CHEMISTRY: A STRUCTURAL VIEW

CHEMISTRY

A structural view

SECOND EDITION

D. R. STRANKS M.Sc., Ph.D., F.R.A.C.I.
Professor of Inorganic Chemistry, University of Adelaide

M. L. HEFFERNAN M.Sc., Ph.D., A.R.A.C.I.
Senior Lecturer in Chemistry, Monash University

K. C. LEE DOW B.Sc. (Hons), B.Ed., A.R.A.C.I.
Senior Lecturer, Centre for the Study of Higher Education, University of Melbourne

P. T. McTIGUE M.Sc., D.Phil., A.R.A.C.I.
Senior Lecturer in Physical Chemistry, University of Melbourne

G. R. A. WITHERS B.Sc.
Senior Chemistry Master, Melbourne Church of England Grammar School

CAMBRIDGE UNIVERSITY PRESS
CAMBRIDGE · LONDON · NEW YORK

Published by the Syndics of the Cambridge University Press
The Pitt Building, Trumpington Street, Cambridge CB2 1RP
Bentley House, 200 Euston Road, London NW1 2DB
32 East 57th Street, New York, NY 10022, USA

Published in Australia, New Zealand, Papua, New Guinea and
Taiwan by Melbourne University Press

Library of Congress catalogue card number: 74–31783

ISBN 0 521 20707 X hard covers
ISBN 0 521 09928 5 paperback

Dewey decimal classification number 541

First edition 1965
Reprinted with corrections 1966
Second edition 1970
Reprinted in monochrome with minor corrections 1974
Reprinted 1977, 1979

Printed and bound in Great Britain at
William Clowes & Sons Limited, Beccles & London

Preface to Second Edition

This text is intended to be an introduction to general chemistry in which the major emphasis is on the physico-chemical principles underlying both inorganic and organic chemistry. No great mathematical sophistication is assumed; in particular, calculus is used only in chapter 12, and then but briefly.

The contents of the book can be summarized under four main headings:

(1) **Chemical stoichiometry.** The mole concept is used exclusively in the quantitative treatment of problems in chemical stoichiometry.

(2) **Chemical bonding.** The idea that energy changes occur during chemical reactions at constant temperature is introduced in a quantitative treatment of thermochemistry. The fact that *all* chemical bonding results from the electrostatic attractions between protons and electrons is stressed, and a primarily structural classification of matter is used. The charge cloud repulsion hypothesis is introduced and used to help predict the shapes of simple molecules and ions. Much care has been given to achieving realistic representations of molecular shapes in the illustrations.

(3) **Chemical equilibrium.** A quantitative treatment of the equilibrium law is developed through a discussion of gas phase, acid-base, complex ion, gas-solid, solubility and oxidation-reduction equilibria.

(4) **Factual chemistry.** Presentation and discussion is closely linked to the principles developed earlier in the text. Three chapters deal with elementary organic chemistry. Ten chapters introduce the periodic table, and discuss the comparative chemistry of the elements of the second and third periods, the eight main groups and the first transition series. There are brief chapters on chemical kinetics and reaction mechanisms.

PREFACE

This book differs in a number of ways from its first edition. The more important changes are

(a) *The introduction and use of the International System of Units (SI).* In accordance with recent trends in many parts of the world, we have introduced and *used* 'SI' units. The authors feel strongly that the wider use of these units must eventually benefit the whole scientific community by facilitating interdisciplinary communication. As far as we know, this is the first chemistry text at this level in which these units have been extensively used. Specific problems raised in the use of these units in chemistry are briefly discussed in chapter 2.

(b) *A substantial increase in the amount of organic chemistry.* A new chapter (19) on hydrocarbons has been inserted, and a discussion of elementary aromatic chemistry introduced.

(c) *The inclusion of a more detailed account of gaseous and heterogeneous equilibria.* Chapter 13 has been substantially rewritten, and chapters 15 and 16 have changed places and been partly rewritten. The changes make possible a consideration of the relation between standard redox potentials and equilibrium constants in chapter 17.

There has also been a number of minor changes. Chapters 3, 5 and 12 have been substantially rewritten; and in the chapter on reaction mechanisms, now chapter 21, the example of the hydrogen-iodine reaction has regretfully been replaced by the carbon dioxide-nitric oxide reaction as an example of a simple bimolecular collision mechanism. J. H. Sullivan (*Journal of Chemical Physics*, **46** (1967), p. 73) has performed a disservice to the authors of elementary texts by his persistence in disproving a mechanism which had virtually reached the status of a chemical dogma. Elementary discussions of the mechanisms of gaseous reactions can never be the same again!

Finally, we have included a short appendix on systematic nomenclature. In the text itself, we have taken a fairly arbitrary line in using IUPAC-recommended names where these are generally acceptable, but retaining those 'trivial' names which are strongly entrenched.

D. R. Stranks
M. L. Heffernan
K. C. Lee Dow
P. T. McTigue
G. R. A. Withers

Acknowledgments

Thanks are due to the following organizations for financial assistance used to defray the secretarial and research expenses of the working group who prepared the manuscript of the first edition of this book: Thomas Baker (Kodak), Alice Baker and Eleanor Shaw Benefactions; Royal Australian Chemical Institute (Victorian Branch); Shell Company of Australia Ltd; Ford Motor Company of Australia Ltd; Broken Hill Proprietary Company Ltd; Imperial Chemical Industries of Australia and New Zealand Ltd; Conzinc Riotinto of Australia Ltd; Kraft Foods Ltd; Australian Paper Manufacturers Ltd.

D. R. Stranks
M. L. Heffernan
K. C. Lee Dow
P. T. McTigue
G. R. A. Withers

Contents

Preface to Second Edition v

Acknowledgments vii

1 The Nature of Chemistry 1

2 Review of Basic Concepts 4

3 Behaviour of Gases 26

4 Stoichiometry 46

5 Chemical Reactions and Heat 70

6 Atomic Structure 95

7 Aggregates of Atoms 118

8 Diatomic Molecules 130

9 Polyatomic Molecules 148

10 Bonding between Molecules 168

11 Bonding into Infinite Arrays 180

12 Rates of Chemical Reactions 196

13 Homogeneous Equilibria 212

14 Acids and Bases 245

15 Complex Ion Equilibria 271

16 Heterogeneous Equilibria 277

17 Oxidation and Reduction 299

18	Some Simple Organic Compounds	327
19	Hydrocarbons	336
20	Functional Group Chemistry	351
21	Mechanisms of Chemical Reactions	374
22	The Periodic Table	383
23	Elements of the Second and Third Periods	398
24	Group I and II Elements	412
25	Group VII Elements	422
26	Group VI Elements	436
27	Group V Elements	451
28	Group IV Elements	464
29	Group III Elements	473
30	Group VIII Elements	479
31	Transition Elements	483
	Table of Relative Atomic Weights	494
	Appendix	496
	Answers	500
	Index	509

1

The Nature of Chemistry

Atoms are the fundamental building blocks of our world; and the very large number of different substances in and on our planet are all aggregates of atoms arranged in an almost endless variety of ways.

The chemist is concerned with studying the way in which atoms aggregate, or stick together, and the properties of such aggregates. In these studies the research chemist performs two functions:

(a) He collects experimental data on the properties and behaviour of pure substances under various conditions.

(b) He attempts, in the light of his experimental observations, to formulate general laws which can satisfactorily *describe* the properties and behaviour of pure substances.

1.1 CHEMISTRY IN THE WORLD

Man's knowledge of chemistry is growing at an ever-increasing rate, and his understanding of the laws governing the behaviour of atoms and aggregates of atoms is undergoing constant review and modification as new facts are discovered. Most of the laws and principles of chemistry which we regard as useful today have been formulated during the past 150 years by physicists and chemists engaged in 'pure' or 'academic' research. In the nineteenth century these workers often carried out their investigations in their own private laboratories; in this century, however, research has become both more organized and more expensive, and fundamental research is most commonly pursued in universities and research institutes.

Parallel to this development of our understanding of the fundamentals of chemistry, there has been accumulated a vast body of knowledge loosely termed 'applied chemistry', which is concerned with those chemicals and chemical reactions which are of economic importance. Chemical industry produces an almost endless array of materials which are essential in modern society, e.g. plastics, fertilizers, pharmaceutical products, pesticides. Chemical manufacturers are continually alert for ways and means of developing new products, and for better methods of producing established products. Because of the pressure of the constant search for new products and the improvement of old products, all large chemical companies maintain research laboratories staffed by highly trained scientists; it is a common practice to maintain a proportion of the staff engaged in 'pure' or 'academic' research in order that the company's scientists can maintain contact with latest developments at the frontiers of chemistry. Indeed, so complex has modern chemical industry become that the very maintenance of efficient production of existing products normally requires the services of highly trained scientists with a sound understanding of the principles of chemistry.

1.2 CHEMISTRY AMONG THE SCIENCES

There is a tendency to think of science as a collection of separate disciplines—physics, chemistry, geology, biochemistry, physiology, botany, zoology and many others. However, all the 'sciences' have a common goal in so far as they are ultimately concerned with the study of matter in all its forms. That is to say, 'science' is a manifestation of man's attempts to understand his material environment. In practice, science is approached in two quite different ways; let us consider these two ways in turn:

(a) If a scientist wishes to investigate matter with a view to formulating general laws about its behaviour, he normally carries out experiments with artificially created systems built up in a laboratory. In this way it is possible to observe the behaviour of essentially simple systems. This approach is that universally used in physics. Chemistry is primarily concerned with one very important aspect of the behaviour of matter—the way in which atoms behave when they cluster together—and chemistry today encompasses an area of science which fifty years ago was the domain of physics, i.e. physics has reached into chemistry, providing it with the general principles with which to understand the ways in which aggregates of atoms behave.

(b) There are many areas of knowledge which deal with systems which, from the physicist's point of view, are hopelessly complicated, e.g. geology, bacteriology, zoology and botany, which deal with the crust of the earth, and the living organisms on the earth. Scientists working with these disciplines are restricted in the extent to which they can simplify their experiments, because of the inherent complexity of their subjects. As science advances, however, the operation of the fundamental laws of physics can be seen, even in quite complex systems. Thus, in recent years, chemistry has reached into botany and zoology to provide a description of the principles of heredity in terms of the chemical nature of certain large molecules in living cells.

Ultimately, there are no sharp boundaries between the various parts of science—the division into physics, chemistry, etc., is simply a convenience forced upon us because of the need for specialization—no one man could possibly 'specialize' in science as a whole, as the human brain has not the capacity to cope with the sheer volume of material he would have to remember. We can draw up a rough flow diagram showing the interconnections between some of the traditional parts of science. It has been drawn as something of an hierarchy with physics at the top, since physics is properly the rigorous study of the fundamental laws governing the behaviour of matter.

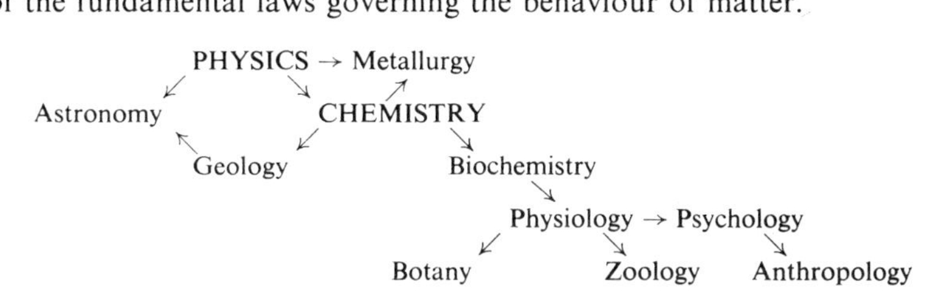

Chemistry, in a sense, is a part of physics; the great importance of chemistry stems from the fact that the world we live on consists of atoms aggregated together in clusters of the utmost complexity. Geology, botany and zoology, the 'earth' sciences, are inevitably becoming increasingly dependent upon chemistry to explain and codify the vast stock of facts they have accumulated over the years. The progressive invasion of chemistry by physics during the past thirty years is a compelling analogy to the present-day invasion of the 'earth' sciences by chemistry. This 'invasion' highlights the importance of physics and chemistry, since both these branches of science involve a study of fundamental principles, which are being increasingly applied in almost all other branches of science.

2

Review of Basic Concepts

Much of our understanding of chemistry depends upon two theories of quite fundamental importance. The first of these is the strongly-held belief that matter may be regarded as composed of individual, discrete entities, which are called *atoms*. The other theory concerns the nature and structure of these atoms. This chapter is directed towards some clarification of these two theories, and towards an interpretation of some of their consequences. In particular, we shall be interested in how the relative masses of the atoms of different elements may be established, and how we can calculate the actual numbers of atoms present in a known mass of a pure substance of known composition.

2.1 THE ATOMIC NATURE OF MATTER

The atomic theory of matter proposes that all matter is composed of aggregates of very large numbers of small particles called atoms. These are extremely small and are indivisible by ordinary chemical means.

This theory is of such basic importance for our understanding of the chemical behaviour of matter that we must consider why we accept it so firmly. The evidence is, in fact, quite indirect, as we cannot directly see atoms in the way that we can see apples or elephants. Although the evidence is indirect, a great mass of it, of quite diverse character, has been accumulated since 1800, and this evidence forms a very convincing case for the acceptance of the theory. We shall see many of these pieces of evidence in this book, but since the theory is of such immediate importance we shall support it by considering two of these lines of evidence now.

(1) Evidence from radioactivity. Certain radioactive elements emit a very high energy beam of radiation called α-radiation. If this beam strikes a zinc sulphide screen, tiny flashes of light (scintillations) are visible, suggesting that a stream of particles is striking the screen. The scintillations provide a means of counting the actual number of these particles. When this radiation is collected the material is found to be the gaseous element, helium. These observations suggest that helium gas is not a continuous fluid but comprises a large number of tiny discrete particles. Other gases, too, are composed of large numbers of small individual particles. In the case of helium, the particles are, in fact, helium atoms.

(2) Evidence from field emission experiments. This is obtained from an instrument called a field emission microscope. It consists of a pointed wire cathode (negative electrode) composed of some high melting metal, such as tungsten. The wire has a very fine tip of radius 1 μm (10^{-6} m) or less. The wire is mounted at the centre of a hollow hemi-spherical anode (positive electrode) coated with some scintillating material such as zinc sulphide. The whole system is mounted in a glass envelope enabling a high vacuum to be maintained within it. If voltages of the order of 10,000 volts are applied between cathode and anode, an image of the tip is formed on the anode, which is produced by the bombardment of the anode by a stream of electrons, in a fashion similar to the formation of an image on a television screen. Electrons are emitted from atoms on the tip of the cathode and are accelerated by the electric field to the anode, where they produce scintillations, thus giving rise to a number of bright spots. Each spot is believed to correspond to an atom on the tip of the cathode. The device produces magnifications of any atoms adsorbed on the tungsten tip of 10^5 or 10^6.

If very small traces of oxygen gas are admitted to the apparatus one can see single spots on the screen, which subsequently split into two spots. It is believed that the oxygen gas is adsorbed on to the tip and the image of oxygen molecules (O_2) is formed on the screen. It seems likely that the events seen on the screen represent oxygen molecules dissociating into oxygen atoms.

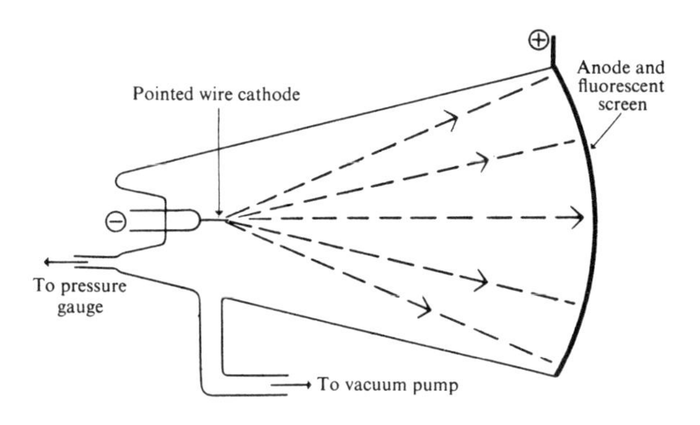

Fig 2.1 Field emission microscope.

EXERCISE
List any other evidence you can find which supports the atomic theory of matter.

The atomic theory successfully accounts for the mass relations observed in chemical compounds, and in the reactions of chemical compounds. The atomic theory was first put forward as a result of a study of the diffusion of gases into one another, but its main support arose from observations of the mass relationships in chemical reactions. It is of interest to consider such a case to test the usefulness of the atomic theory.

A compound of copper and chlorine is known in which the copper and chlorine are united in the ratio by mass 1 : 1.12. This compound is called copper(II) chloride. Another compound, copper(I) chloride, exists, in which the copper and chlorine are united in the ratio by mass 1 : 0.56. The proportion of chlorine in the copper(II) chloride is just twice that in the copper(I) chloride, for a fixed mass of copper. This suggests that, on the atomic scale, the proportion of chlorine to copper atoms in the copper(II) chloride is twice the proportion of chlorine to copper atoms in the copper(I) chloride. Thus, the mass relationships suggest an explanation entirely consistent with the atomic concept. In many other similar pairs of compounds which have been analysed similar simple integral mass ratios have been found. These ratios are also satisfactorily accounted for by the atomic theory.

2.2 THE STRUCTURE OF ATOMS

Just as a wealth of indirect evidence has been accumulated to support the atomic theory of matter, so a wealth of indirect evidence has been accumulated to support the idea that atoms themselves have structure. We must be content to be concerned less with the evidence than with some of the conclusions which have been drawn from it. The picture of atomic structure which has developed may be called the *nuclear model of atomic structure*.

The atom may be pictured as consisting of a very small, positively charged *nucleus*, surrounded by a number of negatively charged *electrons*. The charge on the nucleus of an atom is characteristic for each element and is always a multiple of a fundamental unit of charge, viz. 1.60206×10^{-19} coulomb, For example, the charge on the sodium nucleus is $+17.62266 \times 10^{-19}$ coulomb, which is just eleven times the fundamental unit of charge. Since the atom is electrically neutral the collective charge of the surrounding electrons is $-17.62266 \times 10^{-19}$ coulomb, which is again just eleven times the fundamental unit of charge, but opposite in sign to the nuclear charge.

The radius of an atom may be determined experimentally and is found to be of the order of 10^{-8} cm, whilst the radius of the nucleus is of the order of 10^{-13} cm. *Clearly the electrons occupy most of the volume of the atom.*

EXERCISE
Calculate the ratio of the volume of an atom to the volume of its nucleus, assuming an atomic radius of 10^{-8} cm and a nuclear radius of 10^{-13} cm. Remember that the volume of a sphere depends on the cube of its radius.

The mass of a helium atom nucleus is 6.64626×10^{-24} g. The mass of the electrons in a helium atom is 1.82166×10^{-27} g. Clearly the electrons contribute very little to the total mass of the atom. *Most of the mass of the atom resides in its nucleus.*

EXERCISE
Calculate the percentage of the total mass of the helium atom which is due to its electrons.

2.3 THE STRUCTURE OF THE NUCLEUS

Most nuclei contain two types of particles, viz. *protons* and *neutrons*. Protons carry a charge of $+1.60206 \times 10^{-19}$ coulomb and have a mass of 1.67239×10^{-24} g. Neutrons have no charge and have a mass of 1.67470×10^{-24} g. These particles are held together in nuclei by short-range nuclear forces which are very strong, much stronger than the electrostatic forces encountered in atoms.

Surrounding the nucleus are electrons each bearing a charge of -1.60206×10^{-19} coulomb and having a mass of 9.1083×10^{-28} g. The fundamental unit of charge (1.60206×10^{-19} coulomb) may be represented by the symbol e. The charge on the electron may be represented as $-$e, and the charge on the proton may be represented as $+$e.

The sodium atom has a nucleus containing eleven protons, hence bearing a charge of $+11$e. In the neutral sodium atom the number of electrons must equal the number of protons to maintain electrical neutrality. Thus there are eleven electrons with a total charge of -11e. If an electron is removed from a sodium atom a nett charge of $+$e will be present. This entity is referred to as the sodium *ion* (Na^+). The fluorine atom has a nucleus containing nine protons, hence in the neutral atom we find nine electrons. If an additional electron is inserted, the entity formed will have a net charge of $-$e. The entity is referred to as a fluoride ion (F^-).

The atoms of each element have a characteristic number of protons in their nuclei. This number is called the *atomic number* (Z). We shall see in chapter 6 that the nuclear charge is indirectly responsible for the chemical properties of each element, in that it determines the number of electrons in the neutral atom and these in turn dictate the chemical behaviour of the atom. Definition:

An element is a substance all of whose atoms have the same atomic number.

For a given element the number of neutrons in the nuclei of the different atoms may vary. Definition:

Atoms of the one element which differ in the number of neutrons in the nuclei are called isotopes.

We may refer to the different nuclear species as *nuclides*. The number of protons and neutrons (collectively known as *nucleons*) in the nucleus of a nuclide is referred to as its *mass number* (A). If we represent the neutron number as N, we can write for any nuclide:

$$A = N + Z$$

To represent a nuclide we write the symbol for the element with the atomic number and mass number both on the left of the symbol. Thus we write ${}^{A}_{Z}X$.

EXERCISE
For the cadmium nuclide, ${}^{112}_{48}Cd$, state:
(a) the number of protons;
(b) the number of neutrons;
(c) the number of nucleons;
(d) the number of electrons in the neutral atom;
(e) the number of electrons in the ion, Cd^{2+}.

It would be wrong to infer that protons, neutrons and electrons are the only fundamental particles which constitute atoms. In recent years physicists have discovered large numbers of species which are described as subatomic particles. At present, this number is greater than one hundred. It includes such species as the neutrino, the positron, the μ-meson, the π-meson and the photon. In fact there is considerable doubt now as to whether the terms 'fundamental' or 'elementary' are suitable for this vast array of particles, and it is likely that in the future some deeper organizational pattern or structure will be seen to underlie them.

We simply need to note that the three particles we consider in detail are not the only species which make up atoms but, for chemists, they are the most important ones.

2.4 ISOTOPES

Many naturally-occurring elements have been found to consist of more than one isotope. Thus hydrogen ($Z = 1$) consists mainly of the isotope ${}^{1}_{1}H$, but there is also a small amount of a second isotope ${}^{2}_{1}H$, often called deuterium, and an extremely small amount of a third, radioactive isotope ${}^{3}_{1}H$, often called tritium. Carbon ($Z = 6$) consists mainly of the isotope ${}^{12}_{6}C$, but there is also a much smaller proportion of the isotope ${}^{13}_{6}C$, and a very small amount of the radioactive isotope ${}^{14}_{6}C$. On the one hand, tin consists of ten isotopes; on the other hand, some naturally-occurring elements consist of one type of atom only. Thus fluorine ($Z = 9$) consists entirely of ${}^{19}_{9}F$. For an element such as this, it is not very logical to say that the element has isotopes, but in practice we sometimes say that the element has one isotope only.

Some interesting facts have been noticed about the isotopes of elements. First, elements with odd atomic number have one or two isotopes only. Second, the mass number of a stable nuclide is equal to twice its atomic number, or else greater than twice its atomic number. Isotopes for which this is not true are usually radioactive. Third, the number of neutrons in a nucleus tends to be even. Fourth, elements of even atomic number are generally more abundant in nature than those with odd atomic number. Whilst these facts are quite well known, the explanations for these observations are still very far from clear.

Table 2.1
Some common isotopes

Nuclide	*Relative abundance in nature (%)*	*Atomic number*	*Mass number*	*Neutron number*	*Relative isotopic mass*
$^{1}_{1}H$	99.986	1	1	0	1.00783
$^{2}_{1}H$	0.0145	1	2	1	2.01410
$^{3}_{1}H$	very small	1	3	2	3.01610
$^{10}_{5}B$	19.91	5	10	5	10.01294
$^{11}_{5}B$	80.09	5	11	6	11.00930
$^{12}_{6}C$	98.888	6	12	6	12 exactly
$^{13}_{6}C$	1.112	6	13	7	13.00336
$^{14}_{6}C$	*ca.* 10^{-10}	6	14	8	14.00329
$^{16}_{8}O$	99.76	8	16	8	15.99491
$^{17}_{8}O$	0.04	8	17	9	16.99913
$^{18}_{8}O$	0.20	8	18	10	17.99916
$^{35}_{17}Cl$	75.771	17	35	18	34.96885
$^{37}_{17}Cl$	24.230	17	37	20	36.96590

The relative proportions of nuclides of a given element, whether the element be free or combined, are constant within the precision of normal chemical experiments. In a few cases variations are found. Thus, in the case of lead, the ordinary form consists of isotopes of mass numbers 204, 206, 207 and 208. However, lead from the rare mineral, curite, consists almost entirely of the isotope of mass number 206. This isotope is the end product of the radioactive decay series derived from uranium 238.

2.5 ATOMIC WEIGHT

For chemical purposes it is not necessary to determine the absolute masses of atoms in gram, but only the relative masses of atoms. By ascribing a standard mass to a selected atom, we can establish a relative scale of atomic masses. The atom $^{12}_{6}C$ is chosen as standard and is assigned a mass of exactly 12 units. This particular atom is chosen because, for technical reasons, it is the most convenient practical standard for use in mass spectrometry. The choice of 12 units as the assigned mass is a historical legacy from the early days of the study of chemistry. Then the standard was taken as hydrogen, and the average mass of an atom of naturally-occurring hydrogen was taken as being 1. Since hydrogen has the lightest atoms, this set all other atoms as having relative masses greater than 1, with the average mass of a carbon atom being about 12.

The *mass spectrometer* is used to determine the relative masses of isotopes and their relative abundances. The instrument performs three tasks:
(a) the creation of ions from a sample;

(b) the separation of these ions according to their mass to charge ratios (m/Q); the charge (Q) is always an integral multiple of the fundamental charge (e);

(c) measurement of the relative abundance of the ions and their relative masses; this information, presented graphically, is known as a *mass spectrum*.

Thus different isotopes of a given element may be separated, the masses may be determined, and the relative proportions found.

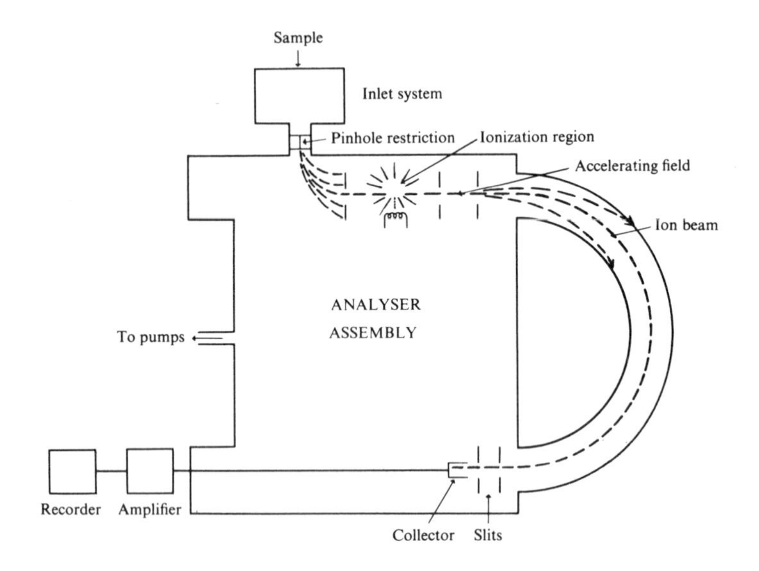

Fig 2.2 Mass spectrometer.

The element must enter the analyser assembly in the form of a gas. Thus liquid and solid elements must first be vaporized; a solid element, or a compound of it, being vaporized upon a heated filament. The now gaseous sample is introduced into an evacuated inlet system and the atoms are allowed to pass through a pin-hole restriction and enter the analyser assembly, which is maintained at a pressure below 10^{-6} mmHg. The atoms are bombarded by electrons from a heated filament and electrons are knocked off atoms producing positive ions. The ions leave the ionization region through slits, and are accelerated by an intense electric field. The rapidly moving ions are diverted into circular paths by a magnetic field parallel to the slits and normal to the ion beam. Each ion follows a circular path of radius r, which for ions of the same charge is proportional to the square root of the mass of the ion, as given by the relation

$$r = \frac{1}{B}\sqrt{\frac{2mV}{Q}}$$

where m kg is the mass of the ion, r m is the radius of curvature, B weber m^{-2} is the magnetic field intensity, Q coulomb is the charge on the ion, and V volt is the accelerating potential.

Ions of different mass are diverted into paths of different radius traversing an arc of 180°. Ions of the appropriate mass pass through resolving slits and strike a collector, giving up their charges which are amplified electronically. The resultant ion currents are recorded on a chart recorder. Positive ion beams of increasing mass are successively brought to focus at the collector by continuously decreasing the accelerating voltage. As each beam sweeps across the collector its intensity is recorded graphically giving a continuous plot of the intensities of successive ion beams. The height of each peak is proportional to the number of positive ions of given mass reaching the collector in unit time and the ion mass is identified on the abscissa by reference to a known standard.

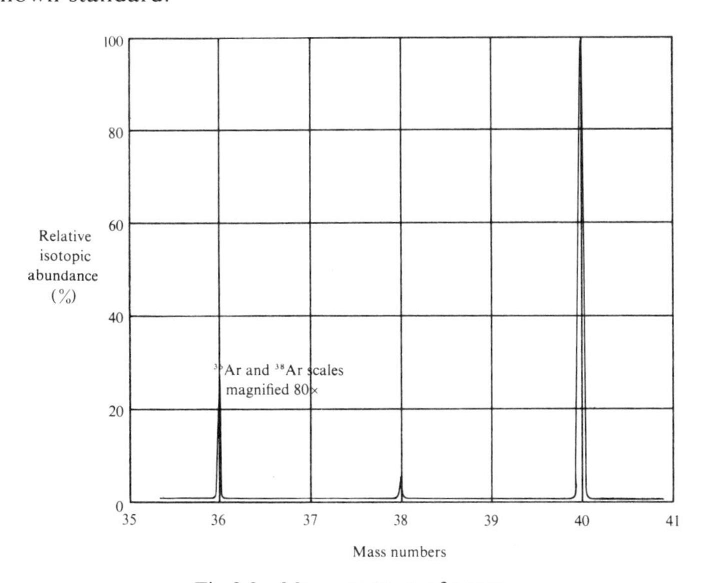

Fig 2.3 Mass spectrum of argon.

Thus we can obtain from the mass spectrum the mass of an isotope relative to the mass of a $^{12}_{6}C$ atom. On this scale the relative mass of the proton is determined to be 1.00783 and that of the neutron may be calculated to be 1.00866 units. Hence the relative mass of the $^{19}_{9}F$ nuclide might be expected to be

$$\{(9 \times 1.00783) + (10 \times 1.00866)\} = 19.1520$$

It is, in fact, found to be 18.9984. The difference is associated with the forces binding nucleons in the nucleus. It is found that the mass difference is related to the *nuclear binding energy*, as given by the mass-energy relation

$$E = mc^2,$$

where E joule is the binding energy, m kg is the mass difference, and c m s^{-1} is the velocity of light.

Notice that since the relative masses of proton and neutron are nearly unity, we would expect the mass number to approximate to the relative isotopic mass. That this is so may be seen from columns four and six in Table 2.1.

We have seen that most natural elements consist of a mixture of isotopes. Since we deal with naturally-occurring elements in chemical reactions it is useful to have a measure of the relative masses of their atoms.

The atomic weight of an element is the weighted mean of the masses of the isotopes of the element relative to the mass of a $^{12}_{6}C$ atom taken as 12 exactly.

Notice that the atomic weight is a ratio and hence has no units.

Thus naturally-occurring nitrogen is composed of 99.633 per cent of $^{14}_{7}N$ and 0.367 per cent of $^{15}_{7}N$, hence the atomic weight is

$$\frac{(99.633 \times 14.00307) + (0.367 \times 15.00011)}{100} = 14.0067$$

The value of the atomic weight is limited by the accuracy of the least accurately known quantity, in this case the relative abundances.

The atomic weight of carbon is slightly in excess of 12 (actually 12.011) because of the presence of slight amounts of the isotope $^{13}_{6}C$.

If we say that the atomic weight of sulphur is 32, we mean that the average mass of a sulphur atom is 32/12 times as great as the mass of a $^{12}_{6}C$ atom. If the atomic weight of oxygen is 16, then on the average, a sulphur atom is twice as heavy as an oxygen atom.

EXERCISE

Calculate the atomic weight of boron if it contains 18.83 per cent of $^{10}_{5}B$ of relative isotopic mass 10.0130, and 81.17 per cent of $^{11}_{5}B$ of relative isotopic mass 11.0093.

2.6 MOLECULAR WEIGHT AND FORMULA WEIGHT

In pure substances atoms are normally found aggregated in some way. The modes of aggregation will be discussed in chapter 7, but we will review some of the important and common modes in this section.

In some substances atoms occur in groups bonded together to form discrete units, called *molecules*. Thus water is composed of the elements hydrogen and oxygen. The atoms of these elements are organized into molecules each consisting of two hydrogen atoms bonded to an oxygen atom. We may represent this molecule by the *molecular formula*, H_2O.

H_2O H–O–H (bent)

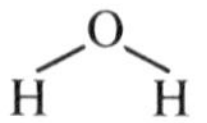

Fig 2.4 Representations of the formula of a water molecule.

If liquid water is vaporized, the water molecules pass from the liquid state to the vapour state unchanged, emphasizing that these molecules behave as discrete units. Because molecules behave as units there must exist forces or bonds holding the atoms together. In chemical reactions these bonds may be broken, thus disrupting the molecules and leading to the formation of new molecules.

Some elements exist as molecules. Thus solid white phosphorus consists of molecules of formula P_4. Each molecule consists of four phosphorus atoms bonded into a discrete unit. We say that this molecule is tetratomic because it contains four atoms. Other elements exist as molecules. Elements having diatomic molecules include oxygen (O_2), nitrogen (N_2), fluorine (F_2), chlorine (Cl_2), bromine (Br_2) and iodine (I_2). Both monoclinic and rhombic sulphur exist as octatomic molecules (S_8). The noble gases exist as monatomic molecules, for example, helium (He), neon (Ne) and argon (Ar). In all of these cases the molecules persist from the solid to the vapour state, although molecules composed of more than one atom may be disrupted into their constituent atoms if the temperature is high enough.

Just as we can assign relative masses to atoms, so we can assign relative masses to molecules.

The molecular weight is the sum of the atomic weights of the atoms as given in the molecular formula.

The molecular weight of water must be

$$\{(2 \times 1.00797) + (15.9994)\} = 18.0153$$

It follows that a molecule of water is about 18/12 times as heavy as a ${}^{12}_{6}C$ atom.

The molecular weight of phosphorus is

$$(4 \times 30.9738) = 123.8952$$

A molecule of phosphorus is about 124/18 times as heavy as a water molecule.

Just as the relative masses of atoms can be determined by mass spectrometry, so too can the relative masses of molecules. When a molecular species is introduced into a mass spectrometer, a number of peaks may be recorded in the mass spectrum. Sometimes it is possible to identify one prominent peak as arising from the removal of just one electron from the original molecule. The peak of this particular singly-charged molecular ion is termed the 'parent peak', if it can be identified unambiguously. In such instances a high precision mass spectrometer enables the relative mass of the ion producing this peak to be found with great accuracy. From this relative mass and the accurately known relative isotopic masses one can deduce the molecular formula of thc compound, and hence the molecular weight. However, if the 'parent peak' escapes detection, or is of low intensity, molecular weight determination by mass spectrometry may be difficult.

Molecular weights of gases may be found by determining gas densities. This method will be discussed in chapter 3, where it will be seen that this method is generally much less accurate than mass spectrometric methods.

In many substances, however, atoms do not aggregate into molecules. Thus, in diamond (a form of carbon) the atoms are not organized into discrete molecules, but are bonded one to another in a network structure. In the compound silicon carbide (SiC), the silicon and carbon atoms are similarly bonded into a network. In metallic elements, and some alloys, we have an array of positive ions in a sea of electrons. Again no discrete molecules may be detected. In sodium chloride we have an array of sodium ions, Na^+, and chloride ions, Cl^-, aggregated together. Again no discrete molecules can be detected.

In the cases discussed we cannot assign a molecular weight since there are no molecules. Instead we can assign a quantity known as the formula weight.

The formula weight is the sum of the atomic weights as given in the simplest (empirical) formula of the substance.

In the case of sodium chloride the simplest formula is NaCl. This means that in the crystal the sodium ions and chloride ions occur in the ratio by number 1 : 1. Thus the formula weight of sodium chloride is

$$(22.9898 + 35.453) = 58.4428$$

In the case of silicon carbide the simplest formula is SiC. This means that in the crystal the silicon atoms and carbon atoms occur in the ratio by number 1 : 1. Thus the formula weight of silicon carbide is

$$(28.086 + 12.0111) = 40.097$$

In the case of water the formula weight is 18.0153, and this is also the molecular weight since water does contain discrete molecules.

2.7 THE MOLE CONCEPT

In measuring the quantity of some material we may choose various systems of measurement. If we wish to buy pomegranates we may buy them by the pound (measuring in terms of mass), or by the basketful (measuring in terms of volume), or by the dozen (measuring in terms of number). In chemistry it is found to be useful to measure quantity by the numbers of particles. A knowledge of the number of particles usually enables us to calculate the relevant masses and volumes. We might choose as our unit any arbitrary number—a dozen, a score, a gross or a quadrillion. It is found to be particularly useful to take as our number quantity the number of carbon atoms in 12.00000 g of the isotope $^{12}_{6}C$. Any amount of substance that contains this number of particles is called a mole of that substance.

The mole is the amount of substance which contains as many elementary units as there are atoms in 12 g of carbon-12.

Notice that in this formal definition, which has been internationally agreed upon, the mole is described as an amount of substance. For convenience, chemists are in the habit of simply referring to the mole as a number. Thus, if we represent the mole by the symbol N_A, we know that 1 mole of oxygen molecules is N_A oxygen molecules. Half a mole of copper atoms is $N_A/2$ copper atoms, and one-third of a mole of emus is $N_A/3$ emus.

We know from mass spectrometry that the atomic weight of fluorine is 19.0. Hence by definition

$$\frac{\text{mass of fluorine atom}}{\text{mass of } ^{12}_{6}\text{C atom}} = \frac{19.0}{12.0}$$

hence

$$\frac{\text{mass of } N_A \text{ fluorine atoms}}{\text{mass of } N_A\ ^{12}_{6}\text{C atoms}} = \frac{19.0}{12.0}$$

But the mass of N_A $^{12}_{6}C$ atoms is 12.0 g, hence the mass of N_A fluorine atoms is 19.0 g. Thus 19.0 g is the mass of 1 mole of fluorine atoms. It may similarly be shown for any element that a mass in gram of the element, numerically equal to the atomic weight, contains 1 mole of atoms. Thus since the atomic weight of copper is 63.5, then 63.5 g of copper must contain 1 mole of copper atoms.

It is also known from mass spectrometry that the molecular weight of oxygen is 32.0. Thus we write

$$\frac{\text{mass of } N_A \text{ oxygen molecules}}{\text{mass of } N_A\ ^{12}_{6}\text{C atoms}} = \frac{32.0}{12.0}$$

It follows that the mass of 1 mole of oxygen molecules is 32.0 g. Similarly it may be shown that for any element or compound composed of molecules, a mass in gram, numerically equal to the molecular weight, contains 1 mole of molecules. Thus since the molecular weight of water is 18.0, then 18.0 g of water contains 1 mole of water molecules. Notice that 1 mole of oxygen molecules must contain 2 mole of oxygen atoms, and 1 mole of water molecules must contain 2 mole of hydrogen atoms and 1 mole of oxygen atoms.

These arguments may be extended to compounds which do not contain molecules. Thus the formula weight of sodium chloride is 58.5, hence 58.5 g of sodium chloride contain 1 mole of sodium ion-chloride ion pairs. Or we might say that 58.5 g of sodium chloride contains 23.0 g of sodium ion, which contains 1 mole of sodium ion, and 35.5 g of chloride ion, which contains 1 mole of chloride ion.

EXAMPLE

(a) What would be the mass of 0.2 mole of phosphorus molecules?
(b) What would be the mass of 0.167 mole of calcium chloride?
(c) How many mole is present in 10.0 g silica, SiO_2?

SOLUTION

(a) 0.2 mole $P_4 = \frac{1}{5}$ of $(4 \times 31.0) = 24.8$ g
(b) 0.167 mole $CaCl_2 = 0.167 \times \{40 + (2 \times 35.5)\} = 18.5$ g
(c) 10.0 g $SiO_2 = 10.0/\{28.1 + (2 \times 16.0)\} = 1/6$ mole

EXERCISE

Ethane has the molecular formula C_2H_6. What is the mass of:
(a) 3.5 mole of ethane;
(b) $5N_A$ molecule of ethane?

As a final example of the mole concept we will consider the charge associated with a mole of electrons. The magnitude of the charge on the electron is 1.60206×10^{-19} coulomb. Hence the magnitude of the charge associated with 1 mole of electrons is $(N_A \times 1.60206 \times 10^{-19})$ coulomb. This quantity has the value 96,491 coulomb and is referred to as 1 *faraday* (F) of charge.

Remember in quoting chemical quantity by number of mole that the elementary unit must be specified, and may be an atom, a molecule, an ion, a radical or an electron, etc.

2.8 THE AVOGADRO CONSTANT

The number of particles in 1 mole is known as the *Avogadro constant.* It may be experimentally determined by a number of widely different methods. We will discuss three of these in this section.

(1) Radioactivity. When radium decays radioactively it emits α-particles (helium nuclei $^4_2He^{2+}$). The presence of an α-particle may be registered by a suitable device, for example, a Geiger counter. Alpha particles rapidly gain electrons and form helium atoms. It is possible to collect α-particles in a thin-walled glass vessel and also record the number of particles collected. The mass of the helium collected can be measured and since the mass of 1 mole of helium is known, the number of helium atoms in 1 mole of helium can be calculated.

EXAMPLE

It is found that 1 gram of radium emits 2.9×10^{15} particles in one day. The volume occupied by the helium formed from these α-particles is 1.1×10^{-3} cm³ at 0°C, 1 atmosphere pressure. This amount of helium has a mass of 1.965×10^{-8} g. Calculate the value of the Avogadro constant.

SOLUTION

1.965×10^{-8} g contains 2.9×10^{15} helium atoms.
One mole of helium is N_A molecules of helium, and because helium is a monatomic gas, it is also N_A helium atoms.

Hence 4.003 g of helium contains N_A atoms of helium. The number of atoms in 4.003 g of helium is

$$\frac{4.003}{1.965 \times 10^{-8}} \times 2.9 \times 10^{15} = 5.9 \times 10^{23} \text{ atoms}$$

The value of the Avogadro constant is 5.9×10^{23}.

(2) Electrolysis. If a solution of silver nitrate is electrolysed, silver ions (Ag^+) are discharged at the negative electrode (cathode) by the process of each silver ion gaining 1 electron. One can determine the number of coulomb required to discharge a given mass of silver and hence the number of coulomb required to discharge 1 mole of silver ions. Since the charge on the electron is known, one can thus calculate the value of the Avogadro constant.

EXAMPLE

It is found that 641 coulomb discharge 0.7168 g of silver. If the fundamental unit of charge is 1.6021×10^{-19} coulomb, calculate the value of the Avogadro constant.

SOLUTION

One mole of silver ions has a mass of 107.87 g, thus 0.7168 g of silver represents 0.7168/107.87 mole of silver ion.
Hence 641.0 coulomb discharges 0.7168/107.87 mole of silver ion. One mole of silver ion is discharged by

$$\frac{641.0 \times 107.87}{0.7168} = 9.649 \times 10^4 \text{ coulomb}$$

But each silver ion requires 1 electron for discharge, hence the number of electrons required to discharge 1 mole of silver ion is

$$\frac{9.649 \times 10^4}{1.6021 \times 10^{-19}} = 6.024 \times 10^{23}$$

The value of the Avogadro constant is 6.024×10^{23}.

(3) Oil film. A film of oleic acid of known mass is spread over a clean water surface. The area of the film may be measured. If the density of oleic acid is known the volume of the film may be calculated, and hence the thickness of the film. It is assumed that the film is 1 molecule thick, hence if the relative dimensions of the molecule are known, or assumed, the volume of a molecule can be calculated. If the mass of 1 mole of oleic acid is known, then the number of molecules per mole can be calculated from the known volume of 1 molecule and the volume of 1 mole.

EXAMPLE

One cm^3 of a solution of oleic acid in hexane (containing 0.53×10^{-4} mole of oleic acid per dm^3 of solution) is added to a water surface. The hexane evaporates leaving an oleic acid film, on the water surface, of area 207 cm^2. The density of oleic acid is 0.891 g cm^3 and the mass of 1 mole is 282 g. If the molecule of oleic acid is assumed to be of cubic shape, calculate the value of the Avogadro constant.

SOLUTION

The number of mole of oleic acid in 1 cm^3 of solution is

$$\frac{1}{1000} \times 0.53 \times 10^{-4} = 0.53 \times 10^{-7}$$

Hence the mass of the oleic acid film is

$$0.53 \times 10^{-7} \times 282 \text{ g}$$

Hence the volume of the oleic acid film (volume = mass/density) is

$$\frac{0.53 \times 10^{-7} \times 282}{0.891} \text{ cm}^3$$

Hence the thickness of the oleic acid film (thickness = volume/area) is

$$\frac{0.53 \times 10^{-7} \times 282}{0.891 \times 207} = 0.81 \times 10^{-7}\ \text{cm}$$

Hence the volume of 1 molecule, assumed to be a cube, is

$$(0.81 \times 10^{-7})^3 = 0.53 \times 10^{-21}\ \text{cm}^3$$

The volume of 1 mole of oleic acid is

$$\frac{282}{0.891} = 317\ \text{cm}^3$$

Hence the number of molecules per mole is

$$\frac{317}{0.53 \times 10^{-21}} = 6 \times 10^{23}$$

The value of the Avogadro constant is 6×10^{23}.

It should be noted that this method is far less accurate than the other two. In particular the assumption of a cubic molecule is not completely valid.

The accepted value for the Avogadro constant, based on the latest experimental determinations, is

$$N_A = 6.0226 \times 10^{23}$$

2.9 GRAVIMETRIC DETERMINATION OF ATOMIC WEIGHTS

Before the development of the mass spectrometer, atomic weights were determined by chemical methods. These involved the reaction of very pure chemicals, weighing to a high order of accuracy, and comparison of the unknown atomic weight against accurately known atomic weights, or against the standard of atomic weight. These methods developed a very high order of accuracy as the result of very careful experimental technique. They could not, however, reach the very high order of accuracy obtained in mass spectrometric determinations.

EXAMPLE

In 1909 Baxter and Tilley determined the atomic weight of silver. They reacted iodine with concentrated nitric acid to form iodic acid

$$I_2 + 10HNO_3 \rightarrow 2HIO_3 + 4H_2O + 10NO_2$$

The iodic acid was heated to form iodine pentoxide

$$2HIO_3 \rightarrow I_2O_5 + H_2O$$

This was reduced with hydrazine to form iodide ion and then reacted with silver ion to form silver iodide

$$3N_2H_4 + I_2O_5 \rightarrow 3N_2 + 2H^+ + 2I^- + 5H_2O$$

$$Ag^+ + I^- \rightarrow AgI$$

In one determination it was found that 8.72163 g of iodine pentoxide yielded a silver iodide precipitate containing 5.63619 g of silver. Calculate the atomic weight of silver.

SOLUTION

From the above equations we see that 1 mole of I_2O_5 yields 2 mole of I^-, and 1 mole of I^- yields 1 mole of AgI.
Hence 1 mole of I_2O_5 yields 2 mole of AgI containing 2 mole of Ag. The mass of 1 mole of I_2O_5 is

$$\{(2 \times 126.90) + (5 \times 15.9994)\} = 333.80 \text{ g}$$

Hence the number of mole of I_2O_5 used is

$$\frac{8.72163}{333.80}$$

Hence the number of mole of Ag reacted is

$$\frac{2 \times 8.72163}{333.80} = 0.052257$$

Let the atomic weight of silver be x, then the number of mole of silver reacted is

$$\frac{5.63619}{x}$$

Thus

$$\frac{5.63619}{x} = 0.052257$$

$$\therefore x = 107.86$$

Other values obtained at this time were 107.868, 107.876, 107.874 and 107.878. The mass spectrometric value is 107.870. Notice that, in the above calculation, the accuracy of the value obtained is limited by the least accurately known piece of data, which in this case is the atomic weight of iodine.

2.10 UNITS FOR PHYSICAL QUANTITIES

There is now an internationally agreed system of symbols and units for most of the physical quantities which are used in chemistry. This system is known as SI (Système International d'Unités), and has been accepted and recommended for use by both the International Unions of Pure and Applied Chemistry and Pure and Applied Physics (IUPAC and IUPAP), as well as many other international organizations representing several disciplines of science and technology. For this reason, SI units have been used in this book as far as has been practicable.

SI admits six basic quantities—although IUPAC and IUPAP have recommended that the mole be added to this list. We shall therefore treat the mole as a basic quantity, and seven basic quantities are shown, together with their names and symbols, in Table 2.2.

Table 2.2

Basic quantities and their symbols (SI)

Physical quantity	*Symbol for quantity*	*Name of SI unit*	*Symbol for SI unit*
Length	l	metre	m
Mass	m	kilogram	kg
Time	t	second	s
Electric current	I	ampere	A
Absolute temperature	T	kelvin	K
Amount of substance	n	mole	mol
Luminous intensity	I_v	candela	cd

Notice that the basic unit of mass, the kilogram, is named in such a way as to suggest that the gram (see Table 2.4) is the fundamental unit of mass. It is expected that the kilogram will eventually be renamed but, until that time, the present name for the unit of mass will be retained.

Each of the basic units has been defined in some suitable arbitrary fashion, and all scientific measurements are ultimately referred to these seven independent arbitrary definitions. There is also a number of additional SI units which are derived from the seven fundamental units. Some of them are shown in Table 2.3.

Table 2.3

Some derived physical quantities and their symbols (SI)

Physical quantity	*Name of SI unit*	*Symbol for SI unit*	*Definition of SI unit*
Energy	joule	J	$kg\ m^2\ s^{-2}$
Force	newton	N	$kg\ m\ s^{-2} = J\ m^{-1}$
Electric charge	coulomb	C	$A\ s$
Electric potential difference	volt	V	$kg\ m^2\ s^{-3}\ A^{-1} = J\ A^{-1}\ s^{-1}$
Electric resistance	ohm	Ω	$kg\ m^2\ s^{-3}\ A^{-2} = V\ A^{-1}$

In dealing with numerical values of SI units, a problem often arises in that very large or very small values occur; e.g. bond lengths in molecules are of the order of 10^{-10} metre. In order to accommodate such extreme values, a set of prefixes, together with their appropriate symbols, has been approved for use with SI units. These are shown in Table 2.4.

Table 2.4

Allowed prefixes for physical quantities (SI)

Fraction	*Prefix*	*Symbol*	*Multiple*	*Prefix*	*Symbol*
10^{-1}	deci	d	10	deka	da
10^{-2}	centi	c	10^{2}	hecto	h
10^{-3}	milli	m	10^{3}	kilo	k
10^{-6}	micro	μ	10^{6}	mega	M
10^{-9}	nano	n	10^{9}	giga	G
10^{-12}	pico	p	10^{12}	tera	T

$$\text{i.e. } 1 \text{ picometre} \equiv 1 \text{ pm} = 10^{-12} \text{ m}$$

$$\text{or } 1 \text{ nanosecond} \equiv 1 \text{ ns} = 10^{-9} \text{ s}$$

The addition of the prefix to a unit effectively forms a new unit, so that

$$1 \text{ dm}^3 \equiv 1(\text{dm})^3 = 10^{-3} \text{ m}^3 \ (\textit{not } 1 \text{ d(m)}^3)$$

$$1 \text{ km}^3 \equiv 1(\text{km})^3 = 10^{9} \text{ m}^3 \ (\textit{not } 1 \text{ k(m)}^3)$$

In general, units formed by the addition of approved prefixes have the same status as the basic SI unit(s) from which they derive.

A number of metric units which have long been in common use by chemists have become redundant as a result of the introduction of SI. Some of these are:

(a) The litre is now replaced by the cubic metre as the fundamental unit of volume. The litre has been redefined so that

$$1 \text{ litre} \equiv 1000 \text{ cm}^3 = 1 \text{ dm}^3$$

Although the litre will doubtless continue to be used by name as a convenient unit of volume, we shall not use it extensively in this book.

(b) The ångström, where by definition

$$1\text{Å} \equiv 10^{-10} \text{ m}$$

The convenience of the ångström lies in the fact that the lengths of chemical bonds and the distances between ions in crystals, are of the order of a few ångström. Again, this unit will doubtless continue to be used by chemists, although its use should progressively wane.

(c) The degree Celsius (°C), which is a commonly-used measure of temperature. The appropriate SI unit is the kelvin, and the Celsius scale is related to the kelvin scale by

$$°\text{C} = (\text{K} - 273.16)$$

We shall frequently quote temperature in °C for convenience.

There remain some widely used physical quantities which have not yet been assigned special names or symbols in SI. One of the most important of these for chemists is concentration. The appropriate SI unit of concentration would be mol m^{-3}; however, as no symbol or name has been given to this quantity we shall continue to use the well-known concentration unit of mole per litre; i.e. we shall use as our unit of concentration the quantity mol dm^{-3}, give it the symbol M, and the name 'molar'.

Finally, we shall point out one minor inconvenient feature of the use of the SI system in chemistry. This involves the calculation of the number of mole of any pure substance from its known mass, and its known atomic or molecular weight. In practice, one calculates this by making use of a table of relative atomic weights, as:

$$\text{number of mole} = \frac{\text{mass of substance in g}}{\text{atomic or molecular weight}}$$

Of course, in using SI units, one should express both the mass of the substance, and the atomic or molecular weight, in kilogram. However, as chemists are so accustomed to using tables of relative atomic weights, we shall continue with their use and in this case, and in this case only, masses will normally be expressed in gram rather than in kilogram.

QUESTIONS

Atomic structure

1. Why is the atomic number of an atom, rather than its mass number, used to identify an element made up of these atoms?
2. For each case name two ions which have the same number of electrons as:
   (a) the helium atom;
   (b) the neon atom;
   (c) the oxide ion.
3. Supply the missing information in the following table:

Atomic number	*Neutron number*	*Mass number*	*Symbol*
5	5	10	$^{10}_{5}B$
8	(a)	17	$^{17}_{8}O$
14	(b)	(c)	$^{28}_{14}Si$
(d)	(e)	(f)	$^{22}_{10}Ne$
34	(g)	78	(h)
30	36	(i)	$^{(j)}_{(k)}Zn$

4. If atoms are bombarded by high energy neutrons, the nucleus of an atom may combine with a neutron, expel a proton, and become the nucleus of a new element, thus for example

$$^{27}_{13}Al + ^{1}_{0}n \rightarrow ^{1}_{1}H + ^{27}_{12}Mg$$

Which isotope, of which element, would be formed in such a reaction for each of the following:

(a) $^{16}_{8}O$ (c) $^{37}_{17}Cl$

(b) $^{31}_{15}P$ (d) $^{85}_{37}Rb$

5. Assuming that the nucleus of a fluorine atom is a sphere of radius 5×10^{-13} cm, and that the mass of a fluorine atom is 3.155×10^{-23} g, calculate the density of matter in a fluorine nucleus in:

   (a) g cm^{-3}

   (b) ton cm^{-3}

   (One pound is equivalent to 454 g.)
6. The chemical atomic weight of rubidium is 85.47. The relative masses of the isotopes are 84.94 and 86.94. Find the relative abundances of these two isotopes in naturally-occurring rubidium.
7. Calculate the atomic weight of each of the following elements from the data given in the table:

Element	*Relative abundance (%)*	*Relative isotopic mass*
(a) Lithium	7.4	6.02
	92.6	7.02
(b) Magnesium	78.8	23.99
	10.2	24.99
	11.0	25.99
(c) Copper	69.1	62.95
	30.9	64.95
(d) Silicon	92.2	27.98
	4.7	28.98
	3.1	29.98
(e) Gallium	60.5	68.95
	39.5	70.95

8. Aniline ($C_6H_5.NH_2$) is subjected to electron bombardment. The ionized products are passed through a mass spectrometer. Traces are obtained for species with charge/mass ratios of 0.01075, 0.01299, 0.1250, 0.02598, 0.06250. Account for these results.

Mole concept

9. Which of the following has the greatest mass:
   (a) 100 g of copper;
   (b) 6 mole of helium;
   (c) 12×10^{23} atoms of silver?
10. Suppose at atom of an element X has a mass of 4.0×10^{-23} g.
    (a) What is the relative atomic weight of X on the scale $^{12}_{6}C = 12.0000$?
    (b) How many mole of X atoms is there in 48 g of X?
    (c) How many gram of X is required to combine with 0.54 mole of Y atoms in a reaction requiring two atoms of Y for every three atoms of X?
11. Iron has an atomic weight of 55.8 and a density of 7.9 g cm^{-3}.
    (a) What is the volume occupied by 1 mole of iron atoms?
    (b) Calculate the average volume occupied by an iron atom.
    (c) If an iron atom is spherical in shape, what is its approximate diameter?
12. In a chemical reaction requiring two atoms of aluminium for every three atoms of oxygen, how many mole of oxygen atoms is required by 2.7 g of aluminium? What mass of oxygen is required?

13. If 0.24 mole of tungsten (W) reacts with 3.84 g of sulphur to form a compound, tungsten sulphide (WS_3), how many gram of the compound can be prepared?
14. (a) How many gram of phosphine (PH_3) in 0.6 mole of PH_3?
    (b) How many mole of P atoms and of H atoms in 0.15 mole of PH_3?
    (c) How many gram of P and of H in 0.2 mole of PH_3?
    (d) How many molecules of PH_3 in 0.5 mole of PH_3?
    (e) How many atoms of P and of H in 0.25 mole of PH_3?
15. How many mole of atoms in:
    (a) 10 g of sodium;
    (b) 10 g of sulphur?
    How many mole of sulphur molecules (S_8) in 10 g of sulphur?
    How many molecules in 10 g of sulphur?
16. How many mole of molecules in:
    (a) 6.4 g of nitric oxide (NO);
    (b) 6.4 of nitrogen dioxide (NO_2)?
17. In 314 g of barium chlorate ($Ba(ClO_3)_2.H_2O$) calculate:
    (a) the number of mole of barium ion (Ba^{2+});
    (b) the number of mole of oxygen atoms;
    (c) the number of mole of chlorate ion (ClO_3^-).
18. If 2×10^{21} molecules of ammonia (NH_3) are removed from 100 milligram of ammonia, how many mole of ammonia is left?
19. One molecule of an unknown compound is found to have a mass of 3.27×10^{-22} g. Find the molecular weight of the compound.
20. One atom of gold is placed on one square of a chess board, two atoms are placed on the second square, four atoms on the third square, and so on. How many gram of gold atoms would be present on the 64th square?
21. A crystal of caesium chloride is an orderly array of caesium ions and chloride ions. The crystal may be regarded as built up of cubes, consisting of one caesium ion and one chloride ion (an ion pair) of edge 4.13×10^{-8} cm. The density of caesium chloride is 3.97 g cm^{-3}. Calculate the number of ion pairs per mole of caesium chloride.
22. What is the mass, in gram, of:
    (a) one calcium atom;
    (b) one lead atom?

Atomic weight determination

23. Cobalt oxide (CoO) contains 78.61 % of cobalt. Calculate the atomic weight of cobalt.
24. Find the atomic weight of nickel if 3.370 g of nickel was obtained by reduction of 4.286 g of the oxide (NiO).
25. 5.030 g of mercury(II) chloride ($HgCl_2$) was decomposed by electrolysis and 3.716 g of mercury was obtained. Find the atomic weight of mercury.
26. 4.150 g of tungsten was burned in chlorine and 8.950 g of tungsten chloride (WCl_6) was formed. Find the atomic weight of tungsten.
27. 1.1658 g of cobalt displaced 2.5826 g of gold from a gold bromide solution. The equation for the reaction is

$$3Co + 2Au^{3+} \rightarrow 3Co^{2+} + 2Au$$

If the atomic weight of gold is 197.0, find the atomic weight of cobalt.

28. 2.668 g of lithium chloride (LiCl) gave, on evaporation with nitric acid, 4.3389 g of lithium nitrate ($LiNO_3$). Calculate the atomic weight of lithium. The equation for the reaction is

$$LiCl + HNO_3 \rightarrow LiNO_3 + HCl$$

29. 7.542 g of bismuth bromide ($BiBr_3$) when reacted with silver nitrate solution yielded 9.486 g of silver bromide. Calculate the atomic weight of bismuth. The equation for the reaction is

$$BiBr_3 + 3AgNO_3 \rightarrow Bi(NO_3)_3 + 3AgBr$$

30. 15.85 g of zinc was dissolved in sulphuric acid and the hydrogen evolved was burned to form 4.336 g of water. From these data calculate the atomic weight of zinc.

3

Behaviour of Gases

It was pointed out in chapter 2 that once a concept of measurement by number of particles has been established we may seek for laws relating number of particles with other simple physical quantities, such as mass, volume, pressure and temperature. In the case of gases quite simple laws may be found making such connections. In fact, historically the mole concept was arrived at largely as a consequence of the study of the mass, volume, pressure and temperature relationships found for gases.

3.1 SOME PROPERTIES OF GASES

Gases show many marked differences in properties from solids and liquids. A study of such properties yields insight into the nature of gases.

(a) If 0.1 g of a gas is placed in a container of volume 1 dm^3 it will fill it completely, and 0.1 g of a gas will also fill a container of volume 1000 dm^3 completely. This would not be true for say 0.1 g of lead. These observations suggest that:

Gases occupy the whole volume of the containing vessel.

(b) One mole of nitrogen (of mass 28.0 g), at 0°C and 1 atmosphere pressure, occupies a volume of 22,400 cm^3. At 0°C and 50 atmosphere pressure the volume is 441 cm^3. One mole of nitrogen, in the liquid state, occupies a volume of 34.6 cm^3 at 1 atmosphere pressure. If the pressure becomes 50 atmosphere the volume becomes 34.5 cm^3. One mole of nitrogen, in the solid state, occupies a volume of 27.2 cm^3 at 1 atmosphere pressure. If the pressure becomes 50 atmosphere the volume is still nearly 27.2 cm^3. The observations suggest that:

Gases are easily compressed compared with their corresponding liquid or solid forms.

(c) Consider an experiment in which we place a gas jar of carbon dioxide above a gas jar of the reddish-brown gas, bromine, with a glass plate separating the two gases. If the plate is removed the brown colour of the bromine spreads evenly throughout the two jars, even though bromine vapour is denser than carbon dioxide. The bromine is said to *diffuse* into the upper vessel. It can also be shown that the carbon dioxide spreads evenly throughout the two vessels. This latter movement is a combination of diffusion and movement due to the weight of the gas. If the positions of the gases are reversed the carbon dioxide diffuses into the upper vessel whilst the bromine moves into the lower vessel, until again the composition is even throughout. The observations suggest that:

If two gases which do not react chemically are mixed in a vessel, the composition of the mixture becomes uniform throughout. We say that gases are miscible.

Liquids may also be miscible, for example, alcohol and water. But they may be immiscible, for example, oil and water. Solids are not normally miscible at room temperature and pressure.

(d) From the data given in property (b), we can calculate the densities of nitrogen in the gaseous, liquid and solid states. In the gaseous state the density of nitrogen, at 0°C and 1 atmosphere pressure, will be

$$\frac{28.0}{22{,}400} = 0.00125 \text{ g cm}^{-3}$$

In the liquid state, the density of nitrogen will be

$$\frac{28.0}{34.6} = 0.81 \text{ g cm}^{-3}$$

In the solid state, the density of nitrogen will be

$$\frac{28.0}{27.2} = 1.03 \text{ g cm}^{-3}$$

These observations suggest that:

The densities of gases are very low compared with the densities of the corresponding liquids and solids.

3.2 KINETIC MOLECULAR THEORY OF GASES

The properties of gases discussed above suggest a theory of gas structure which might account for the observed phenomena. Such a theory has been put forward and is known as the kinetic molecular theory of gases. This theory attempts to account for gas properties in terms of the constituent particles of the gas. It presents a 'micro' picture of gas behaviour. In later chapters some gas properties will be studied from a more general, or 'macro' viewpoint. Let us now consider the postulates of the kinetic theory.

(a) A gas consists of a large number of molecules placed far apart, relative to the size of the molecules. Between the molecules is a perfect vacuum.

We can gain some idea of the spacing if we realize that in 1 cm^3 of a gas there are about 2.7×10^{19} molecules (at 0°C, 1 atmosphere pressure). Thus each molecule is moving, on the average, in a volume of

$$\frac{1}{2.7 \times 10^{19}} = 4 \times 10^{-20} \text{ cm}^3$$

The volume of a molecule is of the order of 10^{-23} cm^3, since the radius of a molecule is of the order of 10^{-8} cm. Hence the average volume in which a molecule moves is about 4000 times greater than the volume of the molecule. Most of the volume of a gas is empty space.

(b) **Molecules are independent of one another.** There are no strong attractive or repulsive forces between molecules.

(c) **Molecules are in a state of rapid motion.** Molecules move with a velocity of the order of the speed of sound, 740 mph, at room temperature. Molecules collide with one another and with the walls of the containing vessel. It has been calculated that in 1 cm^3 of helium gas (at 0°C, 1 atmosphere pressure) there are 14×10^{28} collisions per second. It has been calculated that the distance the helium molecule moves between collisions, under the same conditions, is, on the average, 1.67×10^{-5} cm. In all collisions momentum and kinetic energy are conserved.

(d) **The average kinetic energy of the molecules is a measure of the temperature of the gas.** This postulate will be examined in detail in section 3.7.

Using these postulates we may qualitatively account for the observed properties discussed in section 3.1.

(a) Because molecules are in a state of rapid motion they will fill any container very quickly. The volume of the gas will be limited by the walls of the containing vessel, which restrict the motion of the molecules.

(b) Since the distances between molecules are large compared with their molecular dimensions we would expect a gas to be readily compressible. If we assume that in a liquid or solid the molecules are much closer together we would expect them to be much less compressible.

(c) The mixing of gases is to be expected since distances between molecules are great, permitting one gas to occur in the presence of another. The rapid motion of molecules would lead us to expect that a gas may spontaneously move out of a containing vessel, that is, the gas diffuses.

(d) Since the distances between molecules are large in gases, whereas the molecules are much closer in the liquid and solid states—in fact, virtually in contact—we would expect the density of a gas to be less than the density of the corresponding liquid or solid state.

Thus we can qualitatively account for observed properties of gases using this theory.

We shall see that it accounts both qualitatively and quantitatively for other gas properties we shall meet in this chapter.

3.3 THE GENERAL GAS EQUATION

We have seen that gases exert pressure on the walls of any containing vessel. Pressure has the dimension of force per unit area and will therefore be expressed in units of newton per square metre ($N\,m^{-2}$). In practice, gas pressures are frequently measured with a mercury manometer, and in this case the pressures are normally recorded as the length of the column of mercury supported by the gas pressure. The pressure exerted by the Earth's atmosphere at sea level supports a column of mercury around 760 mm high, and this value has been arbitrarily defined as being the 'standard atmosphere'. The atmosphere, so defined, is sometimes used as a unit of pressure so that

$$1 \text{ atmosphere} \equiv 760 \text{ mmHg}$$

760 mmHg correspond to a pressure of 101,325 $N\,m^{-2}$ so that we have

$$1 \text{ atmosphere} \equiv 760 \text{ mmHg} \equiv 101{,}325 \text{ N m}^{-2}$$
$$\simeq 101.3 \text{ kN m}^{-2}$$

The pressure of a gas might be expected to be a function of the number of collisions of molecules, with unit area of the wall in unit time, and hence to depend on the number of mole of molecules present. Suppose we place 0.01 mole of hydrogen in a vessel such as the one shown in Figure 3.1.

The vessel has a volume of 2 dm^3 and its temperature is maintained at 0°C. The number of mole per unit volume is 0.01/2 or 0.005 mol dm^{-3} and the observed pressure is 85.1 mmHg.

If we repeat this experiment using 0.01 mole of helium, we find that the pressure is again 85.1 mmHg. If we repeat this experiment for a large number of gases we find that:

The pressure exerted by a fixed number of mole of any gas, at constant volume and temperature, is constant, and is independent of the nature of the gas.

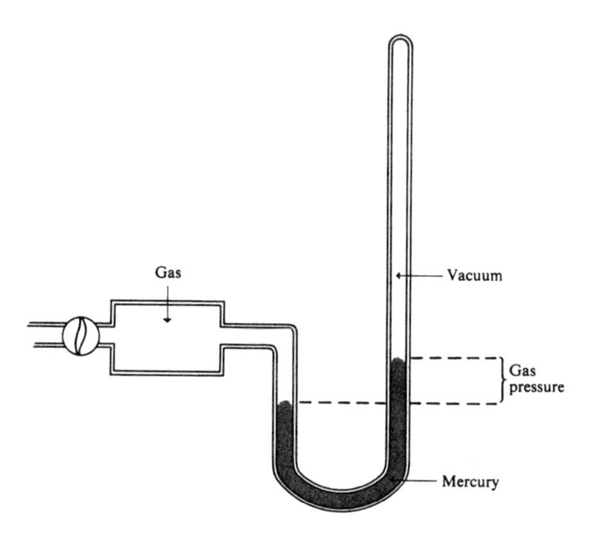

Fig 3.1 Apparatus to measure the pressure of a given number of mole of gas.

Suppose we repeat the above experiments using a vessel of volume 0.5 dm^3, still at 0°C. The number of mole per unit volume is now 0.01/0.5 or 0.02 mol dm^{-3}. Hence the number of mole per unit volume has increased four times (from 0.005 mol dm^{-3} to 0.02 mol dm^{-3}). The observed pressure is now 340.4 mmHg, which is four times as great as before (85.1 mmHg). Repetition of such experiments for different values of the volume (V) and number of mole (n), shows that the pressure is such that:

$$p \propto \frac{n}{V} \tag{1}$$

provided that the temperature is constant. This relation is demonstrated by Figure 3.2, in which pressures are expressed in kN m^{-2}.

EXERCISE
Express the pressures 85.1 mmHg and 340.4 mmHg observed on the mercury manometer in N m^{-2} and kN m^{-2}.

This relationship is consistent with the kinetic theory. If we increase the number of mole per unit volume we should expect that the number of collisions, with unit area of wall surface per unit time, will rise and hence the pressure should rise.

Does the pressure depend on any other factors? One obvious factor to investigate is temperature. Let us take 0.10 mole of a gas in a vessel of volume 2270 cm^3 and a temperature of 0°C. The observed pressure is 100 kN m^{-2}. If

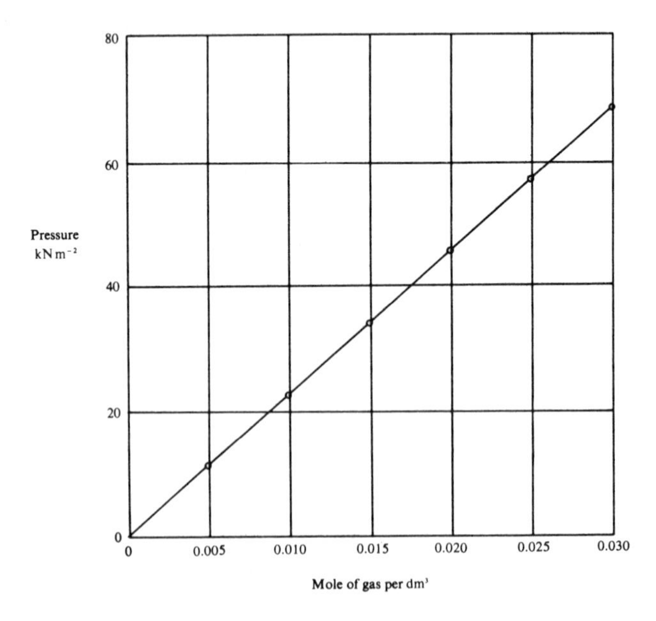

Fig 3.2 Variation of pressure with number of mole of gas per dm^3.

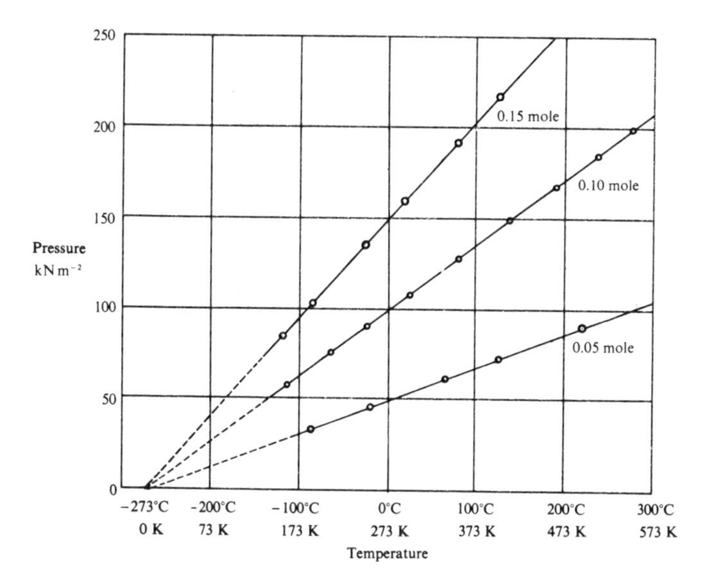

Fig 3.3 Variation of pressure with temperature for various amounts of gas.

the temperature is raised, the pressure is observed to rise, and if the temperature is lowered, the pressure is observed to drop. Plots of pressure versus temperature are shown in Figure 3.3 for three different amounts of gas. In each case, it is apparent that there is a linear relation between pressure and temperature, and inspection of the data shows that

$$p = \text{constant} \times (^\circ\text{C} + 273.16)$$

where the value of the constant depends on the amount of gas present. This simple relation has been used to define an absolute scale of temperature, which is defined as

$$T = {}^\circ\text{C} + 273.16$$

This temperature scale is called the kelvin scale, and units on this scale are called 'kelvin', symbol K. Having defined absolute temperature in this way, we can therefore write for any quantity of gas,

$$p \propto T \tag{2}$$

Combining relations (1) and (2), we can write

$$p \propto \frac{nT}{V}$$

$$\therefore p = \text{constant} \times \frac{nT}{V}$$

The value of the constant will depend only on the units chosen. We choose to measure pressure in $N\,m^{-2}$, volume in m^3, temperature in kelvin, hence we can write

$$pV = nRT$$

where R is a constant known as the *gas constant*. The relation is known as the *general gas equation*. This equation provides us with an algebraic expression by which the pressure, temperature, volume and number of mole of any sample of any gas are related.

EXAMPLE 1

Show, for any number of different gases at the same temperature and at the same pressure, that equal volumes contain an equal number of mole (or equal numbers of molecules).

SOLUTION

For any two gases A and B,

$$p_A V_A = n_A R T_A \quad \text{and} \quad p_B V_B = n_B R T_B$$

$$\text{Since } p_A = p_B, \quad V_A = V_B \quad \text{and} \quad T_A = T_B,$$

and R is a constant, then

$$n_A = n_B$$

This of course holds true for *any* gases, as A and B are any gases. This general relationship is experimentally based, and has been known as *Avogadro's law*.

EXAMPLE 2

Obtain the numerical value of the gas constant R.

SOLUTION

This could be found from experimental data for the pressure, temperature and volume of any quantity of any gas. Selecting one such set of data, it has been found that 0.1 mole of hydrogen at a pressure of 754 mmHg and a temperature of 17°C occupies a volume of 2.40 dm^3.

So

$$T \text{ (in unit of K)} = 290$$

$$p \text{ (in unit of N m}^{-2}\text{)} = 754/760 \times 101{,}325$$

$$V \text{ (in unit of m}^3\text{)} = 2.40 \times 10^{-3}$$

$$\text{As } R = \frac{pV}{nT}$$

$$\therefore \quad R = \frac{(754/760) \times 101{,}325 \times 2.40 \times 10^{-3}}{0.1 \times 290}$$

$$= 8.31$$

When the quantities used are expressed in the units shown, the unit of R is $J\,K^{-1}\,mol^{-1}$. *The value of the gas constant is therefore* 8.31 $J\,K^{-1}\,mol^{-1}$. This is true for any volume of any gas, at any temperature and any pressure. It is a numerical value which should be committed to memory.

EXERCISE

Remembering the unit of J to be kg $m^2 s^{-2}$, and that of N to be kg m s^{-2}, confirm that the unit of R is J K^{-1} mol^{-1}.

EXAMPLE 3

0.05 mole of gaseous argon is maintained in a vessel of volume 500 cm^3 at 25°C. What pressure would it exert?

SOLUTION

$$n = 0.05$$

$$T = 298 \qquad \text{(K)}$$

$$V = 500 \times 10^{-6}\ (\text{m}^3)$$

$$R = 8.31$$

$$\therefore p = \frac{nRT}{V}$$

$$= \frac{0.05 \times 8.31 \times 298}{500 \times 10^{-6}}$$

$$= 248{,}500$$

The pressure is therefore 248,500 N m^{-2}. Notice the importance, when carrying out calculations with the general gas equation, of expressing pressure in N m^{-2}, volume in m^3, temperature in K, and the gas constant as 8.31 J K^{-1} mol^{-1}.

3.4 MOLECULAR WEIGHT DETERMINATION

Suppose we have m g of a gas of molecular weight M, then the number of mole of gas is

$$n = m/M$$

where m is in gram; hence, from the general gas equation

$$pV = \frac{m}{M} \cdot RT$$

$$\therefore M = \frac{mRT}{pV}$$

This relationship enables us to determine M experimentally. Notice that, since the density of the gas is

$$\rho = m/V$$

we can write

$$M = \rho \frac{RT}{p}$$

where ρ is in *gram* per cubic metre. Finally, if we have two gases whose densities have been determined at the same temperature and pressure, then

$$\frac{M_1}{M_2} = \frac{\rho_1}{\rho_2}$$

EXAMPLE

Calculate the molecular weight of a gas if 0.574 g occupies 548 cm³, at 22°C and a pressure of 740 mmHg.

SOLUTION

$$p = \frac{740}{760} \times 101{,}300 = 98{,}800 \text{ N m}^{-2}$$

$$V = 548 \times 10^{-6} \text{ m}^3$$

$$T = 22 + 273 = 295 \text{ K}$$

$$\therefore M = \frac{0.574 \times 8.31 \times 295}{98{,}800 \times 548 \times 10^{-6}}$$

$$= 26.1$$

EXAMPLE

560 cm³ of carbon dioxide weighs 1.00 g at a certain temperature and pressure, whilst 880 cm³ of another gas, at the same temperature and pressure, weigh 1.50 g. Find the molecular weight of the second gas.

SOLUTION

If M is the molecular weight of the second gas and ρ g m^{-3} is its density, then

$$\frac{M}{M_{CO_2}} = \frac{\rho}{\rho_{CO_2}}$$

$$\therefore \frac{M}{44} = \frac{1.50/880}{1.00/560}$$

$$\therefore M = 42$$

The ratio of molecular weights of gases may be determined experimentally using an instrument called a *gas density balance*.

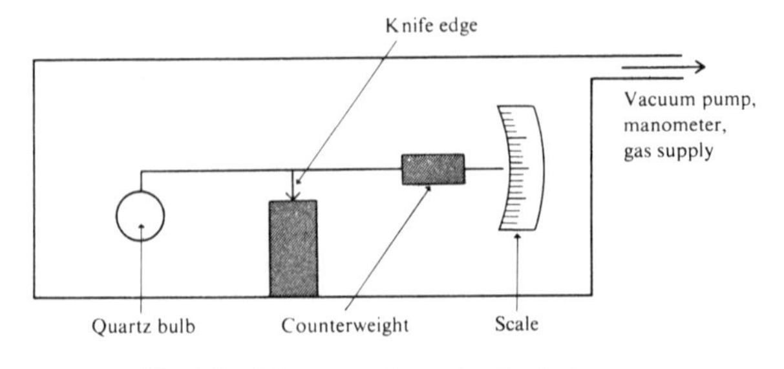

Fig 3.4 Diagram of gas density balance.

A bulb is suspended from a beam which is counterweighted on the other side by a small mass; the whole apparatus is about 10 cm long and 3 cm wide. It is evacuated and a gas of known molecular weight M_1, is slowly admitted; the bulb displaces the gas and the pointer moves downwards. The gas is admitted until the pointer reaches some pre-determined position on the scale, and the gas pressure (p_1) is then recorded. The whole procedure is repeated with a gas of unknown molecular weight, M_2, until the pointer again reaches the pre-determined position, and the pressure (p_2) is again recorded. If ρ_1 g m^{-3} and ρ_2 g m^{-3} are the respective densities of the two gases, we can write for the first gas

$$M_1 = \rho_1 \cdot \frac{RT}{p_1}$$

and for the second gas

$$M_2 = \rho_2 \cdot \frac{RT}{p_2}$$

Since the bulb was equally displaced on each occasion, the densities in both cases must have been equal, hence

$$\rho_1 = \rho_2$$

$$\therefore \frac{M_1}{M_2} = \frac{p_2}{p_1}$$

Thus the molecular weight of the unknown gas may be found. This instrument is very suitable for determining the molecular weights of small samples of gas.

3.5 PARTIAL PRESSURES IN A GAS MIXTURE

Suppose we place n_1 mole of a gas in a vessel of volume V m^3 at a temperature of T K. From the general gas equation, the pressure (p_1 N m^{-2}) exerted by the gas is

$$p_1 = \frac{n_1 RT}{V}$$

If we now add n_2 mole of a second gas to the same vessel, then as the general gas equation is equally applicable to this gas, its pressure (p_2 N m^{-2}) is

$$p_2 = \frac{n_2 RT}{V}$$

For the mixture we can again apply the general gas equation and the total pressure of the mixture (p_T N m^{-2}) is

$$p_T = \frac{(n_1 + n_2)RT}{V}$$

but

$$p_1 + p_2 = \frac{(n_1 + n_2)RT}{V}$$

$$\therefore p_T = p_1 + p_2$$

Hence, if we have a mixture of gases, then provided they do not react, we can state that:

The total pressure of a mixture of gases is the sum of the partial pressures of the constituent gases, at constant temperature.

This is sometimes called the *Law of Partial Pressures*. The *partial* pressure of a gas is the pressure which the gas would exert if it alone occupied the volume of the mixture at the particular temperature.

EXAMPLE 1

250 cm³ of carbon monoxide gas was collected in an inverted jar, over water, at 10°C. The barometric pressure was 765 mmHg. Find the number of mole of carbon monoxide collected.

SOLUTION

The jar will contain water vapour in addition to the gas. At 10°C the pressure due to the water vapour is 9.2 mmHg, provided the gas is saturated with water vapour.

From the law of partial pressures we can write

$$\text{gas pressure} + \text{water vapour pressure} = \text{atmospheric pressure}$$
$$\therefore \text{gas pressure} + 9.2 = 765$$
$$\therefore \text{gas pressure} = 755.8 \text{ mmHg.}$$

From the general gas equation

$$n = \frac{pV}{RT}$$
$$= \frac{(755.8/760) \times 101{,}300 \times 250 \times 10^{-6}}{8.31 \times 283}$$
$$= 0.0107 \text{ mole}$$

EXAMPLE 2

A vessel of volume 500 cm³ contains a gas A, which exerts a pressure of 76.0 mmHg at 0°C. A second vessel of volume 800 cm³ contains a gas B, which exerts a pressure of 152 mmHg. The vessels are connected. What is the total pressure in the system at 0°C?

SOLUTION

For gas A, the volume has increased from 500 cm³ to (500 + 800) = 1300 cm³. Thus the pressure has decreased to

$$\frac{500}{1300} \times 76.0 = 29.2 \text{ mmHg}$$

For gas B, the volume has increased from 800 cm³ to (500 + 800) = 1300 cm³. Thus the pressure has decreased to

$$\frac{800}{1300} \times 152 = 93.5 \text{ mmHg}$$

The total pressure is the sum of the partial pressures and is therefore

$$29.2 + 93.5 = 122.7 \text{ mmHg}$$

ALTERNATIVE SOLUTION

For gas A, in its original volume, we can write

$$p_A V_A = n_A RT$$

$$\therefore n_A = \frac{p_A V_A}{RT}$$

$$= \frac{(76/760) \times 101{,}300 \times 500 \times 10^{-6}}{8.31 \times 273}$$

$$= 0.00223 \text{ mole}$$

For gas B, in its original volume, we can write

$$p_B V_B = n_B RT$$

$$\therefore n_B = \frac{p_B V_B}{RT}$$

$$= \frac{(152/760) \times 101{,}300 \times 800 \times 10^{-6}}{8.31 \times 273}$$

$$= 0.00714 \text{ mole}$$

Hence for the mixture the total number of mole is (0.00223 + 0.00714) = 0.00937 mole.
Hence the total pressure (p_T N m^{-2}), when the gases mix, is

$$p_T = \frac{(n_A + n_B)RT}{(V_A + V_B)}$$

$$= \frac{0.00937 \times 8.31 \times 273}{1300 \times 10^{-6}}$$

$$= 16{,}400 \text{ N m}^{-2}$$

The pressure may also be given as $\frac{16{,}400}{101{,}300} \times 760 = 122.7$ mmHg.

Notice that in using the general gas equation we have been careful to see that the volume is in m^3, the pressure in N m^{-2}, and the temperature in kelvin.

Clearly the first solution is the superior one, in this particular problem, because of its greater simplicity.

3.6 REAL AND IDEAL GASES

The general gas equation was derived from a set of experimentally based proposals about gas behaviour. If gases do not conform to these proposals under all conditions, then the general gas equation will not exactly describe gas behaviour; i.e. it will be only an approximation to the behaviour of real gases. From the general gas equation, it follows that, for a given number of mole, and at constant temperature, $pV = nRT =$ constant. Thus, for a gas obeying the general gas equation, the product $p \times V$ should be independent of the individual values of p and V, and a plot of pV versus p should be a horizontal line. For all real gases, however, it is found that, at pressures around atmospheric, plots of pV versus p deviate noticeably from the horizontal. Data for a number of gases are given in Figure 3.5. It can be seen, for example that, as the pressure rises, carbon dioxide is more compressible than expected, and hydrogen less so. Because real gas behaviour deviates from that predicted by the general gas equation, it has become useful to refer to the hypothetical gas which would obey the general gas equation as an 'ideal' gas, and indeed the general gas equation is sometimes called 'the

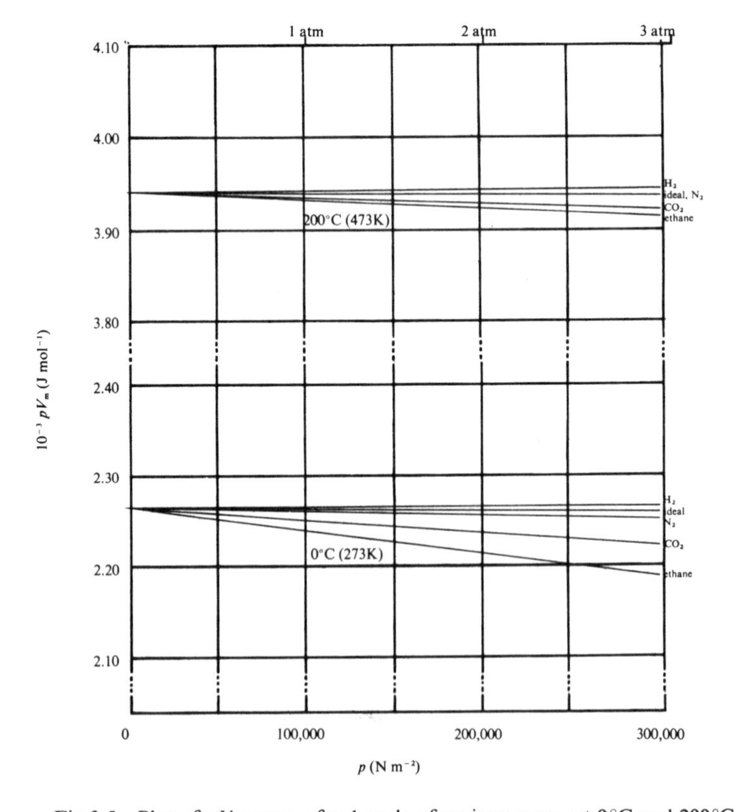

Fig 3.5 Plot of pV versus p for 1 mole of various gases, at 0°C and 200°C.

ideal gas equation'. No real gases ever quite behave ideally, although nearly all gases approximate quite well to ideal gas behaviour at pressures around atmospheric, and gases like hydrogen and helium approximate very closely to ideal behaviour at low pressures and high temperatures.

The growing importance of deviations from ideal behaviour with falling temperature is also shown in Figure 3.5 in which pV is plotted versus p at both 0°C and 200°C. Notice that deviations from ideality are larger at the lower temperature, but that even in the case of the largest deviation shown (ethane), the deviation from ideal behaviour is less than 3 per cent at a pressure of 3 atmospheres. It is sometimes convenient to record the extent of the deviation of the behaviour of a particular gas from ideality by recording the molar volume ($V_m = V/n$) of the gas (i.e. the volume occupied by 1 mole) under some standard set of conditions. Data are usually quoted for a gas pressure of 1 standard atmosphere (101,300 N m^{-2}) and 0°C (273 K), and these conditions are referred to as S.T.P.—standard temperature and pressure. We can calculate the molar volume expected for an ideal gas at S.T.P. from the ideal gas equation, since

$$V_m = \frac{RT}{p} = \frac{8.31 \times 273}{101{,}300} \quad (n = 1)$$

$$= 0.022414 \text{ m}^3 \text{ mol}^{-1} = 22.414 \text{ dm}^3 \text{ mol}^{-1}$$

Such data are shown in Table 3.1.

Table 3.1

Molar volumes of real gases

Gas	*Formula*	*Density (g dm^{-3} at S.T.P.)*	*Molar mass (g)*	*Molar volume (dm^3 at S.T.P.)*
Hydrogen	H_2	0.08989	2.016	22.430
Helium	He	0.1782	4.003	22.46
Ideal gas	—	—	—	22.414
Nitrogen	N_2	1.251	28.013	22.400
Carbon monoxide	CO	1.251	28.011	22.402
Oxygen	O_2	1.429	31.999	22.392
Carbon dioxide	CO_2	1.977	44.010	22.262
Ammonia	NH_3	0.7711	17.031	22.093
Chlorine	Cl_2	3.215	70.906	22.061
Sulphur dioxide	SO_2	2.927	64.063	21.887

It can be seen that the easily liquefiable gases (SO_2, Cl_2 and NH_3) show the greatest deviations from ideal behaviour, while those gases which are difficult to liquefy (H_2, He, N_2, CO, O_2) approximate more nearly to ideal behaviour. The same conclusions can be drawn from a consideration of Figure 3.5.

Real gases show deviations from ideal behaviour for two reasons. First, molecules do have a finite volume, whereas the kinetic theory supposes them to be point particles. Further, the kinetic theory assumes that molecules do not interact with one another. In fact, weak forces do act between molecules, their magnitude falling off sharply as the intermolecular distance increases. It is because of the existence of these forces that gases can be condensed to liquids.

A number of empirical equations have been put forward to account for the pV curves of real gases. One of the most useful of these equations is the Van der Waals equation:

$$\left(p + \frac{a}{V_m^2}\right)(V_m - b) = RT$$

where a is a constant which takes account of the attractive forces between gas molecules, and b is a constant which takes account of the finite size of molecules. Values of a and b have been determined experimentally, and are tabulated for many gases. A more general but completely empirical equation which is useful for dealing with gross deviations from ideal behaviour is the virial equation:

$$pV_m = RT\left(1 + \frac{B}{V_m} + \frac{C}{V_m^2} + \frac{D}{V_m^3} + \ldots\right)$$

where B, C, D are called the second, third, fourth . . . virial coefficients, and these are determined experimentally. They are found to vary with temperature.

3.7 KINETIC MOLECULAR THEORY AND TEMPERATURE

It was mentioned in section 3.2 that gas molecules are in a state of rapid motion. The range of velocities shown by molecules can be experimentally determined. These velocities range widely, from a few molecules which are hardly moving to a few moving at very high velocities with most molecules moving at velocities between these extremes. Molecules will possess kinetic energy due to this motion and the kinetic energy will also range widely. At a given temperature the average kinetic energy of a molecule is

$$\tfrac{1}{2}m\overline{u^2}$$

where m is the mass of a molecule, and $\overline{u^2}$ is the mean of the squares of the velocities of the molecules. In the mathematical development of the kinetic theory, the average of the squares of the velocities $(\overline{u^2})$ is used, rather than the square of the average velocity $(\bar{u})^2$.

As molecules collide, their individual kinetic energies will alter but the overall kinetic energy will remain fixed. If the temperature rises, the individual molecular velocities rise and so does the mean square velocity $(\overline{u^2})$. In fact, it is believed that the absolute temperature (T K) is such that

$$T \propto \tfrac{1}{2}m\overline{u^2}$$

This is, in fact, the fourth postulate of the kinetic theory given in section 3.2.

If two gases are at different temperatures they will have different average kinetic energies. If the gases are mixed the molecules will collide and a redistribution of kinetic energy will take place, until again the mean square velocity is constant for the mixture. The two gases come to the same temperature.

The distribution of molecular velocities or kinetic energies for a gas such as oxygen may be determined experimentally, or calculated. For oxygen at three different temperatures the distribution is shown in Figure 3.6.

For oxygen the mean velocity at 0°C is 461 m s^{-1}. If the temperature rises to 227°C, the mean velocity becomes 624 m s^{-1}. It has increased by 35 per cent or about one-third. If we consider higher velocities, say in excess of 1000 m s^{-1}, we find that at 0°C about 0.3 per cent have such velocities, whilst at 227°C about 5 per cent have such velocities, an increase of seventeen times or 1700 per cent. We shall see the importance of these relative changes when we deal with rates of chemical reactions.

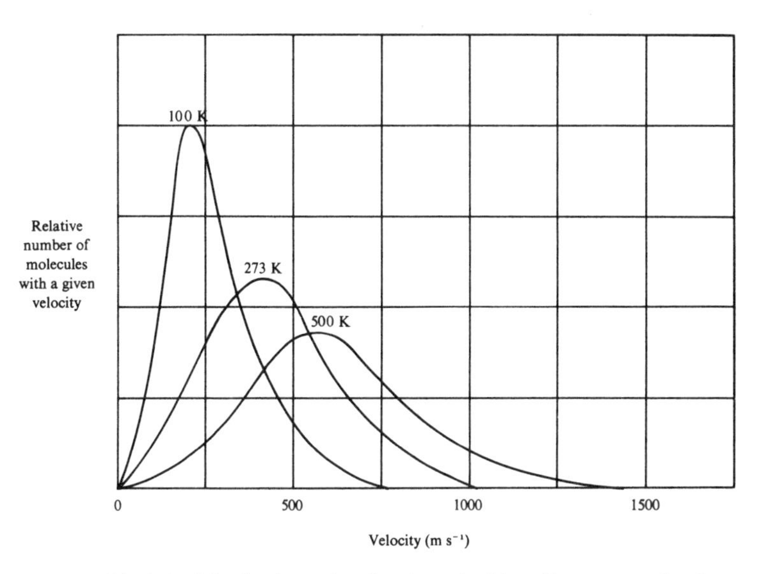

Fig 3.6 Distribution of molecular velocities of oxygen molecules at different temperatures.

For a fixed number of mole of gas (n) at a fixed temperature and constant volume

$$T \propto \tfrac{1}{2}nM\overline{u^2}$$

where M g is the mass of 1 mole. If the temperature rises, then $\overline{u^2}$ rises, hence we should expect that the number of collisions in unit time per unit area of wall surface would rise. This is consistent with the relation between pressure and temperature discussed in section 3.3, namely

$$p \propto T$$

At 0 K pressure is zero, hence the molecules are motionless and kinetic energy is also zero. This is a theoretical extreme for a gas in that all gases (except helium) solidify or liquefy well before the absolute zero could be attained.

3.8 KINETIC MOLECULAR THEORY AND PRESSURE

Consider n mole of gas of volume V m^3, at a constant temperature T K. The pressure of the gas is proportional to the number of collisions in unit time on unit area of wall surface.

The number of collisions will depend on the number of mole per unit volume (n/V) and on the mean molecular velocity (u). We will take as a measure of mean velocity the square root of the mean square velocity,

$$u = (\overline{u^2})^{\frac{1}{2}}$$

Hence the number of collisions in unit time per unit area of wall surface is proportional to $\frac{nu}{V}$,

hence the pressure $\propto \frac{nu}{V}$

The pressure will also depend on the mean momentum of each molecule, mu. Thus the pressure (p) depends on the number of collisions and on the mean momentum, hence

$$p \propto \frac{nu}{V} \times mu$$

or,

$$p \propto \frac{nm\overline{u^2}}{V}$$

But we have stated that the temperature is proportional to the mean kinetic energy,

$$T \propto \tfrac{1}{2}m\overline{u^2}$$

$$\therefore \quad pV = \text{another constant} \times nT$$

We have established in section 3.3, as a result of experiment, the general gas equation

$$pV = nRT$$

In this section we have established this same equation as a direct consequence of the kinetic molecular theory.

QUESTIONS

1. Account for the following observations in terms of the kinetic molecular theory:
   (a) the volume of a gas can be readily reduced by increasing the applied pressure;
   (b) two gases can readily be mixed together;
   (c) the pressure exerted by a gas increases as its temperature is raised;
   (d) 6 gram of nitrogen in a vessel of volume 3 dm^3, at a certain temperature and pressure, exert twice the pressure of 3 gram of nitrogen under the same conditions.

2. The volume occupied by 39.9 g of argon at 1000°C, 760 mmHg pressure, is 104.4 dm^3. Each argon atom can be assigned a radius of 1.54×10^{-8} cm. Calculate the fraction of the gas which is due to the volume of the molecules.
3. State qualitatively the effect on the number of molecular collisions per unit time on unit area of the wall of a vessel, containing 0.1 mole of neon, if the following changes take place:
   (a) the temperature is raised at constant volume;
   (b) the volume is increased at constant temperature;
   (c) 0.05 mole of argon is added at constant temperature and volume;
   (d) 0.05 mole of argon is added at constant temperature and pressure.
4. One dm^3 of ammonia at 25°C, 8×10^4 N m^{-2} pressure is compressed (at constant temperature) so that the pressure becomes 8×10^5 N m^{-2}. The volume is then found to be 96 cm^3. At 5×10^6 N m^{-2} pressure and 25°C, the gas has become a liquid of volume 0.9 cm^3. Calculate the volume the gas would occupy at these two pressures, assuming ideal behaviour. What are the factors leading to non-ideal behaviour of gases? Under what conditions of temperature and pressure will ammonia behave most like an ideal gas?

Law of partial pressures

5. A student collects some nitrogen over water. He finds he has collected 5×10^{-2} mole of nitrogen and 6.83×10^{-4} mole of water vapour. If the total pressure is 762 mmHg, find the pressure due to the water vapour.
6. 250 cm^3 of oxygen is collected over water at 20°C. The total pressure is found to be 768 mmHg and the vapour pressure of water at 20°C is 17.5 mmHg. Calculate the partial pressure of the oxygen. What would this partial pressure become if the volume was reduced to 100 cm^3?
7. A vessel of volume 1500 cm^3 contains nitrogen and a little water. The temperature is 15°C and the total pressure is 750 mmHg. The vapour pressure of water at 15°C is 12.8 mmHg. The volume of the vessel is halved. What will be the new total pressure?
8. A flask contains 2 dm^3 of hydrogen at 3×10^5 N m^{-2} pressure. A second flask contains 6 dm^3 of argon at 1×10^5 N m^{-2} pressure. If the flasks are connected, what will be the final pressure, the temperature remaining constant throughout?
9. One dm^3 of oxygen at 2×10^5 N m^{-2} pressure, 2 dm^3 of argon at 0.5×10^5 N m^{-2} pressure, and 4 dm^3 of helium at 380 mmHg, are passed into a vessel of volume 8 dm^3. What is the final pressure in the vessel, the temperature remaining constant throughout?
10. 1.6 g of oxygen and 0.08 mole of nitrogen are placed in a vessel of volume 150 cm^3 at 10°C. What is the pressure of the mixture?
11. Two gram of argon and 0.4 g of helium are placed in a vessel containing some water at 0°C. The volume of the vessel is 4.3 dm^3 and the total pressure is 599.3 mmHg. Calculate the vapour pressure of ice at 0°C.
12. A vessel, of volume 100 cm^3, contained hydrogen and also a little chloroform liquid. The temperature of the vessel was 20°C and the vapour pressure of chloroform at 20°C is 160.5 mmHg. If the total pressure was 680 mmHg, calculate the number of mole of hydrogen present. What assumptions must be made in this calculation?

General gas equation and mole quantities

13. Calculate the number of mole of hydrogen sulphide (H_2S) in:
    (a) 12 g of hydrogen sulphide;

(b) 6 dm³ of hydrogen sulphide, at S.T.P.;
(c) 6 dm³ of hydrogen sulphide, at 100°C, 737 mmHg.

14. A gas sample contains 0.4 g of helium and 0.4 g of hydrogen. Calculate:
(a) the number of mole of each gas;
(b) the number of molecules of each gas;
(c) the volume of the mixture, at S.T.P., in dm³;
(d) the volume of the mixture, at 27°C, 700 mmHg.

15. A sample of methane (CH_4) has a mass of 20 g. Calculate:
(a) the number of mole of methane;
(b) the number of molecules of methane;
(c) the number of mole of hydrogen atoms;
(d) the volume, in dm³, at 17°C, 720 mmHg.

16. A flask of 1 dm³ contains 0.5 g of hydrogen. A second flask of volume 2 dm³ contains 8.4 g of nitrogen. Both flasks are at 0°C. Which flask contains the greater number of mole of molecules? In which flask is the pressure greater?

17. 0.04 mole of gas is collected over water at 20°C, and 768 mmHg. The water vapour pressure at 20°C is 17.5 mmHg. What is the volume occupied by the gas?

18. A vessel has a volume of 5 dm³. It contains nitrogen at 10°C, and 764 mmHg. It is desired to raise the pressure to 900 mmHg, without altering the volume or temperature. How may this be done?

19. The density of liquid carbon tetrachloride (CCl_4) is 1.594 g cm^{-3}. It is desired to fill a 2-dm³ vessel with the vapour at 100°C, and 769 mmHg. What volume of liquid will be required to provide this vapour?

20. Two glass bulbs have the same volume. One is filled with oxygen, and the other is filled with sulphur dioxide gas, both being at the same temperature and pressure.
(a) Compare the number of mole of each gas.
(b) Compare the number of molecules of each gas.
(c) Compare the number of gram of each gas.
If the temperature of the bulb containing oxygen is doubled, the volume remaining constant, how do the two gases now compare in:
(a) pressure;
(b) number of moles;
(c) number of molecular collisions per unit time on unit area of wall surface?

21. Which gas sample contains more molecules, 600 cm³ of a gas at 600 mmHg, and 2(C, or 550 cm³ of the same gas at 770 mmHg, and 0°C?

22. It is possible, under special conditions, to obtain pressures as low as 10^{-10} mmHg. Calculate the number of molecules in 22.414 dm³ of hydrogen at this pressure and at 10°C.

23. Find the density in g dm^{-3} of chlorine trifluoride gas (ClF_3) at:
(a) S.T.P.;
(b) 20°C, 764 mmHg.

24. The density, at S.T.P., of sulphur dioxide gas is 2.9267 g dm^{-3}. Calculate the mass of 1 mole of the gas. What is the mass of 1 mole of the gas calculated from the known atomic weights? Why are the values different?

25. 0.612 g of a gas occupy a volume of 419 cm³ at 70°C, and 742 mmHg. Calculate the molecular weight of the gas.

26. A certain volume of the gas chlorine monofluoride weighs 1.945 g under certain conditions of temperature and pressure. The same volume of nitrogen under the same conditions weighs 1.000 g. If the atomic weight of nitrogen is known to be 14, calculate the molecular weight of chlorine monofluoride.

27. The ratio of the density of phosphine gas to the density of argon gas, both under the same conditions of temperature and pressure, is 0.85. If the atomic weight of argon is 40, calculate the molecular weight of phosphine.
28. By careful measurement and suitable extrapolation the limiting value for the quantity ρ/p, as p approaches zero, has been found to be 1.2495 for carbon monoxide gas (CO), at 0°C. Calculate, as accurately as the data allow, the molecular weight of carbon monoxide. Compare this with the value found by summing the known atomic weights in the formula.
29. A gas has a density of 1.27 g dm^{-3} at 30°C, and 747 mmHg. Find the mass of 1 mole of the gas.
30. Find the molecular weight of a gas if 250 cm^3 of the dry gas weighed 0.981 g at 30°C, and 710 mmHg.
31. 1.56 dm^3 of a gas is collected over water at 25°C and a barometric pressure of 758 mmHg. If the mass of the gas is 2.02 g and the vapour pressure of water at 25°C is 23.5 mmHg, calculate the molecular weight of the gas.
32. Some liquid carbon tetrachloride (CCl_4) is spread out on a solid surface to form a film of area 100 cm^2, which is believed to be 1 molecule thick. When vaporized the volume of the vapour is 9.18×10^{-4} cm^3 at S.T.P. Given the molecular weight of carbon tetrachloride and the density of the liquid (1.594 g cm^{-3}), calculate the thickness of the film and hence the approximate volume of 1 molecule of carbon tetrachloride, assuming it has a cubic shape.
33. Suppose the atomic weight of $^{12}_{6}C$ had been taken as 100 units, what would this have made the value of the Avogadro constant compared to its value on the scale $^{12}_{6}C = 12.0000$? On the scale $^{12}_{6}C = 100$, how many atoms of $^{12}_{6}C$ would there be in 12 g of $^{12}_{6}C$?
34. What is the molar volume of water under each of the following conditions;
    (a) solid at 0°C, density of ice is 0.915 g cm^{-3};
    (b) liquid at 0°C, density of liquid water is 1.000 g cm^{-3};
    (c) gas at 100°C, density of water vapour (at 100°C, 1 atmosphere pressure) is 5.88×10^{-4} g cm^{-3}.
35. When carbon monoxide and oxygen react they form carbon dioxide. Energy is released in this reaction so that the temperature rises. Two mole of carbon monoxide react with 1 mole of oxygen in a sealed vessel at 4×10^5 N m^{-2} initial pressure and 25°C.
    (a) What is the final pressure after reaction, temperature still at 25°C?
    (b) What is the final pressure after reaction, if the temperature rises to 300°C?
36. At what temperature (in °C) would ethylene (C_2H_4) have the same density at 740 mmHg as has nitrogen at 50°C, and 783 mmHg?

Stoichiometry

The mole concept finds one of its most useful applications in the calculation of the masses and volumes of chemical substances taking part in, or produced by, chemical reactions. From the chemical equation for the reaction we can determine the ratio of moles of reactants and products. Once this ratio is known, we can calculate the mass or volume of some other reactant or product. The study of the quantitative relationships implied by a chemical reaction is called *stoichiometry* (Greek—*stoicheion*—element).

4.1 CHEMICAL FORMULAE

Chemical formulae provide a convenient shorthand method of representing, both qualitatively and quantitatively, the composition of chemical substances. We may distinguish a number of different types of formulae, each one having its own particular use and value.

(1) An empirical formula *shows the elements present in the compound and the ratio of the number of mole of each element present.* Ethane is known to contain carbon and hydrogen united in the mole ratio 1 : 3. The empirical formula is CH_3. In one molecule of ethane the atoms of carbon and hydrogen are united in the ratio 1 : 3.

In ethane we have discrete molecules, but in the case of sodium chloride there are no discrete molecules. This substance is made up of an assembly of sodium ions, Na^+, and chloride ions, Cl^-. Since the compound is electrically neutral the number of sodium ions in the crystal must equal the number of chloride ions in the crystal. The ions occur in the ratio by number 1 : 1 in the crystal. The formula NaCl may be regarded as an empirical formula. In the case of sodium sulphate, the sodium ions and sulphate ions occur in the ratio by number 2 : 1. The empirical formula is Na_2SO_4.

For the compound silicon carbide, the silicon and carbon atoms are bonded to one another, in a network, throughout the crystal. No discrete molecules can be distinguished. The ratio by number of silicon and carbon atoms is 1 : 1. The empirical formula is SiC. Similarly, silicon dioxide is an array of silicon and oxygen atoms bonded in a network, the ratio by number of the atoms being 1 : 2. The empirical formula of the compound is SiO_2.

(2) A molecular formula *shows the actual numbers of atoms of each element in one molecule of the compound.* A molecular formula thus tells us the number of mole of each element in 1 mole of the compound. It is strictly applicable only to substances existing as discrete molecules, hence we would not expect to speak of the molecular formula of sodium chloride, which is made up of ions, or of the molecular formula of silicon carbide, which is made up of atoms bonded together into a network.

The mass of 1 mole of ethane may be determined by gas density measurements, and is 30 g. The mass of 1 mole of CH_3 is 15 g, hence the molecular formula of ethane must be C_2H_6. One molecule of ethane consists of two atoms of carbon and six atoms of hydrogen. One mole of ethane (30 g) contains 2 mole of carbon (24 g) and 6 mole of hydrogen atoms (6 g).

EXERCISE

What are the empirical formulae of these elements and compounds, whose molecular formulae are given: water (H_2O), nitrogen (N_2), hydrogen peroxide (H_2O_2), oxygen (O_2), ozone (O_3), benzene (C_6H_6), propane (C_3H_8), sulphur (S_8)?

(3) A structural formula *shows the way in which atoms are bonded in a molecule and the geometrical arrangement of the atoms.* In phosphorus vapour we can detect phosphorus molecules of empirical formula P. The molecular formula of phosphorus is P_4, and the structural formula is shown in Figure 4.1.

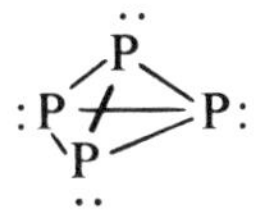

Fig 4.1 Formula of phosphorus vapour.

In the above figure each stroke represents a bond uniting two atoms. Similarly, the empirical formula of ethane is CH_3. The molecular formula of ethane is C_2H_6 and the structural formula is shown in Figure 4.2.

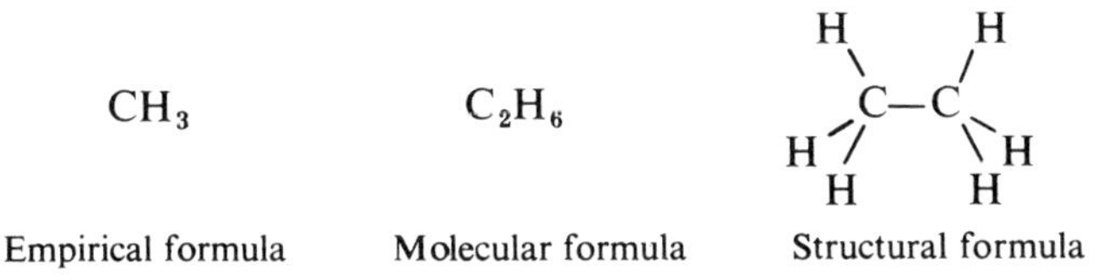

Fig 4.2 Formulae of ethane.

In chapters 8 and 9 we will meet *valence structures* of molecules. Valence structures convey much the same information as structural formulae, but in some cases convey it in a more detailed fashion.

4.2 CALCULATIONS INVOLVING FORMULAE

EXAMPLE 1

Calculate the percentage of fluorine in the compound uranyl fluoride (UO_2F_2).

SOLUTION

One mole of this compound contains 1 mole of uranium atoms, 2 mole of oxygen atoms, and 2 mole of fluorine atoms.

$$\text{The mass of 1 mole} = \{238 + (2 \times 16.0) + (2 \times 19.0)\}$$
$$= 308 \text{ g}$$

One mole of the compound contains $2 \times 19.0 = 38.0$ g of fluorine. Hence the percentage of fluorine

$$= \frac{38}{308} \times 100$$
$$= 12.3 \text{ per cent}$$

EXAMPLE 2

In 1964 the preparation of a complex compound of potassium, molybdenum, and cyanide ($K_xMo_y(CN)_z$) was reported. The percentage composition by mass was found to be K = 25.51 per cent, Mo = 31.54 per cent. Calculate the empirical formula of the compound.

SOLUTION

In 100 g of the compound there would be 25.51 g of potassium, 31.54 g of molybdenum, and 42.95 g of cyanide.
Hence the ratio

mole of potassium : mole of molybdenum : mole of cyanide

$$= \frac{25.51}{39.1} : \frac{31.54}{95.9} : \frac{42.95}{26.0}$$
$$= 0.652 : 0.329 : 1.65$$
$$= 1.98 \;\; : 1.00 \;\; : 5.02$$

Because of the experimental errors involved in the data, the ratio is not exactly integral and must be rounded off to the nearest whole number ratio.

Hence the required empirical formula is $K_2Mo(CN)_5$.

EXAMPLE 3

0.220 g of a volatile compound containing carbon, hydrogen and chlorine yielded on combustion in oxygen 0.195 g of carbon dioxide, and 0.0804 g of water. 0.132 g of the compound yielded, after suitable treatment, 0.3822 g of silver chloride. 0.1089 g of the compound occupied a volume of 37.15 cm^3 at 135°C, 767 mmHg. Calculate the molecular formula of the compound.

SOLUTION

Suppose the formula of the compound is $C_xH_yCl_z$.

On combustion 1 mole of $C_xH_yCl_z$ will yield x mole of CO_2 containing x mole of carbon.

0.220 g of the compound yields on combustion

$$\frac{0.195}{44} = 0.00443 \text{ mole of } CO_2$$

Hence 0.220 *g of the compound contains* 0.00443 *mole of carbon atoms.*

Similarly, on combustion 1 mole of $C_xH_yCl_z$ will yield $y/2$ mole of H_2O containing y mole of hydrogen atoms.

0.220 g of the compound yields on combustion

$$\frac{0.0804}{18} = 0.00445 \text{ mole of } H_2O$$

Hence 0.220 *g of the compound contains* 0.00890 *mole of hydrogen atoms.*

One mole of $C_xH_yCl_z$ will yield z mole of AgCl containing z mole of chlorine atoms.

0.132 g of the compound yield

$$\frac{0.3822}{143.5} = 0.002664 \text{ mole of AgCl}$$

Hence 0.220 g of the compound yield

$$\frac{0.220}{0.132} \times 0.002664 = 0.00444 \text{ mole of AgCl}$$

Hence 0.220 *g of the compound contains* 0.00444 *mole of chlorine atoms.*

The ratio

$$\begin{aligned} &\text{mole of carbon : mole of hydrogen : mole of chlorine} \\ &= 0.00443 : 0.00890 : 0.00444 \\ &= 1 : 2 : 1 \end{aligned}$$

Hence the empirical formula is CH_2Cl.

The mass of 1 mole of $C_xH_yCl_z$ can be obtained from the general gas equation

$$\begin{aligned} M &= \frac{mRT}{pV} \\ &= \frac{0.1089 \times 8.31 \times 408}{767/760 \times 101{,}300 \times 37.15 \times 10^{-6}} \\ &= 97.3 \end{aligned}$$

The mass of 1 mole of CH_2Cl is $(12 + 2 + 35.5) = 49.5$ g

The molecular formula of $C_xH_yCl_z$ must be $C_2H_4Cl_2$.

ALTERNATIVE SOLUTION

An alternative method is to calculate the percentage by weight of each element in the compound and then proceed to the molecular formula. It may be noted that this is a rather less direct solution to the problem than the first solution.

Since 1 mole of CO_2 contains 1 mole of carbon (12.0 g), the mass of carbon in 0.195 g of CO_2 is

$$\frac{12.0}{44.0} \times 0.195 \text{ g}$$

Hence the percentage of carbon in 0.220 g of the compound is

$$\frac{(12.0/44.0 \times 0.195) \times 100}{0.220} = 24.2 \text{ per cent}$$

Similarly the percentage of hydrogen in 0.220 g of the compound is

$$\frac{(2.02/18.0 \times 0.0804) \times 100}{0.220} = 4.06 \text{ per cent}$$

The percentage of chlorine in 0.132 g of the compound is

$$\frac{(35.5/143.5 \times 0.3822) \times 100}{0.132} = 71.7 \text{ per cent}$$

This last percentage could also have been obtained from the two known percentages by subtraction $\{100 - (24.2 + 4.06)\} = 71.7$ per cent. Thus the ratio

$$\begin{aligned} &\text{mole of carbon : mole of hydrogen : mole of chlorine} \\ &= \frac{24.2}{12.0} : \frac{4.06}{1.01} : \frac{71.74}{35.5} \\ &= 2.02 : 4.06 : 2.02 \\ &= 1 : 2 : 1 \end{aligned}$$

Hence the empirical formula is CH_2Cl.
The remainder of the problem is calculated as shown in the first solution.

4.3 CHEMICAL EQUATIONS

Just as a formula is a shorthand description of a compound or element, so also is a chemical equation a shorthand description of a chemical reaction.

We will take as an example the chemical equation for the reaction of xenon tetrafluoride with potassium iodide. This equation is written as follows

$$XeF_4(s) + 4KI(aq) \rightarrow Xe(g) + 4KF(aq) + 2I_2(s)$$

Such a chemical equation tells us:

(a) *The nature of the reactants and products.* By suitable symbols we can indicate the state of each substance. Thus the symbol (s) indicates solid, the symbol (aq) indicates a substance dissolved in water, the symbol (g) indicates gas. Hence the equation indicates that solid xenon tetrafluoride reacts with a solution of potassium iodide to produce xenon gas, iodine solid, and a solution of potassium fluoride. If in a chemical equation the states of reactants and products are unambiguously known, the symbols, indicating state, may be omitted.

(b) *The relative numbers of atoms, ions, or molecules involved in the reaction.* In fact, in the above reaction both potassium iodide and potassium fluoride exist in aqueous solution as ions, so that we can write

$$XeF_4(s) + 4K^+(aq) + 4I^-(aq) \rightarrow Xe(g) + 4K^+(aq) + 4F^-(aq) + 2I_2(s)$$

The potassium ions are common to both sides of the equation and may be omitted. Thus we may write the simplified equation as

$$XeF_4(s) + 4I^-(aq) \rightarrow Xe(g) + 4F^-(aq) + 2I_2(s)$$

This last equation is referred to as an *ionic equation*. It indicates that one molecule of XeF_4 reacts with four iodide ions (I^-), to produce one atom of Xe, four fluoride ions (F^-), and two molecules of I_2.

The equation at the beginning of this section is referred to as a *molecular equation*.

(c) *The relative numbers of mole of substances involved in the reaction.* Thus our molecular equation indicates that 1 mole of xenon tetrafluoride (207.3 g) reacts with 4 mole of potassium iodide (4×166.0 g) to produce 1 mole of xenon (131.3 g), 4 mole of potassium fluoride (4×58.1 g) and 2 mole of iodine (2×253.8 g). The ionic equation indicates that 1 mole of xenon tetrafluoride (207.3 g) reacts with 4 mole of iodide ion (4×126.9 g) to produce 1 mole of xenon (131.3 g), 4 mole of fluoride ion (4×19.0 g) and 2 mole of iodine (2×253.8 g).

If we have 20.73 g of xenon tetrafluoride reacting, then we have 20.73/207.3 or 0.1 mole. We expect this amount to react with 0.4 mole of potassium iodide to produce 0.1 mole of xenon. The mass of 0.1 mole of xenon is 0.1×131.3 g or 13.13 g, and its volume (S.T.P.) is 0.1×22.4 dm^3 or 2.24 dm^3. Since a chemical equation expresses the ratio of mole of reactants and products we could write the equation

$$XeF_4(s) + 4I^-(aq) \rightarrow Xe(g) + 4F^-(aq) + 2I_2(s)$$

in the form

$$\tfrac{1}{4}XeF_4(s) + I^-(aq) \rightarrow \tfrac{1}{4}Xe(g) + F^-(aq) + \tfrac{1}{2}I_2(s)$$

without altering the mole ratio, since

$$\tfrac{1}{4}\text{ mole } XeF_4 : 1 \text{ mole } I^- = 1 \text{ mole } XeF_4 : 4 \text{ mole } I^-$$

An equation thus provides us with considerable information about a reaction. However, it fails to supply some relevant information. It does not:

(a) Indicate the rate of reaction. For example, we have no idea how rapidly the iodine is produced in the above reaction simply by inspecting the equation.

(b) Indicate the conditions necessary to initiate or maintain the reaction. We do not know if it is necessary to use an elevated temperature to initiate the above reaction.

(c) Indicate whether the reaction goes to completion.
(d) Indicate the way in which the particles of reactants actually come together to form the products. There is no indication of the mechanism of the reaction. It is exceedingly unlikely that the reaction

$$XeF_4(s) + 4I^-(aq) \rightarrow Xe(g) + 4F^-(aq) + 2I_2(s)$$

proceeds by the simultaneous collision of four iodide ions with a xenon tetrafluoride molecule. In all probability, a series of simple reactions occurs, eventually yielding the products. If the equations for all of these steps are summed we obtain the above overall stoichiometric equation.

The problems of the rate of reaction, extent of reaction, and mechanism of reaction will be discussed in later chapters.

4.4 STOICHIOMETRIC CALCULATIONS

In the remaining sections of this chapter we will consider calculations based upon the stoichiometry of chemical reactions. In these calculations we will see how the mole concept provides a unifying principle linking apparently dissimilar problem situations.

In performing these calculations we will make two important assumptions:

(a) *That the reactions give complete conversion of reactants to products.* In many chemical reactions it is found that the products themselves react to re-form the reactants. In such systems products and reactants eventually come to a state of *equilibrium*. We will not discuss this condition here, but it will be examined in detail in chapter 13.
(b) *That no side-reactions occur, yielding other than the stated products.* We may be given a problem in which we are required to calculate the mass of carbon dioxide formed when a given mass of carbon burns in air. If we are told that the equation for this reaction is

$$C + O_2 \rightarrow CO_2$$

we can readily calculate the mass of carbon dioxide. However, if we actually perform the experiment we may well find that in practice we obtain less carbon dioxide than the calculated mass. This may be because our given mass of carbon was not only reacting to form carbon dioxide, but also to form some carbon monoxide

$$2C + O_2 \rightarrow 2CO$$

In our calculations we will assume that the stated equation completely describes the reaction, and that no side reactions take place.

In cases where the above assumptions are not valid, the calculated amounts of products would exceed the amounts actually attainable by experiment.

4.5 CALCULATIONS INVOLVING MASSES OF PRODUCTS AND REACTANTS

EXAMPLE 1

The radioactive element technetium (Tc) burns in oxygen to form a bright yellow, volatile heptoxide (Tc_2O_7). Calculate the number of gram of heptoxide formed by the combustion of 5.7 milligram of technetium.

SOLUTION

The equation for the reaction is

$$4Tc + 7O_2 \rightarrow 2Tc_2O_7$$

Hence 4 mole of Tc yields 2 mole of Tc_2O_7

We start with $\frac{5.7 \times 10^{-3}}{98}$ mole of Tc

and this will yield $\frac{2}{4} \times \frac{5.7 \times 10^{-3}}{98}$ mole of Tc_2O_7

Hence the mass of Tc_2O_7 formed will be

$$\left(\frac{2}{4} \times \frac{5.7 \times 10^{-3}}{98}\right) \times \{(2 \times 98) + (7 \times 16)\}$$

$$= 8.96 \times 10^{-3} \text{ g or } 8.96 \text{ mg}$$

EXAMPLE 2

Metallic nickel reacts with carbon monoxide on heating to form liquid nickel tetracarbonyl, $Ni(CO)_4$. If 5.10 g of nickel is heated with 9.31 g of carbon monoxide, calculate the mass of nickel tetracarbonyl formed and the mass of any unreacted substance remaining.

SOLUTION

The equation for the reaction is

$$Ni + 4CO \rightarrow Ni(CO)_4$$

One mole of Ni reacts with 4 mole of CO to form 1 mole of $Ni(CO)_4$.

Initially we have $5.10/58.7 = 0.0869$ mole of Ni, and $9.31/28.0 = 0.3324$ mole of CO. 0.0869 mole of Ni would require $(4 \times 0.0869) = 0.3476$ mole of CO for complete reaction, whilst 0.3324 mole of CO would require $(\frac{1}{4} \times 0.3324) = 0.0831$ mole of Ni for complete reaction. Obviously there is insufficient CO for complete reaction with 0.0869 mole of Ni, hence the 0.3324 mole of CO reacts with 0.0831 mole of Ni to form 0.0831 mole of $Ni(CO)_4$, and $(0.0869 - 0.0831) = 0.0038$ mole of Ni is left unreacted.

The mass of $Ni(CO)_4$ formed is

$$0.0831 \times 170.7 = 14.2 \text{ g}$$

The mass of Ni left unreacted is

$$0.0038 \times 58.7 = 0.223 \text{ g}$$

EXAMPLE 3

56 cm^3 of liquid carbon tetrachloride (density $= 1.594$ g cm^{-3}) reacts with solid tungsten sulphide (WS_3) to form tungsten hexachloride (WCl_6) and carbon disulphide (CS_2).

Calculate the volume of carbon tetrachloride required to form 10.0 g of tungsten hexachloride.

SOLUTION

The equation for the reaction is

$$2WS_3 + 3CCl_4 \rightarrow 2WCl_6 + 3CS_2$$

Three mole of CCl_4 yields 2 mole of WCl_6
We are required to form 10.0/396.9 = 0.0252 mole of WCl_6

Hence we require $\frac{3}{2} \times 0.0252$

$$= 0.0378 \text{ mole of } CCl_4$$

Hence the mass of CCl_4 required is

$$0.0378 \times 154.0 \text{ g}$$

Since volume = mass/density, the volume required is

$$\frac{0.0378 \times 154.0}{1.594}$$

$$= 3.65 \text{ cm}^3$$

Notice that *in all examples we have assumed that mass is conserved.* For example, we have assumed in Example 1 that the mass of technetium remains unchanged in the reaction. This is almost an axiom of stoichiometry and is known as the *Law of Conservation of Mass.* In chapter 2 however, it was pointed out that mass and energy are interconvertible and in chemical changes some mass may be destroyed and appear in the form of energy. In such chemical reactions the quantity of matter changed to energy would be very small to account for any observed energy release; in fact, such a quantity of matter is well beyond the possibility of detection by normal methods of weighing. In chemistry the law of conservation of mass may be accepted as being true within normal limits of precision.

There is not necessarily a conservation of mole of reactant and product molecules. In Example 2 we see that there is a total of 5 mole of reactants, but only 1 mole of product. However, a number of mole of atoms are conserved—there are 9 mole of atoms of reactants and 9 mole of atoms of products. Notice also in Example 2 that not all of each reactant was used, but for every mole of nickel 4 mole of carbon monoxide reacted, thus preserving the mole ratio prescribed by the equation.

4.6 CALCULATIONS INVOLVING MASSES AND VOLUMES OF REACTANTS AND PRODUCTS

EXAMPLE

A student wishes to collect 500 cm³ of hydrogen at 15°C, 762 mmHg barometric pressure. He decides to prepare the gas by reacting aluminium with hydrochloric acid and collecting the gas by downward displacement of water in an inverted, graduated jar. Calculate the minimum mass of aluminium that would be needed in this preparation. The aqueous vapour pressure at 15°C is 12.8 mmHg.

SOLUTION

The equation for the reaction is

$$2Al + 6HCl \rightarrow 2AlCl_3 + 3H_2$$

or

$$2Al + 6H^+ \rightarrow 2Al^{3+} + 3H_2$$

We shall calculate the number of mole of hydrogen to be collected, using the general gas equation.

The pressure of the hydrogen is found by subtracting the pressure due to the water vapour from the atmospheric pressure. This is because the atmospheric pressure is equal to the sum of the pressures exerted by the hydrogen gas and by the water vapour (law of partial pressures).

Hence the pressure of hydrogen is

$$762 - 12.8 = 749.2 \text{ mm}$$
$$= 749.2/760 \times 101{,}325 \text{ N m}^{-2}$$

The volume of hydrogen is

$$500 \text{ cm}^3 = 500 \times 10^{-3} \text{ dm}^3 = 500 \times 10^{-6} \text{ m}^3$$

Since $pV = nRT$

$$\therefore \quad n = pV/RT$$
$$= \frac{749.2/760 \times 101{,}325 \times 500 \times 10^{-6}}{8.31 \times 288}$$
$$= 0.0208 \text{ mole}$$

From the equation we see that 2 mole of Al yield 3 mole of H_2.

Hence $(2/3 \times 0.0208) = 0.01386$ mole of Al yield 0.0208 mole of H_2.

The mass of aluminium required is

$$0.01386 \times 27.0 = 0.374 \text{ g}$$

4.7 CALCULATIONS INVOLVING VOLUMES OF GASEOUS REACTANTS AND PRODUCTS

EXAMPLE 1

In 1962 it was discovered that at 400°C gaseous xenon tetrafluoride (XeF_4) reacts with hydrogen to form xenon and gaseous hydrogen fluoride. Assuming constant pressure and temperature, calculate the volumes of xenon and hydrogen fluoride produced if 3.8 dm³ of xenon tetrafluoride reacts with excess hydrogen at 400°C.

SOLUTION

The equation for the reaction is

$$XeF_4 + 2H_2 \rightarrow Xe + 4HF$$

From the general gas equation, since temperature and pressure are constant, we see that

$$\text{number of mole of gas } (n) \propto \text{volume } (V)$$

One mole of XeF_4 yields 1 mole of Xe and 4 mole of HF. Hence 3.8 dm³ XeF_4 yields 3.8 dm³ of Xe and $(4 \times 3.8) = 15.2$ dm³ of HF.

The simple volume relationships apparent in the above calculations were discovered experimentally by Gay-Lussac in 1808. He summarized his observations in the *Law of Combining Volumes.*

When gases react they do so in simple proportions by volume, and the volume of any gaseous product also bears a simple ratio to that of the reacting gases, all volumes measured at the same temperature and pressure.

EXAMPLE 2

10 cm³ of a hydrocarbon (C_xH_y) is exploded with excess oxygen, of volume 80 cm³. The volume of the mixture after reaction is 60 cm³. If this mixture is treated with potassium hydroxide, the volume becomes 30 cm³. If the volumes are all measured at S.T.P., calculate the formula of the hydrocarbon.

SOLUTION

One mole of C_xH_y will form x mole of CO_2 requiring x mole of O_2, since 1 mole of CO_2 contains 1 mole of O_2. One mole of C_xH_y will form $y/2$ mole of H_2O, since 1 mole of water contains 2 mole of hydrogen atoms, and this will require $y/4$ mole of O_2, since 1 mole of water contains half a mole of oxygen molecules.

After the explosion the mixture had a volume of 60 cm³, being made up of unreacted oxygen and carbon dioxide. $(60 - 30) = 30$ cm³ of carbon dioxide was absorbed by the potassium hydroxide, leaving 30 cm³ of unreacted oxygen.

Hence the volume of oxygen which reacted was $(80 - 30) = 50$ cm³. The equation for the combustion may be written as

$$C_xH_y + (x + y/4)O_2 \rightarrow xCO_2 + (y/2)H_2O(l)$$

From experiment

$$10\text{ cm}^3 \quad 50\text{ cm}^3 \qquad\qquad 30\text{ cm}^3 \quad 0\text{ cm}^3$$

also

$$10\text{ cm}^3 \quad (x + y/4)10\text{ cm}^3 \quad 10x\text{ cm}^3 \quad 0\text{ cm}^3$$

Hence we can equate volumes

$$10x = 30$$

$$\therefore \qquad x = 3$$

$$(x + y/4).10 = 50$$

$$\therefore \qquad y = 8$$

Hence the formula of the hydrocarbon is C_3H_8.

Notice that the volume of water formed in the reaction is taken as zero, because the volume of liquid water is negligible compared with the volume of the gaseous reactants and product at 0°C and 1 atmosphere pressure.

4.8 REACTIONS IN SOLUTION

In investigating chemical reactions we are usually interested in the amounts of substances reacting or produced. One method of determining the amount of a particular substance is to react it completely with a known amount of another substance and then calculate, from the stoichiometric equation, the quantity of the first substance. This procedure is often carried out in solution. Solutions can be readily prepared containing accurately known amounts of a solute. Such solutions are called *standard solutions*. If standard solutions are used we can then accurately measure volumes, containing known amounts of substances, using pipettes and burettes. This procedure is often more convenient than weighing procedures.

We may choose any convenient units for the amounts of solute and solution. In stoichiometric procedures it is convenient to use mole quantities for the solute.

A molar solution is one which contains 1 *mole of solute in* 1 *dm*^{3} *of solution.*

The molarity of a solution is the number of mole of solute in 1 *dm*^{3} *of solution.*

The symbol M represents the molarity of the solution. A 2M HCl solution contains 2 mole of HCl in 1 dm^3 of solution, and a 0.5M NaOH solution contains 0.5 mole of NaOH in 1 dm^3 of solution. Such a solution may be prepared by taking a 1-dm^3 *standard flask*. Such a flask contains an accurately known volume at a given temperature. To this flask is added 0.5 mole of NaOH, or $(0.5 \times 40) = 20$ g of NaOH. Water is added until the volume of the solution formed is 1 dm^3. In fact, one would have to add 1001 cm^3 to achieve this. Notice that the volume of water required to form 1 dm^3 of solution is not necessarily 1 dm^3. The concentration of this solution is 0.5 mole dm^{-3}, or 20 g dm^{-3}. Since the sodium hydroxide is present in the form of ions, the concentration of OH^- is also 0.5 mole dm^{-3}, and the concentration of Na^+ is also 0.5 mole dm^{-3}.

If one wished to make up a 0.1M solution of acetic acid one would place in the standard flask $(0.1 \times 60) = 6$ g of acetic acid, and then add water to form 1 dm^3 of solution. We would then say that the *analytical concentration* of acetic acid was 0.1 mole dm^{-3}. However, the actual concentration of acetic acid molecules in solution is 0.0988 mole dm^{-3}, because some of the acetic acid reacts with water to form acetate ions and hydrogen ions, and in this case 0.0012 mole of acetic acid has so reacted. The analytical concentration of a solute is not necessarily the same as the actual concentration of the solute in the solution.

Other measures of concentration of solutions include the number of gram of solute per 100 g of solvent, and the number of gram of solute per 100 g of solution. A solution containing 3 g of hydrogen peroxide in 100 g of solution is said to be a 3 per cent H_2O_2 solution. A more important measure is the number of mole of solute per kg of solvent. This is described as the *molality* of the solution.

Because none of these units is as convenient in volumetric calculations they will not be used in this textbook. The molal scale is however very frequently used in other physico-chemical measurements.

EXAMPLE 1

How many mole of sulphuric acid are there in 250 cm^3 of 0.20M H_2SO_4? How many gram of H_2SO_4 are present?

SOLUTION

1000 cm^3 of solution contains 0.20 mole of H_2SO_4
Hence 250 cm^3 of the solution contains

$$\frac{250}{1000} \times 0.20 = 0.05 \text{ mole of } H_2SO_4$$

The mass of H_2SO_4 in 250 cm^3 is $0.05 \times 98 = 4.9$ g

EXAMPLE 2

What mass of $BaCl_2.2H_2O$ is required to prepare 1.5 dm^3 of 0.15M Ba^{2+} solution? What is the number of mole of Ba^{2+} and Cl^- in 1.5 dm^3 of this solution?

SOLUTION

One dm^3 of 0.15M Ba^{2+} solution contains 0.15 mole of Ba^{2+}. Hence 1.5 dm^3 of 0.15M Ba^{2+} solution contains $(1.5 \times 0.15) = 0.23$ mole of Ba^{2+}.

Since there are 2 mole of chloride ion present for every 1 mole of barium ion, the number of mole Cl^- present is $(2 \times 0.23) = 0.46$ mole of Cl^-.

One mole of $BaCl_2.2H_2O$ yields in solution 1 mole of Ba^{2+}. Hence 0.23 mole of Ba^{2+} is released by 0.23 mole of $BaCl_2.2H_2O$. Hence the mass of $BaCl_2.2H_2O$ required is $(0.23 \times 244.3) = 56$ g of $BaCl_2.2H_2O$.

EXAMPLE 3

A solution of nitric acid has a density of 1.42 g cm^{-3}, and contains 72.0 per cent by mass of nitric acid. Calculate the molarity of the solution.

SOLUTION

In 100 g of solution there is 72.0 g of HNO_3.
Hence in $100/1.42 = 70.4$ cm^3 of solution there is $72.0/63.0 = 1.14$ mole of HNO_3.
Hence in 1000 cm^3 of solution there is

$$\frac{1000}{70.4} \times 1.14 = 16.2 \text{ mole of } HNO_3$$

The solution is 16.2M.

4.9 REACTIONS INVOLVING ACIDS AND ALKALIS

If we wish to determine the concentration of a solution of an acid, we may react it with a known volume of an alkali of known concentration. The reaction is continued until it is just complete. We then say that the *equivalence point* has been reached. To detect this equivalence point we may use an indicator or some electrical method. The point at which the indicator changes colour is called the *endpoint*. The endpoint may be a good or bad approximation to the equivalence point. The latter is fixed for a given reaction but the endpoint may vary with experimental circumstances. The volume of acid added is noted and one can then calculate its concentration from the stoichiometric equation. To determine the concentration of alkali one simply reverses the procedure. This method is an example of the procedure known as *titration*.

EXAMPLE 1

50 cm^3 of a sulphuric acid solution exactly reacts with 20 cm^3 of 0.2M NaOH. Calculate the molarity of the sulphuric acid solution.

SOLUTION

The equation for the reaction is

$$2NaOH + H_2SO_4 \rightarrow Na_2SO_4 + 2H_2O$$

Two mole of NaOH reacts with 1 mole of H_2SO_4

Hence $\frac{20 \times 0.2}{1000}$ mole of NaOH reacts with $\frac{\frac{1}{2}(20 \times 0.2)}{1000}$ mole of H_2SO_4. If the molarity of the H_2SO_4 solution is x, then the number of mole of H_2SO_4 used is $\frac{50 \times x}{1000}$

Hence
$$\frac{50 \times x}{1000} = \frac{\frac{1}{2}(20 \times 0.2)}{1000}$$

$$\therefore \quad x = 0.04$$

The solution is 0.04M

ALTERNATIVE SOLUTION

The ionic equation for the reaction is

$$H^+ + OH^- \rightarrow H_2O$$

One mole of OH^- reacts with 1 mole of H^+

Hence $\frac{20 \times 0.2}{1000}$ mole of OH^- reacts with $\frac{20 \times 0.2}{1000}$ mole of H^+

But every mole of H_2SO_4 yields 2 mole of H^+ in solution

$$H_2SO_4 \rightarrow 2H^+ + SO_4^{2-}$$

Hence if the molarity of the H_2SO_4 solution is x with respect to H_2SO_4, it must be $2x$ with respect to H^+

Thus the number of mole of H^+ reacted is $\frac{50 \times 2x}{1000}$

Hence $$\frac{50 \times 2x}{1000} = \frac{20 \times 0.2}{1000}$$

$$\therefore \quad x = 0.04$$

The solution is 0.04M

EXAMPLE 2

If v cm³ of 0.10M HCl is added to $2v$ cm³ of 0.30 HNO_3, what volume of 0.20M NaOH is required to react with the mixture?

SOLUTION

The equations for the reactions are

$$HCl + NaOH \rightarrow NaCl + H_2O$$
$$HNO_3 + NaOH \rightarrow NaNO_3 + H_2O$$

Hence $\frac{0.10 \times v}{1000}$ mole of HCl reacts with $\frac{0.10 \times v}{1000}$ mole of NaOH

Also $\frac{0.30 \times 2v}{1000}$ mole of HNO_3 reacts with $\frac{0.30 \times 2v}{1000}$ mole of NaOH

The total number of mole of NaOH required is $\left(\frac{0.10 \times v}{1000} + \frac{0.30 \times 2v}{1000}\right)$

If the volume of NaOH required is V cm³, then the number of mole of NaOH required is $\frac{0.20 \times V}{1000}$

Hence $$\frac{0.20 \times V}{1000} = \left(\frac{0.10 \times v}{1000} + \frac{0.30 \times 2v}{1000}\right)$$

$$\therefore V = 3.5v \text{ cm}^3$$

EXAMPLE 3

To 20.0 cm³ of 1.00M HCl is added 0.1216 g of magnesium. The excess acid is neutralized by reaction with 50.0 cm³ of 0.200M NaOH. Calculate the atomic weight of magnesium.

SOLUTION

The equations for the reactions are

$$Mg + 2HCl \rightarrow MgCl_2 + H_2$$
$$NaOH + HCl \rightarrow NaCl + H_2O$$

One mole of magnesium reacts with 2 mole of HCl. Let the mass of 1 mole of magnesium be A g. Then, $0.1216/A$ mole of magnesium reacts with $0.2432/A$ mole of HCl. Hence the number of mole of HCl left unreacted is

$$\left(\frac{20.0 \times 1.00}{1000} - \frac{0.2432}{A}\right)$$

But the number of mole of NaOH required to neutralize this is $\dfrac{50.0 \times 0.200}{1000}$

Hence
$$\left(\frac{20.0 \times 1.00}{1000} - \frac{0.2432}{A}\right) = \frac{50.0 \times 0.200}{1000}$$
$$\therefore A = 24.3$$

4.10 REACTIONS INVOLVING PRECIPITATION

EXAMPLE

To 40.0 cm^3 of 1.00M $AgNO_3$ is added 20.0 cm^3 of 0.500M $AlCl_3$. What is the molarity of the resulting silver nitrate solution?

SOLUTION

The equation for the reaction is

$$AlCl_3(aq) + 3AgNO_3(aq) \rightarrow Al(NO_3)_3(aq) + 3AgCl(s)$$

One mole of $AlCl_3$ reacts with 3 mole of $AgNO_3$. Hence $\dfrac{20.0 \times 0.500}{1000}$ mole of $AlCl_3$ reacts with $\dfrac{3 \times 20.0 \times 0.500}{1000}$ mole of $AgNO_3$. But we started with $\dfrac{40.0 \times 1.00}{1000}$ mole of $AgNO_3$, hence the number of mole of $AgNO_3$ remaining is

$$\left(\frac{40.0 \times 1.00}{1000} - \frac{3 \times 20.0 \times 0.500}{1000}\right) = 0.0100$$

The volume of the mixture after reaction is (40.0 + 20.0) = 60.0 cm^3. Hence 60.0 cm^3 contains 0.0100 mole of unreacted $AgNO_3$. Hence 1000 cm^3 contains

$$\frac{1000 \times 0.0100}{60.0} = 0.167 \text{ mole of } AgNO_3$$

The solution is 0.167M.

ALTERNATIVE SOLUTION

The ionic equation for the reaction is

$$Ag^+(aq) + Cl^-(aq) \rightarrow AgCl(s)$$

Hence 1 mole of Ag^+ reacts with 1 mole of Cl^-. But 1 mole of $AlCl_3$ yields, in solution, 3 mole of Cl^-. Hence $\dfrac{20.0 \times 0.500}{1000}$ mole of $AlCl_3$ yields $\dfrac{3 \times 20.0 \times 0.500}{1000}$ mole of Cl^- which will react with $\dfrac{3 \times 20.0 \times 0.500}{1000}$ mole of Ag^+. From this point the calculation proceeds as shown above.

4.11 REACTIONS INVOLVING OXIDATION-REDUCTION

EXAMPLE 1

17.2 cm^3 of a 0.120M $FeCl_3$ solution is exactly reduced by 37.2 cm^3 of a $SnCl_2$ solution. Find the molarity of this solution.

SOLUTION

The ionic equation for the reaction is

$$2Fe^{3+}(aq) + Sn^{2+}(aq) \rightarrow 2Fe^{2+}(aq) + Sn^{4+}(aq)$$

Two mole of Fe^{3+} reacts with 1 mole of Sn^{2+}. Hence $\frac{17.2 \times 0.120}{1000}$ mole of Fe^{3+} reacts with $\frac{\frac{1}{2} \times 17.2 \times 0.120}{1000}$ mole of Sn^{2+}.

If the molarity of the Sn^{2+} is x, then the number of mole of Sn^{2+} which reacts is $\frac{37.2 \times x}{1000}$

Hence
$$\frac{37.2 \times x}{1000} = \frac{1}{2} \times \frac{17.2 \times 0.120}{1000}$$

$$\therefore x = 0.0277$$

The solution is 0.0277M.

EXAMPLE 2

0.156 g of potassium dichromate ($K_2Cr_2O_7$) is dissolved in 100 cm^3 of water in a standard flask. The solution is made up to 500 cm^3 by the addition of water. What is the molarity of an iron(II) sulphate solution of which 24.6 cm^3 react exactly with 20.0 cm^3 of the potassium dichromate solution? What is the concentration, in mole dm^{-3} and g dm^{-3}, of the solution referred to: (a) Fe^{2+}, (b) $FeSO_4.7H_2O$?

SOLUTION

The ionic equation for the reaction is

$$Cr_2O_7^{2-} + 14H^+ + 6Fe^{2+} \rightarrow 2Cr^{3+} + 7H_2O + 6Fe^{3+}$$

One mole of $K_2Cr_2O_7$ reacts with 6 mole of Fe^{2+} ion (or 6 mole of iron(II) sulphate). The molarity of the $K_2Cr_2O_7$ solution is

$$\frac{2 \times 0.156}{294} = 0.00106M$$

Hence $\frac{20.0 \times 0.00106}{1000}$ mole of $K_2Cr_2O_7$ exactly reacts with $6 \times \frac{20.0 \times 0.00106}{1000}$ mole of Fe^{2+}.

If the molarity of the Fe^{2+} ion is x, then the number of mole of iron(II) sulphate reacted is $\frac{24.6 \times x}{1000}$

Hence $$\frac{24.6 \times x}{1000} = 6 \times \frac{20.0 \times 0.00106}{1000}$$

$$\therefore x = 0.00517$$

The solution is 0.00517M.

The concentration of both Fe^{2+} and $FeSO_4.7H_2O$ is 0.00517 M. The concentration with respect to Fe^{2+} is 0.00517×55.8

$$= 0.289 \text{ g dm}^{-3}$$

The concentration with respect to $FeSO_4.7H_2O$ is 0.00517×277.8

$$= 1.44 \text{ g dm}^{-3}$$

QUESTIONS

Equations

1. The reaction between hydrogen and oxygen to form water may be represented

$$H_2 + \tfrac{1}{2}O_2 \rightarrow H_2O$$

but not $$2H + O \rightarrow H_2O$$

Explain why the last equation is considered to be incorrect.

2. In 1962 it was discovered that xenon reacted with fluorine to produce a fluoride of xenon, which existed as a solid at 25 °C, but at temperatures above 200 °C it existed as a gas. At 400 °C this xenon fluoride reacted quantitatively with hydrogen to produce xenon and hydrogen fluoride. Outline suitable chemical and physical procedures which should demonstrate that the reaction of the xenon fluoride with hydrogen occurs according to the equation

$$XeF_4 + 2H_2 \rightarrow Xe + 4HF$$

and that the molecular formula of the xenon fluoride is XeF_4.

Formulae

3. Calculate the percentage by mass of each element in the following compounds:

(a) Al_2O_3 (c) $MgCl_2.6H_2O$

(b) $Cu(OH)_2$ (d) $Fe_2(SO_4)_3$

4. A compound has the following composition by mass: 32% carbon, 6.7% hydrogen, 42.6% oxygen, and 18.7% nitrogen. Find the empirical formula of the compound.
5. One gram of a compound containing only carbon and hydrogen gave on combustion 1.635 g of water and 2.995 g of carbon dioxide. Calculate the empirical formula of the compound.
6. Two gram of phosphorus is burned in air and forms 4.582 g of an oxide. This quantity of the oxide is found to react with exactly 1.744 g of water to give 6.326 g of a compound of hydrogen, phosphorus and oxygen. Determine the empirical formula of the oxide and of the other compound.
7. 5.60 g of copper(II) chloride ($CuCl_2$) combines with 4.22 g of ammonia to form a compound. What is its empirical formula?
8. A compound of carbon and hydrogen is found by analysis to contain 92.26% of carbon. This substance is a gas at 200 °C, 1.013×10^5 N m^{-2} pressure, and under these conditions has a density of 2.012 g dm^{-3}. Calculate the molecular formula of the compound.

9. A gaseous compound of carbon and hydrogen contains 92.26% carbon. When 1.373 g of the gas is collected over water at 25°C, and a barometric pressure of 770 mmHg, it is found to occupy a volume of 1.308 dm^3. What is the molecular formula of the compound?
   The vapour pressure of water at 25°C is 23.8 mmHg.
10. When 1.000 g of a compound, known to contain only carbon, hydrogen, and oxygen, is burned in air, 1.910 g of carbon dioxide and 1.173 g of water are formed. The vapour of this compound is 1.64 times as dense as nitrogen under the same conditions of temperature and pressure. What is the molecular formula of the compound?
11. A compound has a percentage composition by mass: sulphur 23.7%, chlorine 52.6%, and oxygen 23.7%. When 0.3375 g of the compound was vaporized the volume occupied at 100°C, 770 mmHg, was 75.7 cm^3. Find the molecular formula.
12. A compound contains carbon, hydrogen, oxygen, and nitrogen. On analysis it was found that 0.3006 g of the compound, on combustion, gave 0.1804 g of water and 0.3530 g of carbon dioxide. A further 0.2500 g of the compound evolved 38.76 cm^3 of nitrogen at 13°C, 768 mmHg, on decomposition. The molecular weight of the compound was found to be 75.00. Find the molecular formula of the compound.

Mass-mass calculations

13. If 3.72 g of element X exactly reacts with 4.80 g of oxygen to form a compound whose molecular formula is shown, from other experiments, to be X_4O_{10}, what is the atomic weight of X?
14. What masses of barium sulphate and calcium sulphate must be mixed so that 1000 g of a mixture contains equal numbers of mole of each?
15. Zinc and nitric acid react according to the equation

$$4Zn + 10HNO_3 \rightarrow 4Zn(NO_3)_2 + N_2O + 5H_2O$$

   Calculate for each case which reagent is in excess, and by how many mole:
   (a) 0.231 mole Zn and 1.000 mole HNO_3
   (b) 0.672 mole Zn and 0.817 mole HNO_3
   (c) 0.500 mole Zn and 0.500 mole HNO_3
16. What mass of silver chloride will be precipitated when a solution containing 2.884 g of silver nitrate is added to one containing 5.000 g of sodium chloride? Will any sodium chloride remain unreacted? If so, how many gram?
17. Potassium dichromate reacts with potassium chloride, in the presence of concentrated sulphuric acid, to form the deep red liquid chromyl chloride (CrO_2Cl_2) of density 1.935 g cm^{-3}

$$K_2Cr_2O_7 + 4KCl + 3H_2SO_4 \rightarrow 2CrO_2Cl_2 + 3K_2SO_4 + 3H_2O$$

   Calculate
   (a) the mass of potassium chloride required to form 6.00 g of chromyl chloride;
   (b) the volume of chromyl chloride liquid formed from 10.0 g of potassium dichromate.
18. Titanium tetrachloride ($TiCl_4$) is a liquid, of density 1.76 g cm^{-3}. It reacts with water to form titanium dioxide (TiO_2) and hydrogen chloride. Calculate the mass of titanium dioxide formed on complete reaction of 100 cm^3 of titanium tetrachloride with water.

19. When solid lead dioxide (PbO_2) is heated, it forms solid lead monoxide (PbO) and oxygen. Heating solid barium peroxide (BaO_2) yields solid barium monoxide (BaO) and oxygen. A mixture of lead dioxide and barium peroxide was heated until both decompositions were complete. If the initial mass of the mixture was 15.00 g and the final mass was 13.80 g, what mass of lead dioxide was present in the original mixture?
20. A mixture of calcium and magnesium carbonates of mass 1.84 g was strongly heated until no further loss in mass was detected. The residue had a mass of 0.900 g. What percentage by mass of the mixture was calcium carbonate?

Mass-volume calculations

21. Some magnesium was partially oxidized in air. The residue was dissolved in excess hydrochloric acid and displaced 3.00 dm^3 of hydrogen at 26°C, 752 mmHg. Calculate the mass of unoxidized magnesium present. Water vapour pressure at 26°C is 25.2 mmHg.
22. (a) What mass of aluminium sulphide is required for the production of 6.70 dm^3 of hydrogen sulphide, at S.T.P., according to the equation

$$Al_2S_3 + 6HCl \rightarrow 2AlCl_3 + 3H_2S$$

    (b) What mass of hydrochloric acid solution, containing 36.0% by weight of HCl, would be required for the above reaction?
    (c) What volume of gaseous HCl, at S.T.P., would have to be dissolved in water to give sufficient hydrochloric acid to liberate 1.00×10^{21} molecules of hydrogen sulphide?
23. How many cm^3 of oxygen gas, at S.T.P., can be made from 100.0 milligram of sodium peroxide (Na_2O_2) according to the equation

$$2Na_2O_2 + 2H_2O \rightarrow 4NaOH + O_2$$

24. How many cm^3, at S.T.P., of carbon dioxide can be produced from 5.00 cm^3 of liquid heptane (C_7H_{16}) on combustion of the heptane in oxygen, forming carbon dioxide and water? The density of liquid heptane is 0.684 g cm^{-3}.
25. (a) How many gram of manganese dioxide is needed to prepare 612 cm^3 of chlorine at 30°C, 765 mmHg, according to the equation

$$MnO_2 + 4HCl \rightarrow MnCl_2 + 2H_2O + Cl_2$$

    (b) How many mole of hydrogen chloride is needed in the reaction?
26. What volume of hydrogen sulphide (at 25°C, 756 mmHg) is needed to precipitate all the antimony in 100 g of antimony chloride ($SbCl_3$) as antimony sulphide (Sb_2S_3)?
    The equation for the reaction is

$$2SbCl_3 + 3H_2S \rightarrow Sb_2S_3 + 6HCl$$

27. 20.0 g of copper was heated in a current of chlorine and then weighed, 30.7 g of copper and copper(II) chloride being obtained. Calculate the volume of chlorine used in the experiment, measured at 20°C, 763 mmHg.
28. 50.0 cm^3 of an aqueous solution of hydrogen peroxide is decomposed catalytically according to the equation

$$2H_2O_2 \rightarrow 2H_2O + O_2$$

    If 580 cm^3 of dry oxygen, measured at 22°C, 767 mmHg, is obtained, find the mass of hydrogen peroxide per dm^3 of solution.

Volume-volume calculations

29. What is the percentage increase in volume when nitrogen dioxide (NO_2) completely decomposes into nitric oxide (NO) and oxygen?
30. What volume of chlorine gas, measured at 30°C, 756 mmHg, can be obtained by reaction of 5.70 dm^3 of hydrogen chloride with 50.0 dm^3 of oxygen, both volumes measured at 30°C, 756 mmHg? The equation for the reaction is

$$4HCl + O_2 \rightarrow 2H_2O + 2Cl_2$$

31. Ammonia reacts with oxygen to form nitric oxide (NO) and steam. 57.0 dm^3 of ammonia is reacted with 140 dm^3 of oxygen, both volumes measured at 200°C, 4×10^4 N m^{-2}. Which gas does not completely react and what volume of this gas is left at 200°C, 4×10^4 N m^{-2}?
32. Butane (C_4H_{10}) gas is burned in oxygen to form carbon dioxide and water. All volumes are measured at S.T.P.
    (a) How many dm^3 of oxygen is required to produce 3.00 dm^3 of carbon dioxide?
    (b) If 16.0 dm^3 of oxygen is used, how many dm^3 of butane will be burned?
    (c) If 10.0 dm^3 each of butane and oxygen are reacted, how many dm^3 of carbon dioxide is produced?
33. When 1 dm^3 of a gaseous compound of carbon and hydrogen is burned in excess oxygen, 2 dm^3 of carbon dioxide and 3 dm^3 of water vapour are produced, all volumes measured at 126°C, 1×10^5 N m^{-2}. What is the molecular formula of the compound?
34. A mixture of oxygen and hydrogen sulphide, each at a partial pressure of 400 mmHg, in a closed vessel at 150°C, is made to react according to the equation

$$2H_2S + 3O_2 \rightarrow 2H_2O + 2SO_2$$

    What is the final total pressure at 150°C, the volume remaining constant?
35. 18.32 cm^3 of a pure, gaseous hydrocarbon (C_xH_y) was mixed with 80.00 cm^3 of oxygen and burned producing 36.64 cm^3 of carbon dioxide, and leaving 15.88 cm^3 of oxygen unreacted. If the conditions were S.T.P. throughout, find the formula of the hydrocarbon.
36. 20 cm^3 of a hydrocarbon was mixed with 140 cm^3 of oxygen (more than enough to react). The mixture was sparked to start the combustion and after completion of the reaction the volume was found to be 110 cm^3. This was reduced to 70 cm^3 following treatment of the mixture with potassium hydroxide. If the conditions were S.T.P. throughout, find the formula of the hydrocarbon.
37. 25 cm^3 of a mixture of hydrogen, methane, and carbon dioxide is exploded with 25 cm^3 of oxygen. The total volume decreased to 17.5 cm^3. On treatment with potassium hydroxide the volume further decreased to 7.5 cm^3. If the conditions are S.T.P. throughout, find the volume composition of the mixture.

Molarity introductory exercises

38. What mass of solute is needed to prepare 1 dm^3 of each of the following solutions:
    (a) 0.60M sodium hydroxide
    (b) 0.50M sulphuric acid
    (c) 2.00M iron(III) chloride
    (d) 0.100M barium hydroxide
    (e) 2.00M sodium carbonate ($Na_2CO_3.10H_2O$)

39. What mass of solute is present in each of the following solutions:
    (a) 500 cm^3 of 6.00M HCl
    (b) 2.0 dm^3 of 0.40M H_3PO_4
    (c) 5.00 dm^3 of 0.500M $CaCl_2$
    (d) 100 cm^3 of 0.0500M $Ca(OH)_2$
    (e) 127 cm^3 of 0.400M $AgNO_3$?
40. What mass of solute is needed to prepare:
    (a) 140 cm^3 of 0.560M KCl
    (b) 2.80 dm^3 of 1.04M $Cu(NO_3)_2$
    (c) 70 cm^3 of 0.20M KOH
    (d) 310 cm^3 of 0.240M HNO_3
    (e) 400 cm^3 of 0.850M $CuSO_4$ from $CuSO_4.5H_2O$
41. To prepare 200 cm^3 of 0.40M HCl what volume of 10M acid would be required?
42. 150 cm^3 of 2.0M KNO_3 is mixed with 70 cm^3 of 1.5M K_2SO_4. What is the molarity of each salt in the final solution?
43. If 50 cm^3 of 3.0M H_3PO_4 is diluted to 800 cm^3, how many mole of solute are present in 40 cm^3 of the dilute solution?
44. A 186 g sample of a dilute aqueous solution of hydrogen peroxide liberated 2.80 dm^3 of oxygen, at S.T.P., after complete decomposition into water and oxygen. Calculate the concentration of the hydrogen peroxide in:
    (a) per cent by mass;
    (b) molarity (assume that the density of the solution is 1 g cm^{-3}).
45. A solution of hydrogen chloride in water contains 20.15% HCl by weight and has a density of 1.015 g cm^{-3}. It is boiled at 760 mmHg pressure, and has a constant boiling point of 110°C. What is the molarity of this constant boiling point acid?

Acid-base titrations

46. (a) How many cm^3 of 0.480M H_2SO_4 is required to neutralize 26.0 cm^3 of 0.612M NaOH?
    (b) What is the molarity of a solution of H_2SO_4 if 42.7 cm^3 is required to exactly titrate 27.5 cm^3 of 0.612M NaOH?
    (c) To what volume must 400 cm^3 of 0.612M NaOH be diluted to yield a 0.500M solution?
47. Some anhydrous sodium carbonate solid completely reacts with 20.0 cm^3 of 0.250M HCl. Find the mass of sodium carbonate required.
48. In each of the cases given below state which reactant is in excess and by how many mole it is in excess.
    (a) 1.0 g of magnesium reacts with 10 cm^3 of 2.0M HCl
    (b) 0.70 g of Na_2CO_3 reacts with 10 cm^3 of M HCl
    (c) 0.70 g of Na_2CO_3 reacts with 10 cm^3 of M H_2SO_4
    (d) 5.50 g of $BaCO_3$ reacts with 100 cm^3 of 0.600M HCl
49. What volume of 0.10M H_2SO_4 would be required to neutralize a mixture of 0.500 g of NaOH and 0.800 g of KOH?
50. 50 cm^3 of 0.10M HNO_3 is mixed with 60 cm^3 of 0.10M $Ca(OH)_2$. What volume of 0.05M H_2SO_4 is required to neutralize the mixture?
51. 578 cm^3 of dry hydrogen chloride gas, at 25.0°C, 1.50×10^5 N m^{-2}, is dissolved to make 200 cm^3 of solution. This is added to 400 cm^3 of 0.100M $Ba(OH)_2$. How many cm^3 of 2.00M HCl is needed to titrate the resulting solution to the endpoint?

52. Solutions of KOH and HNO_3 are standardized as follows. 26.8 cm^3 of the acid neutralizes 25.1 cm^3 of the hydroxide and 41.2 cm^3 of the hydroxide is used to titrate 2.00 g of benzoic acid. Benzoic acid has a molecular weight of 122.1 and reacts with KOH according to the following equation:

$$C_6H_5.CO.OH + KOH \rightarrow C_6H_5.CO.OK + H_2O$$

In titrating the benzoic acid the endpoint was passed and 2.60 cm^3 of HNO_3 solution was required to complete the titration. Calculate the molarity of the HNO_3 solution and of the KOH solution.

53. One gram of an acid (HA) of unknown formula is dissolved in water and 20.72 cm^3 of 0.588M NaOH solution is added. The reaction may be represented

$$HA + NaOH \rightarrow NaA + H_2O$$

7.82 cm^3 of 0.510M HCl is required to neutralize the excess hydroxide. What is the molecular weight of HA?

54. A sample of $Ba(OH)_2.8H_2O$ has lost some of its water of crystallization. Two gram of the solid is dissolved in water and requires 35.2 cm^3 of 0.620M HCl to titrate to the endpoint. How many gram of water would the 2-gram sample have to absorb to form pure $Ba(OH)_2.8H_2O$?
55. 0.930 g of a hydrated form of sodium carbonate exactly reacts with 150 cm^3 of 0.100M HCl. Calculate the number of molecules of water of crystallization present.

Precipitation titrations

56. A solution contains 2.50 g of ammonium sulphate. How many cm^3 of 0.120M $BaCl_2$ solution would be required to completely precipitate all the sulphate ion from the solution?
57. Excess hydrochloric acid was added to 10.0 cm^3 of $AgNO_3$ solution and 1.00 g of AgCl was obtained in the reaction. What was the molarity of the $AgNO_3$ solution?
58. Excess hydrogen sulphide was passed through 200 cm^3 of copper(II) chloride solution and 1.877 of copper(II) sulphide precipitated. Calculate the molarity of the copper(II) chloride solution.
59. How many cm^3 of 0.150M $AgNO_3$ solution is required to precipitate all the chloride ion from a sample of rock salt, of mass 0.300 g, and containing 98.5% NaCl?
60. What mass of AgCl can be obtained from 500 cm^3 of 1.00M $AgNO_3$ solution, if 16.0 cm^3 of 0.200 KCl solution is added?
61. 30.0 cm^3 of 0.200M $Fe_2(SO_4)_3$ solution is added to 30.0 g of anhydrous barium chloride. Calculate the mass of the reactant left in excess and also the mass of barium sulphate formed in the reaction.
62. A mixture known to contain NaCl and KCl only weighs 2.000 g. The mixture is dissolved in water and all of the chloride is precipitated as silver chloride by the addition of 50.00 cm^3 of 0.6000M $AgNO_3$. What fraction of the mixture is NaCl?
63. A certain organic compound is found by analysis to contain 31.9% carbon and 5.30% hydrogen. Analysis shows that the only other element present is chlorine. Find the empirical formula. The compound is heated with sodium and the chlorine present in 1.50 g of the compound is converted to chloride ion. How many cm^3 of 0.450M $AgNO_3$ is required to react with the chloride ion? When a 1.50 g sample of the compound is converted to vapour it is found to occupy 410 cm^3 at 100°C, 762 mmHg. Find the molecular formula of the compound.

Oxidation-reduction titrations

64. Concentrated nitric acid contains 68% by mass of HNO_3. If this solution is 15M, how many dm^3 of the concentrated acid is needed to react with 160 g of copper?

The equation for the reaction is

$$Cu + 4HNO_3 \rightarrow Cu(NO_3)_2 + 2H_2O + 2NO_2$$

or

$$Cu + 4H^+ + 2NO_3^- \rightarrow Cu^{2+} + 2H_2O + 2NO_2$$

65. How many gram of silver metal will react with 2.000 dm^3 of 10.00M HNO_3 according to the equation

$$Ag + 2HNO_3 \rightarrow AgNO_3 + H_2O + NO_2$$

66. How many cm^3 of 0.0500M $KMnO_4$ is required to oxidize 1.00 g of anhydrous iron(II) sulphate, in dilute acid solution, according to the equation

$$MnO_4^- + 5Fe^{2+} + 8H^+ \rightarrow Mn^{2+} + 5Fe^{3+} + 4H_2O$$

67. 0.420 g of $KMnO_4$ is exactly reduced by 40.0 cm^3 of an acidified solution containing Fe^{2+} ion. Find the molarity of the Fe^{2+} ion.

68. If x g of pure iron is dissolved completely in excess dilute sulphuric acid, what volume of a $KMnO_4$ solution, containing 3.16 g of $KMnO_4$ per dm^3, is required for complete reaction with the iron solution?

69. 0.812 g of pure iron is dissolved in dilute sulphuric acid and the solution is titrated with a solution of potassium dichromate, containing 12.0 g of $K_2Cr_2O_7$ per dm^3. What volume of the potassium dichromate solution is required? The equation for the reaction is

$$Cr_2O_7^{2-} + 6Fe^{2+} + 14H^+ \rightarrow 2Cr^{3+} + 6Fe^{3+} + 7H_2O$$

70. 10.0 cm^3 of 0.200M H_2O_2 exactly reduces 5.00 cm^3 of a $KMnO_4$ solution. Find the molarity of the $KMnO_4$ solution and its concentration in g dm^{-3}. The equation for the reaction is

$$2MnO_4^- + 5H_2O_2 + 6H^+ \rightarrow 2Mn^{2+} + 5O_2 + 8H_2O$$

71. 1.70 g of an impure sample of anhydrous oxalic acid ($H_2C_2O_4$) completely reacts with 4.34 cm^3 of 1.20M $KMnO_4$. Find the percentage by mass of oxalic acid in the sample. The equation for the reaction is

$$2MnO_4^- + 5H_2C_2O_4 + 6H^+ \rightarrow 2Mn^{2+} + 10CO_2 + 8H_2O$$

72. Mercury reduces potassium dichromate in the presence of hydrochloric acid. In an experiment so conducted that one compound only of mercury was produced, 1 gram of the metal reduced 50 cm^3 of M/60 $K_2Cr_2O_7$. From the data decide whether a mercury(I) (Hg_2^{2+}) or mercury(II) (Hg^{2+}) compound was formed. The relevant equations are

$$3Hg + Cr_2O_7^{2-} + 14H^+ \rightarrow 3Hg^{2+} + 2Cr^{3+} + 7H_2O$$

$$6Hg + Cr_2O_7^{2-} + 14H^+ \rightarrow 3Hg_2^{2+} + 2Cr^{3+} + 7H_2O$$

5

Chemical Reactions and Heat

Present-day living conditions are greatly dependent on the ready availability of energy in its various forms. In our homes, our transport and in industry, energy is widely used. Within our own bodies, energy is vital. For industrial uses, fuels like coal or oil are burned to obtain energy, whilst in our bodies energy is derived, for instance in our muscles, from the hydrolysis of large phosphate-containing molecules. All these energy-yielding processes are chemical reactions. This is in itself sufficient reason to enquire as to the relation between chemical reactions and energy. In addition, we shall see in later chapters how a study of the energy changes involved in reactions can lead to a better understanding of many of the fundamental processes in chemistry.

In this chapter we will be concerned primarily with heat energy and chemical reactions. Such a study is often described as *thermochemistry*.

5.1 HEAT EFFECTS IN CHEMICAL REACTIONS

A few test tube experiments will quickly confirm that chemical reactions are often accompanied by heat changes. If some pellets of solid sodium hydroxide or a few cm^3 of concentrated sulphuric acid are added to water in a test tube, heat energy is released. The temperature of the test tube and its contents rises. Later, we will discuss this release of heat energy as arising from the chemical reactions involved in the dissolution process.

A chemical reaction which generates heat energy is described as an exothermic reaction.

Examples of other exothermic processes include the neutralization of aqueous sodium hydroxide with aqueous hydrochloric acid (and most other acid-base neutralizations), the hydration of anhydrous white $CuSO_4$ crystals to yield blue $CuSO_4.5H_2O$ crystals, and the condensation of steam to form liquid water. All these examples of exothermic processes involve the formation of chemical bonds of various types and strengths.

A chemical reaction which absorbs heat energy is described as an endothermic reaction.

There are many examples of endothermic reactions. If crystalline sodium thiosulphate ('hypo') is added to water in a test tube, the test tube feels distinctly cool as heat energy is absorbed from the surroundings including the hand. Likewise, the precipitation of magnesium carbonate from aqueous magnesium chloride and sodium carbonate solutions is an example of the many precipitation reactions which are endothermic. Another general type of process which is endothermic is the evaporation of a liquid. A dramatic illustration of this is the freezing of a water film on the outside of a test tube by the rapid evaporation of ether from inside the test tube.

If crystalline sodium bromide is dissolved in a test tube of water, it is difficult to detect any temperature change. We might be tempted to describe this reaction as 'thermoneutral', i.e. a reaction in which heat energy is neither absorbed nor evolved. In fact more precise measurements would reveal that the dissolution process is very slightly exothermic. It is interesting to note that as the precision and sensitivity of temperature measurements have steadily improved in recent years, so the number of reactions which are *exactly* thermoneutral has steadily fallen. One process which is almost exactly thermoneutral is the mixing of very dilute aqueous solutions of electrolytes like sodium chloride and potassium chloride. The absence of a heat change accords with the belief that these electrolytes consist of cations and anions which are essentially independent of each other in dilute solution.

Before we can discuss any further the relation between heat changes and chemical reactions, it is necessary to define suitable units of energy and to consider an experimental method for measuring these heat changes quantitatively.

5.2 UNITS OF ENERGY

To assist in understanding the nature of energy, let us first consider how electricity is generated and used. In a power station, fuel is burned to yield heat which then raises steam. This steam performs mechanical work in driving a turbine which turns a rotor in a magnetic field and electrical energy is thereby produced. The electrical energy is then converted back to heat when the consumer throws a switch and allows electric current to flow through a resistance. This cycle of operations has interconverted three forms of work—thermal, mechanical and electrical—and the original source of energy was the chemical oxidation of a fuel.

Of course during this cycle, energy will be 'lost' through inefficient processes like friction in turbine bearings, incomplete heat transfer and resistive losses in the electrical circuits. The eventual fate of all forms of work is heat.

Very precise experiments can make due allowance for such energy losses and it has been established that, if such allowances are made, a given amount of mechanical work will always produce a definite quantity of either thermal energy or electrical work. A quantity of one form of work has been found to be equivalent to a definite quantity of every other form of work. It is therefore possible to define all forms of work in terms of a single phenomenon which we call *energy*.

The fundamental unit of energy is the joule.

The joule is defined in terms of mechanical work, and 1 joule is equal to the work done when a force of 1 newton (1 kg m s^{-2}) moves through a distance of 1 metre. However, in the laboratory, the most convenient way of generating a known quantity of energy is by electrical means. Thus if a steady current of 1 ampere is passed through a resistance, such that a potential difference of 1 volt is maintained between the ends of the resistance, heat energy is produced at the rate of 1 joule per second. So we may write:

$$1 \text{ joule} = 1 \text{ volt} \times 1 \text{ ampere} \times 1 \text{ second}$$

Heat has not always been recognized as a form of energy, and was originally measured in terms of a unit called the calorie, which was approximately equal to the amount of heat required to raise the temperature of one gram of water by 1 °C. The calorie is now defined in terms of the joule, and

$$1 \text{ calorie} \equiv 4.1840 \text{ joule}$$

Table 5.1

Units of energy and their equivalence

Unit	*Abbreviation*	*Equivalent units*
1 joule	J	0.23901 cal
1 kilojoule	kJ	0.23901 kcal or 10^3 J
1 calorie	cal	4.1840 J
1 kilocalorie	kcal	4.1840 kJ or 4.1840 × 10^3 J

It must be stressed that the fundamental unit of energy is the joule, and this unit will be used throughout this book. However, it remains true that a majority of chemists still prefer to use the older calorie as a convenient measure of heat energy, and it is necessary for us to be reasonably familiar with both units. Today, the most accurate measurements of the heats of chemical reactions are quoted in joule, because of the ease with which calorimeters may be electrically standardized (see section 5.3) and it is for this reason that we have used the joule in this text. In reading another book which uses calories, it is necessary to multiply the number of calories by the factor 4.184 (4 as a rough rule) to express the amount of heat energy in terms of the joule.

The amount of heat energy involved in many chemical reactions is quite large in terms of joule and it is convenient to define the larger unit, the kilojoule where

$$1 \text{ kilojoule} \equiv 10^3 \text{ joule}$$

Similarly, 1 kilocalorie $\equiv 10^3$ calorie

Units of energy, and their inter-relations, are summarized in Table 5.1.

5.3 THE MEASUREMENT OF HEAT CHANGES

(1) A reaction in solution

It is important to appreciate the difference between temperature and heat energy. The temperature of a body is the quantitative expression, relative to an arbitrary zero point, of the 'hotness' of that body. The temperature of a body indicates the direction of heat transfer between that body and its surroundings. For instance, any material at 40°C will always transfer heat energy to its surroundings at 20°C. However, no nett transfer of heat occurs between 1 kg of iron at 40°C and 1 g of iron at 40°C when these two masses come in contact. Temperature is therefore independent of the mass of the material.

The preliminary test tube experiments described in Section 5.1 served to demonstrate that chemical reactions are accompanied by heat changes. The amount of heat energy evolved or absorbed during a reaction is measured in an instrument described as a *calorimeter*. The design of a calorimeter depends mainly on the type of reaction under study. Figure 5.1 illustrates the essential features of a calorimeter suitable for measuring heat changes occurring when a solid or a liquid is dissolved in a solvent such as water.

The reaction is carried out in a vacuum-jacketed Dewar vessel designed to reduce heat losses from the vessel to a minimum. A definite mass of water is placed in the vessel and the chemical under study is contained in a small glass vessel which can be broken open when the reaction is to be started. A stirrer assists in the rapid mixing of the contents and the establishment of rapid thermal equilibrium. The rise or fall in temperature is registered by

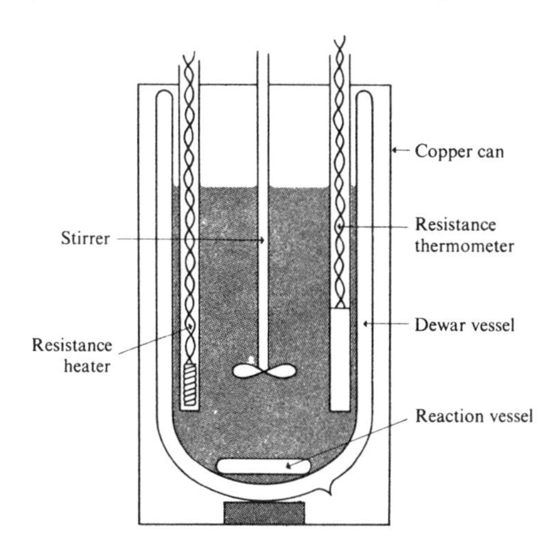

Fig 5.1 A calorimeter for measuring heats of solution.

either a mercury-in-glass thermometer or, as here, by means of a more sensitive platinum resistance thermometer. The relation between the amount of heat energy released inside the calorimeter and the consequent temperature rise must first be established by an independent calibration. This is often performed by dissipating a known amount of electrical energy in the resistance heater within the calorimeter, and noting the resultant temperature rise. Some typical experimental results for the dissolution of solid sodium nitrate in water will now be used to illustrate the operation of the calorimeter described above:

(a) *Electrical standardization of the calorimeter.* 250 g water was placed in the calorimeter with a sealed glass vessel containing 0.0132 mole of solid $NaNO_3$. The temperature of the calorimeter was observed until a steady value of 20.368 °C was established. Then a constant current of 1.51 amp was passed through the calibrating resistance for a period of 50.1 s. The potential difference across the resistance was 6.05 volt. The temperature of the calorimeter rose from 20.368 °C to 20.755 °C.

Electrical energy dissipated

$$= 1.51 \text{ A} \times 6.05 \text{ V} \times 50.1 \text{ s}$$

$$= 457.7 \text{ J} = 458 \text{ J}$$

$$\text{Temperature rise} = (20.755 - 20.368) = 0.387\,°\text{C}$$

Hence the release of 458 J heat energy causes a rise in temperature of the calorimeter and contents of 0.387 °C. A temperature rise of 1.00 °C would correspond to the release of 458/0.387 = 1180 J heat energy. This is the *calibration factor* for the calorimeter referring to 250 g water and 0.0132 mole of solid $NaNO_3$ in the sealed glass vessel.

(b) *The heat of solution.* The glass vessel was next smashed and the solid $NaNO_3$ dissolved with stirring. The temperature in the calorimeter fell from 20.755 to 20.521 °C.

$$\text{Temperature fall} = 20.755 - 20.521 = 0.234\,°\text{C}$$

Using the calibration factor of 1180 J per degree obtained in (a), we can calculate the heat energy change corresponding to this fall in temperature.

$$\text{Heat energy change} = 1180 \times 0.234 = 277 \text{ J}$$

Hence the dissolution of 0.0132 mole of $NaNO_3$ in 250 g of water would require the *absorption* of 277 J of heat energy from the surroundings in order to maintain the calorimeter plus contents at constant temperature. Thus we can write for the constant temperature process

$$0.0132 \text{ mole NaNO}_3(\text{s}) + \text{H}_2\text{O(l)} \rightarrow 0.0132 \text{ mole NaNO}_3(\text{aq})$$

$$q = +277 \text{ J}$$

where q is the heat energy change for the reaction, and the positive sign indicates that heat must be *added to* the system in order to maintain it at constant temperature. Thus the reaction is endothermic—it absorbs heat. The (aq) symbol on the right hand side of the equation indicates that the $NaNO_3$ is in aqueous solution.

(2) A reaction in the gas phase

It should be noted that the calorimeter described in the above experiment was not sealed off from the atmosphere, and consequently reactions will necessarily be carried out at atmospheric pressure. Calorimeters designed for studying reactions between gases must be sealed, and such reactions are therefore studied under constant volume conditions. If gases are evolved or used up during a chemical reaction, it is found that the heat changes measured under constant pressure conditions differ slightly from those measured under constant volume conditions. For this reason it is customary for all heat changes to be referred to constant pressure conditions, and all values quoted in this book will refer specifically to constant pressure conditions. It may be assumed that all measurements made under constant volume conditions have been appropriately corrected when necessary.

A calorimeter suitable for studying the heat changes during a chemical reaction between gases is shown in Figure 5.2. This is known as a 'bomb' calorimeter. In making thermochemical measurements, the calorimeter bomb is initially evacuated, and known amounts of the gases which are to react are admitted. The calorimeter (bomb plus water-bath) is then calibrated electrically in the same way as described in the previous example. Finally, reaction inside the bomb is initiated, e.g. by sparking the mixture; the temperature rise caused by the reaction is then recorded, and the heat

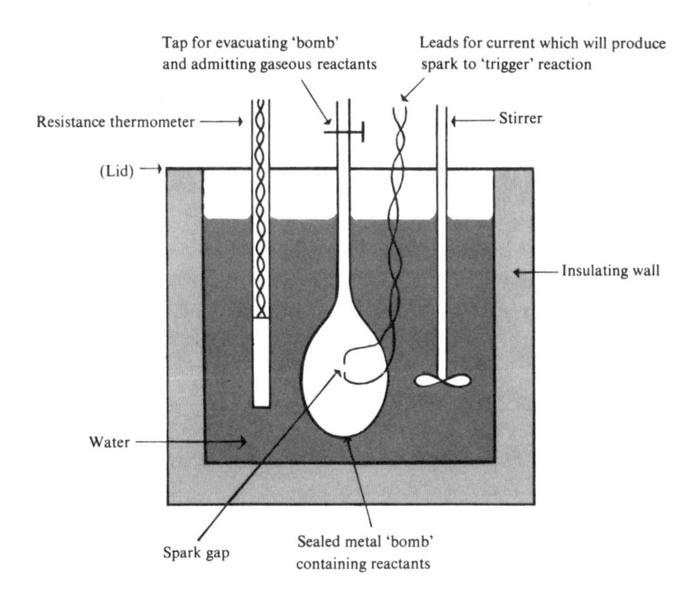

Fig 5.2 'Bomb' calorimeter for gaseous reactants.

energy released by the reaction may thus be calculated. We will now describe a typical experiment to determine the heat energy released during the burning of carbon monoxide in oxygen.

(a) *Electrical standardization of the calorimeter.* The bomb consisted of a steel vessel of total volume 2000 cm³, immersed in 10 kg of water held in an insulated container.

0.045 mole of carbon monoxide gas and 0.045 mole of oxygen gas were admitted to the bomb. The bomb was sealed, and the temperature of the whole calorimeter was noted to be 24.505°C. A current of 2.2 ampere under a potential difference of 10 volt was allowed to pass for 300 seconds through the resistance heater, and the temperature of the bath was observed to rise to 24.655°C. Thus we have temperature rise, $\Delta T = 24.655 - 24.505 = 0.150°C$.

This temperature rise was brought about by an energy input of

$$2.2 \text{ amp} \times 10 \text{ volt} \times 300 \text{ seconds}$$
$$= 6600 \text{ joule}$$

Therefore the calibration factor of the calorimeter plus contents is

$$\frac{6600}{0.150} = 44{,}000 \text{ J °C}^{-1}$$

(b) *The heat of reaction.* The mixture inside the bomb is exploded by the discharge of a tiny electric spark, which serves to trigger the reaction between gaseous carbon monoxide and gaseous oxygen to produce gaseous carbon dioxide. Notice that we initially put into the calorimeter 0.045 mole of carbon monoxide and 0.045 mole of oxygen. Since only 1 mole of oxygen is used up for every 2 mole of carbon monoxide, we will have when reaction is completed 0.045 mole of carbon dioxide together with 0.0225 mole of oxygen left in excess; i.e. only 0.0225 mole of oxygen took part in the reaction.

When reaction is complete, the temperature of the calorimeter is observed to rise from 25.255°C to 25.543°C. Thus we have the temperature rise,

$$\Delta T = 25.543 - 25.255 = 0.288°C$$

and from the calibration factor determined in (a) we can calculate the energy released by the chemical reaction as

$$0.288 \times 44{,}000 = 12{,}670$$
$$= 12.67 \text{ kJ}$$

Clearly we would have to remove 12.67 kJ of heat energy from the calorimeter in order to maintain it at the temperature of 25.255°C. We can represent this fact symbolically as:

$$0.045 \text{ mole } CO(g) + 0.0225 \text{ mole } O_2(g) \rightarrow 0.045 \text{ mole } CO_2(g)$$
$$q = -12.67 \text{ kJ}$$

The negative sign indicates that the reacting system will *lose* energy to the surroundings in maintaining a constant temperature; i.e. the reaction is exothermic.

5.4 HEAT CHANGES AND STOICHIOMETRY

Having established a suitable means (calorimetry) of accurately measuring the heat energy changes accompanying chemical reactions, we shall now see if there is any relation between the amount of substance involved in a chemical reaction and the amount of heat energy released or absorbed. In Table 5.2 we have recorded the heats absorbed during the dissolution of varying amounts of sodium nitrate in 250 g of water.

Table 5.2

Heat absorbed on dissolution of solid $NaNO_3$ in 250 g water

Mass $NaNO_3$ dissolved	*No. of mole $NaNO_3$*	*Heat absorbed (joule)*	*Heat* absorbed *per mole of $NaNO_3$ dissolved (joule)*	q (*kJ*)
0.373	0.0044	92	21,000	+21.0
0.560	0.0066	139	21,000	+21.0
1.120	0.0132	277	21,000	+21.0
1.680	0.0198	417	21,000	+21.0
1.865	0.0220	460	21,000	+21.0

Average heat energy change on dissolution of $NaNO_3$ in water at constant temperature = $+21.0$ kJ mol^{-1}.

We can draw an important conclusion from the data of Table 5.2, namely, that the amount of heat energy absorbed when sodium nitrate is dissolved in water is directly proportional to the number of mole of sodium nitrate dissolved, i.e. the more sodium nitrate dissolved, the more heat absorbed.

In Table 5.3 we have recorded data for the carbon monoxide plus oxygen reaction, for the combustion in excess oxygen of varying amounts of gaseous carbon monoxide.

Table 5.3

Heat evolved on combustion of carbon monoxide in excess oxygen to give carbon dioxide

Mole of CO(g) burned	*Heat evolved (joule)*	*Heat* evolved *per mole of CO(g) burned (joule)*	q (*kJ*)
0.009	2,535	283,000	−283
0.018	5,070	283,000	−283
0.027	7,605	283,000	−283
0.036	10,140	283,000	−283
0.045	12,670	283.000	−283

Average heat energy change per mole of carbon monoxide burned at constant temperature = -283 kJ mol^{-1}.

From these data we can draw the conclusion that the amount of heat evolved by burning carbon monoxide in excess oxygen is proportional to the amount of carbon monoxide consumed.
Analogous conclusions apply to every chemical reaction.

EXERCISE
0.05 mole of gaseous methane is completely burnt in excess oxygen to yield gaseous carbon dioxide and liquid water. 44.5 kJ of heat energy was evolved.
(a) Write a balanced chemical equation for the reaction occurring.
(b) Calculate the amount of heat energy that would be evolved by the combustion of 1 mole of methane.

5.5 HEAT CONTENT

In expressing quantitative results of the sort discussed in section 5.4, it is customary for chemists to use a quantity called 'heat content' or 'enthalpy', to which is given the symbol H. This quantity is defined in such a way that the heat energy change, q, for any reaction, *measured at a constant temperature and pressure*, gives the change in heat content accompanying the reaction, i.e.

$$q \text{ (heat energy change)} = \Delta H \text{ (heat content change)}$$

The definition of the heat content, H, arises from the application to chemistry of that fundamental law of physics, the Law of Conservation of Energy, which states that energy can neither be created nor destroyed. Although it is now known that mass and energy are interconvertible, it is a fact of experience that in chemical changes such effects lie well outside the limits of detection, and the simple statement of the law given above is valid within the limits of experimental detection.

Energy is usually given the symbol U, and a simple practical mathematical statement of the law of conservation of energy is

$$\Delta U = q + w$$

In physical terms this says that the energy of any body or system can be increased either by
(a) the *addition* of heat energy, q, or
(b) the performance of work, w, *on* the body or system. The work may be in any of a number of forms, e.g. mechanical work or electrical work.

Conversely, the energy of a body can be reduced by the removal of heat energy from the body, or the performance of work *by* the body. From the chemist's point of view, the heat energy change, q, is by far the most important way in which the energies of chemical systems change, since, as we have already seen, heat is released or absorbed during the occurrence of almost all chemical reactions. There is however one comparatively unimportant situation in which mechanical work *may* be done during a chemical change, and that is in those cases when the *volume* of a chemical system changes appreciably during a reaction. Since any chemical system will normally be subjected to some external pressure, e.g. the atmosphere, a volume change will involve the movement of a force (since pressure is force per unit area) through some definite distance, and hence mechanical work will be done.

Now volume changes are only significant in gaseous reactions in which there is a change in the total number of mole of gas present, e.g. in the reaction

$$2H_2(g) + O_2(g) \rightarrow 2H_2O(l)$$

where the number of mole of gas present drops from 3 to zero. If we allow for the effect on q of the small work term accompanying such volume changes, we get:

$$q = \Delta U - w$$

where w is the mechanical work done *on* the chemical system in changing its volume. Mechanics tells us that

$$w = -p\Delta V$$

at constant pressure, where ΔV is the volume change and p is the pressure, so that

$$q = \Delta U + p\Delta V \text{ at constant pressure.}$$

It has thus been decided to arbitrarily define

$$H = U + pV$$

so that $$\Delta H = \Delta U + p\Delta V \text{ at constant pressure}$$
$$= q \qquad \text{at constant pressure}$$

i.e. H is defined in such a way that

$$q \text{ (heat energy change)} = \Delta H \text{ (heat content change)}$$

at constant pressure.

In practice, we must emphasize that the difference between ΔH and ΔU in almost all chemical reactions, even when volume changes *are* involved, is barely outside the usual experimental error. For example, the heat evolved per mole of carbon monoxide burned in oxygen has been measured in a bomb calorimeter as -281.6 kJ at constant volume. This corresponds to the change in internal energy (ΔU) for the reaction:

$$CO(g) + \tfrac{1}{2}O_2(g) \rightarrow CO_2(g)$$

The value of ΔH for this reaction may then be calculated by evaluating the $p\Delta V$ term for this particular reaction.

For the reaction of 1 mole of CO we have

$$p\Delta V = \Delta nRT = -\tfrac{1}{2} \times 8.31 \times 300$$
$$= -1.2 \text{ kJ}$$
$$\therefore \Delta H = \Delta U + p\Delta V$$
$$= -281.6 - 1.2$$
$$= -282.8$$

Thus the $p\Delta V$ term leads to a correction to this particular measured heat energy change of less than 0.5 per cent in converting it to ΔH.

5.6 THE ASSIGNMENT OF HEAT CONTENT CHANGES TO REACTIONS

The use of the ΔH notation may be illustrated in the cases of the two examples given in section 5.4. For the dissolution of $NaNO_3$ in water it was apparent that the reaction absorbed heat to the extent of 21,000 J per mole of $NaNO_3$ dissolved. That is, if we wanted to go from the starting materials ($NaNO_3$(s) and water) to the final state ($NaNO_3$, aq), the chemical system would have had to gain 21,000 J of heat energy from the surroundings per mole of $NaNO_3$ dissolved, *in order to maintain constant temperature.*

This may be written symbolically:

$$NaNO_3(s) + aq \rightarrow NaNO_3(aq) \qquad \Delta H = +21{,}000\ J$$

This means that 1 *mole* of $NaNO_3$(s), when dissolved in an indeterminate but large excess of water, denoted by 'aq', will gain 21,000 J of heat in forming 1 *mole* of $NaNO_3$(aq) at the same temperature. Analogously we could write

$$2NaNO_3(s) + aq \rightarrow 2NaNO_3(aq) \qquad \Delta H = +42{,}000\ J$$

$$0.5\ NaNO_3(s) + aq \rightarrow 0.5\ NaNO_3(aq) \qquad \Delta H = +10{,}500\ J$$

where we are dissolving 2, and 0.5 mole of $NaNO_3$(s) respectively.

These heat content changes represent changes on going from a solid to a solution. The reverse process would be one of crystallization of a solid from solution, and it follows that the heat changes will also be reversed, so

$$NaNO_3(aq) \rightarrow NaNO_3(s) + aq \qquad \Delta H = -21{,}000\ J$$

This process would give out heat to the surroundings and would therefore be exothermic.

The thermochemistry of the carbon monoxide plus oxygen reaction may be represented in a similar fashion to the sodium nitrate dissolution. It was clear in section 5.4 that when 1 mole of carbon monoxide reacted with 0.5 mole of oxygen to give carbon dioxide, 283,000 J of heat energy had to be released to the surroundings in order to maintain the chemical system at constant temperature. This would be symbolically represented as:

$$CO(g) + \tfrac{1}{2}O_2(g) \rightarrow CO_2(g) \qquad \Delta H = -283{,}000\ J$$

and further,

$$2CO(g) + O_2(g) \rightarrow 2CO_2(g) \qquad \Delta H = -566{,}000\ J$$

and in reverse,

$$CO_2(g) \rightarrow CO(g) + \tfrac{1}{2}O_2(g) \qquad \Delta H = +283{,}000\ J$$

This latter is a process we cannot observe directly at ordinary temperatures, but the equation expresses well the fact that 1 mole of carbon monoxide gas and 0.5 mole of oxygen gas has a much higher heat content than has 1 mole of gaseous carbon dioxide.

It should finally be noted that in both the examples chosen, no temperature was specified. In the absence of any indications to the contrary, it is normally assumed that measurements are made at room temperatures (about 300 K).

More precise indications are not necessary, as it is found in practice that ΔH is almost independent of temperature over a small temperature range (say 100 K).

EXERCISE

Given that the combustion of 0.01 mole of hydrogen in excess oxygen to yield liquid water results in the evolution of 2860 J of heat energy, calculate ΔH for the following reactions:

(a) $H_2(g) + \frac{1}{2}O_2(g) \rightarrow H_2O(l)$

(b) $H_2O(1) \rightarrow 2H_2(g) + O_2(g)$

(c) $0.001 H_2(g) + 0.0005 O_2(g) \rightarrow 0.001 H_2O(l)$

5.7 THE MANIPULATION OF HEAT CONTENT CHANGES

One important consequence of the way in which H and ΔH have been defined (see section 5.5) is that every chemical substance may be assigned a definite heat content that will depend only on the temperature and pressure of the substance. This means that we can speak of a substance having a definite heat content in the same way as it has a definite weight, or a definite volume under a given set of physical conditions. We can make use of this property of H by recognizing that the heat content change, ΔH, for *any* process, will depend only on the initial and final states in the process, and *not* on the way in which the process is carried out.

A simple illustration of this is seen on considering the heat changes associated with the melting, vaporization and sublimation of water. The melting of ice is described by the equation:

$$H_2O(s) \rightarrow H_2O(l); \; \Delta H_{fus} = +6.00 \text{ kJ}$$

where ΔH_{fus} is the *latent heat of fusion* of water. Similarly vaporization of water is expressed by the equation:

$$H_2O(l) \rightarrow H_2O(g); \; \Delta H_v = +44.01 \text{ kJ}$$

where ΔH_v is the *latent heat of vaporization* of water. We can therefore imagine that ice may be converted to vapour by the sequence:

$$H_2O(s) \rightarrow H_2O(l) \rightarrow H_2O(g)$$

where it is first melted, and then the resulting liquid water is vaporized. Alternatively, instead of this two-stage process, we can vaporize ice directly without going via the liquid, as in

$$H_2O(s) \rightarrow H_2O(g); \; \Delta H_s = ?$$

where ΔH_s is the *heat of sublimation* of ice. We therefore have two different ways in which we can go from ice to water vapour, and this may be represented in the following form:

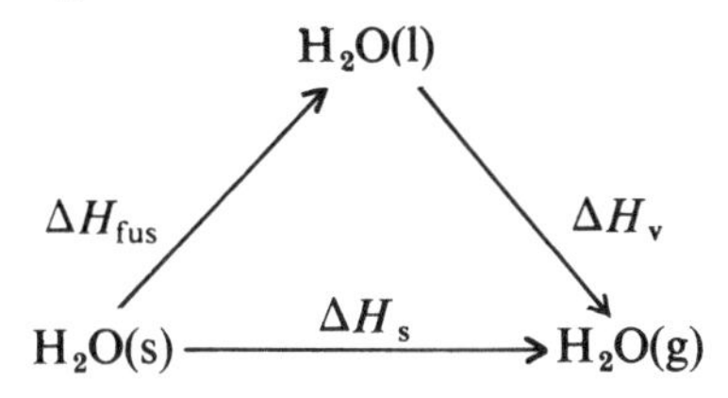

Since we know that there is a definite heat content difference between $H_2O(g)$ and $H_2O(s)$, the ΔH for the process

$$H_2O(s) \rightarrow H_2O(g)$$

will be the same irrespective of whether we carry out the process directly, or go via the liquid state, i.e.

ΔH_s must be equal to the sum of ΔH_{fus} and ΔH_v, and

$$\Delta H_s = \Delta H_{fus} + \Delta H_v = (+6.00 + 44.01)$$
$$= +50.01 \text{ kJ}$$

An alternative means of compounding ΔH values is simply to treat the thermochemical equations as algebraic equations, and manipulate them accordingly. Thus

A. $H_2O(s) \rightarrow H_2O(l)$ $\qquad \Delta H_A = +6.00$ kJ

B. $H_2O(l) \rightarrow H_2O(g)$ $\qquad \Delta H_B = +44.01$ kJ

Adding (A) and (B) we obtain

C. $H_2O(s) + H_2O(l) \rightarrow H_2O(l) + H_2O(g)$ $\qquad \Delta H_C = \Delta H_A + \Delta H_B$

The $H_2O(l)$ cancels and we simplify to

C. $H_2O(s) \rightarrow H_2O(g)$ $\qquad \Delta H_C = +50.01$ kJ

EXERCISE

The latent heat of sublimation of solid sulphur trioxide is 67.2 kJ mol^{-1} and its latent heat of fusion is 20.3 kJ mole^{-1}. Calculate the latent heat of vaporization of liquid sulphur trioxide.

A more complex example involving the combination of two ΔH values to get a third is provided by the reaction

$$CO_2(g) + H_2(g) \rightarrow CO(g) + H_2O(g)$$

This reaction cannot be studied directly, but we can measure the following ΔH values:

A. $H_2(g) + \frac{1}{2}O_2(g) \rightarrow H_2O(g)$ $\qquad \Delta H_A = -242.0$ kJ

B. $CO(g) + \frac{1}{2}O_2(g) \rightarrow CO_2(g)$ $\qquad \Delta H_B = -283.0$ kJ

The required reaction may be obtained by subtracting (B) from (A) giving:

C. $H_2(g) + \frac{1}{2}O_2(g) - CO(g) - \frac{1}{2}O_2(g) \rightarrow H_2O(g) - CO_2(g)$

$$\Delta H_C = \Delta H_A - \Delta H_B$$

or, simplifying:

$$H_2(g) + CO_2(g) \rightarrow CO(g) + H_2O(g) \qquad \Delta H_C = +41.0 \text{ kJ}$$

Another example is provided by the calculation of ΔH for the reaction:

A. $C(s) + 2H_2(g) \rightarrow CH_4(g)$

given

B. $C(s) + O_2(g) \rightarrow CO_2(g)$ $\quad \Delta H_B = -394.0$ kJ

C. $H_2(g) + \frac{1}{2}O_2(g) \rightarrow H_2O(l)$ $\quad \Delta H_C = -242.0$ kJ

D. $CH_4(g) + 2O_2(g) \rightarrow CO_2(g) + 2H_2O(l)$ $\quad \Delta H_D = -803.0$ kJ

These equations may be combined to eliminate O_2, CO_2 and H_2O, by multiplying equation (C) by 2, giving

B. $C(s) + O_2(g) \rightarrow CO_2(g)$ $\quad \Delta H_B = -394.0$ kJ

2C. $2H_2(g) + O_2(g) \rightarrow 2H_2O(l)$ $\quad 2\Delta H_C = -484.0$ kJ

D. $CH_4(g) + 2O_2(g) \rightarrow CO_2(g) + 2H_2O(l)$ $\quad \Delta H_D = -803.0$ kJ

These may be combined to give ΔH_A as follows:

$$B + 2C - D$$

$$= C(s) + O_2(g) + 2H_2(g) + O_2(g) - CH_4(g) - 2O_2(g)$$

$$\rightarrow CO_2(g) + 2H_2O(l) - CO_2(g) - 2H_2O(l)$$

$$\Delta H_A = \Delta H_B + 2\Delta H_C - \Delta H_D$$

$$= -394 - 484 + 803$$

and simplifying we have:

$$C(s) + 2H_2(g) \rightarrow CH_4(g) \qquad \Delta H_A = -75.0 \text{ kJ}$$

EXERCISE

Calculate ΔH for the reaction

$$2S(s) + 3O_2(g) \rightarrow 2SO_3(g)$$

given

$S(s) + O_2(g) \rightarrow SO_2(g)$; $\quad \Delta H = -297.0$ kJ

$2SO_2(g) + O_2(g) \rightarrow 2SO_3(g)$; $\quad \Delta H = -196.0$ kJ

5.8 HEATS OF FORMATION

The energy content of any chemical substance will be compounded of the potential energy inherent in the electrical and nuclear interactions of the constituent atoms of the substance, together with the kinetic energies of motion of the atoms and individual electrons and nuclei. It is impossible to evaluate exactly all these various contributions to the total energy of a substance, particularly the huge energies locked away in atomic nuclei. For the chemist who wants some numerical measure of the energy or, alternatively, heat content, which is nearly identical with the energy, it is therefore necessary to choose some arbitrary energy zero on which to base some numerical heat content scale. A convenient practical measure has been the adoption of a universally agreed convention as to the definition of a quantity known as the standard heat of formation.

The standard heat of formation of a compound is defined as the heat change accompanying the formation of 1 mole of this compound from its constituent elements, with each substance in its standard state at a specified temperature.

For our purposes 25°C, or 298 K, will always be used as the reference temperature. The standard state of a compound refers to the normal state in which the compound exists at 298 K and at atmospheric pressure. Thus H_2, Cl_2, HCl and CO_2 are all gaseous in their standard states, Br_2, Hg and H_2SO_4 are all liquids, whilst I_2, P_4, and Pb are all solids. In cases where allotropic forms of a compound exist, the standard state is that of the stable form, e.g. graphite is the standard state of carbon at 298 K and one atmosphere pressure.

An immediate consequence of the definition of standard heat of formation is that *the standard heat of formation of all elements is zero*, since no heat change is involved when they are formed from themselves in their standard states.

The standard heat of formation of a compound is assigned the symbol ΔH_f° where the superscript ° indicates the standard state and the subscript f refers to the formation reaction. Consequently we may write for some typical elements:

$$\Delta H_f^\circ(H_2, g) = 0 \qquad \Delta H_f^\circ(Br_2, l) = 0 \qquad \Delta H_f^\circ(I_2, s) = 0$$

In thinking of heat effects in chemical reactions, we have seen in section 5.5 that it is convenient to visualize each compound as having a definite heat content or 'enthalpy'. The standard heat of formation is a measure of this heat content, the value being expressed relative to an arbitrary zero for the constituent elements. The question of whether heat will be evolved or absorbed during a reaction is determined by the combined heat content of the products relative to that of the reactants.

Consider the formation of carbon dioxide from carbon and oxygen:

$$\text{C (graphite)} + O_2(g) \rightarrow CO_2(g) \qquad \Delta H = -393.4 \text{ kJ}$$

where, by definition, $\Delta H = \Delta H_f^\circ(CO_2, g) = -393.4$ kJ
Figure 5.3 represents the *relative* heat contents of the products and reactants for this reaction, and

$$\begin{aligned}\Delta H &= (\text{heat content } CO_2) - (\text{heat content C} + \text{heat content } O_2)\\ &= \Delta H_f^\circ(CO_2, g) - \{\Delta H_f^\circ(\text{C, graph}) + \Delta H_f^\circ(O_2, g)\}\\ &= -393.4 - (0 + 0) = -393.4 \text{ kJ}\end{aligned}$$

We see that heat is released because the heat content of carbon dioxide is lower than the combined heat contents of graphite and oxygen. If we visualize the reaction on the atomic scale, the conversion of carbon atoms in graphite

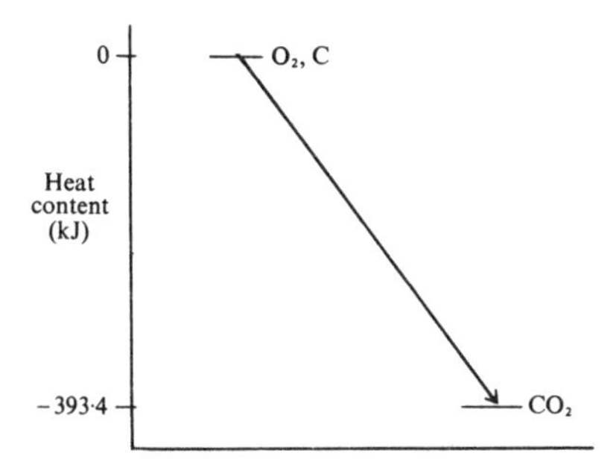

Fig 5.3.

and oxygen molecules to carbon dioxide molecules releases some of the potential energy associated with electron arrangements in the graphite and molecular oxygen. This released potential energy is converted to kinetic energy of the CO_2 molecules and the temperature of the CO_2 molecules rises. This interconversion of potential energy and kinetic energy is analogous to the interconversion of potential energy to kinetic energy when a ball falls to the ground from a height. Just as work is required to lift the ball back to its former position, so work, in the form of heat energy, must be supplied to CO_2 molecules to convert them into graphite and oxygen. This heat energy is stored in the graphite and oxygen as potential energy, which is available for release in a suitable reaction.

The opposite situation arises with nitrogen dioxide when it is formed from its elements:

$$\tfrac{1}{2}N_2(g) + O_2(g) \rightarrow NO_2(g) \qquad \Delta H_f^\circ = +33.7 \text{ kJ}$$

Figure 5.4 shows that the heat content of NO_2 is higher than that of nitrogen plus oxygen. The heat content change in this reaction is:

$$\Delta H = \Delta H_f^\circ(NO_2, g) - \{\tfrac{1}{2}\Delta H_f^\circ(N_2, g) + \Delta H_f^\circ(O_2, g)\}$$
$$= +33.7 - (0 + 0) = +33.7 \text{ kJ}$$

In this case, kinetic energy of the reactant N_2 and O_2 molecules must be converted to potential energy stored within the NO_2 molecule.

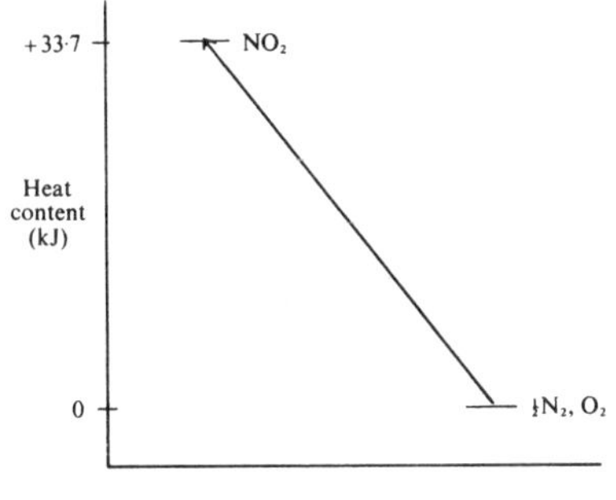

Fig 5.4.

The standard heats of formation of many compounds are now known. Only a few of these ΔH_f° values may be measured by direct calorimetry and it is often necessary to combine the ΔH values for a number of reactions in order to obtain a desired ΔH_f°. A simple example is the determination of the standard heat of formation of diamond. The stable allotrope of carbon at room temperature is graphite, so we require

A. $\quad C\text{ (graphite)} \rightarrow C\text{ (diamond)} \qquad \Delta H_A = ?$

This reaction cannot be studied directly but both allotropes may be burned in oxygen thus:

B. $\quad C\text{ (graphite)} + O_2(g) \rightarrow CO_2(g) \qquad \Delta H_B = -393.4\text{ kJ}$

C. $\quad C\text{ (diamond)} + O_2(g) \rightarrow CO_2(g) \qquad \Delta H_C = -395.3\text{ kJ}$

Then we have

$$\begin{aligned}\Delta H_A &= \Delta H_B - \Delta H_C \\ &= +1.9\text{ kJ}\end{aligned}$$

Hence the standard heat of formation of diamond, ΔH_f°(diamond) is 1.9 kJ.

A more complex example is provided by the determination of the heat of formation of ethanol (ethyl alcohol). Clearly, we cannot hope to measure ΔH directly for the reaction:

A. $\quad 2C(s) + 3H_2(g) + \frac{1}{2}O_2(g) \rightarrow CH_3.CH_2.OH(l)$

However, we can obtain ΔH values for the reactions:

B. $\quad C(s) + O_2(g) \rightarrow CO_2(g) \qquad \Delta H_B = -393\text{ kJ}$

C. $\quad 2H_2(g) + O_2(g) \rightarrow 2H_2O(l) \qquad \Delta H_C = -572\text{ kJ}$

D. $\quad 2CH_3.CH_2.OH(l) + 7O_2(g) \rightarrow 4CO_2(g) + 3H_2O(l)$

$$\Delta H_D = -2736\text{ kJ}$$

The ΔH values for these reactions may be combined as described in the previous section, and we obtain

$$\begin{aligned}\Delta H_A &= \Delta H_f^\circ(C_2H_5OH, l) \\ &= 2\Delta H_B + \tfrac{3}{2}\Delta H_C - \tfrac{1}{2}\Delta H_D \\ &= -276\text{ kJ}\end{aligned}$$

Table 5.4 lists some ΔH_f° values for some typical compounds, some of which have positive ΔH_f° values, whilst others have negative values.

Data in Table 5.4 may be used to deduce heats of reaction. For example, consider the reaction:

$$HCl(g) + \tfrac{1}{2}I_2(s) \rightarrow \tfrac{1}{2}Cl_2(g) + HI(g) \qquad \Delta H = ?$$

Table 5.4

Standard heats of formation of selected substances

Element or compound	ΔH_f° (*kJ mol*$^{-1}$)	*Element or compound*	ΔH_f° (*kJ mol*$^{-1}$)	*Eelment or compound*	ΔH_f° (*kJ mol*$^{-1}$)
Na(s)	0	N_2(g)	0	C (graphite)	0
NaF(s)	−571	HN_3(g)	+292	C (diamond)	+1.9
NaCl(s)	−411	NH_3(g)	−46.1	CO(g)	−115
NaBr(s)	−359	P(s)	0	CO_2(g)	−393
NaI(s)	−286	PH_3(g)	+9.25	NO(g)	+90.1
I_2(s)	0	AsH_3(g)	+41.0	NO_2(g)	+33.7
F_2(g)	0	H_2O(l)	−286	N_2O_4(g)	+9.65
Cl_2(g)	0	H_2S(g)	−20.2	S (rhombic)	0
Br_2(g)	0	Se(s)	0	S (monoclinic)	+0.34
IF(g)	−94.0	H_2Se(g)	+85.6	SO_2(g)	−296
ICl(g)	+17.5	Te(s)	0	CH_4(g)	−74.9
IBr(g)	+40.9	H_2Te(g)	+153	SiH_4(g)	−61.9
HF(g)	−268	CaO(s)	−635	Si(s)	0
HCl(g)	−92.4	CaF_2(s)	−1210	C_2H_2(g)	+226
HBr(g)	−36.2	$CaCl_2$(s)	−795	C_2H_4(g)	+52.4
HI(g)	+25.9	$CaSO_4$(s)	−1430	C_2H_6(g)	−84.4
H_2(g)	0	O_2(g)	0		

where the heat of reaction is unknown. The heat of reaction is given by the expression:

$$\begin{aligned}\Delta H &= (\text{heat content of products}) - (\text{heat content of reactants})\\ &= [\tfrac{1}{2}\Delta H_f^\circ(Cl_2, g) + \Delta H_f^\circ(HI, g)] - [\Delta H_f^\circ(HCl, g) + \tfrac{1}{2}\Delta H_f^\circ(I_2, s)]\\ &= [0 + (+25.9)] - [(-92.4) + 0]\\ &= +118.3 \text{ kJ}\end{aligned}$$

This reaction is therefore endothermic. In fact, this reaction cannot be studied directly in a calorimeter because it proceeds quite slowly.

A rather different example is the burning of ammonia in pure oxygen. Under appropriate conditions, the reaction is:

$$4NH_3(g) + 3O_2(g) \rightarrow 2N_2(g) + 6H_2O(g) \quad \Delta H = ?$$

and
$$\begin{aligned}\Delta H &= [2\Delta H_f^\circ(N_2, g) + 6\Delta H_f^\circ(H_2O, g)] - [4\Delta H_f^\circ(NH_3, g) + 3\Delta H_f^\circ(O_2, g)]\\ &= [0 + (6 \times -242)] - [(4 \times -46.1) + 0]\\ &= -1268 \text{ kJ}\end{aligned}$$

However, in the presence of a glowing platinum wire acting as a catalyst, ammonia will burn to nitrogen dioxide as follows:

$$4NH_3(g) + 7O_2(g) \rightarrow 4NO_2(g) + 6H_2O(g)$$

For this reaction we can again calculate ΔH from known ΔH_f° values.

$$\begin{aligned}\Delta H &= [4\Delta H_f^\circ(NO_2, g) + 6\Delta H_f^\circ(H_2O, g)] - [4\Delta H_f^\circ(NH_3, g) + 7\Delta H_f^\circ(O_2, g)]\\ &= [(4 \times 33.7) + (6 \times -242)] - [(4 \times -46.1) + 0]\\ &= -1133 \text{ kJ}\end{aligned}$$

5.9 BOND ENERGIES

The definition that the standard heats of formation of elements in their standard states are zero is a very convenient device for deducing heats of formation of compounds. However, this has avoided the quite fundamental issue of the 'strength' of chemical bonds in molecules like H_2. Before a satisfactory theory of chemical bonding is developed, it is necessary to have accurate measurements of the energy involved in holding the hydrogen atoms together. This energy is visualized as the energy necessary to pull these two hydrogen atoms apart and to separate them by a sufficiently large distance where the attractive forces are effectively zero:

$$H_2(g) \rightarrow H(g) + H(g) \qquad \Delta H = ?$$

This process has been followed in the reverse sense by first dissociating hydrogen molecules to atoms in a radio-frequency discharge tube and measuring the heat of recombination in a suitable calorimeter:

$$H(g) + H(g) \rightarrow H_2(g) \qquad \Delta H = -432\ \text{kJ}$$

The value of ΔH is therefore $+432$ kJ, indicating that the dissociation reaction requires the expenditure of a large amount of energy. It is usual to describe this as the *bond dissociation energy*, D, where D (H—H) = 432 kJ is a quantitative measure of the strength of the H—H bond. This energy is actually that required to dissociate 1 mole of hydrogen molecules into atoms. We can write analogously

$$Cl_2(g) \rightarrow 2Cl \qquad \Delta H = +239\ \text{kJ}$$

where 239 kJ is the energy required to dissociate 1 mole of gaseous chlorine molecules into 2 mole of chlorine atoms. The energy required to dissociate 1 molecule of chlorine into 2 chlorine atoms would be

$$E = (\Delta H/N_A)$$

where N_A is the Avogadro constant, i.e.

$$E = (239{,}000/6.023 \times 10^{23})\ \text{J molecule}^{-1}$$

$$= 3.97 \times 10^{-19}\ \text{J molecule}^{-1}$$

Because of the convenient magnitude of the energies involved, it is customary to express bond dissociation energies in units of energy per mole of the molecules dissociated.

A wide range of experimental techniques is available to determine bond dissociation energies of molecules. All these techniques essentially determine the minimum amount of energy required to dissociate a molecule. One technique varies the energy of light incident on the molecule. From the absorption properties of the molecule, it is possible to deduce when dissociation occurs. For example, the minimum frequency required to dissociate the gaseous iodine molecule into atoms can be determined:

$$I_2(g) + h\nu \rightarrow I(g) + I(g) \qquad \Delta H = \text{D (I—I)}$$

The energy of a light quantum is given by the basic expression:

$$E = h\nu = D\,(I\text{—}I)$$

where h is a universal constant called Planck's Constant, ν is the frequency of the incident light and D (I—I) is the bond dissociation energy of the molecule. Some bond dissociation energies are given in Table 5.5.

Table 5.5

Bond dissociation energies of some diatomic molecules

Molecule	*Bond dissociation energy (kJ)*	*Molecule*	*Bond dissociation energy (kJ)*
O_2	490	CO	1072
H_2	432	NO	626
F_2	151	HF	561
Cl_2	239	HCl	428
Br_2	190	HBr	362
I_2	149	HI	295

Let us consider the successive dissociation of bonds in methane which can only be approximately determined:

$$CH_4(g) \rightarrow CH_3(g) + H(g) \qquad D(CH_3\text{—}H) = 427\ kJ$$

$$CH_3(g) \rightarrow CH_2(g) + H(g) \qquad D(CH_2\text{—}H) = 371\ kJ$$

$$CH_2(g) \rightarrow CH(g) + H(g) \qquad D(CH\text{—}H) = 523\ kJ$$

$$CH(g) \rightarrow C(g) + H(g) \qquad D(C\text{—}H) = 344\ kJ$$

It is often difficult to separate experimentally these successive dissociation processes, and in these circumstances it is usual to quote the mean bond energy E where, in this example, E is one-quarter of the heat required for the process:

$$CH_4(g) \rightarrow C(g) + 4H(g) \qquad \Delta H = 1665\ kJ$$

Thus $E(CH_4) = \Delta H/4 = 416$ kJ. In this book, however, we shall use only bond dissociation energies of diatomic molecules.

QUESTIONS

(*In these calculations it is important to specify the sign of* ΔH *values in your answers.*)

1. State which of the following reactions are exothermic and which are endothermic:

(a) $I_2(s) + Cl_2(g) \rightarrow 2ICl(g) \qquad \Delta H = +35\ kJ$

(b) $N_2(g) \rightarrow 2N(g) \qquad \Delta H = +946\ kJ$

(c) $Ag^+(aq) + 2NH_3(aq) \rightarrow Ag(NH_3)_2^+(aq) \qquad \Delta H = -111\ kJ$

(d) $H^+(aq) + HC_2O_4^-(aq) \rightarrow H_2C_2O_4(aq) \qquad \Delta H = +0.54\ kJ$

(e) $Si(s) + 2H_2(g) \rightarrow SiH_4(g) \qquad \Delta H = -61.9\ kJ$

If each of these reactions were conducted in a thermally insulated vessel, state, in each case, whether the temperature of the contents would be higher or lower after the reaction was complete.

2. Considering some of the reactions listed above, calculate the heat changes which would occur when the following quantities of reagents are allowed to react completely:
   (a) 25.4 g solid iodine and 7.1 g gaseous chlorine;
   (b) 25.4 g solid iodine and 3.55 g gaseous chlorine;
   (c) 12.7 g solid iodine and 3.55 g gaseous chlorine;
   (d) 12.7 g solid iodine and 7.1 g gaseous chlorine;
   (e) 100 cm^3 0.1M $AgNO_3$ and 100 cm^3 0.2M aqueous ammonia;
   (f) 100 cm^3 0.1M $AgNO_3$ and 50 cm^3 0.4M aqueous ammonia;
   (g) 200 cm^3 0.05M $AgNO_3$ and 100 cm^3 0.2M aqueous ammonia;
   (h) 200 cm^3 0.05M $AgNO_3$ and 100 cm^3 0.4M aqueous ammonia;
   (i) 2.8 g solid silicon and 0.4 g gaseous hydrogen;
   (j) 2.8 g solid silicon and 4 dm^3 gaseous hydrogen, measured at S.T.P.
3. Calculate the heat changes involved in the reaction when:
   (a) 4.50 g gaseous ICl is completely decomposed to solid iodine and gaseous chlorine;
   (b) 5 dm^3 of gaseous SiH_4 (measured at S.T.P.) is completely decomposed to solid silicon and gaseous hydrogen. What mass of silicon would be produced in this reaction? If the hydrogen gas produced in this reaction were collected in a 20-dm^3 vessel at 25°C, calculate the pressure of hydrogen inside this vessel.
4. Calculate the heat change involved when 40 g sodium sulphate decahydrate crystallizes from water, given the information:

$$Na_2SO_4.10H_2O(s) + aq \rightarrow Na_2SO_4(aq) \qquad \Delta H = +79.0 \text{ kJ}$$

5. How much heat energy is required to remove 17 g ammonia, from an aqueous solution, into the gas state, given the information:

$$NH_3(g) + aq \rightarrow NH_3(aq) \qquad \Delta H = -35.4 \text{ kJ}$$

6. How much heat energy is either absorbed or evolved when 2.24 dm^3 of water vapour, at 25°C and 760 mmHg, is condensed to liquid water at the same temperature? Use the information:

$$H_2O(l) \rightarrow H_2O(g) \qquad \Delta H = +44.0 \text{ kJ}$$

What volume of liquid water is formed in this case? Is it possible to condense water vapour at 25°C?

7. When solid white phosphorus is burned in excess oxygen the following reaction occurs:

$$P_4(s) + 5O_2(g) \rightarrow P_4O_{10}(s) \qquad \Delta H = -3005 \text{ kJ}$$

When gaseous phosphorus is burned in excess oxygen, at the same temperature as before, the reaction is:

$$P_4(g) + 5O_2(g) \rightarrow P_4O_{10}(s) \qquad \Delta H = -3018 \text{ kJ}$$

Calculate the heat change for the process:

$$P_4(s) \rightarrow P_4(g)$$

8. 1.60 g rhombic sulphur was burned in excess oxygen to form gaseous sulphur dioxide and 14.800 kJ of heat energy was evolved. When 1.60 g monoclinic sulphur was burned, 14.817 kJ of heat energy was evolved. Calculate the heat change involved when 1.60 g rhombic sulphur is completely converted to 1.60 g monoclinic sulphur. Hence calculate the heat change for the reaction:

$$S\ (\text{rhombic}) \rightarrow S\ (\text{monoclinic})$$

9. A gaseous mixture of 100 g oxygen and 100 g hydrogen is sparked to form water, according to the equation

$$2H_2(g) + O_2(g) \rightarrow 2H_2O(l)$$

(a) What mass of water is formed?
(b) How many kilojoule of heat would be evolved?
Use the following data:

$$2H_2(g) + O_2(g) \rightarrow 2H_2O(g) \qquad \Delta H = -485\ \text{kJ}$$

$$H_2O(g) \rightarrow H_2O(l) \qquad \Delta H = -44\ \text{kJ}$$

10. Calculate ΔH for the process

$$4NH_3(g) + 5O_2(g) \rightarrow 4NO(g) + 6H_2O(g)$$

given:

$$N_2(g) + 3H_2(g) \rightarrow 2NH_3(g) \qquad \Delta H = -92.2\ \text{kJ}$$

$$2H_2(g) + O_2(g) \rightarrow 2H_2O(g) \qquad \Delta H = -484\ \text{kJ}$$

$$N_2(g) + O_2(g) \rightarrow 2NO(g) \qquad \Delta H = +180.2\ \text{kJ}$$

11. A mixture of propane and oxygen, each initially at a partial pressure of 50 kN m^{-2}, is reacted in a closed vessel at 300°C as shown:

(A) $$C_3H_8(g) + 5O_2(g) \rightarrow 3CO_2(g) + 4H_2O(g)$$

(a) What is the final total pressure at this temperature, the volume remaining constant?

(b) Given: $3C(s) + 4H_2(g) \rightarrow C_3H_8(g) \qquad \Delta H = -104\ \text{kJ}$

$$C(s) + O_2(g) \rightarrow CO_2(g) \qquad \Delta H = -393\ \text{kJ}$$

$$2H_2(g) + O_2(g) \rightarrow 2H_2O(g) \qquad \Delta H = -484\ \text{kJ}$$

calculate ΔH for the reaction (A) above.

(Notc: 10^5 N m^{-2} $\equiv$ 10^2 kN m^{-2})

12. Calculate ΔH for the process:

$$HCl(g) \rightarrow H(g) + Cl(g)$$

given:

$$Cl_2(g) \rightarrow 2Cl(g) \qquad \Delta H = 242\ \text{kJ}$$

$$H_2(g) \rightarrow 2H(g) \qquad \Delta H = 432\ \text{kJ}$$

$$H_2(g) + Cl_2(g) \rightarrow 2HCl(g) \qquad \Delta H = -185\ \text{kJ}$$

13. Calculate ΔH for each of the reactions:
(a) $2FeCl_3(s) \rightarrow 2Fe(s) + 3Cl_2(g)$
(b) $2FeCl_3(s) \rightarrow 2FeCl_2(s) + Cl_2(g)$
(c) $3FeCl_2(s) \rightarrow 2FeCl_3(s) + Fe(s)$
given that

$$Fe(s) + Cl_2(g) \rightarrow FeCl_2(s) \qquad \Delta H = -341 \text{ kJ}$$

$$Fe(s) + \tfrac{3}{2}Cl_2(g) \rightarrow FeCl_3(s) \qquad \Delta H = -400 \text{ kJ}$$

14. Calculate ΔH for the process:

$$2Al(s) + 3Cl_2(g) \rightarrow Al_2Cl_6(s)$$

given:

$$2Al(s) + 6HCl(aq) \rightarrow Al_2Cl_2(aq) + 3H_2(g) \qquad \Delta H = -1003 \text{ kJ}$$

$$H_2(g) + Cl_2(g) \rightarrow 2HCl(g) \qquad \Delta H = -184 \text{ kJ}$$

$$HCl(g) + aq \rightarrow HCl(aq) \qquad \Delta H = -72.4 \text{ kJ}$$

$$Al_2Cl_6(s) + aq \rightarrow Al_2Cl_6(aq) \qquad \Delta H = -643 \text{ kJ}$$

15. Calculate the heat of formation of solid calcium hydroxide, i.e. ΔH for the process

$$Ca(s) + O_2(g) + H_2(g) \rightarrow Ca(OH)_2(s)$$

given

$$H_2(g) + \tfrac{1}{2}O_2(g) \rightarrow H_2O(l) \qquad \Delta H = -286 \text{ kJ}$$

$$CaO(s) + H_2O(l) \rightarrow Ca(OH)_2(s) \qquad \Delta H = -64 \text{ kJ}$$

$$Ca(s) + \tfrac{1}{2}O_2(s) \rightarrow CaO(s) \qquad \Delta H = -635 \text{ kJ}$$

16. Hydrazine is often used as a rocket fuel, its heat of combustion being:

$$N_2H_4(l) + O_2(g) \rightarrow N_2(g) + 2H_2O(l) \qquad \Delta H = -622 \text{ kJ}$$

What would be the expected heat of reaction if fluorine gas were used as oxidant instead of oxygen?
Assume the products to be $N_2(g)$ and $HF(g)$.
Use the data:

$$\tfrac{1}{2}H_2(g) + \tfrac{1}{2}F_2(g) \rightarrow HF(g) \qquad \Delta H = -269 \text{ kJ}$$

$$H_2(g) + \tfrac{1}{2}O_2(g) \rightarrow H_2O(l) \qquad \Delta H = -286 \text{ kJ}$$

17. A small amount of pure sulphuric acid is sealed into a glass ampoule and placed in a calorimeter containing exactly 100 cm^3 of water at 25.00°C. A heating coil is present in the calorimeter and a current of 2 amp, under a potential difference of 6 volt, is passed for 24 s. The temperature in the calorimeter rises to 25.62°C. The ampoule of sulphuric acid is now broken under conditions of continuous stirring and the temperature rises to 26.10°C.
A 25.00 cm^3 aliquot of the sulphuric acid solution is titrated against a 0.05 M sodium hydroxide solution. 22.9 cm^3 of the sodium hydroxide solution are required to just neutralize the acid.
(a) Write chemical equations for the reactions occurring when sulphuric acid is brought into contact with water.

(b) Determine ΔH for the reaction

$$H_2SO_4(l) \rightarrow H_2SO_4(aq)$$

(c) How much heat would be evolved if 10 g of sulphuric acid were dissolved in excess water?

18. The standard heats of formation of acetylene gas (C_2H_2) and carbon dioxide gas are, respectively, +226 and −393 kJ mol^{-1}. Calculate the heat change involved when 0.1 mole of gaseous acetylene is completely burned to CO_2 and water vapour. What use is made of this reaction in industry?

19. Gaseous ethylene (C_2H_4) may be 'hydrogenated' to produce gaseous ethane (C_2H_6) as follows:

$$C_2H_4(g) + H_2(g) \rightarrow C_2H_6(g)$$

Calculate the heat change involved in this reaction, given that the standard heats of formation of ethylene and ethane are +52.4 and −84.4 kJ mol^{-1} respectively. Calculate the heat change involved when 6.5 dm^3 of gaseous ethylene is reacted with 4.6 dm^3 of gaseous hydrogen, both volumes being expressed at S.T.P.

20. A motorist fills his petrol tank with 'Super-Hep' motor fuel, which consists of pure pentane (C_5H_{12}). Later the motorist prefers 'Rocket-O' motor fuel which is a mixture of 1 mole of pentane and 1 mole of butane (C_4H_{10}). Calculate the heat evolved when 0.01 mole of each fuel is exploded as a gas in the car cylinder, with excess air, to yield carbon dioxide and water vapour. If the pentane fuel only burned to carbon monoxide and water vapour, calculate the loss in heat energy when 0.01 mole pentane is burned to CO rather than to CO_2. The standard heats of formation are as follows:

$$C_4H_{10}(g) : \Delta H_f^\circ = -124 \text{ kJ mol}^{-1}$$

$$C_5H_{12}(g) : \Delta H_f^\circ = -145 \text{ kJ mol}^{-1}$$

$$H_2O(g) : \Delta H_f^\circ = -242 \text{ kJ mol}^{-1}$$

$$CO_2(g) : \Delta H_f^\circ = -393 \text{ kJ mol}^{-1}$$

$$CO(g) : \Delta H_f^\circ = -115 \text{ kJ mol}^{-1}$$

21. Calculate the heat change for the direct synthesis of gaseous methanol

$$C(s) + 2H_2(g) + \tfrac{1}{2}O_2(g) \rightarrow CH_3OH(g)$$

given that the heat of combustion of methanol is

$$CH_3OH(g) + 1\tfrac{1}{2}O_2(g) \rightarrow CO_2(g) + 2H_2O(g) \qquad \Delta H = -676 \text{ kJ}$$

and that the standard heats of formation of $CO_2(g)$ and $H_2O(g)$ are −393 and −242 kJ mol^{-1}, respectively.

22. The following series of reactions of sulphur dioxide may be observed:

$$SO_2(g) + aq \rightarrow SO_2(aq) \qquad \Delta H = -35.50 \text{ kJ}$$

$$SO_2(aq) + aq \rightarrow HSO_3^-(aq) \qquad \Delta H = -19.20 \text{ kJ}$$

$$HSO_3^-(aq) + aq \rightarrow SO_3^{2-}(aq) \qquad \Delta H = +3.73 \text{ kJ}$$

Calculate the heat change involved when an aqueous solution of 10 g sodium sulphite is treated with excess acid and the sulphite completely converted into gaseous sulphur dioxide.

23. The belles of St Trinian's have just encapsulated their classics mistress in a rocket in preparation for her guided space flight. Elspeth, the head girl, has the choice of two rocket fuels, either 50 kg gaseous diborane (B_2H_6) or 100 kg gaseous benzene (C_6H_6). These fuels burn in excess oxygen as follows:

$$B_2H_6(g) + 3O_2(g) \rightarrow B_2O_3(s) + 3H_2O(g)$$

$$C_6H_6(g) + 7\tfrac{1}{2}O_2(g) \rightarrow 6CO_2(g) + 3H_2O(g)$$

On behalf of Elspeth, calculate which of the two samples of fuel will generate the greater amount of heat energy. Standard heats of formation, expressed in kJ mol^{-1} are: $B_2H_6(g)$: +31.4; $B_2O_3(s)$: −1270; C_6H_6 : +83.9 : $CO_2(g)$: −393; $H_2O(g)$: −242.

6

Atomic Structure

We are already familiar with the idea that an atom consists of a small, massive, positively charged nucleus, surrounded by electrons. It seems that the chemical properties of an atom are determined by the way in which the electrons surrounding the nucleus are organized. It is therefore necessary to study in some detail the ways in which the electrons are arranged around nuclei.

6.1 ELECTROSTATIC FORCES BETWEEN CHARGED PARTICLES

A fundamental generalization about charged particles is that particles of *opposite* charge *attract* one another, and particles of *like* charge ***repel*** one another. This statement about attractive and repulsive forces may be expressed mathematically as follows:

If there are two charges Q_1 and Q_2 coulomb where Q_1 and Q_2 may be either positive or negative, separated by a distance r_{12} metre, then the force F newton which acts on *each* of these particles is given by the expression:

$$F = \frac{1}{4\pi\varepsilon_0} \cdot \frac{Q_1 Q_2}{\varepsilon r_{12}^2}$$

ε_0 is a constant termed the permittivity of free space, and ε is the dielectric constant of the medium between the charges. $\varepsilon = 1$ for a vacuum, and is very close to unity in air.

If Q_1 and Q_2 are both positive, or both negative, then F will be positive. By convention we refer to a positive force on each charged particle as a repulsion. If Q_1 and Q_2 have opposite signs, then F will be negative. By convention we refer to a negative force on each charged particle as an attraction.

If we attempt to bring two like charges closer together, i.e. to make r_{12} smaller, work must be done against the repulsive force operating on each particle. On the other hand, if we bring unlike charges together, work is actually done by the particles, since they can come together spontaneously, and an external agency would have to do work to pull them apart again. From mechanics, it is known that work done on a system goes to *increase* the potential energy of the system. If the system does work itself, then the potential energy of the system is *reduced.*

If we take a system of two infinitely separated charges as having zero potential energy, then the potential energy, V, of two charges Q_1 and Q_2 coulomb, separated by r_{12} metre, is given by

$$V = \frac{1}{4\pi\varepsilon_0} \cdot \frac{Q_1Q_2}{\varepsilon r_{12}}$$

where V is given in joule. Thus if r_{12} = a,

$$V_a = \frac{1}{4\pi\varepsilon_0} \cdot \frac{Q_1Q_2}{\varepsilon a}$$

and V_a may be interpreted as the work which must be done by an external force when the charges Q_1 and Q_2 are brought together from $r_{12} = \infty$ to r_{12} = a. This work is equal to the gain in potential energy of the system of

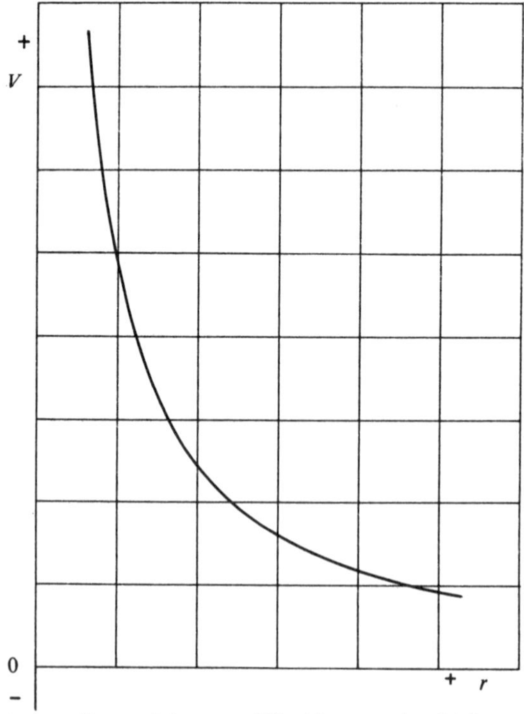

Fig 6.1 Variation of potential energy (V) with separation (r) for two like charges.

two charges, and will be positive for two like charges, and negative for two unlike charges. We shall consider the way in which V depends upon r_{12}, and the signs of Q_1 and Q_2, by taking two cases separately.

(a) *Q_1 and Q_2 have same sign*; when Q_1 and Q_2 are either both positive or both negative, work must be done by an external agency to bring the charges together against the repulsive forces between them.

By doing this work, the potential energy of the system of two charges is accordingly raised. The variation of potential energy with the distance between the charges is shown in Figure 6.1.

(b) *Q_1 and Q_2 have opposite signs*; in this case, the two charges attract each other. Thus, the system can do work on an external agency as the two charges approach each other, and the potential energy of the system falls. The variation of potential energy with the distance between the charges is shown in Figure 6.2.

6.2 ATOMIC ENERGY LEVELS

Hydrogen molecules may be bombarded with high energy electrons by passing an electric discharge through a tube containing gaseous hydrogen at a fairly low pressure. Sufficient energy can be provided by collision with high energy electrons to break up hydrogen molecules into hydrogen atoms;

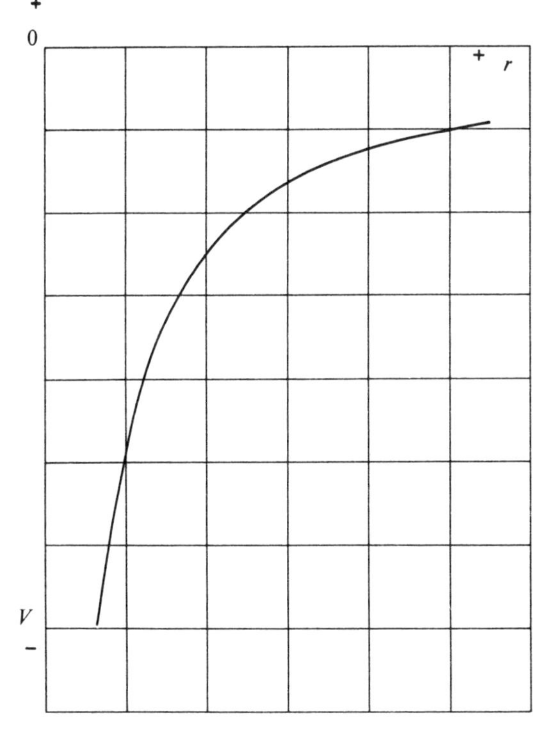

Fig 6.2 Variation of potential energy (V) with separation (r) for two unlike charges.

the hydrogen atoms thus formed may then be energized by further collisions with high energy electrons. The nett effect is that tiny concentrations may be built up of hydrogen atoms with a good deal of excess internal energy. Many of these energetic or 'excited' hydrogen atoms rapidly lose their energy again in collisions with the walls of the container holding the gas, but some of the 'excited' atoms give out their energy as electromagnetic radiation—visible light, ultraviolet light, etc.

The light which is emitted by excited hydrogen atoms may be analysed with an instrument known as a spectrograph. This device, by using a prism, breaks up the light from the atoms into its component wavelengths (or frequencies), and by recording the various frequencies on a photographic plate an emission spectrum is obtained. Part of the emission spectrum of atomic hydrogen is shown on Figure 6.3, and it is quite clear that only certain quite distinct frequencies of radiation have been emitted by the hydrogen atoms.

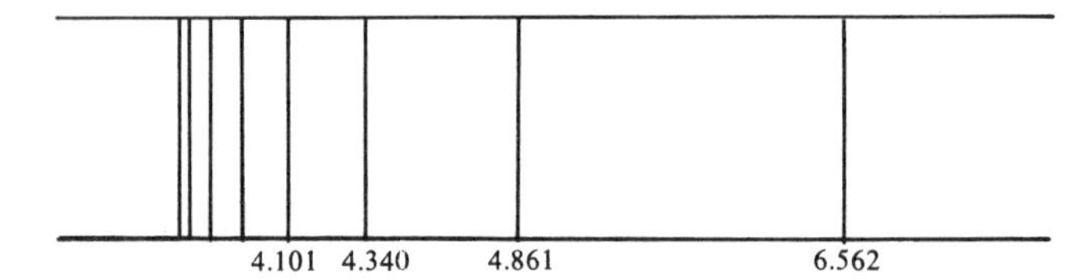

Fig 6.3 Part of the emission spectrum of the hydrogen atom. The numbers attached to the emission lines give the wavelengths of the lines in metre $\times$ 10^7. All the lines in this spectrum are due to transitions from the 2nd, 3rd, 4th etc. excited levels down to the 1st excited level.

Einstein showed in 1905 that light of a definite wavelength consisted of a stream of identical photons, each photon bearing a definite packet, or 'quantum' of energy. Thus:

$$E = h\nu$$

where E is the amount of energy carried by one photon, ν is the frequency of the radiation, and h is a universal constant known as Planck's constant. All of this means, then, that excited hydrogen atoms give out their excess internal energy only as discrete packets or 'quanta'.

This feature of the emission spectrum of hydrogen puzzled physicists for many years around the beginning of this century. It seemed that their experimental results could not be explained unless it was assumed that a hydrogen atom could only have certain discrete amounts of internal energy. By making this assumption, the emission spectrum could be understood as arising from 'transitions' from one internal energy level to another, with the difference in energy between the two levels being released as a photon of a frequency decided by the expression:

$$E = E_2 - E_1 = h\nu$$

where E_2 and E_1 are the energies of the two energy levels involved; hence

$$\nu = E/h$$

Thus the existence of discrete energy levels in the hydrogen atom placed a restriction on the frequencies of radiation which the atom could emit.

Unfortunately, the whole idea of energy levels was impossible to accommodate in the framework of nineteenth-century physics. Newton's laws of motion contain the implicit assumption that a material body is perfectly free to have any amount of energy we choose to give it. For example, we may give a golf ball a great deal of energy by striking it (squarely) with a driver, or alternatively we may give it only a little energy by tapping it gently with a putter. We may also give it any amount of energy we choose between

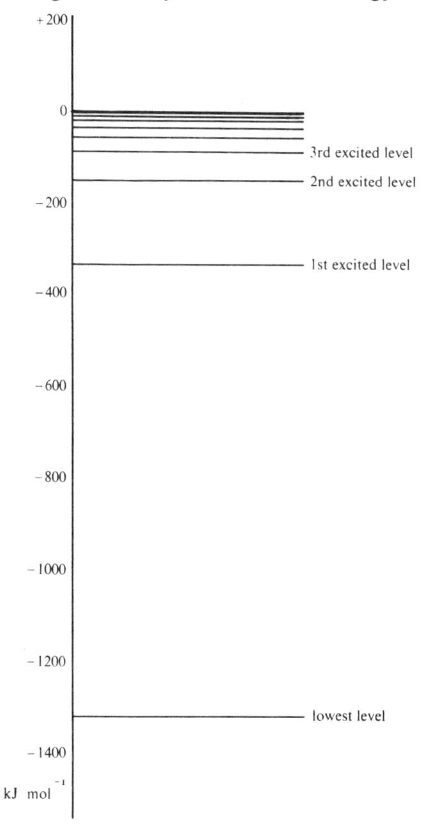

Fig 6.4 Energy level diagram for the hydrogen atom.

these two extremes by choosing the appropriate club and applying the appropriate amount of force. The idea that our ball may be restricted, as is the hydrogen atom, to only quite definite amounts of energy (say driver energy *or* putter energy in the case of the golf ball) seems quite fantastic to us.

Thus, although it was possible in the early 1900s to draw up an energy level diagram for the hydrogen atom, no-one was quite clear as to the significance of such a diagram. The emission spectrum of hydrogen could be rationalized simply by assuming the existence of a set of energy levels for the atom, and the 'lines' in the spectrum corresponded to 'transitions' between any two such levels. Such an energy level diagram for the hydrogen atom is shown in Figure 6.4.

The emission spectra of a good number of many-electron atoms had also been studied at about this time, and these spectra also could only be interpreted by assuming that atoms were able to have only certain discrete amounts of internal energy. Thus the emission spectra of all atoms were seen as arising from 'transitions' between atomic energy levels.

6.3 THE BOHR-RUTHERFORD ATOM

In 1911 Rutherford provided the idea of an atom consisting of a positively charged nucleus, surrounded by electrons, and Bohr used this model of an atom to explain the anomaly of the discrete internal energy levels in the hydrogen atom. In 1913 Bohr introduced the idea that electrons rotated around the nucleus of an atom in definite circular orbits, and that a definite orbit corresponded to a definite electronic energy. The hydrogen atom consists of a proton and electron, and, bearing in mind that they are of opposite charge, we can see that the closer the proton and electron come, the lower (more negative) will be their attractive potential energy. Thus Bohr stated that an electron in a circular orbit of small radius would have a lower (more negative) energy than an electron in an orbit of large radius. Bohr further postulated that only certain orbits were possible for electrons. Since it can be shown from classical mechanics that the energy of an electron in a circular orbit around a proton is determined solely by the radius of the orbit, this postulate of Bohr amounted to saying that there were only certain possible (or 'allowed') energy levels for the electron in the hydrogen atom. Bohr's main ideas may be summarized:

(a) A hydrogen atom possesses internal energy due to the motion of its electron in a circular orbit around the proton.
(b) Only certain orbits are allowed for an electron near a proton. This implies that the atom may have only certain discrete internal energies.
(c) The emission spectrum of a hydrogen atom is due to its electron jumping from a high energy orbit to a low energy orbit, and simultaneously emitting a photon with energy equal to the energy difference between the two orbits.

Bohr's picture of the hydrogen atom is often generalized to larger atoms in which it is assumed that the many electrons of a large atom move in definite orbits around the nuclei. Such pictures are not very satisfactory and they will not be used in this text.

6.4 QUANTUM MECHANICS

The Bohr theory outlined in section 6.3 was developed specifically to deal with the problem of the existence of discrete energy levels in the hydrogen atom, in which task it succeeded. However, Bohr's theory could not be satisfactorily generalized to other more complicated systems (e.g. the sodium atom), since attempts to do so resulted in theoretical predictions about the emission spectrum of such atoms which were at variance with experimental facts. The problem of the existence of atomic energy levels remained, and was not satisfactorily solved until the development of modern quantum mechanics (sometimes called wave mechanics) by Schroedinger in 1926. Quantum mechanics is in reality a completely new set of 'Laws of Motion'. Just as Newton's 'Laws of Motion' apply to the movements of large bodies (such as golf balls, or planets), so the laws of quantum mechanics apply to the motion of small particles such as protons, electrons, neutrons, photons, atoms, etc. The laws of motion expressed in terms of quantum mechanics usually involve us in solving very complicated mathematical equations—indeed, except for very simple systems scientists cannot even now solve the equations exactly—and in this text we shall consider only a few of the results obtained by solving these equations. It is worth noting, however, that quantum mechanics, strictly speaking, applies to the motion of *all* particles—electrons, golf balls or planets. However, when we apply quantum mechanics to bodies which are much more massive than atoms (i.e. golf balls or planets) it turns out that we obtain results identical with those obtained by applying Newton's Laws of Motion. Put in another way:

The laws of quantum mechanics reduce to Newton's Laws of Motion when we are dealing with bodies which are considerably heavier than individual atoms.

This is fortunate, since it means that astronomers may still safely use Newton's Laws of Motion for dealing with the movements of the stars and planets. It is only when we come to deal with tiny particles like protons, neutrons and electrons that the more complex equations of quantum mechanics must be used.

6.5 QUANTUM MECHANICS AND ATOMIC STRUCTURE

When modern quantum mechanics is applied to the hydrogen atom, the theory successfully predicts the correct energy levels for the atom. Indeed, in some ways the picture obtained is similar to that given by Bohr in so far as the atomic energy levels are seen as allowed electronic energy levels; the electron is still pictured as moving about under the influence of the proton, in such a way that the total energy of the electron may have only certain definite values. However, the idea of definite orbits for the electron is completely gone. The electron must now be visualized as moving within a region of space surrounding the nucleus. Such a region of space is called an *orbital*, or in this case an *atomic orbital*. We may define:

The electron in the hydrogen atom moves in a definite region of space which is called an atomic orbital.

Figure 6.4 represents the energy level diagram for the hydrogen atom as determined experimentally. The predictions of quantum mechanics are identical with the experimental facts. Furthermore, according to the theory, for each 'allowed' electronic energy level in the hydrogen atom, there corresponds a definite orbital for the electron to move in.

The theory predicts that the 'allowed' orbitals can be regarded as being arranged in shells around the nucleus. It is customary to number the shells outwards from the nucleus, i.e. the first shell is closest to the nucleus, the second shell is outside the first shell, the third shell is outside the second shell and so on. Each shell is made up of a definite number of orbitals, and the shells, together with the numbers of orbitals they contain, are listed in Table 6.1. A shell of number n is made up of n^2 orbitals.

Table 6.1

Shell no. (n)	*Symbol*	*No. of orbitals* (n^2)
1	K	1
2	L	4
3	M	9
4	N	16
5	O	25

Notice that the shells have been indicated by letters, which we shall use as symbols for the shells. Thus the first shell is called the K-shell, and consists of a single orbital; the second shell is called the L-shell and has four orbitals; etc.

The shells of orbitals we have just discussed are the possible orbitals, corresponding to 'allowed' energy levels, for the hydrogen atom. Put in another way, these orbitals (Table 6.1) are the 'allowed' orbitals for an electron near a proton. We shall now assume that a completely analogous picture is valid for any nucleus. Thus, if, for example, we consider the nucleus of a sodium atom with its eleven positive charges, we assume that the allowed orbitals for the eleven electrons moving around this nucleus are arranged in shells in just the same way as are the orbitals around the proton. Naturally the allowed energies of the electrons in the sodium atom must be different from the allowed electronic energies in the hydrogen atom, since the sodium nucleus has a larger charge and other electrons are also present. We are simply assuming that the allowed orbitals in sodium, and indeed in any atom, are grouped into shells in the same way as are the allowed orbitals in hydrogen.

This assumption allows us to form a satisfactory picture of the electronic structure of many-electron atoms, if we make the additional assumption (sometimes called the Pauli Principle) that there may be no more than two electrons in any given atomic orbital. Thus the maximum possible number of electrons in any given atomic shell will be just twice the number of orbitals in the shell. These numbers are given in Table 6.2.

6.6 ATOMIC GROUND STATES

The total energy of a given electron in an isolated atom is strictly constant, and may be expressed as a sum of its kinetic energy of motion and its electrostatic potential energy. The total electronic energy of a many-electron atom

Table 6.2

Shell no. (n)	*Symbol*	*No. of orbitals* (n^2)	*Max. no. of electrons* ($2n^2$)
1	K	1	2
2	L	4	8
3	M	9	18
4	N	16	32
5	O	25	50

is then the sum of all the individual electronic energies of the atom. The simplest of all atoms is the one-electron hydrogen atom. We have seen that there is a set of possible energy levels for this atom, and the lowest of these is at -1310 kJ mol^{-1}. When a hydrogen atom has this energy, its *average* potential energy has its lowest possible value. At ordinary temperatures, virtually all hydrogen atoms have this energy of -1310 kJ mol^{-1}, and are said to be in the electronic ground states. In general, there will be a set of possible energy levels for any atom, and an atom which is in its state of lowest possible electronic energy is said to be in its electronic ground state. At room temperatures almost all atoms are in their electronic ground states.

6.7 IONIZATION ENERGY

For an isolated hydrogen atom in the gaseous state, the ground state energy is -1310 kJ mol^{-1}, if we consider an infinitely separated proton and electron as our zero of energy, i.e.

$$H(g) \rightarrow H^+(g) + e(g) \qquad \Delta H = +1310 \text{ kJ}$$

This means that we must supply 1310 kJ of energy to remove the electrons from a mole of hydrogen atoms in their electronic ground states. Put in another way, we could say that the ionization energy of normal hydrogen atoms is 1310 kJ mol^{-1}. This is equivalent to an ionization energy of $(-1{,}310{,}000/6.023 \times 10^{23})$ J atom^{-1}, however, as we have noted earlier with bond dissociation energies it is the practice to quote such quantities as so many energy units per mole as well as in energy units per atom.

For a sodium atom, the ionization energy is given by

$$Na(g) \rightarrow Na^+(g) + e(g) \qquad \Delta H = +494 \text{ kJ}$$

We may formally define:

The ionization energy of an atom or ion is the minimum amount of energy required to remove an electron from that atom or ion where the whole process is understood to occur in the gas phase.

Notice that in any of these ionization processes, it is the electron of highest available energy which is removed.

We have seen that the product of the ionization of a sodium atom is a sodium ion. This ion may be further 'ionized' by the removal of an electron thus

$$Na^+(g) \rightarrow Na^{2+}(g) + e(g) \qquad \Delta H = +4560 \text{ kJ}$$

The ΔH value for this process is called the *second* ionization energy of the sodium atom. Indeed, by the successive removal of electrons, we can obtain eleven successive ionization energies of sodium, corresponding to the removal of all eleven electrons from the Na^{11+} nucleus. These data are shown in Table 6.3.

Table 6.3

No. of ionization energy of sodium	*Process*	ΔH ($kJ\ mol^{-1}$)	$I(eV)$
1st	$Na(g) \rightarrow Na^{+}(g) + e(g)$	494	5.12
2nd	$Na^{+}(g) \rightarrow Na^{2+}(g) + e(g)$	4,560	47.3
3rd	$Na^{2+}(g) \rightarrow Na^{3+}(g) + e(g)$	6,900	71.5
4th	$Na^{3+}(g) \rightarrow Na^{4+}(g) + e(g)$	9,540	98.9
5th	$Na^{4+}(g) \rightarrow Na^{5+}(g) + e(g)$	13,400	139
6th	$Na^{5+}(g) \rightarrow Na^{6+}(g) + e(g)$	16,700	173
7th	$Na^{6+}(g) \rightarrow Na^{7+}(g) + e(g)$	20,200	209
8th	$Na^{7+}(g) \rightarrow Na^{8+}(g) + e(g)$	25,500	264
9th	$Na^{8+}(g) \rightarrow Na^{9+}(g) + e(g)$	28,900	300
10th	$Na^{9+}(g) \rightarrow Na^{10+}(g) + e(g)$	141,000	1,460
11th	$Na^{10+}(g) \rightarrow Na^{11+}(g) + e(g)$	160,000	1,700

The following three points should be noticed.

(a) The ionization energy measures the amount of energy supplied in order to remove an electron. The lower the energy of the electron in the atom, the more *negative* will be the electronic energy; thus the *lower* the energy of the electron we are removing, the *higher* will be the corresponding ionization energy.

(b) The ionization energies of sodium increase consistently from first to eleventh. Such an increase is to be expected since, when we take an electron from an atom or ion, the nett positive charge on the species is increased by one unit. The removal of the *next* electron thus requires the expenditure of a little more energy since we are now taking an electron away from a greater positive charge than previously.

(c) In the third column in the table, the ionization energies (I) are quoted in electron volts. An electron volt is the amount of energy acquired by an electron when it is accelerated through a potential difference of 1 volt. The unit is frequently used, since the actual numbers involved are not as inconveniently large as those quoted in kJ. Note that 1 electron volt is equal to $e \times V$ where e is the electronic charge, and V is an electric potential of 1 volt. The charge on the electron is 1.602×10^{-19} coulomb, so

$$1\ \text{eV} = 1.602 \times 10^{-19}\ \text{J}$$

in molar quantities, this is equivalent to

$$\begin{aligned} & 1.602 \times 10^{-19} \times 6.023 \times 10^{23}\ \text{J mol}^{-1} \\ & = 96.51\ \text{kJ mol}^{-1} \\ & = 23.03\ \text{kcal mol}^{-1} \end{aligned}$$

The energy unit used in this text is the joule, however, and we shall continue with its use.

The experimental determination of ionization energies is quite difficult, since it requires an interpretation of the emission spectrum of the appropriate atom or ion. The ionization energy of a sodium atom can be obtained exactly from an interpretation of the emission spectrum of atomic sodium. The ionization energy of, say, Na^{5+}, can be determined exactly from an interpretation of the emission spectrum of the Na^{5+} ion. This particular ion can only be produced in tiny concentrations in a gas discharge tube, where its spectrum has been observed and recorded. Its ionization energy may then be deduced from its spectrum. As the charge on these ions of sodium increases, the ions become increasingly difficult to produce in appreciable quantities and the ionization energies for ions of very high charge are frequently not known.

Data are available for all the ionization energies of the elements with atomic numbers one to eleven (hydrogen to sodium). These ionization energies are given in Table 6.4 in kJ mol^{-1}.

Table 6.4

Ionization	H	He	Li	Be	B	C	N	O	F	Ne	Na
First	1,310	2,370	519	900	799	1,090	1,400	1,310	1,680	2,080	494
Second		5,220	7,310	1,760	2,420	2,390	2,850	3,390	3,360	3,950	4,560
Third			11,800	14,900	3,660	4,600	4,560	5,310	6,070	6,150	6,900
Fourth				20,900	25,000	6,230	7,450	7,450	8,410	9,290	9,540
Fifth					32,600	37,800	9,460	11,000	11,000	12,100	13,400
Sixth						46,900	53,100	13,300	15,100	15,100	16,700
Seventh							64,000	71,000	17,900	20,000	20,200
Eighth								84,000	91,600	23,000	25,500
Ninth									106,000	115,000	28,900
Tenth										130,000	141,000
Eleventh											160,000

6.8 EVIDENCE FOR THE EXISTENCE OF K-, L- AND M-SHELLS

Let us now examine the data assembled in Table 6.4 in order to discover if any regularities can be discerned in the successive ionization energies of the atoms listed. In particular, we shall look for evidence of a shell structure in the arrangements of electrons around the atoms. The ionization energy data tell us the energies required to remove electrons, one at a time from an atom. These energies rise steadily due to the increase in positive charge on the ions as we remove successive electrons. However, if we look at the data on a graph (Figure 6.5) we can see that at certain points there are discontinuities in the plot of ionization energy against charge on the ion.

We may note the following features:

(a) In the case of sodium the first electron is comparatively easy to remove (+494 kJ); the second electron requires about eight times as much

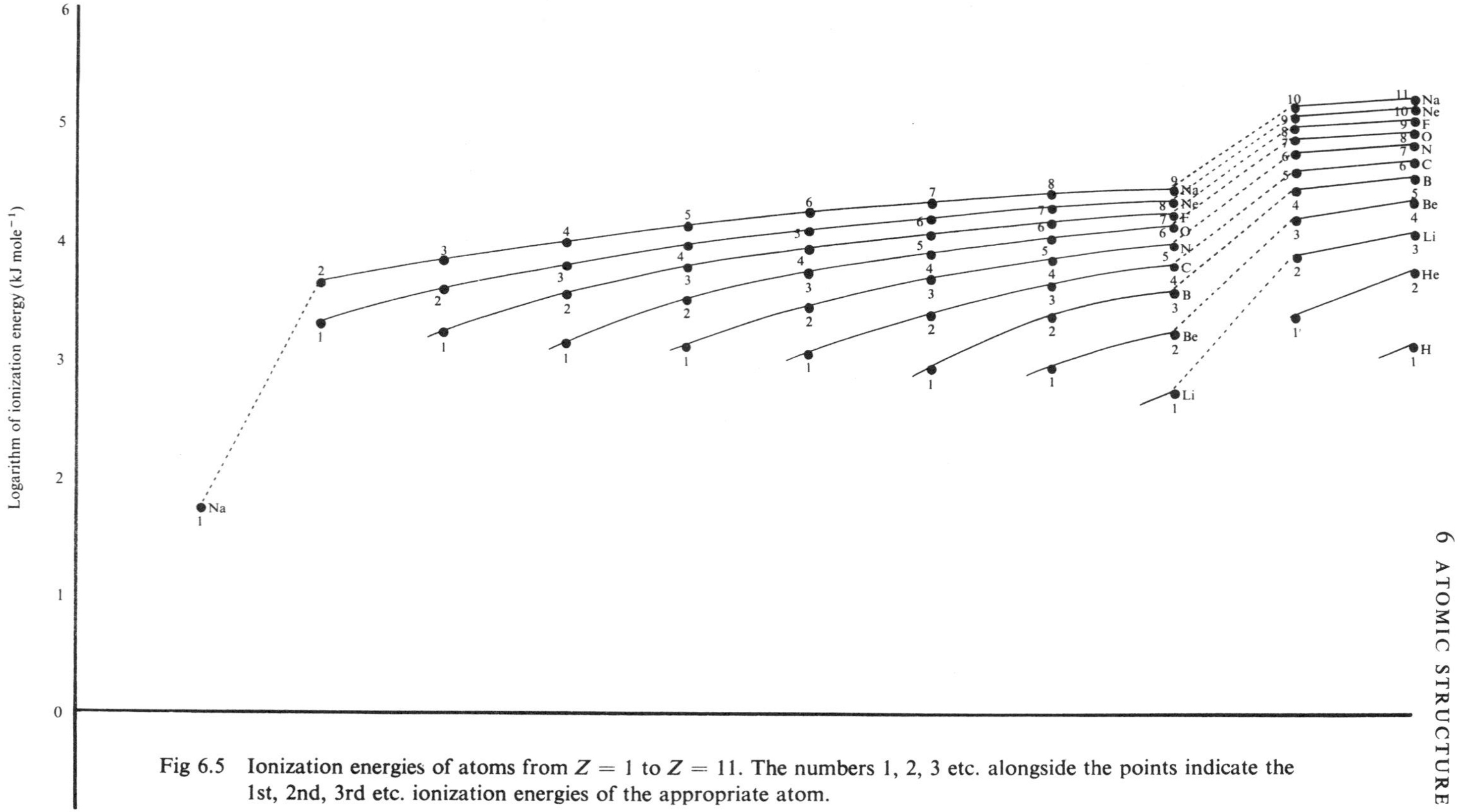

Logarithm of ionization energy (kJ mole^{-1})

Fig 6.5 Ionization energies of atoms from $Z = 1$ to $Z = 11$. The numbers 1, 2, 3 etc. alongside the points indicate the 1st, 2nd, 3rd etc. ionization energies of the appropriate atom.

energy as this to remove it (+4560 kJ), but the next seven electrons are removed with successive increments of only about 50 per cent in the energy required. However, removal of each of the last two electrons requires about five times as much energy as that required by the third last (ninth) electron. Thus there are these two large discontinuous jumps in the plot of ionization energy against charge for sodium, one between the first and second electrons, the other between the ninth and tenth.

(b) For all the atoms shown, the *last two* electrons removed require significantly more energy than all the other electrons.

If we make the general assumption that electrons with the lowest energy (i.e. *highest* ionization energy) will tend to spend their time closer to the nucleus, we might conclude that in all the atoms shown in Figure 6.5 two electrons were closer to the nucleus than all the others. Inspection of the curve for sodium suggests that a further group of eight electrons fall next in order of distance from the nucleus, and a third electron then lies outside the groups of two and eight. This then provides circumstantial evidence for the existence of shells of electrons in these atoms. In sodium we would identify the two lowest energy electrons as K-shell electrons, the next eight as L-shell electrons, and the eleventh as an M-shell electron.

6.9 ENERGY LEVELS IN MANY-ELECTRON ATOMS

In section 6.5, on the basis of quantum mechanics, we introduced the idea that electrons are arranged in shells around nuclei. In section 6.8, circumstantial experimental evidence supporting this general picture was presented. In this section we will enquire as to whether electrons in a particular shell of a given atom all have exactly the same energies, or whether there may be energy differences between electrons in the same shell. It turns out that there is both theoretical and experimental evidence to suggest that electrons in a particular shell of a many-electron atom do *not* all necessarily have the same energy. Although the experimental evidence on this point is quite clear, it is impossible to discuss here, since it involves a quite complicated interpretation of the emission spectra of atoms. In this section therefore we will simply present the essential conclusions reached about the energy levels of atoms.

We have seen that the shells in which electrons move are made up of groups of atomic orbitals; thus there is one orbital in the K-shell, four orbitals in the L-shell, nine orbitals in the M-shell, etc. Experimental and theoretical evidence shows that for all atoms other than hydrogen, the four orbitals in the L-shell do not all have the same energy. For these atoms, the L-shell has two different energy levels. The lower level consists of one orbital, the higher level of three other orbitals which are themselves equivalent. The M-shell has three different energy levels; the lowest level consists of one orbital, the middle level of three orbitals, and the highest level of five orbitals. These findings about the energy levels of shells may be generalized:

(a) The lowest energy level within a shell consists of one orbital, which is usually called an *s-orbital*, or sometimes an *s-subshell*.
(b) The second lowest energy level within a shell consists of a group of three identical orbitals, which are called *p-orbitals*. The group of three p-orbitals is sometimes called a *p-subshell*.
(c) The third lowest energy level of a shell consists of a group of five identical orbitals, which are called *d-orbitals*. The group of five d-orbitals is sometimes called a *d-subshell*.

All the available energy levels within the various shells of an atom are summarized in Table 6.5.

Table 6.5

Shell symbol	*Shell number* (n)	*No. of sub-shells*	*Type of energy level*	*No. of orbitals in energy level*	*Total no. of orbitals* (n^2)
K	1	1	s	1	1
L	2	2	s	1	4
			p	3	
M	3	3	s	1	
			p	3	9
			d	5	
N	4	4	s	1	
			p	3	
			d	5	16
			f	7	
O	5	5	s	1	
			p	3	
			d	5	25
			f	7	
			g	9	

There are a number of interesting regularities in this table:
(a) If n is the shell number, the number of subshells in the shell is equal to n.
(b) The number of orbitals in the successive subshells labelled s, p, d, f, g,— follows the arithmetic series 1, 3, 5, 7, 9,—.
(c) The total number of orbitals, in a shell of number n, is n^2.

It is now possible for us to specify a particular atomic energy level by writing down the number of the shell in which the level appears, followed by the particular subshell in which the level appears, e.g. 3d refers to the energy level which is the d-subshell of the third, or M-shell. More particularly, we might speak of a 3d electron, which would mean an electron in the d-subshell of the third, or M-shell. Similarly a 5s electron would be an electron in the s-subshell of the fifth or O-shell.

As a general principle, within a given shell, the order of subshell energies is

$$s < p < d < f < g \ldots$$

This ordering is based on both theoretical and experimental considerations.

A more subtle question arises, however, viz. do the energy levels of a particular shell 'overlap' the energy levels of a neighbouring shell? For example, is it possible that the 5s level could have a lower energy than the 4d level? It turns out that this is quite possible, and the actual order of energies for the energy level of a many-electron atom is given by

$$1s < 2s < 2p < 3s < 3p < 4s < 3d < 4p < 5s < 4d < 5p < 6s < 4f < 5d \ldots$$

This is shown diagrammatically in Figure 6.6.

In the higher energy levels, this ordering is subject to minor fluctuations with atomic number, e.g. for a few atoms 3d < 4s, or 4d < 5s. These are relatively minor variations, however, and discussion of these effects is beyond the scope of this book.

6.10 ELECTRONIC STRUCTURES OF ATOMS —THE PAULI PRINCIPLE

We can use the atomic energy level scheme from Figure 6.5 to write down the electronic structure of all the atoms known to man, simply from the knowledge of the atomic number of the atom. To do this, one final principle must be formally enunciated. This is the *Pauli Exclusion Principle*, which is based simply on experimental observation, that is:

There can never be more than two electrons in any atomic orbital. This means that any atomic orbital can hold 0, 1 or 2 electrons, but never more than two.

We are now in a position to build up the electronic structures of some atoms. Let us take first the sodium atom, and imagine that we have initially the nucleus of a sodium atom, with its eleven positive charges, Na^{11+}. We have to add eleven electrons to this nucleus to make a normal sodium atom. If we add the electrons one at a time, they will go successively into the orbitals of lowest available energy, remembering that we can put no more than two electrons in any one orbital. The orbital of lowest energy is the 1s orbital, and we therefore put our first two electrons there. The next two go into the 2s orbital, the next six into the 2p subshell (three orbitals) and the eleventh into the 3s orbital. The electron structure we would write is

$$1s^2.2s^2.2p^6.3s^1$$

where the superscript numbers show the number of electrons in each energy level. This is read as 'one-s-two, two-s-two, two-p-six, three-s-one'.

To take a more complex example, let us consider the iron atom. This has an atomic number of 26, and hence we have to add twenty-six electrons to the Fe^{26+} nucleus. The first two electrons go into a 1s orbital, the next eight fill the L-shell, and the next eight the 3s and 3p subshells. The orbital next lowest in energy is not 3d, but 4s, so the next two electrons go into the 4s orbital. The last six electrons then go into the 3d level, and the final electronic structure for Fe is

$$1s^2.2s^2.2p^6.3s^2.3p^6.3d^6.4s^2$$

i.e. the atom has a partly filled 3d subshell.

An important corollary of the Pauli Principle is that there is an upper limit to the number of electrons which can ultimately be fitted into a particular atomic shell. We saw in section 6.9 that the number of orbitals in a shell of number n is given by n^2. Since we can have no more than two electrons in any one orbital, the maximum number of electrons we can have in a shell must be $2n^2$. This maximum number of electrons we can have in the various shells is shown in Table 6.6.

Table 6.6

Shell symbol	K	L	M	N	O	
Shell no. (n)	1	2	3	4	5	(n)
No. of orbitals (n^2)	1	4	9	16	25	(n^2)
Max. no. of electrons ($2n^2$)	2	8	18	32	50	($2n^2$)

6.11 THE PERIODIC CLASSIFICATION

When the electronic structures of all known atoms are written down, it is noticed that characteristic outer-shell electron configurations keep recurring. Thus we find the following electronic configurations.

Atom	*Symbol*	*Electronic configuration*
Lithium	Li	$1s^2.\underline{2s^1}$
Sodium	Na	$1s^2.2s^2.2p^6.\underline{3s^1}$
Potassium	K	$1s^2.2s^2.2p^6.3s^2.3p^6.\underline{4s^1}$
Rubidium	Rb	$1s^2.2s^2.2p^6.3s^2.3p^6.3d^{10}.4s^2.4p^6.\underline{5s^1}$
Caesium	Cs	$1s^2.2s^2.2p^6.3s^2.3p^6.3d^{10}.4s^2.4p^6.4d^{10}.5s^2.5p^6.\underline{6s^1}$
Fluorine	F	$1s^2.\underline{2s^2.2p^5}$
Chlorine	Cl	$1s^2.2s^2.2p^6.\underline{3s^2.3p^5}$
Bromine	Br	$1s^2.2s^2.2p^6.3s^2.3p^6.3d^{10}.\underline{4s^2.4p^5}$
Iodine	I	$1s^2.2s^2.2p^6.3s^2.3p^6.3d^{10}.4s^2.4p^6.4d^{10}.\underline{5s^2.5p^5}$

The atoms from lithium to caesium all have the outer-shell electronic configuration s^1. These atoms also all have very similar chemical properties and are known collectively as the alkali metals. The atoms from fluorine to iodine all have the outer-shell electronic configuration s^2p^5. These atoms also have very similar chemical properties, and are known collectively as the halogens.

It is a matter of general experience that atoms with similar outer-shell electron configurations have similar chemical properties, hence if we can arrange all known atoms in some systematic fashion according to their

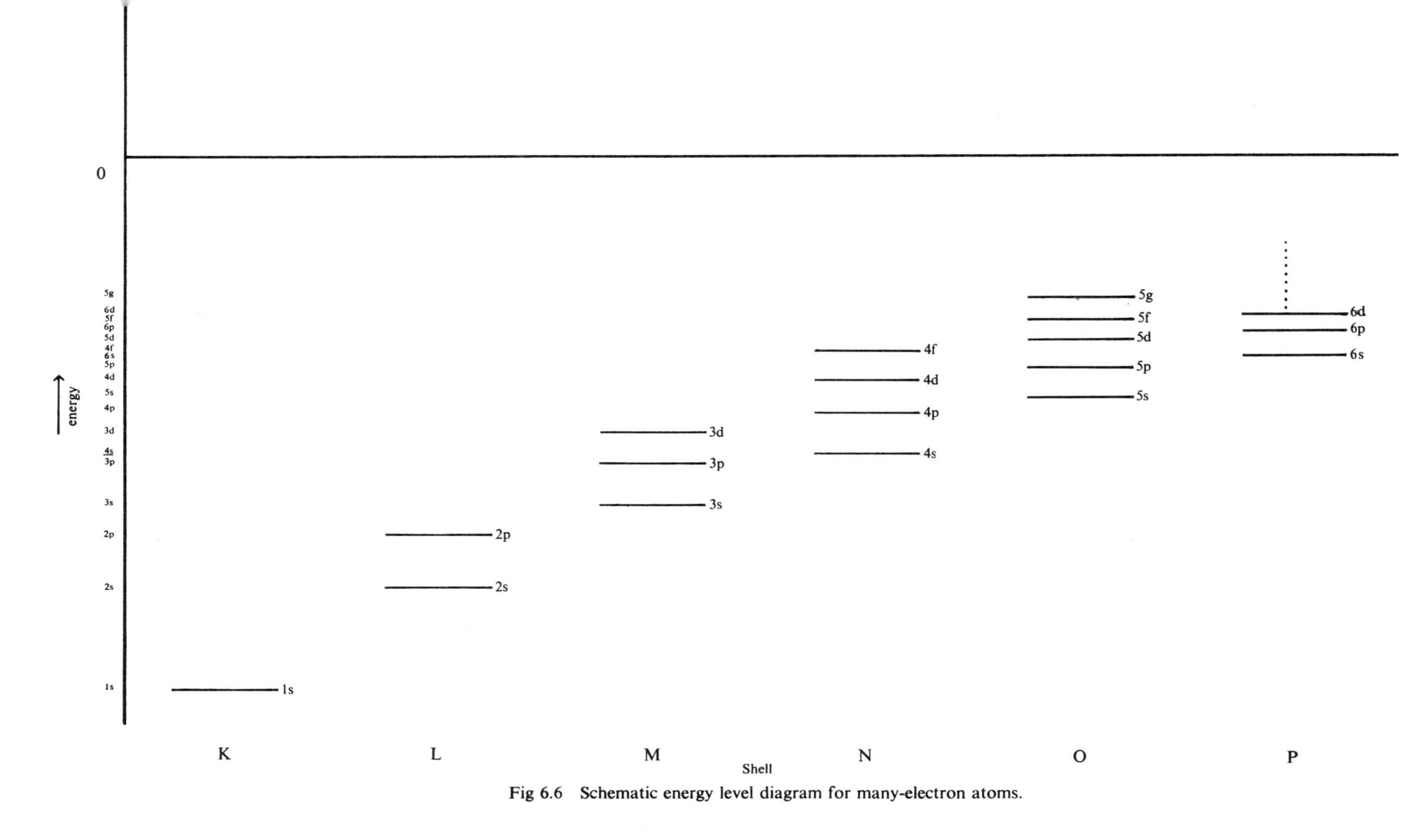

Fig 6.6 Schematic energy level diagram for many-electron atoms.

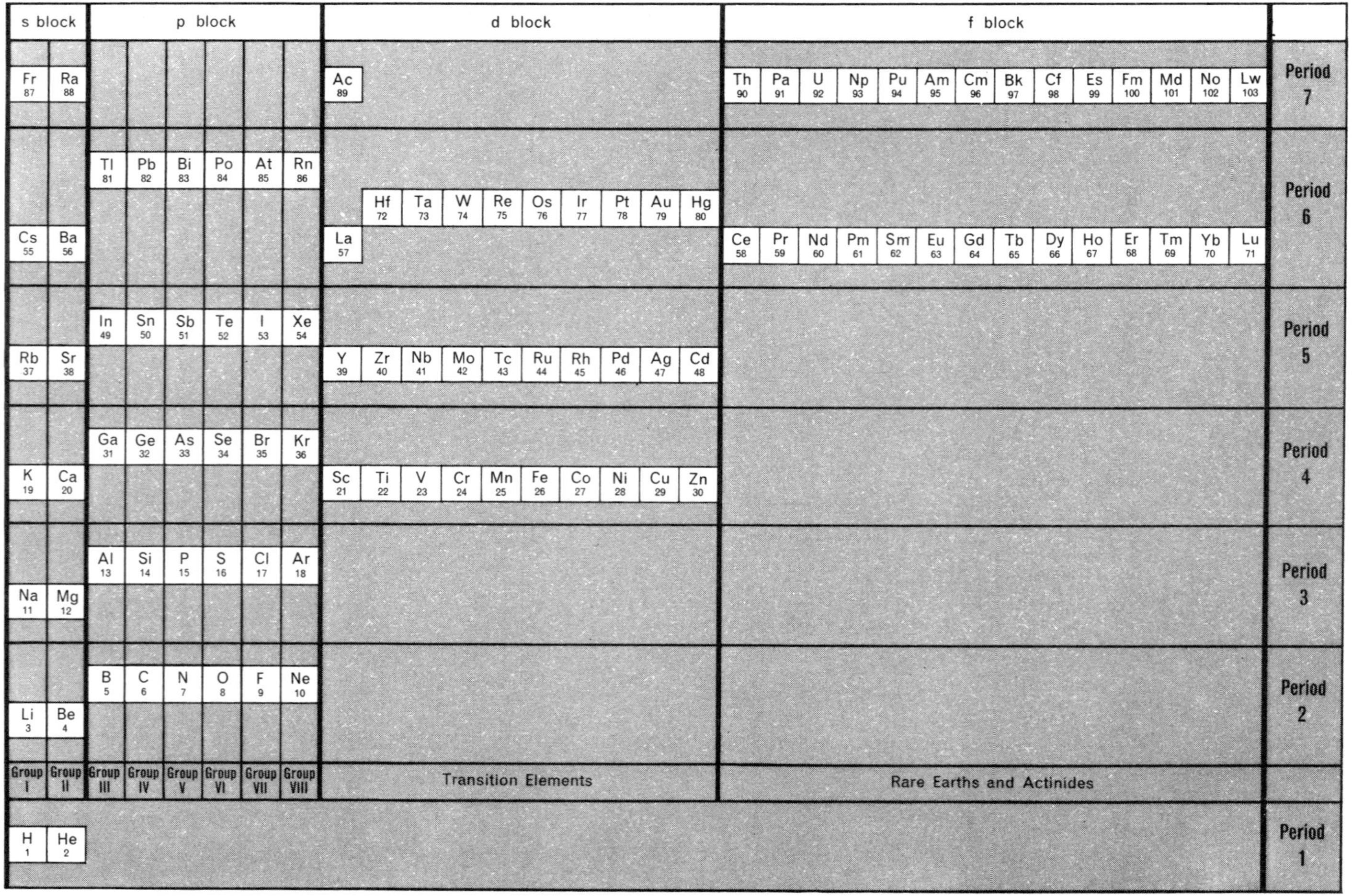

s block	p block	d block	f block	
Fr 87, Ra 88		Ac 89	Th 90, Pa 91, U 92, Np 93, Pu 94, Am 95, Cm 96, Bk 97, Cf 98, Es 99, Fm 100, Md 101, No 102, Lw 103	Period 7
Cs 55, Ba 56	Tl 81, Pb 82, Bi 83, Po 84, At 85, Rn 86	La 57; Hf 72, Ta 73, W 74, Re 75, Os 76, Ir 77, Pt 78, Au 79, Hg 80	Ce 58, Pr 59, Nd 60, Pm 61, Sm 62, Eu 63, Gd 64, Tb 65, Dy 66, Ho 67, Er 68, Tm 69, Yb 70, Lu 71	Period 6
Rb 37, Sr 38	In 49, Sn 50, Sb 51, Te 52, I 53, Xe 54	Y 39, Zr 40, Nb 41, Mo 42, Tc 43, Ru 44, Rh 45, Pd 46, Ag 47, Cd 48		Period 5
K 19, Ca 20	Ga 31, Ge 32, As 33, Se 34, Br 35, Kr 36	Sc 21, Ti 22, V 23, Cr 24, Mn 25, Fe 26, Co 27, Ni 28, Cu 29, Zn 30		Period 4
Na 11, Mg 12	Al 13, Si 14, P 15, S 16, Cl 17, Ar 18			Period 3
Li 3, Be 4	B 5, C 6, N 7, O 8, F 9, Ne 10			Period 2
Group I, Group II	Group III, Group IV, Group V, Group VI, Group VII, Group VIII	Transition Elements	Rare Earths and Actinides	
H 1, He 2				Period 1

Fig 6.7 Periodic table of the elements.

electronic structures, we might also succeed in arranging them according to their chemical properties.

It is possible to construct a table of the atoms in which is shown the arrangement of electrons in the known atomic energy levels. Such tables are called *Periodic Tables*. One such table is shown in Figure 6.7, wherein the atoms are arranged in such a way that the position of an atom indicates how far the process of filling up the available energy levels has progressed. Elements with similar outer-shell configurations appear in vertical columns. This table stresses particularly the arrangement of atomic energy levels since the lowest energy levels appear at the bottom of the table, and the highest energy levels at the top of the table.

In practice, the so-called 'long' form of the periodic table is the most convenient for comparing chemical properties. In this table, the atoms are arranged sequentially in order of increasing atomic number, and again atoms of the same electronic configuration appear in vertical columns.

I	II											III	IV	V	VI	VII	VIII
1 H																	2 He
3 Li	4 Be											5 B	6 C	7 N	8 O	9 F	10 Ne
11 Na	12 Mg											13 Al	14 Si	15 P	16 S	17 Cl	18 Ar
19 K	20 Ca	21 Sc	22 Ti	23 V	24 Cr	25 Mn	26 Fe	27 Co	28 Ni	29 Cu	30 Zn	31 Ga	32 Ge	33 As	34 Se	35 Br	36 Kr
37 Rb	38 Sr	39 Y	40 Zr	41 Nb	42 Mo	43 Tc	44 Ru	45 Rh	46 Pd	47 Ag	48 Cd	49 In	50 Sn	51 Sb	52 Te	53 I	54 Xe
55 Cs	56 Ba	57 La	72 Hf	73 Ta	74 W	75 Re	76 Os	77 Ir	78 Pt	79 Au	80 Hg	81 Tl	82 Pb	83 Bi	84 Po	85 At	86 Rn
87 Fr	88 Ra	89 Ac															

58 Ce	59 Pr	60 Nd	61 Pm	62 Sm	63 Eu	64 Gd	65 Tb	66 Dy	67 Ho	68 Er	69 Tm	70 Yb	71 Lu
90 Th	91 Pa	92 U	93 Np	94 Pu	95 Am	96 Cm	97 Bk	98 Cf	99 Es	100 Fm	101 Md	102 No	103 Lw

Fig 6.8 Periodic table of the elements.

Atoms in the vertical columns labelled with Roman numerals belong to the *main groups* of the periodic table. Thus the elements lithium to francium belong to group I, while the elements fluorine to astatine belong to group VII. The horizontal rows are all numbered with Arabic numerals, and are called *periods*. Hydrogen and helium constitute the first period, lithium to neon the second period and so on. Special allowance is made for the filling of the d-subshells with the 10-atom wide blocks of atoms in the centre of the table, labelled 'transition elements'. Shown below the table are the rare earth and actinide series of atoms arising from the filling of the 4f and 5f subshells.

Notice that the table falls naturally into four parts:

(a) The two main groups on the extreme left of the table, in which the atoms have outer-shell electron configurations s^1 and s^2 respectively. This 'block' of the table is sometimes called the *s-block*, and all the atoms in the block have a half-filled or filled s-orbital for an outer shell.

(b) The six main groups on the extreme right of the table, in which the atoms have outer-shell configurations from s^2p^1 to s^2p^6. This block arises from the systematic filling of the orbitals of a p-subshell, and is sometimes called the *p-block*.

(c) The 10-atom-wide block in the centre of the table, with electronic configurations . . . d^1s^2 to . . . $d^{10}s^2$. This block arises from the systematic filling of the five orbitals of a d-subshell, and the elements in this block are called *transition elements*.

(d) The two 14-atom-wide sets of atoms, usually shown below the table, which arise from the systematic filling of the seven orbitals of the 4f or 5f subshells. These atoms are known as *rare earths* when the 4f subshell is involved, and *actinides* when the 5f subshell is involved.

6.12 ATOMIC ORBITALS

In this chapter to date our main concern has been with electronic energy levels in atoms. We must now turn our attention to the actual way in which electrons move around the nuclei of atoms. We have rejected the Bohr picture of electrons moving in definite orbits around the nucleus and replaced it with the idea of a region of space, known as an orbital, in which the electron may move. The simplest orbital of all is the 1s orbital of the hydrogen atom. An electron in the 1s orbital of a hydrogen atom will be moving around a proton with the minimum allowed total energy (– 1310 kJ) for the hydrogen atom. In order to find the actual position of the electron relative to the proton at any instant of time, let us imagine that we have a tiny sub-atomic camera with which we can take 'photographs' of hydrogen atoms. Fixing our gaze on one particular atom we might take a photograph and find:

Fig 6.9 Instantaneous 'photograph' of a hydrogen atom.

where ⊕ represents the proton, and the dot represents the electron. Simultaneous photographs of five different atoms might yield:

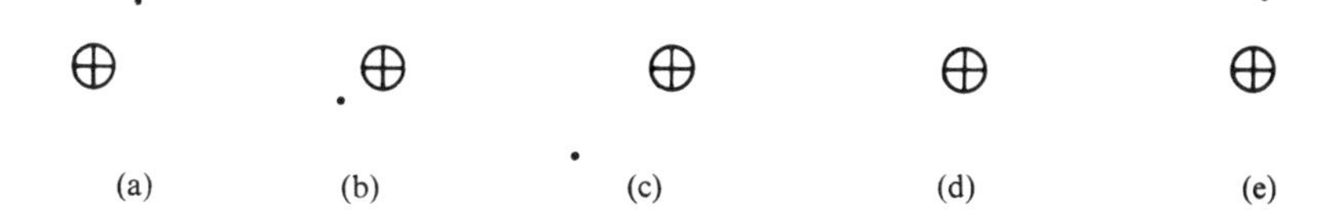

Fig 6.10 Instantaneous 'photographs' of five hydrogen atoms.

If we were to take photographs of ten thousand hydrogen atoms, and scan them, we would be left with the impression that the 1s electron was moving more or less randomly around the proton. If we superimposed various numbers of these photographs, we would see something like:

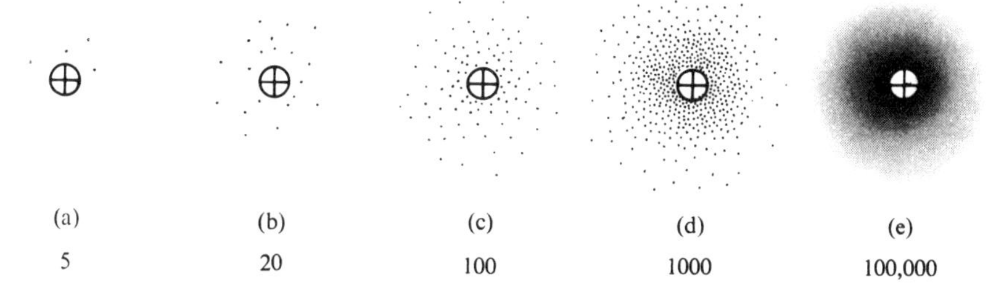

Fig 6.11 Instantaneous 'photographs' of hydrogen atoms superposed, (a) 5 photographs, (b) 20 photographs, (c) 100 photographs, (d) 1000 photographs, (e) 100,000 photographs.

That is, the electron seems to be moving randomly in a spherical region of space around the nucleus, but the region of space in which it moves has no sharp boundaries. The effect of the motion is to make the single electron in a hydrogen atom behave *as if it were* a cloud of negative charge surrounding the nucleus.

The density of the charge cloud at any point gives an indication of the amount of time spent by the electron at that point. For any individual atom it is never possible to state precisely where the electron is or how fast it is moving. The best we can do is to define a mathematical probability of finding

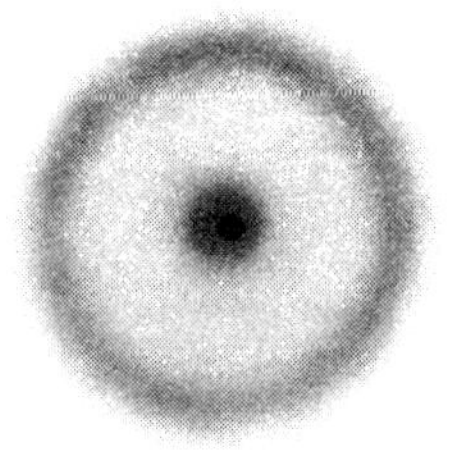

Fig 6.12 Charge cloud picture of a 2s electron in a hydrogen atom.

an electron at a particular point in space; the mathematical probability is at any point closely related to the density of the charge cloud at that point.

In the following chapters, we shall find the idea of a charge cloud very useful for representing the state of an electron in an orbital. We have just noted that an electron in the 1s atomic orbital of hydrogen behaves as a spherical charge cloud. An electron in a 2s atomic orbital also behaves as a spherical charge cloud, but in this case the cloud is a sort of annular shell around the nucleus. A cross-section of the charge cloud of a 2s electron is shown in Figure 6.12.

The charge clouds of electrons in p and d orbitals are not spherical in shape, although filled p and d subshells have an overall spherical shape. However, our main concern in later chapters will be with orbitals in molecules rather than orbitals in atoms. When atoms form molecules, the distinctions between s, p and d orbitals become blurred, and the charge clouds of electrons in molecules can, for our purposes, always be regarded as approximately spherical in shape. This matter will be taken up in chapter 8.

QUESTIONS

1. State the law describing the force between two electrically charged particles. Calculate the force in newton between the following pairs of particles:
   (a) two protons at a distance of 10^{-8} cm apart;
   (b) a proton and an electron at a distance of 10^{-8} cm apart;
   (c) a helium nucleus and an electron at a distance of 10^{-8} cm apart;
   (d) two helium nuclei at a distance of 10^{-8} cm apart;
   (e) two helium nuclei at a distance of 10^{-9} cm apart.

$$\varepsilon_0 = 8.855 \times 10^{-12} \text{ coulomb}^2 \text{ joule}^{-1} \text{ m}^{-1}$$

$$e \text{ (electronic charge)} = 1.602 \times 10^{-19} \text{ coulomb}$$

2. Write down a formula for the potential energy of a system of two electric charges.
   (a) Calculate the potential energies of the five systems defined in question 1, expressing your answers in both joule, and kJ mol^{-1}.
   (b) Draw a graph showing how the potential energy of a system of two protons varies with their distance apart. (Use a scale covering the separation range from 10^{-9} to 10^{-7} cm; use energy units of kJ mol^{-1}.)
3. A proton undergoes collision with a lithium nucleus. The path of the proton is shown in the diagram

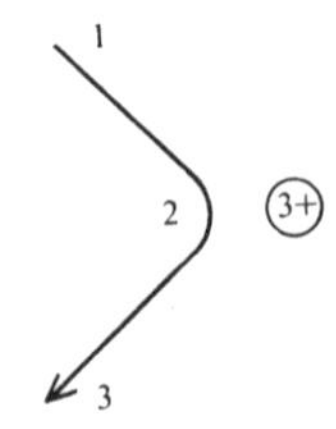

At which of the three points (1, 2, 3) is: (a) the kinetic energy decreasing; (b) the potential energy greatest; (c) the kinetic energy least; (d) the potential energy decreasing?

4. What is meant by the term 'an emission spectrum' of an element? What is the significance of the fact that these spectra consist of a series of discrete lines?
5. List the chief assumptions of the Bohr theory. In what respects did it prove inadequate? What modifications were introduced by quantum mechanics?
6. Write a paragraph describing the distinction between the terms 'orbit' and 'orbital' as used to describe the motion of electrons in atoms.
7. What is meant by the term 'the ground state' of an atom? What happens when an atom in an excited state returns to the ground state?
8. The first ionization energy for beryllium is 897 kJ mol^{-1}, the second ionization energy is 1756 kJ mol^{-1}, and the third ionization energy is 14,830 kJ mol^{-1}. How much energy must be expended per mole of beryllium atoms to remove a total of three electrons?
9. The electron configuration for lithium ($Z = 3$) may be represented $1s^2.2s^1$. Represent the electron configuration of atoms with the following atomic numbers: (a) 1; (b) 7; (c) 21; (d) 23; (e) 28; (f) 31; (g) 45; (h) 53; (i) 64; (j) 92.
10. For each set of elements given, write down the electron configuration and state which elements belong to the same group: (a) $Z = 19, 11, 3, 16$; (b) $Z = 8, 15, 16, 34$; (c) $Z = 9, 5, 31, 14, 22$; (d) $Z = 10, 28, 2, 18, 11$.
11. Name the element having the following electron configuration:
    (a) $1s^2 2s^2.2p^6.3s^1$; (b) $1s^2.2s^2$; (c) $1s^2.2s^2.2p^6.3s^2.3p^6.4s^1$;
    (d) $1s^2.2s^2.2p^6.3s^2.3p^1$; (e) $1s^2.2s^2.2p^2$.
12. Write the electronic configurations of atoms with the following atomic numbers: 12; 36; 42; 56; 62; 68; 74; 77. State which groups of the periodic tables these atoms belong to.
13. By reference to an energy level diagram show why the first period of the periodic table has two elements, whilst the second period has eight elements.
14. Account for the fact that the third period of the periodic table has eight elements and not eighteen.
15. Is it possible for the outer electron of the potassium atom to be in a 4p or a 3d subshell? Under what conditions, if any, could this be achieved?

7

Aggregates of Atoms

7.1 ATOMIC AGGREGATES

All solid objects are composed of atoms 'stuck together' in various ways. We may explain this observation by making two assumptions:

(a) there are forces of attraction between atoms tending to pull them together;
(b) a collection of atoms will tend to arrange itself so that its potential energy is a minimum.

In real systems, the tendency of atoms to arrange themselves in a configuration of minimum potential energy is opposed by the disruptive effects of any kinetic energy possessed by the atoms. Let us take an example. The configuration of minimum potential energy for a collection of iodine atoms is as diatomic molecules packed into a solid, which is the structure of a crystal of iodine at room temperature. The energy required to convert a mole of crystalline iodine to a mole of molecular iodine in the gas phase (sublimation) is given by:

$$I_2(s) \rightarrow I_2(g) \qquad \Delta H = +64.5 \text{ kJ}$$

This result tells us how much energy we must supply in order to overcome the attractive forces holding the molecules into the crystal. Consider now a closed vessel holding a few crystals of iodine. At 0 K, the iodine molecules will be so arranged that the system will have the configuration of minimum potential energy. If we add energy to the iodine crystals and hence raise their temperature, the iodine molecules in the crystal will all obtain varying small amounts of energy. The energy obtained by an individual molecule will cause it to 'shake' around in its position in the crystal. However, not all the iodine molecules will have the same energy; some will have only a little energy, others will have a good deal (compare the energy distribution of the particles in an ideal gas, see chapter 3). Indeed a few of the iodine molecules will obtain sufficient energy to break loose from the lattice and exist as free molecules in the gas phase. By progressively raising the temperature, the fraction of the iodine existing in the gas phase will steadily increase until eventually, at a sufficiently high temperature, enough energy will have been acquired by the iodine to convert all of it to free gaseous molecules. Note that, even for gaseous iodine, the configuration of lowest potential energy is still the solid crystal. The iodine exists as a gas because the vigorous motion of the molecules will not permit the crystal to form. Consider now what happens if we increase the temperature of our vessel of now gaseous iodine. Diatomic iodine molecules can be separated into iodine atoms, if we can provide sufficient energy. Thus, as we saw in chapter 5:

$$I_2(g) \rightarrow 2I(g) \qquad \Delta H = +149 \text{ kJ}$$

This is a good deal more energy than we had to expend in breaking up the iodine crystal. However, by raising the temperature of our tube of iodine to about 1000 K, appreciable numbers of iodine molecules will acquire enough kinetic energy for atoms to fly apart, and some iodine will now exist as free iodine atoms. In a suitable container we would cause almost all of the iodine molecules to break apart by heating to about 2500 K. Again we must stress that the configuration of lowest potential energy for a collection of iodine atoms at 2500 K is still the solid crystal made up of iodine molecules. At 2500 K, however, the iodine has so much kinetic energy that not only has the molecular crystal become disrupted, but the diatomic molecules have themselves been torn apart by the violence of molecular motions.

It should be noted that the disruptive effects of kinetic energy always cause atoms to tend to more ‘disordered’ arrangements. Just as energy can be measured quantitatively (in joule), so too ‘disorder’ can be measured quantitatively, in terms of a quantity known as entropy. A quantitative discussion of entropy is beyond the scope of this book.

We may sum up by saying that the *actual* state of aggregation of any group of atoms at any given temperature is determined by the interplay of two equally important factors, viz.

(a) the tendency to a potential energy minimum;
(b) the disruptive effects of kinetic energy.

7.2 CHEMICAL BONDING

In the following chapters we shall often speak of atoms being bonded together, or of chemical bonds existing between atoms. We saw in section 7.1 that two iodine atoms stick together tenaciously into a diatomic molecule, and we say therefore that a chemical bond exists between the atoms. The strength of this bond may be measured by the amount of energy required to separate one mole of iodine molecules into two moles of iodine atoms. This quantity of energy is called the dissociation energy of the bond between the atoms:

$$I_2(g) \rightarrow 2I(g) \qquad \Delta H = +149 \text{ kJ}$$

As an example of very weak bonding between atoms, consider solid argon, which exists at low temperatures as a close-packed array of argon atoms.

The energy required to overcome the weak forces holding these atoms together is given by:

$$Ar(s) \rightarrow Ar(g) \qquad \Delta H = +6.6 \text{ kJ}$$

This is a sublimation process, and 6.6 kJ is the energy which must be supplied to produce separated argon atoms in the gas phase from the solid crystal. As a contrast to this weak bonding consider tungsten metal, which consists of a closely-packed array of tungsten atoms:

$$W(s) \rightarrow W(g) \qquad \Delta H = +845 \text{ kJ}$$

Obviously tungsten atoms are very firmly bound in the solid. The strength of bonding between atoms can vary from a few kilojoule per mole to about 1000 kilojoule per mole. This is a wide range, and chemists find it useful to distinguish between weak and strong bonding:

(a) weak bonding; any bonding involving energies *less* than about 100 kJ mol^{-1} we call weak bonding. This is exemplified by the bonding between iodine molecules in the iodine crystal.

(b) strong bonding; the term strong bonding refers to bonding involving energies *greater* than about 100 kJ mol^{-1}. This is characteristic of the bonding between atoms in molecules, such as iodine atoms in the I_2 molecule; or metal atoms in a crystal of metal.

To conclude this section, we shall anticipate a later chapter by briefly answering the question: What is the nature of the attractive forces which are responsible for bonding between atoms? Scientists studying the problem have arrived at a remarkably simple answer, viz.

The only important forces between atoms or groups of atoms are electrostatic in nature.

That is, *all* bonding between atoms is due to the electrostatic attraction between electrons and nuclei, which outweighs the combined electron-electron and proton-proton repulsions.

7.3 A TRADITIONAL CLASSIFICATION OF ATOMIC AGGREGATES

A traditional classification of aggregates of atoms is into the three states: solid, liquid and gas. Some properties characteristic of these three states are set out below:

Table 7.1

Solid	*Liquid*	*Gas*
Particles 'fixed' in an orderly array	Particles partially free to move relative to each other hence disordered	Particles completely free to move relative to each other hence disordered
Particles close together hence high density compared with corresponding gas	Particles close together hence high density compared with corresponding gas	Particles far apart hence low density compared with corresponding liquid and solid

In this table, the term 'particles' may refer to atoms, molecules or ions, depending on the precise nature of the substance under consideration.

This solid-liquid-gas classification is of limited usefulness in a discussion of chemical bonding, and for this latter purpose we shall use a detailed classification based on the strength of chemical bonding between atoms. We shall use two basic divisions:

(a) strong bonding of atoms into discrete molecules;
(b) strong bonding of atoms into 'infinite' arrays.

The former classification can apply to all three states of matter, while the latter classification applies only to solids and liquids, since infinite arrays of atoms (e.g. the diamond lattice) can have no existence in the gas phase.

Notice that liquids are the most complex of the three 'traditional' states of matter. They are disordered like gases, but unlike gases the particles in a liquid are close together and may interact strongly. In this latter respect, liquids are like solids, although they lack the simplifying effects of the orderliness of the solid state. Since our most powerful single aid in the study of chemical bonding is a knowledge of the structures of strongly-bonded entities, we shall concentrate our attention almost solely on the gas phase and the solid state. We do this because we have at our disposal powerful experimental tools for determining the structures both of discrete molecules in the gas phase, and of solid lattices. Bonding in liquids may usually be inferred from studies on related gases and solids.

7.4 A NEW CLASSIFICATION OF ATOMIC AGGREGATES

In section 7.3 we presented our new classification of atomic aggregates, using the terms 'strong' bonding in doing so. We had:

(a) *Strong bonding of atoms into discrete molecules.*
(b) *Strong bonding of atoms into infinite arrays.*

Let us consider these classifications separately:

(1) Strong bonding of atoms into discrete molecules. All substances in the gas phase exist as discrete molecules or as atoms. Hydrogen, oxygen, water and carbon dioxide are all examples of substances which form discrete molecules. Furthermore they continue to exist as such in the liquid and solid states, in which the discrete molecules are weakly bonded together. It is most important to realize, however, that not all molecules which exist in the gas phase will persist in the liquid and solid forms of the substance. Thus the diatomic molecule Na_2 has been observed in the gas phase. Solid sodium, however, is a metallic crystal in which no molecular units can be detected. Similarly, when Al_2Cl_6 molecules condense to form a solid, a solid of empirical composition $AlCl_3$ results in which no discrete molecules can be detected. The shapes of some typical molecules are shown in Figure 7.1.

(2) Strong bonding of atoms into infinite arrays. No discrete molecules can be detected in substances exhibiting this structural form, which is thus characteristic of the solid state. We have to consider three subclasses of bonding into infinite arrays:

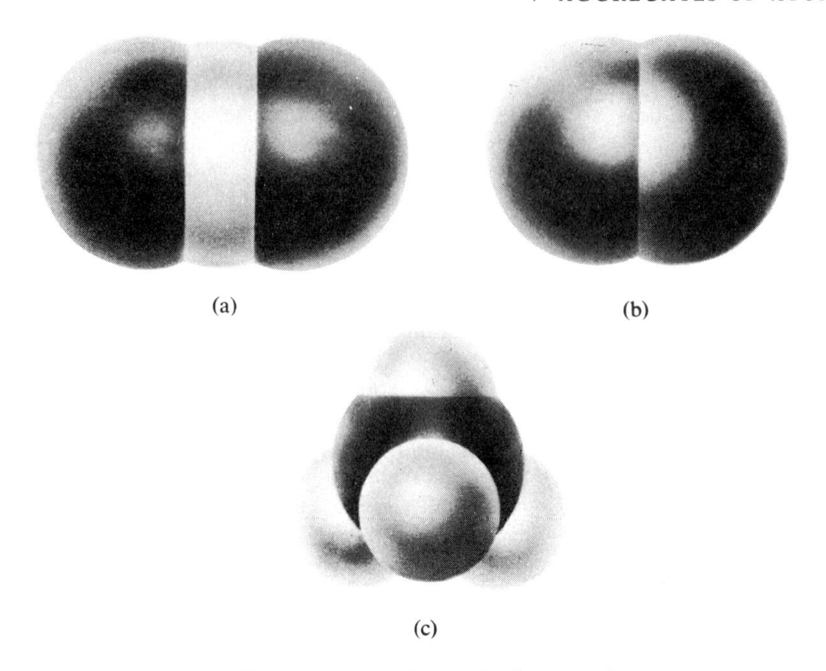

Fig 7.1 Shapes of some simple molecules
(a) CO_2 (carbon dioxide), (b) F_2 (fluorine), (c) CH_4 (methane).

(a) *Strong bonding in three dimensions*; *network lattices*. Examples of this type of structural unit are sodium chloride and diamond. In the solid state both these materials have no structural unit other than the macroscopic crystal.

(b) *Strong bonding in two dimensions*; *weak bonding in one dimension*; *layer lattices*. Many compounds such as aluminium chloride ($AlCl_3$) and cadmium iodide (CdI_2) have structural units which show separation into layers. Thus solid CdI_2 consists of layers made up of infinite double sheets of iodine atoms, cemented together by cadmium atoms. These complex layers are then stacked one on top of the other, with comparatively weak bonding holding the layers together. Each layer has the empirical composition CdI_2.

(c) *Strong bonding in one dimension*; *weak bonding in two dimensions*; *chain lattices*. This is a comparatively rare structural form. It is exemplified by copper(II) chloride, in which infinite chains lie parallel, adjacent chains being held together by comparatively weak forces.

7.5 EXPERIMENTAL DETERMINATION OF THE STRUCTURES OF ATOMIC AGGREGATES

(1) Structures of discrete molecules in the gas phase

Electron diffraction is a direct experimental technique which is generally useful for the determination of the structures of gaseous particles. When a

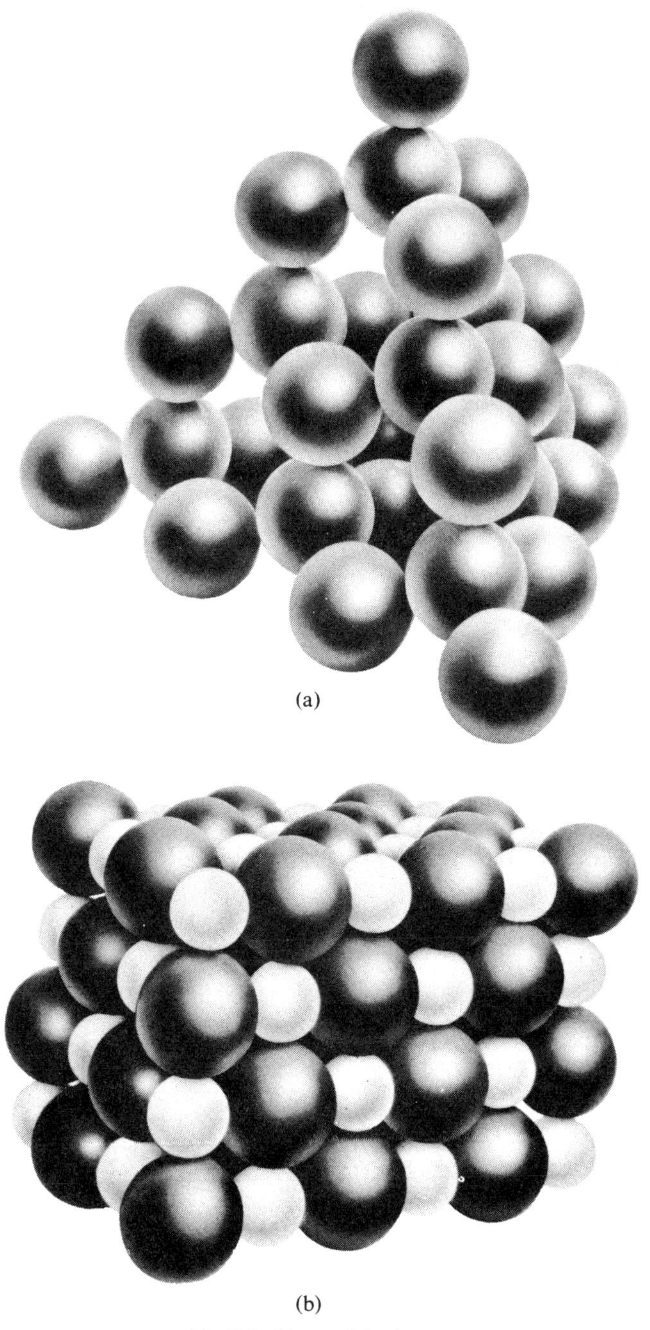
(a)

(b)

Fig 7.2 Network lattices
(a) diamond (C), (b) sodium chloride (NaCl).

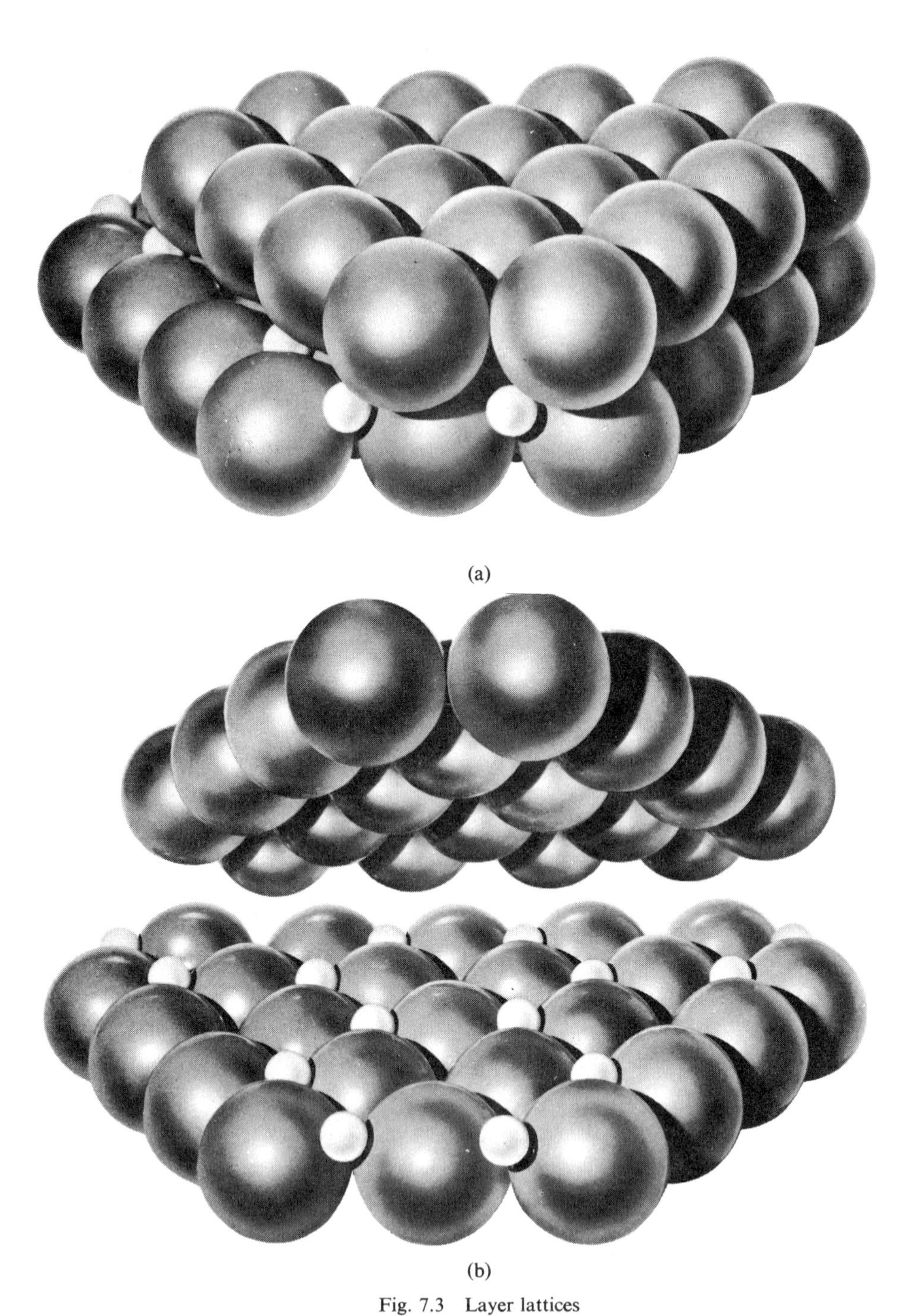

(a)

(b)

Fig. 7.3 Layer lattices
(a) aluminium chloride ($AlCl_3$), (b) aluminium chloride with the layer split into its component sheets, showing Al units in position.

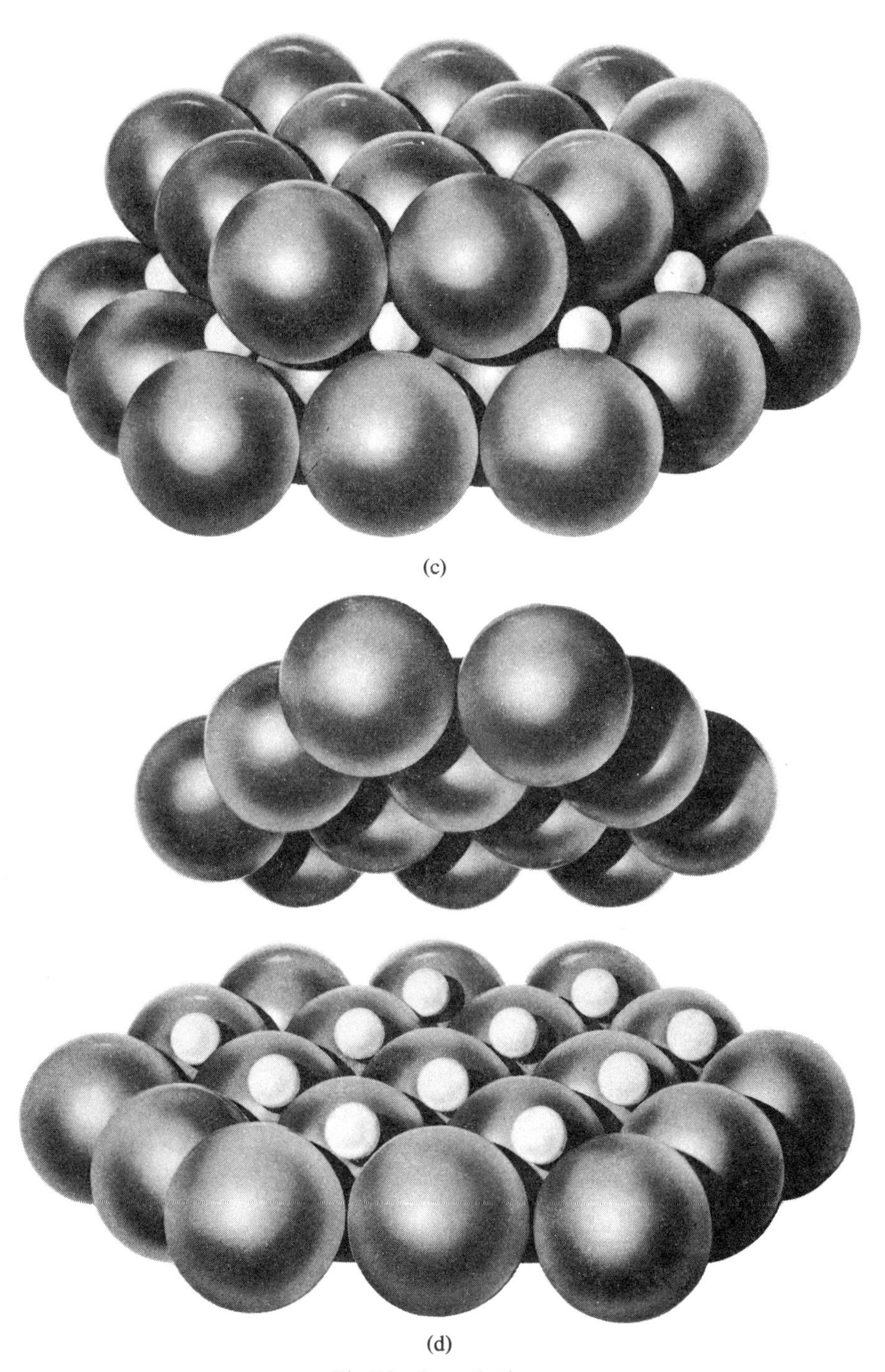

(c)

(d)

Fig 7.3 Layer lattices.
(c) cadmium iodide (CdI_2), (d) cadmium iodide with the layer split into its component sheets, showing Cd units in position.

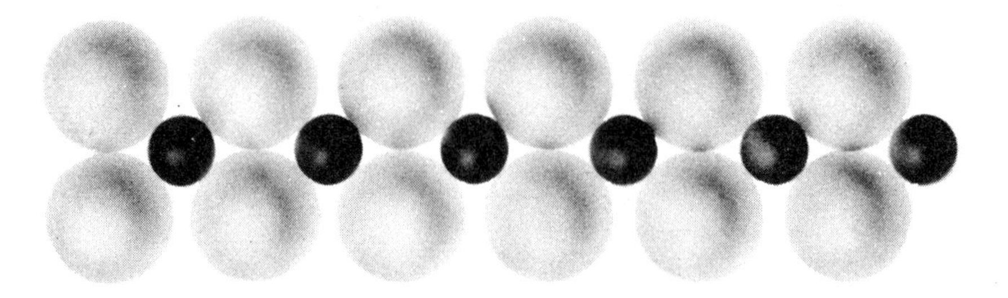

Fig 7.4 Chain lattice; copper(II) chloride ($CuCl_2$).

beam of high energy electrons is passed through a gas at fairly low pressure, the nuclei of the atoms in gaseous molecules scatter the electrons. Figure 7.5 shows a typical experimental arrangement.

The whole system is evacuated and a beam of electrons accelerated to about 40,000 volts is passed through. The gas to be investigated is admitted continuously through a jet in one side of the chamber and is rapidly condensed against a cold surface on the other. The electron beam passes through the gas, is partially scattered, and strikes a photographic plate at the bottom of

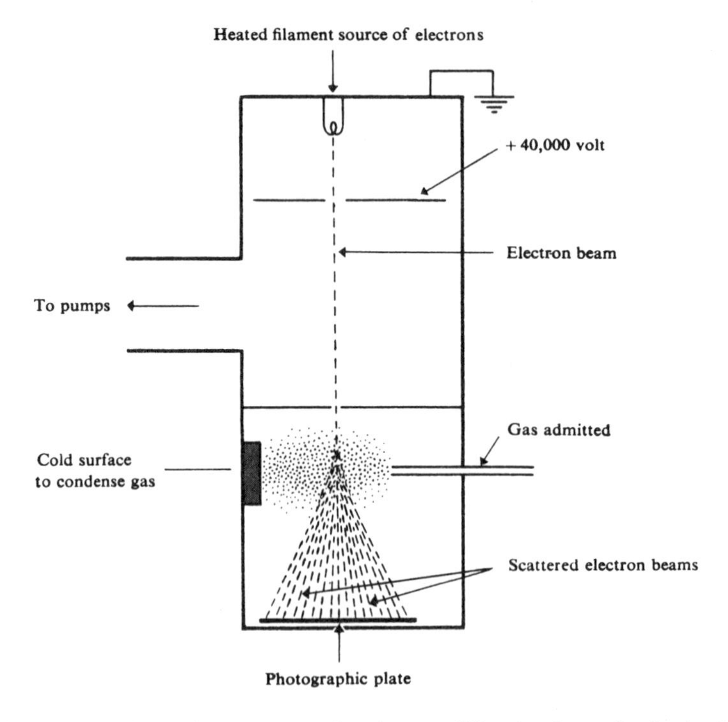

Fig 7.5 Experimental arrangement for electron diffraction by molecules in the gas phase.

the chamber. A photograph is obtained which shows a series of concentric rings, each ring corresponding to a different angle of scattering of the electrons. It is this photograph which provides all the experimental data from which the structure of the molecule may be determined. The angles of scattering of the electrons depend in a complicated way on all the internuclear distances in the molecule. By assuming some definite structure (called a trial structure) for the molecule, an 'expected' diffraction pattern, corresponding to the assumed structure, may be built up by calculation; then, by a process of trial and error, 'expected' patterns are calculated from trial structures until a pattern is found which agrees with the experimentally-found pattern. The assumed structure from which the successful 'expected' pattern was derived is then taken as the actual structure of the molecule. The calculations involved in preparing 'expected' diffraction patterns are very laborious and are normally carried out with the aid of fast electronic computers.

There are a number of other important experimental techniques in common use for determining the structures of molecules in the gas phase. These are the spectroscopic methods which depend upon an interpretation of the way in which the molecules absorb infra-red or microwave radiation.

(2) Structures of solid crystals

The vast majority of known solid structures have been determined by the application of X-ray diffraction techniques. We can observe the scattering of a constant wavelength beam of X-rays by a single crystal of a solid (see Figure 7.6) in which the X-rays are scattered by the electrons surrounding the atomic nuclei in the solid and the *angle* of scattering is determined by the distances between the atomic particles in the solid.

A diffraction pattern is obtained as a series of spots on a photographic plate. By assuming some definite structure for the solid, an 'expected' diffraction pattern may be obtained by calculation, usually with the aid of a high speed electronic computer. By a process of trial and error, 'expected' diffraction patterns are calculated from trial structures until a trial structure is found which correctly reproduces both the positions and intensities of the spots of the experimentally determined diffraction pattern. This trial structure is then

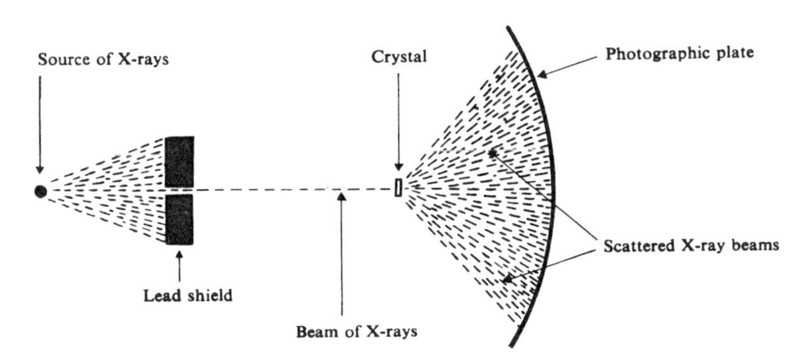

Fig 7.6 Experimental arrangement for X-ray diffraction by solid crystals.

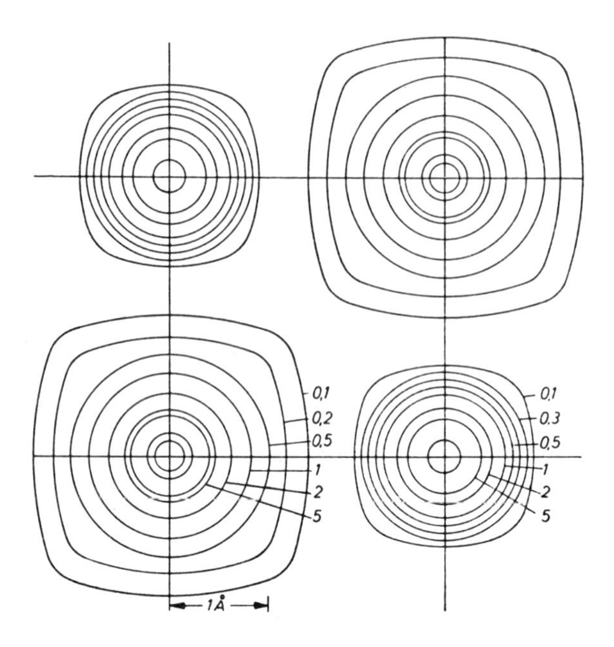

Fig 7.7 Electron density contour diagram for section of sodium chloride lattice. The sodium and chloride ions may be clearly distinguished by their size difference.

assumed to represent the actual arrangement of particles in the solid.

Note that the arrangement of particles in crystals containing either discrete molecules or atoms bonded into infinite arrays may be determined with this technique. X-ray data are frequently presented in the form of electron density contour diagrams, and a typical example of such a diagram is shown in Figure 7.7. Notice how the X-ray scattering data enable us to distinguish entities with different numbers of electrons, by the closeness of the contour lines. Except in very favourable cases, hydrogen atoms, with only one electron, are virtually impossible to detect by means of X-ray diffraction.

Neutron and electron diffraction are also being used increasingly as additional tools in the experimental determination of solid structures.

QUESTIONS

1. Describe qualitatively the structure of a crystal of iodine, of liquid iodine, and of gaseous iodine.
2. The density of solid oxygen, at $-252°C$, is 1.426 g cm^{-3}; the density of liquid oxygen, at $-183°C$, is 1.118 g cm^{-3}; and the density of gaseous oxygen, at S.T.P., is 0.001428 g cm^{-3}. In terms of the structure of the three phases explain why there is such a sharp change in density from the liquid to the gaseous phase.
3. Write down a structural classification of atomic aggregates, and give one example of each structural form.

4. The statement is sometimes made: 'All systems tend to a potential energy minimum.' The minimum potential energy of a collection of carbon dioxide molecules is achieved when the molecules are packed into a solid crystal; yet at room temperature carbon dioxide is a gas, not a solid. Does this mean that the statement about potential energy is incorrect?
5. It is sometimes difficult to decide the precise nature of the bonding between the particles in a liquid. Suggest reasons why this is so.

8

Diatomic Molecules

A good deal of chemistry deals with the properties of molecules, and we shall be concerned in this chapter with the operation of the forces which bond atoms together into molecules. We shall initially restrict ourselves to diatomic molecules—i.e. those containing only two atoms. As we have noted in chapter 7, the forces involved in all chemical bonding are electrostatic in nature. Our task then is to look for the laws governing the operation of these electrostatic forces in order to explain, for example, why it is that two hydrogen atoms can be strongly bonded into a diatomic molecule, whereas helium atoms show no such tendency.

8.1 THE HYDROGEN MOLECULE

The simplest stable molecule known is the hydrogen molecule H_2, which consists of two protons a distance of 0.9×10^{-8} cm apart, and two electrons. When we bring together two hydrogen atoms they form a stable entity H_2, and a good deal of energy is released. Conversely, if we take a hydrogen molecule, work must be done in order to break it apart into its two component atoms:

$$H_2(g) \rightarrow 2H(g) \qquad \Delta H = +432 \text{ kJ}$$

So strong is this interaction between two hydrogen atoms that we speak of a *bond* existing between the atoms, and the energy required to break this bond (432 kJ mol^{-1}) is called the *bond dissociation energy* for the bond in question.

All of this means that two hydrogen atoms exert strong attractive forces on each other, and the atoms will therefore tend to stick together as a pair. If we have a collection of free hydrogen atoms, the forces of attraction between them ensure that the atoms very rapidly pair off to form hydrogen molecules, while heat energy is simultaneously released. Helium atoms, on the other hand, show no tendency to pair off into He_2 molecules, and therefore there can be no strong forces of attraction between helium atoms. In this and succeeding sections, we must thus provide answers to two questions, viz.:

(a) What is the nature of the strong forces of attraction between two hydrogen atoms?

(b) Why do not similar forces of attraction exist between two helium atoms?

Let us return briefly to the hydrogen atom itself. This atom consists of a proton which forms the nucleus of the atom, and an electron moving around the proton, and bound to it by electrostatic attraction. If we took a 'photograph' of a hydrogen atom at some instant of time we would see, typically:

Fig 8.1 Instantaneous 'photograph' of a hydrogen atom.

It is quite easy for us to perceive in this system of proton plus electron that an electrostatic attractive force will tend to hold together the two oppositely charged particles. Consequently it is not surprising to us that a hydrogen atom, consisting of a proton and a *moving* electron, is stable and does not spontaneously break up to give a free proton and a free electron. Indeed, as we have alredy seen, we must supply energy (i.e. do work) to separate the proton and the electron of a normal hydrogen atom:

$$H(g) \rightarrow H^+(g) + e(g) \qquad \Delta H = +1310 \text{ kJ}$$

Our hydrogen molecule is a more complicated system than a hydrogen atom, since the molecule consists of *two* protons and *two* electrons (i.e. *two* hydrogen atoms).

Let us now imagine that we could take a 'photograph' of a hydrogen molecule at some distance of time when we might see:

Fig 8.2 Instantaneous 'photograph' of a hydrogen molecule.

There are six different electrostatic forces to consider. There are two repulsive forces, due to repulsion between the two protons, and repulsion between the two electrons. Counteracting these repulsive forces there are four attractive forces, since each of the two electrons is attracted to each of the two protons. In the configuration shown, a calculation based on the laws of electrostatics expressed in section 6.1 shows that the nett attractive forces outweigh the nett repulsive forces.

However, the configuration of charges shown above cannot be the only one possible for the four particles in a hydrogen molecule, since in the molecule the electrons are in continuous motion. Detailed calculations have been made of the actual energy of the system of two protons and two *moving* electrons. We *know* experimentally that such a system, which is in fact a hydrogen molecule, has a lower energy than two well-separated hydrogen atoms, since:

$$H_2(g) \rightarrow 2H(g) \qquad \Delta H = +432\,kJ$$

The detailed calculations, which are very complicated, were first made by Coolidge and James in 1933. These workers were able to show that the electrostatic forces operating between protons and electrons are quite sufficient to explain why it is that two hydrogen atoms attract each other. Indeed, Coolidge and James were able to correctly calculate the ΔH value for the reaction

$$H_2(g) \rightarrow 2H(g)$$

by applying Coulomb's Law of Electrostatics to protons and electrons, subject only to the limitations of the same quantum mechanics which had been used successfully to calculate the energy levels of the hydrogen atom. We can therefore state a most important conclusion:

The forces binding two hydrogen atoms together into a hydrogen molecule are entirely electrostatic in nature.

The calculations of Coolidge and James have also provided us with some information about the motion of the electrons in the hydrogen molecule. As we might have anticipated, the electrons do not trace out any definite orbits about the protons. If we took a series of 'photographs' of a hydrogen molecule we would see many different instantaneous configurations of the particles, e.g.:

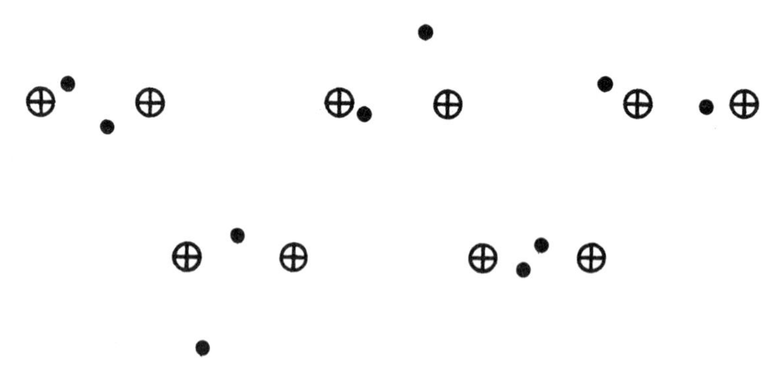

Fig 8.3 Instantaneous 'photographs' of five hydrogen molecules.

The main features we would notice if we inspected a great number of such photographs would be:

(a) the two protons, because of their relatively greater masses, remain, except for small periodic fluctuations, at the same distance apart:
(b) the electrons move rapidly around the nuclei in a more or less random fashion, but tend to be found in the region of space between the two nuclei more frequently than elsewhere.

We saw for the hydrogen atom in section 6.12 that the random motion of an electron made it behave as if it were a charge cloud. The same effect is noticed with the electrons in the hydrogen molecule, and **the two electrons in the hydrogen molecule behave as if they were a charge cloud encompassing the two protons.**

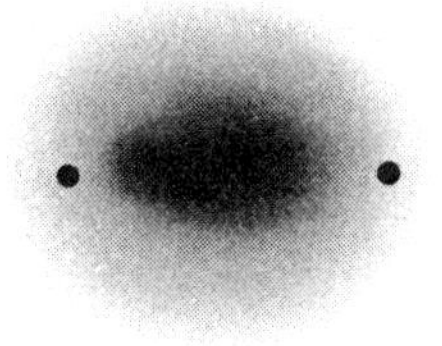

Fig 8.4 Charge cloud picture of hydrogen molecule.

The density of the charge cloud is seen to be greatest between the two nuclei, and in terms of our discussion in section 6.12 we would say that the two electrons were spending most of their time between the two nuclei. This is as expected, since it is in this region between the protons that the electrons experience the strongest forces of attraction, due to the simultaneous effects of both protons. This region of space in which the electrons are moving is called an orbital, and just as we labelled an orbital in an atom an atomic orbital, so

an orbital in a molecule is called a molecular orbital.

In the hydrogen molecule, as we have seen, the orbital encompasses the region between the two protons. Thus the hydrogen molecule consists essentially of two protons being pulled together by the electrostatic attraction for the two electrons moving in the region between them.

8.2 HYDROGEN MOLECULE-IONS AND THE PAULI PRINCIPLE

The Pauli Principle applies to electrons in molecules just as it applies to electrons in atoms, i.e. no molecular orbital can contain more than two electrons; it may have 0, 1 or 2 electrons, but no more. The molecular orbital in the hydrogen molecule contains just two electrons, and hence can contain no more. We could, however, imagine the *removal* of an electron from H_2, yielding H_2^+. This ion has been detected as a short-lived species in a gas discharge tube, and its bond dissociation energy has been determined:

$$H_2^+(g) \rightarrow H^+(g) + H(g) \qquad \Delta H = +255 \text{ kJ}$$

Thus the system of two protons and one electron forms a stable molecule because the electrostatic attraction of two nuclei for one electron becomes greater than the proton-proton repulsion. Indeed, the bond dissociation energy of H_2^+ is about one-half that of the hydrogen molecule, in which there are two electrons:

Table 8.1

Molecule	*No. of electrons*	*Bond dissociation energy (kJ mol^{-1})*
H_2^+	1	255
H_2	2	432

On viewing the data, the unwary might be tempted to suppose that the addition of a third electron to H_2, making H_2^-, would lead to a further increase in the bond dissociation energy. However, the Pauli Principle ensures that this cannot happen. The molecular orbital in the region between the two protons of the hydrogen molecule can contain no more than two electrons. Indeed, this molecular orbital may be considered as arising from the sharing by the two hydrogen atoms of their atomic K-shells. The K-shell of *any* atom is limited to two electrons, and this shared K-shell is similarly limited. As a consequence, a third electron would have to go into an orbital which was *not* between the two protons and would therefore tend to *raise* the potential energy of the system. Not surprisingly the ion H_2^- has never been detected.

Similar arguments apply to the diatomic helium molecule, He_2. A helium atom has the electronic configuration $1s^2$, i.e. a filled K-shell. When two helium atoms are brought together, it is initially impossible for them to share their K-shells since this would place a total of four electrons in the shared K-shell, and this is forbidden by the Pauli Principle. It is necessary therefore to push two of the electrons into a region of space outside the shared K-shell. However, energy must be supplied to do this, and this energy is just about equal to the lowering in energy obtained by having two electrons in a shared K-shell. The nett effect, therefore, is that two helium atoms do not form a covalent bond.

8.3 THE COVALENT BOND

The picture developed in sections 8.1 and 8.2 of the bonding between two hydrogen atoms suggests that the sharing of one or two electrons by two nuclei leads to a nett bonding between the nuclei due to the combined effect of a number of electrostatic interactions. This sort of chemical bonding was first recognized by G. N. Lewis in 1916, and was called by him *covalent bonding*. The *covalent bond* in the hydrogen molecule may be represented diagrammatically as:

$$H : H \qquad \text{or} \qquad H\text{—}H$$

The two electrons in the region of space between the two protons are usually called 'bonding electrons' and we use two dots or a straight line as depicted to represent this pair of electrons. We shall see in succeeding sections how the ideas of sections 8.1 and 8.2 may be generalized to a much wider array of molecules, so we shall summarize at this point the most important features of the covalent bond:

(a) there must be a shared pair of electrons in the region of space between the two nuclei. These electrons are called *bonding* electrons;
(b) the only forces operating are electrostatic in origin.

We have not considered any means of representing the one-electron bond in H_2^+. We shall write

$$(H \cdot H)^+$$

using a single dot to represent the single electron. One-electron bonds are extremely rare in chemical compounds, since atoms or molecules with half-filled orbitals are usually extremely reactive. Nearly all stable molecules possess an even number of electrons, arranged in pairs, and hence the only important covalent bonds we shall meet will be electron pair bonds.

8.4 DIATOMIC MOLECULES OF THE ALKALI METALS

At room temperatures the alkali metals, such as lithium and sodium, normally exist as metallic solids. However, at high temperatures the metals can be vaporized and the vapour contains a mixture of metal atoms in equilibrium with diatomic molecules, thus:

$$2Li(g) \rightleftharpoons Li_2(g)$$

The electronic configuration of a lithium atom is $1s^2.2s^1$; the atom has a filled K-shell, and a half-filled 2s-subshell. When two lithium atoms come together, the single 2s electron from each atom is attracted into a region of space between the two nuclei, and a covalent bond similar to that in H_2 is formed. The K-shell of each lithium atom is filled since it contains 2 electrons, consequently there can be no sharing of any part of the K-shells; a charge cloud picture of the Li_2 molecule shows the two K-shells remaining virtually intact in the molecule. The 2s electrons, on the other hand, being shared between the two atoms, pass into the region of space between the two nuclei, giving rise to a covalent bond:

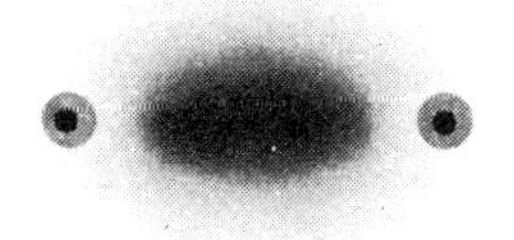

Fig 8.5 Charge cloud picture of Li_2.

The K-shell electrons, which have remained virtually unaffected by the bond formation, are usually called *inner-shell* electrons.

A sodium atom has the electronic configuration $1s^2.2s^2.2p^6.3s^1$; it has filled K- and L-shells, and a single electron in a 3s atomic orbital. When two sodium atoms approach one another, the K- and L-shells, being filled, remain essentially intact; the 3s electrons from each atom are shared between the two atoms, and go into the region of space between the two nuclei, again giving rise to a covalent bond. The charge cloud picture of the Na_2 molecule looks very similar to that for Li_2:

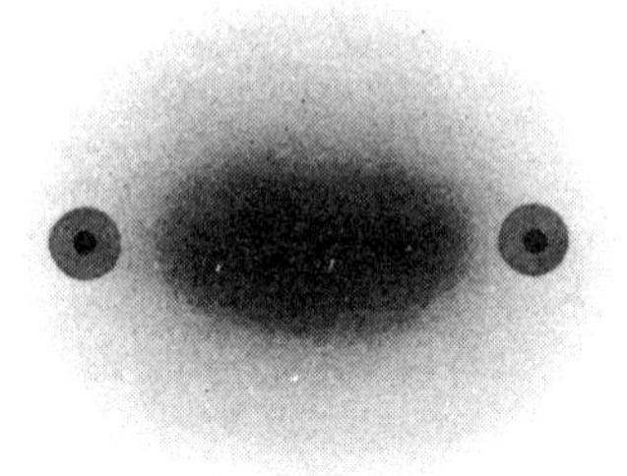

Fig 8.6 Charge cloud picture of Na_2.

These two examples of Li_2 and Na_2 illustrate an important principle, viz.:
In general, only outer-shell electrons are important in the formation of covalent bonds.

We can see this principle illustrated again with the molecule K_2. The potassium atom has the electronic structure $1s^2.2s^2.2p^6.3s^2.3p^6.4s^1$., i.e. it has a filled K- and L-shell; in the M-shell, the s- and p-subshells are filled, and in the outer shell we have a single electron in a 4s atomic orbital. When two potassium atoms come together, the K- and L-shells remain unaffected, and the single electrons in the outer shell are shared, go into the region of space between the two nuclei, and form a covalent bond,

K—K or K : K

Notice that in each of the diatomic molecules Li_2, Na_2 and K_2, there are but two bonding electrons.

8.5 DIATOMIC MOLECULES OF THE HALOGENS

A fluorine atom has the electronic configuration $1s^2.2s^2.2p^5$; it has a filled K-shell, and an L-shell which is one electron short of completion. In all, we have a total of seven electrons in the outer (L) shell. We might be tempted to suggest that two fluorine atoms could come together and share all their outer-shell electrons, thus forming a very strong covalent bond. Unfortunately for fluorine atoms, however, the Pauli Principle prevents them from doing this. The L-shell of an atom can hold no more than eight electrons, hence if a fluorine atom is to share any of its L-shell electrons with another atom it can do so by accepting and sharing one electron only, thus filling its L-shell, and forming a single covalent bond. Thus when two fluorine atoms come together, the K-shells remain virtually unaffected in the final molecule, and a fluorine molecule may be represented:

$$:\underset{\cdot\cdot}{\overset{\cdot\cdot}{F}}:\underset{\cdot\cdot}{\overset{\cdot\cdot}{F}}: \quad \text{or} \quad :\underset{\cdot\cdot}{\overset{\cdot\cdot}{F}}\text{—}\underset{\cdot\cdot}{\overset{\cdot\cdot}{F}}:$$

where we have used the double dot notation for a pair of electrons.

Chlorine has the electronic configuration $1s^2.2s^2.2p^6.3s^2.3p^5$; it has filled K- and L-shells, a filled 3s subshell, and the 3p subshell just one electron short of completion. By accepting and sharing one electron, a chlorine atom can fill its 3p-subshell and form a single covalent bond, as:

$$:\underset{\cdot\cdot}{\overset{\cdot\cdot}{Cl}}\text{—}\underset{\cdot\cdot}{\overset{\cdot\cdot}{Cl}}:$$

Notice that the M-shell of the chlorine atoms has eight electrons and is not filled; there is still an unfilled d-subshell in this shell. Bromine and iodine also form diatomic molecules which may be represented:

$$:\underset{\cdot\cdot}{\overset{\cdot\cdot}{Br}}\text{—}\underset{\cdot\cdot}{\overset{\cdot\cdot}{Br}}: \quad \text{and} \quad :\underset{\cdot\cdot}{\overset{\cdot\cdot}{I}}\text{—}\underset{\cdot\cdot}{\overset{\cdot\cdot}{I}}:$$

8.6 CHARGE CLOUDS IN THE FLUORINE MOLECULE

Returning to the fluorine molecule

$$:\underset{\cdot\cdot}{\overset{\cdot\cdot}{F}}\text{—}\underset{\cdot\cdot}{\overset{\cdot\cdot}{F}}:$$

we can see that there are two classes of electrons in the outer shells of the atoms in this molecule:

(a) *bonding electrons*, which are in the region of space between the two nuclei, and which are 'shared' by both atoms. The orbitals in which bonding electrons move are sometimes referred to as *bonding orbitals*.

(b) *non-bonding electrons*, which are those outer-shell electrons which take no part in bonding; they are not in the region of space between the two nuclei, and hence are not shared by both atoms. A pair of such non-bonding electrons is often called a 'lone pair'. The orbitals in which non-bonding electrons move are sometimes referred to as *non-bonding orbitals*.

When we come to consider the charge cloud representation of a fluorine molecule, we must recall that electrons move in definite regions of space which we have called orbitals. Furthermore, no more than two electrons can be in any particular orbital at the same time. This has been clearly seen in the molecules H_2, Li_2, Na_2, K_2 and F_2, in all of which we have a pair of bonding electrons in a 'bonding orbital' between the two nuclei. In the case of F_2, however, we also have in the outer shell non-bonding electrons, which are not between the nuclei. These electrons must move in orbitals, and thus can also be represented by charge clouds. The question that arises is, what are the shapes of the charge clouds of electrons in non-bonding orbitals? In all the cases we have examined in this chapter, the bonding electrons have been in a somewhat distorted spherical orbital. It turns out that we can explain a good number of the properties of molecules if we assume that non-bonding electrons also move in roughly spherical orbitals. Thus, in general:

We may represent the charge clouds of both bonding and non-bonding outer-shell electrons as being roughly spherical in shape.

The next question we must ask is, how are these spherical charge clouds arranged in space? A simple principle, enunciated by Sidgwick and Powell in 1939, is:

The approximately spherical charge clouds within a particular electronic shell will stay as far away from each other as possible.

This is to be expected because of the electrostatic repulsion between the charge clouds. To take a simple example of this, consider four identical charge clouds around some central nucleus or core. There are many possible ways of arranging these charge clouds around the core, e.g. they may be arranged in a plane:

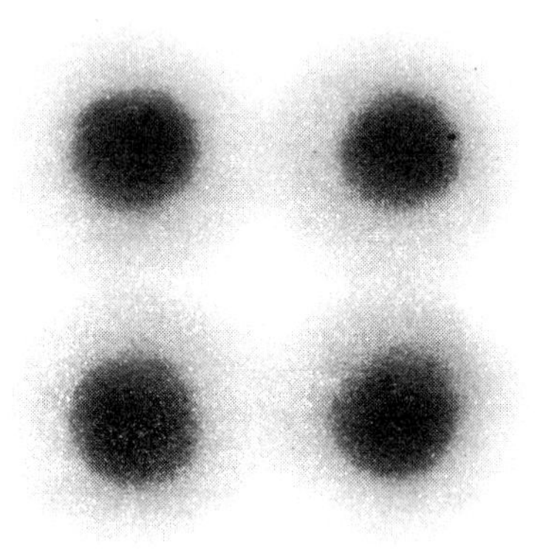

Fig 8.7 Planar arrangement of four charge clouds.

The arrangement of lowest electrostatic potential energy, however, will be that in which the charge clouds are as far away from each other as they can get, and for four identical charge clouds the required arrangement is tetrahedral:

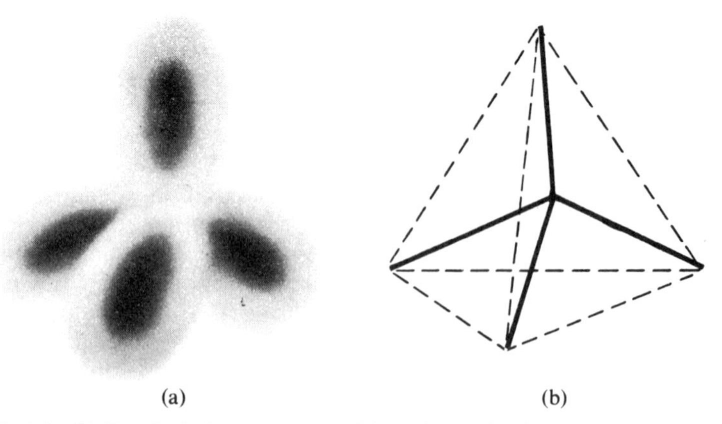

Fig 8.8 (a) Tetrahedral arrangement of four charge clouds, (b) A tetrahedron.

If we apply these ideas to the fluorine molecule:

$$:\underset{\cdot\cdot}{\overset{\cdot\cdot}{F}}-\underset{\cdot\cdot}{\overset{\cdot\cdot}{F}}:$$

we will have one charge cloud for each pair of electrons. Obviously there is a total of seven electron pairs to consider, one shared bonding pair, and six unshared non-bonding pairs, i.e. there are four pairs of electrons in the L-shell of each fluorine atom, with one pair (the bonding pair) common to both atoms. If we require that the charge clouds about a given fluorine atom must get as far away from each other as they can, we obtain:

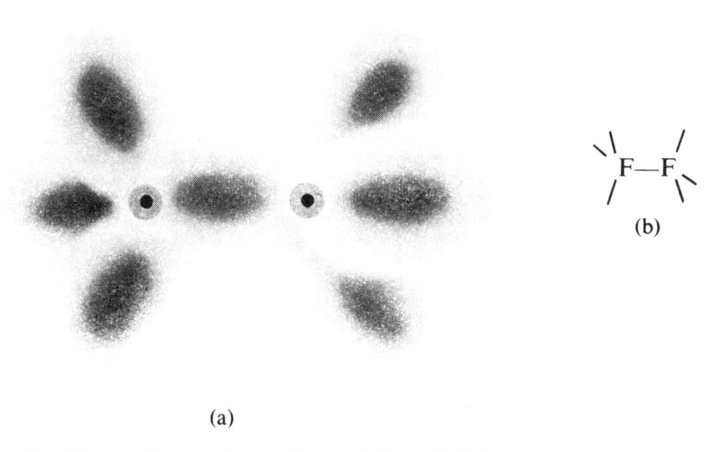

Fig 8.9 (a) Charge clouds picture of F_2, (b) Valence structure.

where there is a roughly tetrahedral arrangement of electron pairs around each fluorine atom. We have also included a more diagrammatic representation where we have used straight lines to represent both bonding and non-bonding pairs of electrons. The lines are drawn along the line between the centre of the charge cloud and the nearest nucleus, and illustrate well the arrangement of the electron pairs in space. Although the arrangement of four charge clouds around the fluorine atoms in F_2 is essentially tetrahedral it is not exactly so because the bonding and non-bonding charge clouds are not identical; we will have more to say about these slight deviations in the next chapter.

8.7 MOLECULES WITH SINGLE BONDS

To date, we have considered the formation of covalent bonds in which we have one pair of electrons responsible for the bonding. Such a bond is called a *single covalent bond*, or an *electron pair bond*. We shall summarize the diagrammatic representations of some molecules with single covalent bonds:

(1) Charge cloud representations. Charge cloud diagrams are used when we wish to emphasize the details of the electron distribution in a molecule. In Figure 8.10 we have charge cloud representations of the hydrogen molecule and the lithium diatomic molecule.

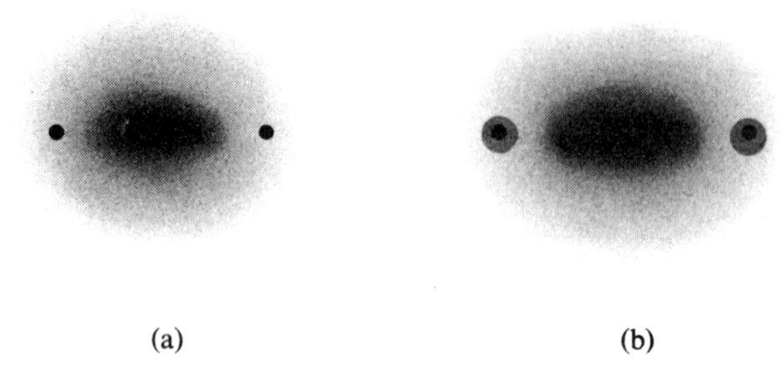

Fig 8.10 Charge cloud pictures: (a) H_2, (b) Li_2.

(2) Valence structures. The charge cloud representations are useful but are far too clumsy for normal use. In the valence structure representations of molecules, the nature of an atom is indicated by its chemical symbol, e.g. H, Li, Na, F. Outer-shell electrons only are represented, and electron pairs may be represented either as two dots, or as a stroke, e.g.:

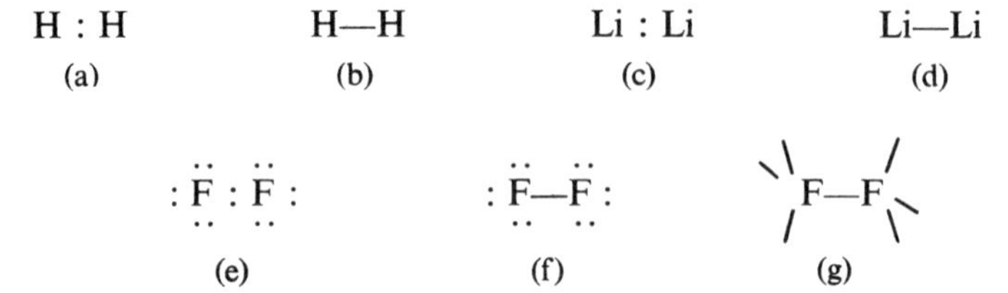

Fig 8.11 Valence structures of H_2 (a) and (b); Li_2 (c) and (d); F_2 (e), (f) and (g).

The last valence structure shown for fluorine (g) is useful if we wish to emphasize the arrangement in space of the non-bonding electron pairs. An alternative use of the stroke for indicating non-bonding pairs is

$$|\overline{\underline{F}}—\overline{\underline{F}}|$$

This last notation is useful when the directional properties of the non-bonding pairs need less emphasis. Finally, it should be noted that when interest lies only in the bonding electrons of a particular molecule, the non-bonding outer-shell electrons may be omitted altogether, e.g.:

F—F

All the notations shown in this section for writing valence structures are in general use, and we shall use them as convenient in the following chapters.

8.8 MOLECULES WITH MULTIPLE BONDS

When two atoms share a pair of electrons, a single covalent bond is formed. If *two* pairs of electrons were shared by two atoms, we would call the resulting bond a *double bond*, and if *three* pairs of electrons were shared, we would have a *triple bond*.

The oxygen atom has the electronic configuration $1s^2.2s^2.2p^4$; it has a filled K-shell, and is two electrons short of a filled L-shell. Thus the atom might accept and share *two* electrons from some other atom, forming a double bond with that atom. In doing so, the number of electrons in the L-shell of the oxygen atom would reach eight, the maximum allowed by the Pauli Principle. Thus when two oxygen atoms come together, the K-shell remains unaltered, and four electrons pass into a bonding region between the two nuclei. Because of the Pauli Principle, two orbitals are required for these four electrons. The actual arrangement of the charge clouds is governed by our charge cloud repulsion principle, and we obtain:

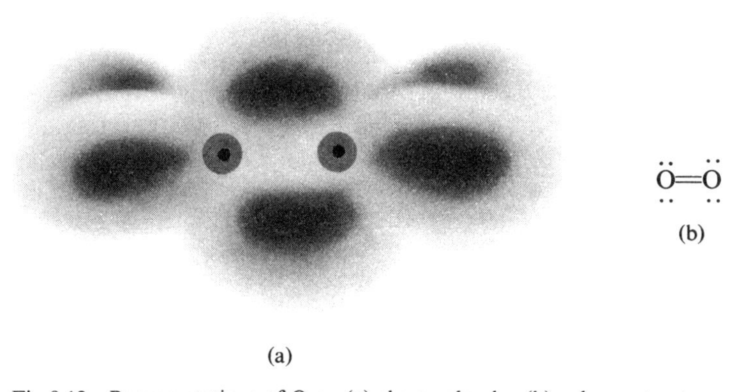

Fig 8.12 Representations of O_2: (a) charge cloud, (b) valence structure.

where there are two bonding charge clouds between the nuclei, and a total of four lone pairs. The arrangement of charge clouds around each oxygen atom is tetrahedral. This picture gives us a satisfactory description of the bonding in the oxygen molecule. Unfortunately, it does not give an adequate description of certain magnetic properties of this molecule, but the satisfactory resolution of this difficulty is beyond the scope of this book; the above picture of a double bond is quite satisfactory for our purposes. Notice that we have used two lines between the nuclei to indicate that two electron pairs are bonding pairs.

The nitrogen atom has the electronic structure $1s^2.2s^2.2p^3$; it has a filled K-shell, and the L-shell is three electrons short of completion. By accepting and sharing three electrons from another atom, the nitrogen atom L-shell becomes filled, and a triple bond is formed. Thus when two nitrogen atoms come together, six electrons in all go into the bonding region between the two nuclei, in three different charge clouds:

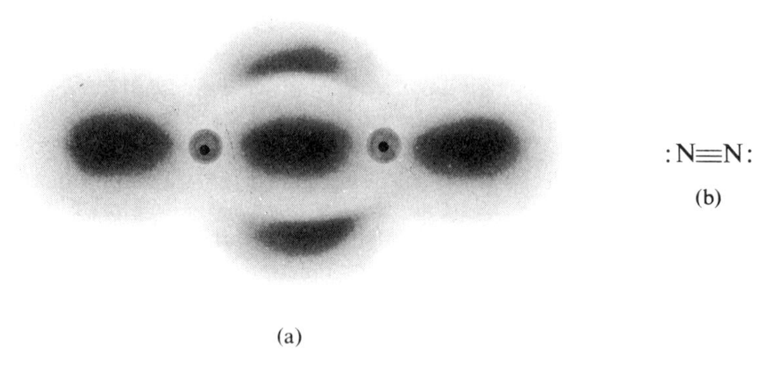

Fig 8.13 Representations of N_2: (a) charge cloud, (b) valence structure.

This is a satisfactory picture of a triple, or 6-electron bond. Note the essentially tetrahedral arrangement of charge clouds around each nitrogen atom. In the valence structure representation of a triple bond, we use three strokes between the nuclei to indicate that three electron pairs are involved in the bond.

8.9 HETERONUCLEAR DIATOMIC MOLECULES

All the diatomic molecules we have dealt with to date have been made up of two identical nuclei. Such molecules are called *homonuclear diatomic molecules*. A diatomic molecule made up from two different nuclei is called a *heteronuclear diatomic molecule*. Consider, for example, the molecule HF. This molecule is formed when a hydrogen atom ($1s^1$) and a fluorine atom ($1s^2.2s^2.2p^5$) come together. These atoms share a pair of electrons, thus bringing about the filling of the K-shell of the hydrogen, and the L-shell of fluorine, giving:

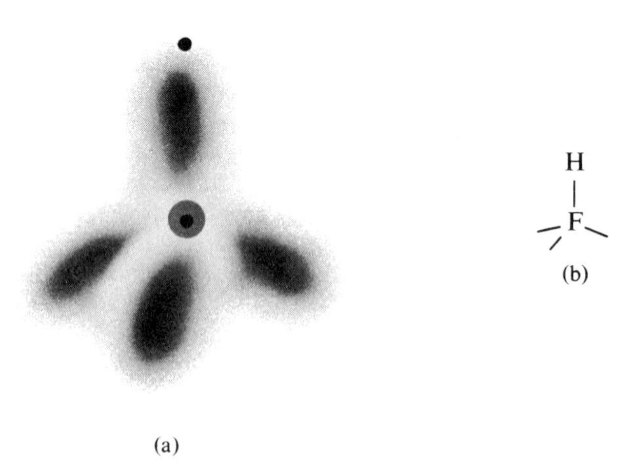

Fig 8.14 Representations of HF: (a) charge cloud, (b) valence structure.

In the charge cloud picture, notice the tetrahedral arrangement of the charge clouds around the fluorine atom. The molecules HCl, HBr, and HI have essentially similar charge cloud pictures.

We can also form diatomic molecules which are ions. The ion O^- has the electronic configuration $1s^2.2s^2.2p^5$, the same as fluorine, and may therefore form one covalent bond. The O^- ion does not normally occur on its own in any chemical compounds, but the OH^- ion may be regarded as arising when a covalent bond is formed between the O^- ion and a hydrogen atom. The resultant molecule-ion has an electronic structure very similar to the HF molecule, and is commonly known as the hydroxide ion.

$$\overset{\ominus}{:}\ddot{\underset{\cdot\cdot}{O}}\text{—H}$$

Its charge cloud representation is very similar to that of hydrogen fluoride, and it occurs in many ionic solids (e.g. NaOH). In writing the valence structure for the hydroxide ion, we have shown the charge of -1 as residing on the oxygen atom. This is done so that we can keep an account of the number of covalent bonds which the atom can form. An uncharged oxygen atom with electronic configuration $1s^2.2s^2.2p^4$ is two electrons short of a completed L-shell, and can form *two* single covalent bonds, as in the water molecule, or one double covalent bond as in the oxygen diatomic molecule. An oxygen atom to which an electron has been added (O^-) can form one covalent bond only, since it is but one electron short of a completed L-shell. Obviously, an oxygen atom to which two electrons have been added has the same electronic configuration as a neon atom, and can form no covalent bonds at all.

8.10 BOND POLARITY IN HETERONUCLEAR DIATOMIC MOLECULES

A new problem which arises in any heteronuclear diatomic molecule, such as HF, is that the atoms at each end of the bond will, in general, have different electron-attracting powers. Thus, in the HF molecule it appears that the fluorine atom attracts electrons more strongly than does the hydrogen atom, and the shared pairs of electrons spend more of their time closer to the fluorine atom than to the hydrogen atom. The molecule is thus sometimes represented:

$$\overset{\delta+}{\text{H}}\text{—}\overset{\delta-}{\text{F}}$$

with small positive and negative charges ($\delta+$, $\delta-$) indicating that the electrons have, on the average, shifted away from the nucleus of the hydrogen atom towards the nucleus of the fluorine atom. Such a molecule, in which a pair of bonding electrons is shared unequally between two atoms, is said to be *polarized.*

An extreme example of polarization is provided by the diatomic molecules of the alkali halides. The alkali halides exist at room temperatures as ionic solids in which no discrete molecules can be detected. At high temperatures, however, these substances can be vaporized, and alkali metal halide molecules have been detected in the vapour. We could write valence structures for these diatomic molecules:

$$\mathrm{Na{-}\ddot{\underset{..}{Cl}}\!:} \qquad \mathrm{K{-}\ddot{\underset{..}{Cl}}\!:} \qquad \mathrm{Cs{-}\ddot{\underset{..}{F}}\!:}$$

However, the electron-attracting power of the halogen is so great, relative to that of the alkali metal, that the alkali metal atom in these molecules gives up an electron almost completely to the halogen atom, and the molecules behave as 'ion-pairs' with valence structures:

$$\overset{+}{\mathrm{Na}}\,\overset{-}{\mathrm{Cl}} \qquad \overset{+}{\mathrm{K}}\,\overset{-}{\mathrm{Cl}} \qquad \overset{+}{\mathrm{Cs}}\,\overset{-}{\mathrm{F}}$$

This situation in which different atoms have different electron-attracting powers is quite general, and chemists have devised a semi-quantitative scale to measure differences in electron-attracting power. This scale is known as the electronegativity scale.

The electronegativity of an atom is a numerical measure of the electron-attracting power of an atom.

Various electronegativity scales have been devised, and it is reassuring to find that, although the actual numerical values of electronegativity differ from one scale to another, the *order* of electronegativities of all the atoms is just about the same on all the scales. Thus fluorine is rated as the most electronegative atom on all scales, and caesium as the least electronegative (excluding francium). A very simple scale, proposed by Mulliken, defines the electronegativity M_X of an atom X as

$$M_X = \tfrac{1}{2}(I + A)$$

where I and A are the ionization energy and electron affinity respectively of the atom X. I and A are defined by the thermochemical equations:

$$X(g) \rightarrow X^+(g) + e(g) \qquad \Delta H = I$$

$$X^-(g) \rightarrow X(g) + e(g) \qquad \Delta H = A$$

i.e. the electron affinity of an atom X is the energy required to form the gaseous atom of X from its singly-charged negative ion in the gas phase. Thus the electron affinity is the energy *released* when an electron is added to a gaseous X atom. Unfortunately, Mulliken's electronegativity scale cannot be written for most atoms because so many electron affinities are not known. An alternative electronegativity scale due to Pauling is therefore used. Pauling's electronegativity scale is obtained empirically from a consideration of experimentally-determined bond dissociation energies. When Mulliken's electronegativity scale can be compared with Pauling's, the two scales are in agreement as to the relative ordering of the electronegativities of the atoms. We therefore quote Pauling's scale, since values are available for most of the atoms in which we are interested.

We may note that the electronegativities *increase* as one goes from left to right across the periodic table and that they *decrease* as one descends the table vertically from elements of lower to higher atomic number.

The *increase* in electronegativity on going from left to right across the periodic table may be understood as follows. As one goes from left to right

H 2.1							He –
Li 1.0	Be 1.5	B 2.0	C 2.5	N 3.0	O 3.5	F 4.0	Ne –
Na 0.9	Mg 1.2	Al 1.5	Si 1.8	P 2.1	S 2.5	Cl 3.0	Ar –
K 0.8	Ca 1.0	Ga 1.6	Ge 1.8	As 2.0	Se 2.4	Br 2.8	Kr –
Rb 0.8	Sr 1.0	In 1.7	Sn 1.8	Sb 1.9	Te 2.1	I 2.5	Xe –
Cs 0.7	Ba 0.9	Tl 1.8	Pb 1.8	Bi 1.9	Po 2.0	At 2.2	Rn –
Fr 0.7	Ra 0.9						

Fig 8.15 Electronegativities of the atoms on Pauling's scale.

across the table within a particular period, the charge on the nucleus is increasing by one unit at the same time as one electron is added to the outer shell. Thus if we consider the second period, all the atoms have a filled K-shell; as electrons are added to the L-shell the nuclear charge is simultaneously increased as we go from lithium to neon. The effective positive charge experienced by the outer-shell electrons in these atoms is not the full nuclear charge, since the nucleus is well shielded by the inner K-shell electrons If this shielding were very efficient, we might suppose that the charge experienced by the outer-shell electrons would be the nuclear charge, *less* the total charge on the inner-shell electrons. This charge is often called the 'core charge' of the atom, and it increases across a period, e.g. in the second period:

Table 8.2

Atom	*Electron configuration*	*(Nuclear charge Z)*	*Core charge (Z − 2)*
Li	$1s^2.2s^1$	+3	+1
Be	$1s^2.2s^2$	+4	+2
B	$1s^2.2s^2.2p^1$	+5	+3
C	$1s^2.2s^2.2p^2$	+6	+4
N	$1s^2.2s^2.2p^3$	+7	+5
O	$1s^2.2s^2.2p^4$	+8	+6
F	$1s^2.2s^2.2p^5$	+9	+7
Ne	$1s^2.2s^2.2p^6$	+10	+8

We shall regard the 'core charge' of an atom as giving us a rough guide to the charge experienced by the outer-shell electrons of the atom. The increase in 'core charge' from lithium to fluorine is in qualitative accord with observed electronegativity trends.

The general *decrease* in electronegativity on going down a group is due to a number of factors. One important factor is the increasing size of the atoms, which means that the outer-shell electrons are moving around a successively larger 'core' as we go down a group. It follows that these outer shell electrons are successively further removed from the nucleus, and are thus not attracted to it so strongly.

8.11 THE STRENGTHS OF COVALENT BONDS

Inspection of some experimentally-determined values of bond dissociation energies of diatomic molecules shows that covalent bonding in general falls into the category we have called 'strong bonding'. That is, it usually requires more than 100 kJ mol^{-1} to break a covalent bond. Data for some molecules are given in Table 8.3.

Table 8.3

Bond dissociation energies of some diatomic molecules (kJ mol^{-1})

Molecule	H_2	F_2	Cl_2	Br_2	I_2	Li_2	Na_2	O_2
Bond dissociation energy	432	151	239	190	149	105	72	490
Molecule	N_2	HF	HCl	HBr	HI	NaCl	KCl	
Bond dissociation energy	946	561	428	362	295	410	423	

Although most covalent bonds have dissociation energies greater than 100 kJ mol^{-1}, there are a few exceptions such as Na_2. However, Na_2 is a most uncommon molecule, and the covalent bonds in virtually all the molecules which exist in appreciable numbers at room temperatures can safely be classed as 'strong bonds'. Double and triple bonds are appreciably stronger than most single bonds, and this is well illustrated by the bond dissociation energies of the oxygen and nitrogen molecules.

QUESTIONS

1. What is the essential difference between a molecular orbital and an atomic orbital?
2. What energy condition must exist if a chemical bond is to form between two atoms as they are brought together?
3. Write a paragraph describing how it is that two hydrogen atoms attract each other, yet two helium atoms do not attract each other.
4. Suggest valence structures for the following diatomic molecules:

$$Na_2, Rb_2, Br_2, I_2, O_2, N_2.$$

5. Fluorine exists as a stable diatomic molecule F_2. Neon, however, shows no tendency to form diatomic molecules. Write a paragraph showing how the Pauli Principle helps us to understand why neon does not exist as Ne_2 molecules.
6. Suggest valence structures for the following entities:

$$Cl_2, K_2, HCl, HI, BrCl, OCl^-$$

7. Suggest valence structures for the following entities:

$$OH^-, SO, ICl, S_2, HS^-$$

8. It is frequently stated that helium does not form any stable chemical compounds. However, the ion HHe^+ has been observed in gas discharge tubes. Write a paragraph explaining why this species can exist, while the molecule HHe is unknown. Would you expect neon to form a molecule-ion HNe^+?
9. Suggest a valence structure for carbon monoxide (CO). Can you suggest more than one possible valence structure?
10. Suggest a valence structure for nitric oxide (hint—since this molecule has an odd number of electrons, there must be a half-filled orbital somewhere in the molecule).
11. The atomic number of selenium is 34:
    (a) write down the electronic structure of selenium;
    (b) to which group in the periodic table does selenium belong?
    (c) would you expect molecules to exist in vaporized selenium? Write valence structures for any molecules you suggest.
12. What is meant by the term electronegativity? What relation is there between electronegativity and polarity of bonds?
13. Indicate the probable trend in polarity of the following diatomic molecules: Cl_2, NaCl, ClF and HCl.
14. Would one expect the inner two-electron charge cloud surrounding the nucleus to be smaller or larger in the beryllium atom than in the lithium atom?
15. The first ionization energies for the elements in Group I of the periodic table are Li 520, Na 496, K 419, Rb 403, Cs 375 kJ mol^{-1}. Suggest an explanation for this variation.
16. The first ionization energies of the atoms of the second period of the periodic table increase regularly from lithium to neon. Suggest a reason why this is so.

9

Polyatomic Molecules

The formation of covalent bonds in polyatomic molecules is governed by the same principle which we have seen established for diatomic molecules. However, a new factor now arises: whereas there is only one possible shape for a diatomic molecule, there is always more than one possible shape for a polyatomic molecule. We have said a good deal about the way in which electron pairs are arranged about an atom. We must be careful not to confuse the arrangement of the electronic charge clouds in a molecule with the *shape* of the molecule. The shape of a molecule is defined by the way in which the *nuclei of the atoms* in the molecule are arranged in space, and the arrangement of the nuclei is a property of the molecule which has to be experimentally determined by the methods discussed in chapter 7. It must be realized that the shapes of *all* the molecules and ions discussed in this chapter have been experimentally determined. Our task will be to *interpret* these known shapes in terms of general theoretical principles, and, in particular, the arrangement of charge clouds in the molecule.

If we consider the water molecule H_2O, we could imagine two possible shapes, viz.:

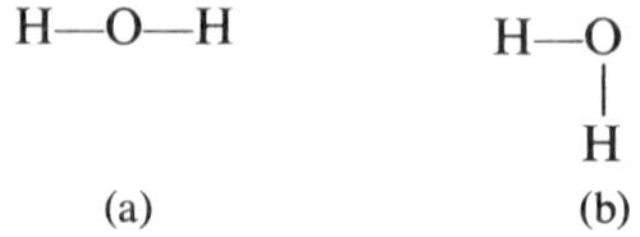

```
H—O—H          H—O
                 |
                 H
 (a)            (b)
```

Fig 9.1 Possible shapes for H_2O: (a) linear, (b) bent.

Experiment shows that the water molecule is bent, and in this chapter we will examine a theoretical principle which is successful in predicting the shapes of such simple molecules. This principle, already introduced in chapter 8, is the Sidgwick-Powell rule:

Charge clouds in the outer shell of an atom in a molecule stay as far away from each other as possible.

The use of this rule for predicting molecular shapes was fully developed by Gillespie and Nyholm in 1957.

9.1 METHANE, AMMONIA AND WATER

The outer shell of the second period atoms, carbon, nitrogen and oxygen, is an atomic L-shell into which no more than eight electrons can ever be fitted. This means that atoms of this period can never have more than four electron pairs in their outer shells, although they may have less than four pairs.

A carbon atom has the electronic configuration $1s^2.2s^2.2p^2$; it has four electrons in its outer shell. By accepting and sharing four electrons from four other atoms (one electron from each), a carbon atom will obtain eight electrons in its L-shell, and form four covalent bonds. This is the absolute maximum number of covalent bonds which carbon can form. Thus in a compound such as CH_4, there are four pairs of electrons in the outer shell of the carbon atom, and we have the charge cloud representation:

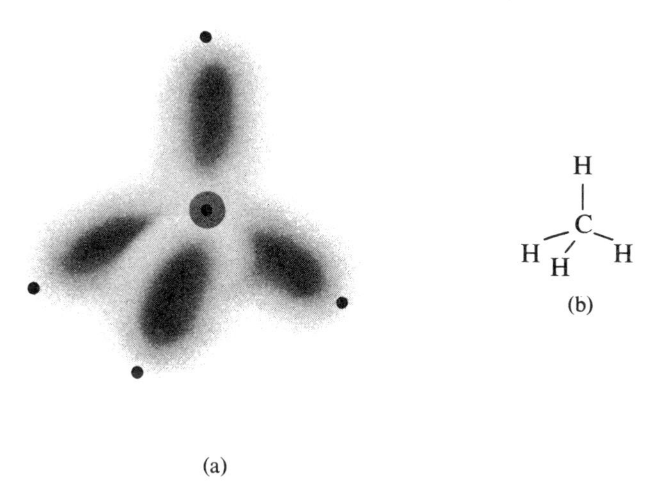

Fig 9.2 Representations of CH_4
(a) charge cloud, (b) valence structure.

In this molecule the four bonding charge clouds in the L-shell of carbon have taken up the expected tetrahedral arrangement so as to minimize the electrostatic potential energy. Notice that the inner or K-shell electrons remain unaffected by the formation of the four covalent bonds.

A nitrogen atom has the electronic configuration $1s^2.2s^2.2p^3$, and hence has five outer-shell electrons. By accepting and sharing three electrons from three other atoms (one electron from each), a nitrogen atom can acquire a filled L-shell, and simultaneously form three covalent bonds. There will thus be three pairs of bonding electrons around the nitrogen atom, and the remaining two electrons on the nitrogen atom will be a non-bonding pair. If these four charge clouds take up the expected tetrahedral arrangement around the nitrogen atom, we obtain the charge cloud picture of ammonia:

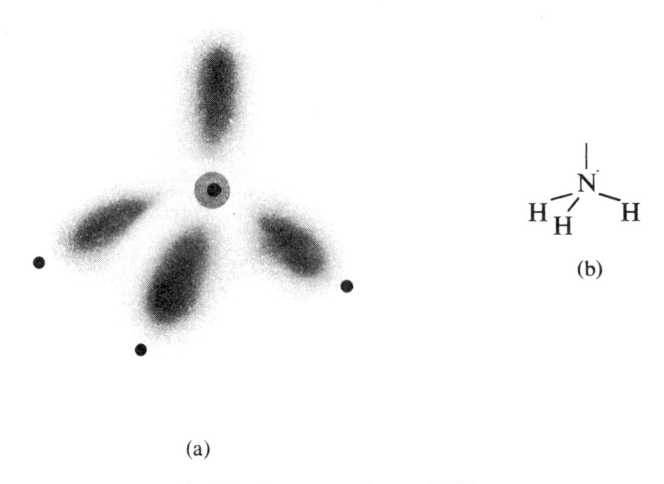

Fig 9.3 Representations of NH_3
(a) charge cloud, (b) valence structure.

The ammonia molecule has the expected pyramidal shape. In practice, although the arrangement of the charge clouds is almost tetrahedral, it is not exactly so since all four charge clouds are not identical. The non-bonding charge cloud seems to spread and take up a little more room than a bonding charge cloud, and $H\hat{N}H = 106° 45'$, which is slightly less than the tetrahedral value of methane in which $H\hat{C}H = 109° 28'$.

An oxygen atom has the electronic configuration $1s^2.2s^2.2p^4$. By accepting and sharing two electrons from two other atoms (one electron from each), the oxygen atom can form two single covalent bonds. It will thus have two bonding electron pairs and two non-bonding pairs in its L-shell. The essentially tetrahedral arrangement of the charge clouds around the oxygen atom leads to the resultant V-shaped arrangement of the nuclei. $H\hat{O}H = 104° 27'$, which is smaller than the ideal tetrahedral angle. As with ammonia, the non-bonding pairs take up a little more room than the bonding pairs.

We may conclude that the charge cloud repulsion hypothesis, which predicts an essentially tetrahedral arrangement of the four electron pairs in the L-shells of carbon, nitrogen and oxygen atoms in the molecules CH_4, NH_3 and OH_2, respectively, accounts well for the observed shapes of these

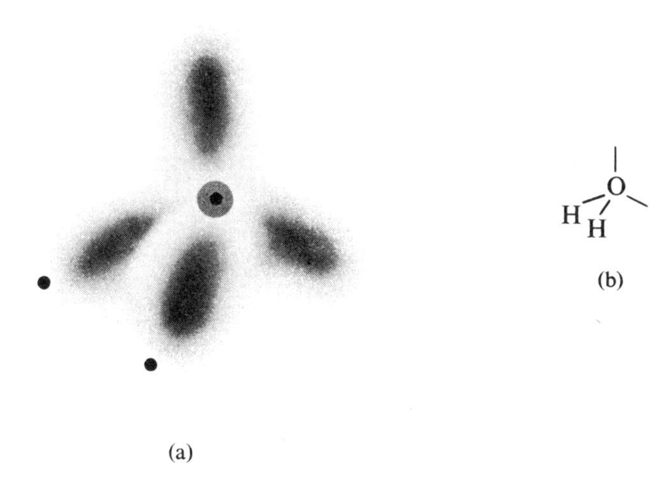

Fig 9.4 Representations of H_2O
(a) charge cloud, (b) valence structure.

molecules. One point worth noting about the three molecules is that they all contain exactly the same number of electrons, viz. ten; two K-shell electrons in the core of carbon, nitrogen or oxygen, and eight outer-shell electrons. Such a set of molecules is called an *isoelectronic set*. A fourth molecule, isoelectronic with methane, ammonia and water, is hydrogen fluoride. Valence structures for all four of these molecules are shown in Figure 9.5, so as to emphasize the common tetrahedral arrangement of electron pairs in the outer shell in each case.

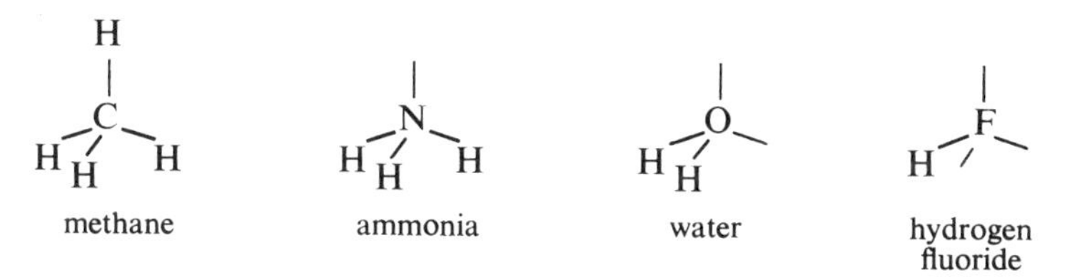

Fig 9.5 Valence structures for the isoelectronic set CH_4, NH_3, OH_2, FH.

9.2 THE AMMONIUM, BOROHYDRIDE AND HYDRONIUM IONS

Two species with the same electronic configuration as a carbon atom are the ions N^+ and B^-. Both have the configuration $1s^2.2s^2.2p^2$ with four outer-shell electrons. By analogy with carbon, we would expect these ions to be capable of forming four covalent bonds, and give rise to tetrahedral molecule-ions. The ions N^+ and B^- do not normally occur as isolated ions, but they can form four covalent bonds and hence give rise to the molecule-ions NH_4^+ and BH_4^-, with the expected tetrahedral arrangement of the bonding charge clouds around the central nucleus. Both ions are known

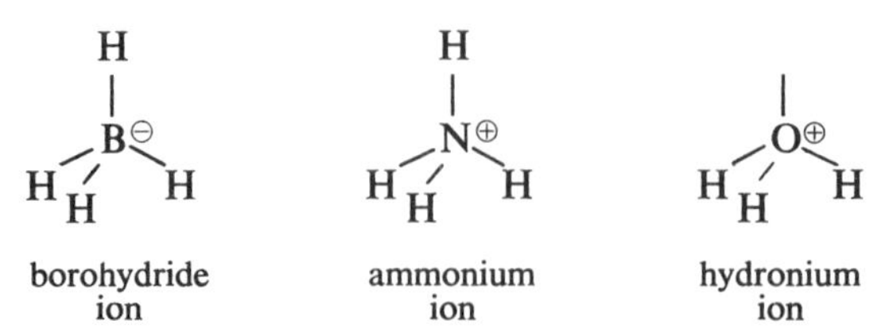

Fig 9.6 Valence structures for the isoelectronic set BH_4^-, NH_4^+, OH_3^+.

in ionic solids (e.g. NH_4Cl, ammonium chloride; and $LiBH_4$, lithium borohydride).

Another well known ion is the hydronium ion, H_3O^+. The species O^+ has the same electronic configuration as a nitrogen atom, viz. $1s^2.2s^2.2p^3$, and, like nitrogen, can form three covalent bonds. This ion has four charge clouds arranged tetrahedrally in the L-shell of oxygen, and is found to have the expected pyramidal shape. It occurs in ionic solids such as hydronium perchlorate ($H_3O.ClO_4$). Valence structures for the isoelectronic set BH_4^-, NH_4^+, OH_3^+ are shown in Figure 9.6.
Notice in the representations for these ions that we have indicated the charges as residing on the boron atom in BH_4^-, and the nitrogen atom in NH_4^+. This is necessary, since in order to form four covalent bonds an atom or ion must have four outer-shell electrons, hence boron must act as B^- in BH_4^- and nitrogen as N^+ in NH_4^+. Similarly, in order to form three covalent bonds, oxygen must have the electronic configuration of an uncharged nitrogen atom, i.e. O^+. Some ions and molecules which show the progressive change in the number of covalent bonds formed by an atom as its charge changes are shown in Figure 9.7. The charges on the atoms in these molecule-ions are frequently referred to as 'formal charges'.

9.3 BORON TRIFLUORIDE AND BERYLLIUM HYDRIDE

Boron, with the electron configuration $1s^2.2s^2.2p^1$, has only three electrons in its outer shell. We might expect it to form a hydride BH_3, but this molecule has never been isolated (see chapter 22). However, boron does form a trifluoride BF_3, in which each of the outer-shell electrons is shared with an L-shell electron of fluorine, giving three bonding pairs in the L-shell of boron. According to our repulsion hypothesis, the three charge clouds should arrange themselves at the corners of an equilateral triangle, since this represents the arrangement of lowest electrostatic energy for three charge clouds. In accord with expectation, BF_3 is a planar triangular molecule, with $F\hat{B}F = 120°$.

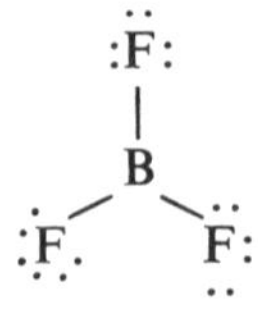

Beryllium has the electronic configuration $1s^2.2s^2$, and although BeH_2

Element	*Formal charge*	*Electronic configuration*	*Name of molecule or ion*	*Valence structure*	*No. of co-valent bonds*
Nitrogen	+1 (N^+)	$1s^2.2s^2.2p^2$	ammonium (NH_4^+)	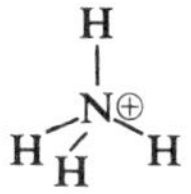 H above; H H H around N⊕	4
(Electronic structure of neutral atom $1s^2.2s^2.2p^3$)	0 (N)	$1s^2.2s^2.2p^3$	ammonia (NH_3)	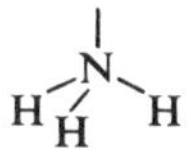 H H H around N	3
	−1 (N^-)	$1s^2.2s^2.2p^4$	amide (NH_2^-)	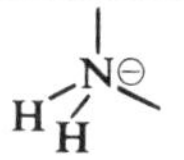 H H around N⊖	2
	−3 (N^{3-})	$1s^2.2s^2.2p^6$	nitride (N^{3-})	$:\ddot{N}:^{3-}$	0
Oxygen	+1 (O^+)	$1s^2.2s^2.2p^3$	hydronium (H_3O^+)	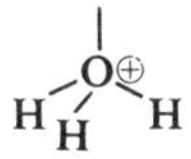 H H H around O⊕	3
(Electronic structure of neutral atom $1s^2.2s^2.2p^4$)	0 (O)	$1s^2.2s^2.2p^4$	water (H_2O)	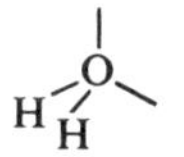 H H around O	2
	−1 (O^-)	$1s^2.2s^2.2p^5$	hydroxide (OH^-)	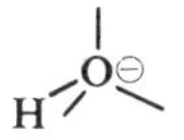 H around O⊖	1
	−2 (O^{2-})	$1s^2.2s^2.2p^6$	oxide (O^{2-})	$:\ddot{O}:^{2-}$	0

Fig 9.7 Covalent bonds formed by nitrogen and oxygen atoms with various formal charges.

is an ionic solid at room temperatures it may be vaporized at higher temperatures, and BeH_2 molecules can be detected in the gas phase. When a beryllium atom forms two covalent bonds there will be only two charge clouds in the L-shell of the beryllium atom. The arrangement of lowest electrostatic energy for these clouds will be linear:

$$H—Be—H$$

Beryllium hydride molecules are, in fact, linear.

9.4 SOME COMPOUNDS OF PHOSPHORUS AND SULPHUR

To date we have considered only atoms whose outer shells have been atomic K- or L-shells. Let us consider some atoms of the third period whose outer shells are M-shells. You will recall that we have been restricted to an upper limit of two electrons (or one pair) in an atomic K-shell, and eight electrons (or four pairs) in an atomic L-shell. The theoretical upper limit for an M-shell is eighteen electrons (or nine pairs). This upper limit of eighteen electrons has never been observed in the outer shell of an atom in a molecule, although atoms with outer shells which are atomic M-shells may certainly accommodate up to twelve electrons (or six pairs) therein.

Consider the phosphorus atom, which has an electronic configuration $1s^2.2s^2.2p^6.3s^2.3p^3$, i.e. a filled K- and L-shell, a filled 3s-subshell, and a p-subshell which is three electrons short of completion in the M-shell. There is also an empty d-subshell in the M-shell. By accepting and sharing three electrons, a phosphorus atom can form three single covalent bonds and give rise to a total of four electrons pairs in the M-shell, e.g. in the stable phosphine molecule, PH_3:

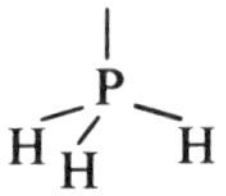

$$\mathrm{H\hat{P}H} = 93°\ 50'$$

In this molecule there are three bonding and one non-bonding pair in the M-shell of phosphorus, and the s- and p-subshells of the shell are filled. Sulphur, with electronic configuration $1s^2.2s^2.2p^6.3s^2.3p^4$, behaves analogously in forming H_2S:

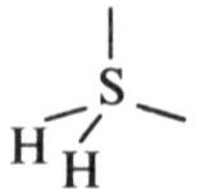

$$\mathrm{H\hat{S}H} = 92°\ 20'$$

except that in this case there are two bonding pairs and two non-bonding pairs in the M-shell. These two atoms are not, however, *limited* to four electron pairs in their outer shells, as were nitrogen and oxygen. Thus phosphorus forms, as well as the expected trifluoride PF_3, a pentafluoride PF_5 in which all five electrons of the phosphorus M-shell appear to be shared in the formation of five covalent bonds. This means that in the PF_5 molecules there are ten electrons (or five electron pairs) in the phosphorus M-shell. Now when five charge clouds are arranged around a central nucleus, the arrangement of minimum electrostatic energy is that of a trigonal bipyramid; three charge clouds in a triangular plane, and one charge cloud above and one below the plane, as shown in Figure 9.8.

Encouragingly, the structure of PF_5 is a trigonal bipyramid:

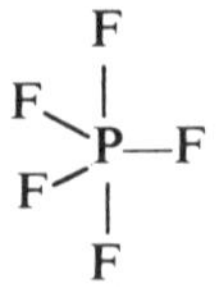

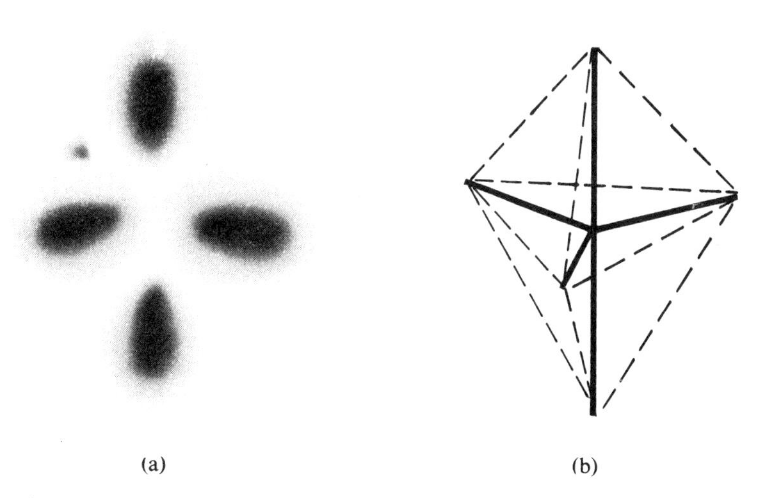

Fig 9.8 (a) Five charge clouds arranged in a trigonal bipyramid, (b) a trigonal bipyramid.

A related case is exemplified by the SF_6 molecule, in which all six M-shell electrons of sulphur are shared in forming six covalent bonds. This means that there is a total of twelve electrons (or six pairs) in the sulphur M-shell. The most stable arrangement of six identical charge clouds around a central nucleus is that of a regular octahedron; four pairs in a plane, and one pair above and one pair below the plane, as shown in Figure 9.9. As expected, the SF_6 molecule has the shape of a regular octahedron:

```
     F
 F\  |  /F
    S
 F/  |  \F
     F
```

9.5 SOME GENERAL RULES FOR COVALENT BONDING

The discussions of the previous sections have established that our hypothesis concerning the mutual repulsion of charge clouds appears to be useful for predicting the shapes of simple polyatomic molecules containing single bonds. It has become apparent that the actual number of electron pairs which we can get into the outer shell of an atom depends upon which atomic shell (K, L, M, etc.) forms the outer shell of the atom in question. Thus, atoms with an atomic L-shell for an outer shell can have no more than four electron pairs in that shell. Atoms with an atomic M-shell for an outer shell may, in principle, have up to nine electron pairs therein, although in practice this number has never been observed. In chemical compounds of such atoms, experience shows us that four, five or six electron pairs are quite commonly observed in the outer shells. The situation is similar for atoms with outer

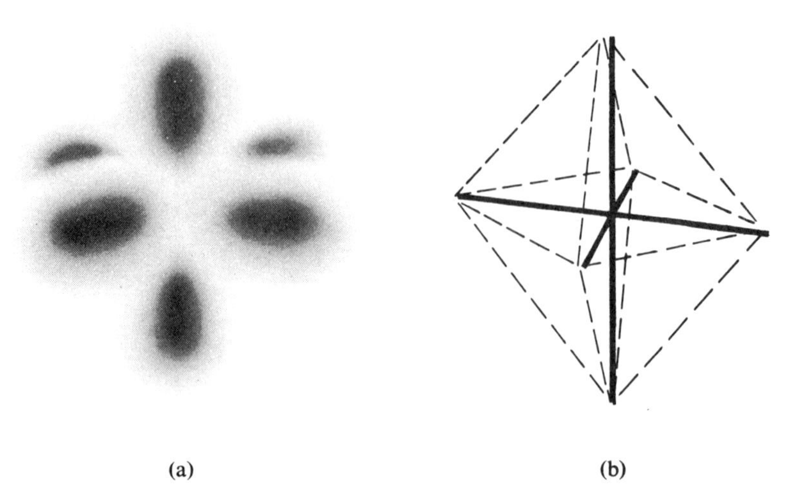

Fig 9.9 (a) Six charge clouds arranged in an octahedron, (b) an octahedron.

N- and O-shells; again four, five and six pairs are fairly commonly observed in these outer shells, although seven and eight pairs also occasionally occur.

The fairly common occurrence of four electron pairs in the outer shells of atoms in molecules, gave rise to the original *octet rule* of G. N. Lewis. Lewis suggested that the number of covalent single bonds formed by an atom could be worked out by assuming that atoms shared electrons in such a way that they obtained just eight electrons in their outer shells, hydrogen (two electrons) being the one exception to the rule. We now know that this rule is obeyed strictly only by atoms whose outer shell is an L-shell. Atoms with outer M-, N- and O-shells do in fact have just eight outer-shell electrons in many of their compounds, but they are not rigidly bound to an octet rule; such atoms quite commonly have ten, twelve and even fourteen electrons in their outer shells in a great number of their compounds. Furthermore, even for the second period

Table 9.1

Outer shell	*Atoms with this outer shell*	*Theoretical maximum no. of electron pairs in shell*	*Maximum observed no. of electron pairs in shell*	*Commonly observed no. of electron pairs in shell*
K	1st Period (H, He)	1	1	1
L	2nd Period (Li → Ne)	4	4	4
M	3rd Period (Na → Ar)	9	7	4, 5, 6
N	4th Period (K → Kr)	16	8	4, 5, 6
O	5th Period (Rb → Xe)	25	8	4, 5, 6

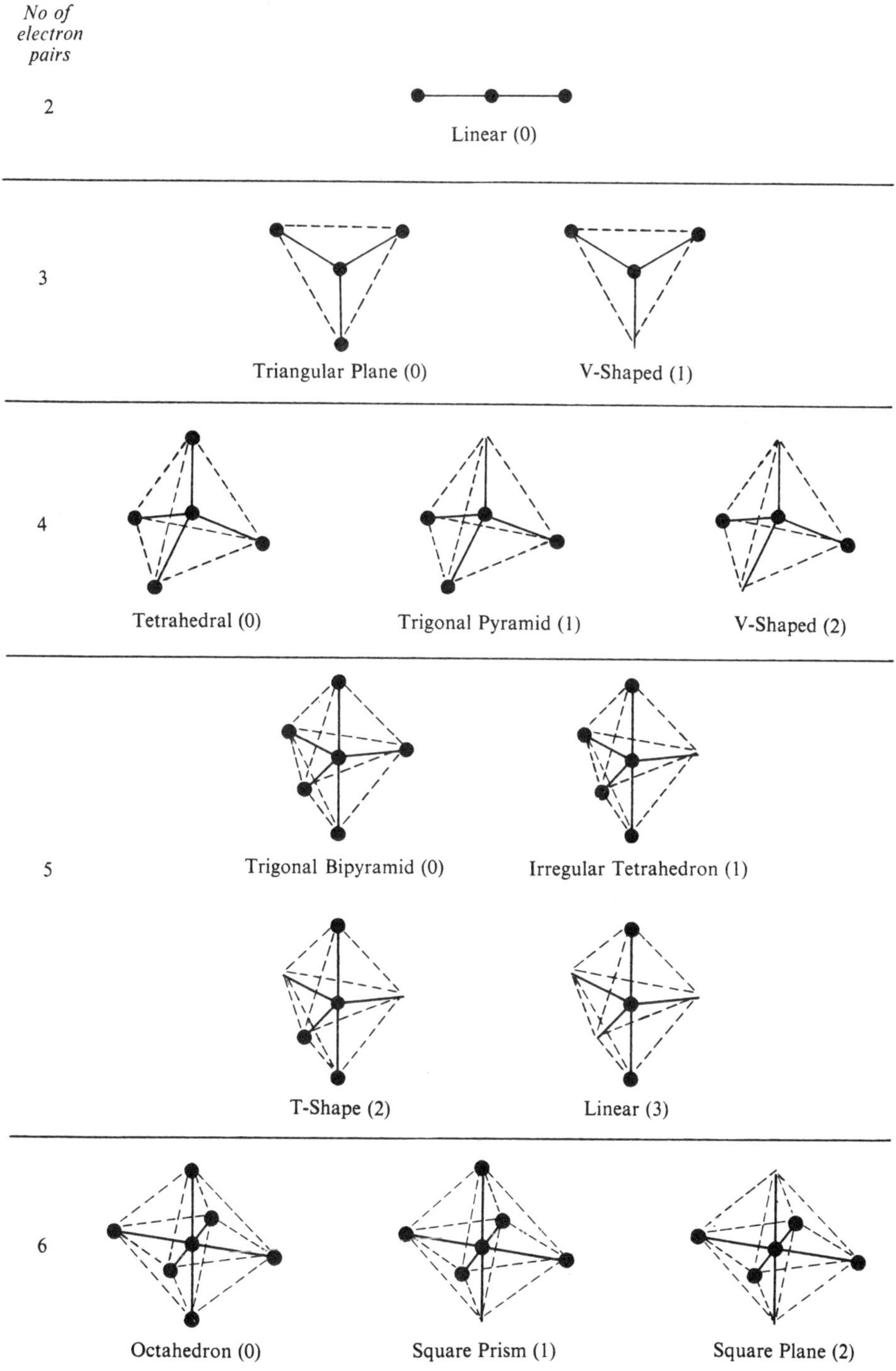

No of electron pairs

2

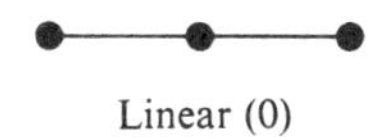

Linear (0)

3

Triangular Plane (0) V-Shaped (1)

4

Tetrahedral (0) Trigonal Pyramid (1) V-Shaped (2)

5

Trigonal Bipyramid (0) Irregular Tetrahedron (1)

T-Shape (2) Linear (3)

6

Octahedron (0) Square Prism (1) Square Plane (2)

Fig 9.10 Shapes of molecules (non-transitional elements). (Number of lone pairs shown after description of shape.)

atoms with outer L-shells, molecules of these atoms exist (Li_2, BeH_2, BF_3) in which there are *fewer* than eight electrons in the atomic outer shell, so that the 'octet rule', even for these atoms, must be seen as fixing an *upper limit* rather than an *absolute value* to the number of electrons in the atomic outer shell. The foregoing discussion is summarized in Table 9.1. The data in the two right-hand columns are based on experimental evidence, and some of the values given are doubtful since they depend upon an interpretation of the electronic structure of a molecule, based on its observed physical properties, including its shape. The reason that the maximum theoretical number of outer-shell electrons is never reached for the M-, N- and O-shells is almost certainly because of the size limitation on the number of atoms which can be fitted around some central atom. Thus a chlorine atom, with an outer M-shell, could never form the maximum possible number of nine covalent bonds, since it would be geometrically impossible to fit nine atoms around a chlorine atom, and make nine strong covalent bonds at the same time. Atoms with outer N- and O-shells are generally somewhat larger in size, and occasionally accommodate as many as eight atoms around themselves in forming covalent bonds. Even these large atoms do this only rarely, and compounds in which atoms form more than six covalent bonds are not very common.

9.6 SHAPES OF POLYATOMIC MOLECULES

The methods developed in this chapter for the prediction of molecular shapes may now be generalized. In Figure 9.10 we show the various shapes which may be derived from molecules containing central atoms with two, three, four, five and six electron pairs.

In each case, the charge clouds due to the electron pairs take up the arrangement of lowest electrostatic potential energy by getting as far away from each other as they can. In some cases, the shape of a molecule is not predicted unambiguously by our simple electron-pair repulsion hypothesis. Consider the ClF_3 molecule. There are seven electrons in the M-shell of chlorine, and by forming three electron pair bonds, one with each of three fluorine atoms, the number of electrons in the chlorine M-shell becomes ten, or five pairs. These five pairs take up the expected trigonal bipyramidal arrangement, and the shape of the molecule is determined by the way in which the three fluorine atoms are arranged around the chlorine atom. If we were free to choose *any* three of the five pairs around chlorine for bonding pairs, we would have three possible shapes for the ClF_3 molecule:

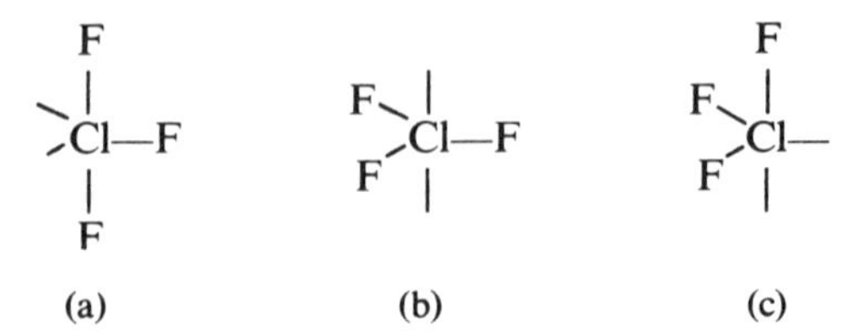

Fig 9.11 Possible valence structures for ClF_3: (a) T-shape, (b) plane, (c) distorted pyramid.

In fact (a) represents the actual shape of the molecule.

A similar ambiguity crops up in molecules with six electron pairs around a central atom. Consider the ICl_4^- anion. This anion may be considered as arising when the I^- ions form four covalent bonds, one with each of four chlorine atoms. The I^- ion has eight electrons in the iodine O-shell. When I^- forms four single covalent bonds, the number of electrons in the O-shell becomes twelve (six pairs), and these six electron pairs arrange themselves octahedrally around the iodine. Since only four of these pairs will be bonding pairs, there are two different possible ways of arranging the four chlorine atoms around the I^- ions, viz.:

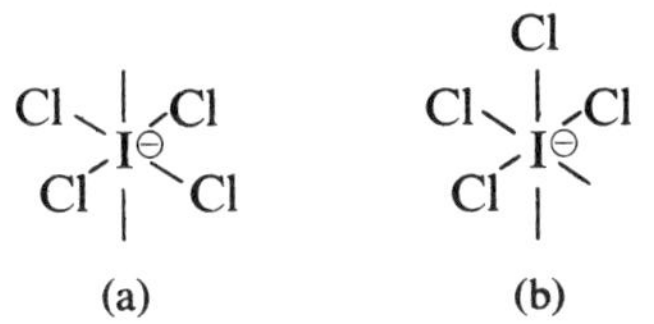

(a) (b)

Fig 9.12 Possible valence structures for ICl_4^-: (a) plane, (b) distorted tetrahedron.

Structure (a) represents the actual structure of this ion.

Table 9.2

The arrangements of electron pairs in valence shells and the shapes of molecules and ions (non-transitional elements)

Total no. of electron pairs	*Arrangement of electron pairs*	*No. of bonding pairs*	*No. of non-bonding (lone) pairs*	*Shape of molecule*	*Examples*
2	Linear	2	0	Linear	$HgCl_2$ (gas), $Ag(CN)_2^-$
3	Triangular plane	3	0	Triangular plane	BCl_3
		2	1	V-shape	$SnCl_2$ (gas)
4	Tetrahedron	4	0	Tetrahedron	CH_4, PCl_4^+
		3	1	Trigonal pyramid	NH_3
		2	2	V-shape	H_2O, F_2O
5	Trigonal bipyramid	5	0	Trigonal bipyramid	PCl_5 (gas)
		4	1	*Irregular tetrahedron	$TeCl_4$
		3	2	*T-shape	ClF_3
		2	3	*Linear	ICl_2^-, I_3^-
6	Octahedron	6	0	Octahedron	SF_6, PCl_6^-
		5	1	Square pyramid	IF_5
		4	2	*Square	ICl_4^-

In both the ambiguous cases quoted above, it is possible to correctly predict the actual shapes of the molecules by appealing to the secondary principle that repulsion is greater between two lone pairs than between a lone pair and a bonding pair, which in turn is greater than the repulsion between two bonding pairs. The application of this additional principle can be rather difficult, however, and we will not pursue the topic further. Some detailed examples of the application of our electron-pair repulsion hypothesis are given in Table 9.2. The cases marked with an asterisk are those where prediction of the actual molecular shapes depends upon the application of the secondary principle relating to the relative magnitude of the repulsions between bonding and non-bonding pairs.

9.7 ETHANE, HYDRAZINE AND HYDROGEN PEROXIDE

The molecules examined to date have consisted essentially of a central atom bonded to a number of other atoms, and the shapes of the molecules have been determined by the number of electron pairs in the outer electronic shell of the central atom. We shall now consider some molecules which are slightly more complex. Ethane has the formula C_2H_6, and the observed shape of the molecule is consistent with the existence of four tetrahedrally-arranged bonding pairs in the L-shell of each carbon atom. So we have the charge cloud diagram as shown in Figure 9.13. Hydrazine (N_2H_4) is a molecule with an almost identical electron distribution. The charge cloud

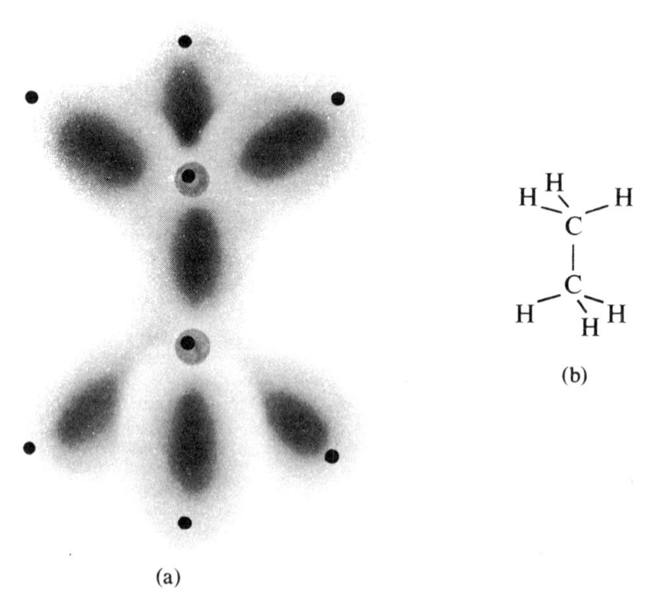

Fig 9.13 Representations of C_2H_6
(a) charge cloud, (b) valence structure.

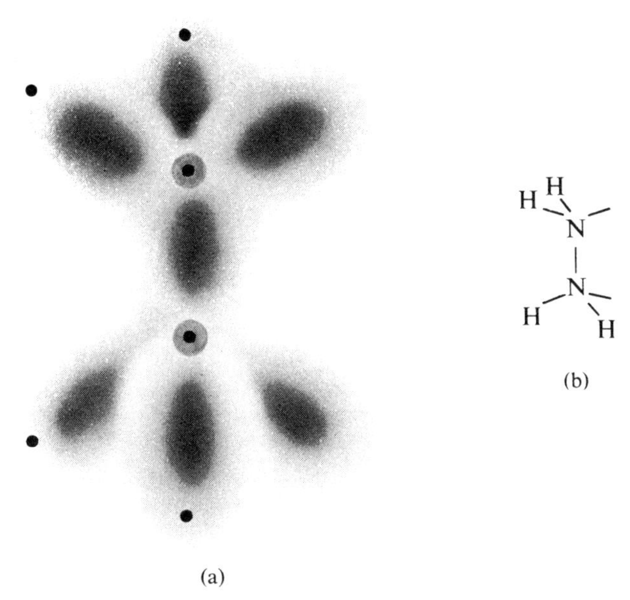

Fig 9.14 Representations of N_2H_4
(a) charge cloud, (b) valence structure.

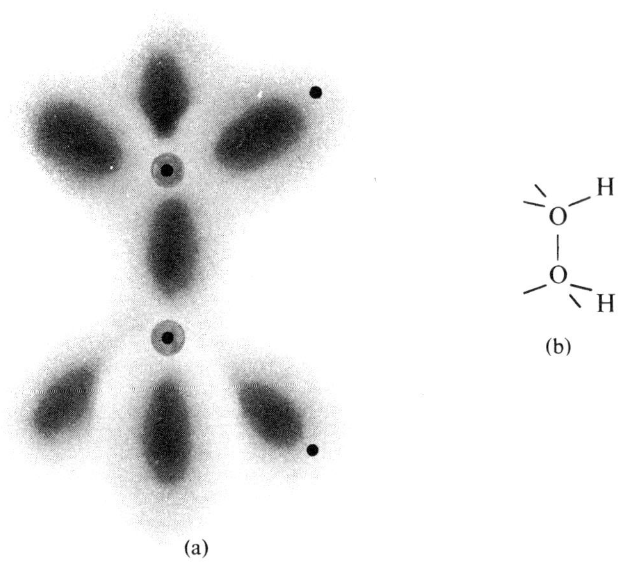

Fig 9.15 Representations of H_2O_2
(a) charge cloud, (b) valence structure.

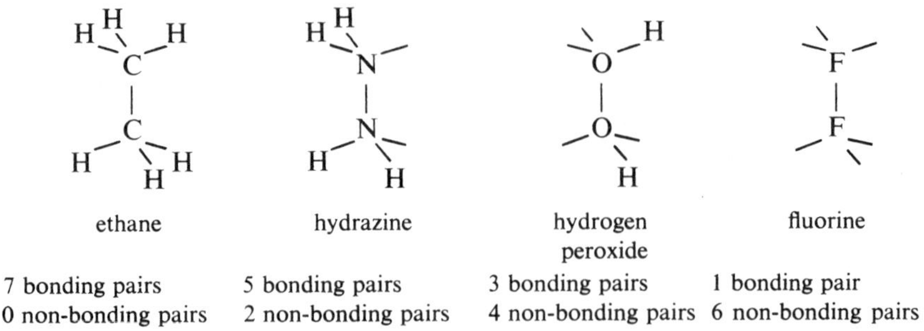

ethane	hydrazine	hydrogen peroxide	fluorine
7 bonding pairs	5 bonding pairs	3 bonding pairs	1 bonding pair
0 non-bonding pairs	2 non-bonding pairs	4 non-bonding pairs	6 non-bonding pairs

Fig 9.16 Valence structures for the isoelectronic set C_2H_6, N_2H_4, O_2H_2, F_2.

diagram for hydrazine is shown in Figure 9.14 with a tetrahedral arrangement of charge clouds around each nitrogen atom. Finally, hydrogen peroxide (H_2O_2) also has this same electron distribution with four charge clouds arranged tetrahedrally around each oxygen atom as shown in Figure 9.15. The relationships between the electronic structures of the three molecules discussed in this section can be seen in the isoelectronic series in Figure 9.16.

9.8 POLYATOMIC MOLECULES AND IONS WITH MULTIPLE BONDS

The prediction of the shapes of polyatomic molecules with multiple bonds follows similar lines to those developed in the earlier sections of the chapter.

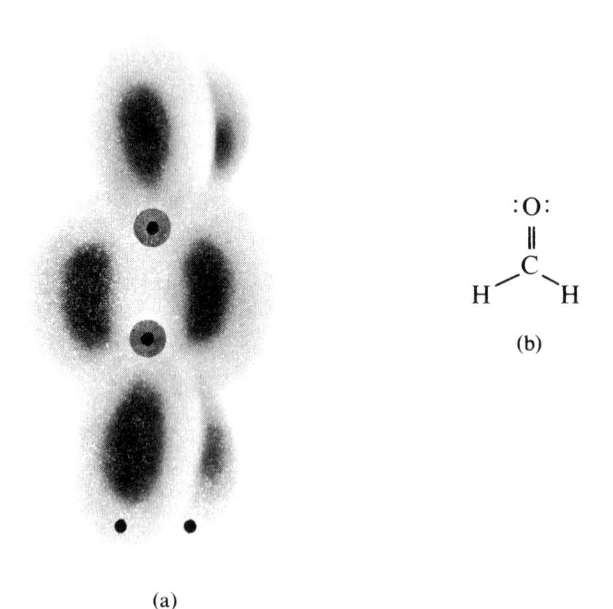

Fig 9.17 Representations of H.CHO
(a) charge cloud, (b) valence structure.

One of the simplest polyatomic molecules with a double bond is formaldehyde H.CHO. The molecule is planar, a shape which is consistent with the tetrahedral arrangement of charge clouds around the carbon and oxygen atoms; the electron distribution in formaldehyde is almost identical to that in another molecule with a double bond, ethylene, as shown in Figures 9.17 and 9.18. This molecule is also planar, and has a tetrahedral distribution of charge clouds about each carbon atom. Other molecules with multiple bonds and tetrahedral charge cloud arrangements are acetylene, a linear molecule with a triple bond, and carbon dioxide, a linear molecule with two double bonds.

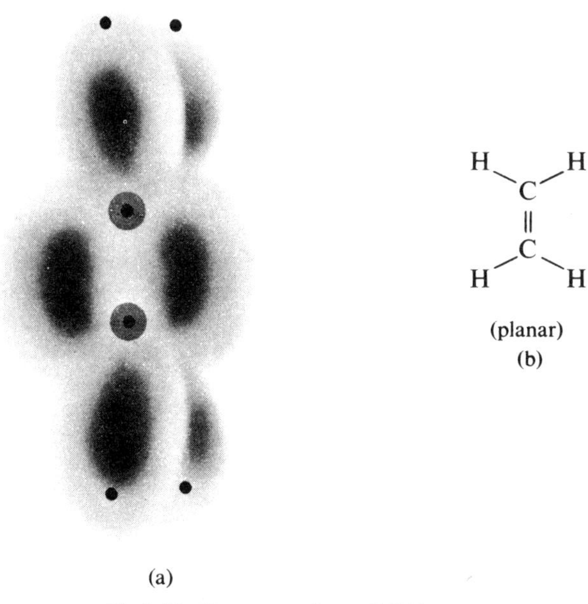

Fig 9.18 Representations of C_2H_4
(a) charge cloud, (b) valence structure.

$H—C\equiv C—H$

(lincar)
(a)

$\ddot{O}=C=\ddot{O}$

(linear)
(b)

Fig 9.19 Representations of
(a) acetylene (C_2H_2), (b) CO_2.

We can also satisfactorily predict the shapes of small molecule-ions such as carbonate. Valence structures for the isoelectronic carbonate and nitrate ions are shown in Figure 9.20. Both ions are planar.

planar (a)   planar (b)

Fig 9.20 Valence structures for (a) NO_3^-, (b) CO_3^{2-}.

In the valence structure for the carbonate ion, there are two single bonds and one double bond to the central carbon atom. The two single bonds are between the carbon atom and O^- ions, which, as we saw in section 8.9, have the same electronic configuration as the fluorine atom, and can form one covalent bond. In the valence structures of the nitrate ion, the nitrogen acts as the N^+ ion, which has the same electronic structure as a carbon atom and can form four covalent bonds. N^+ forms one double bond to an oxygen atom, and two single bonds, each to an O^- ion. As noted previously, the O^- ion can form one covalent bond. The nett charge on the ion is the algebraic sum of the formal charges on the atoms and is $(-1-1+1) = -1$.

The molecules and ions just discussed have multiple bonds formed by atoms possessing a tetrahedral arrangement of four charge clouds, and all the molecules have shapes in accord with our charge cloud repulsion hypothesis. It is also possible to successfully predict the shapes of molecules containing multiple-bonded atoms with five, six or seven charge clouds in their outer shells. In these cases our task of predicting shapes is simplified if we assume that we can treat a four-electron bond, for repulsion purposes, in the same way as a two-electron bond or a lone pair, i.e. we shall assume:

Bonding pairs, bonding 'quartets', and lone pairs all get as far away from each other as possible in an atomic outer shell.

Let us apply this principle to the molecules, sulphur dioxide and sulphur trioxide, which are shown in Figure 9.21. The predictions of our subsidiary hypothesis are satisfactory in these two cases, and are also satisfactory for the sulphite and sulphate ions:

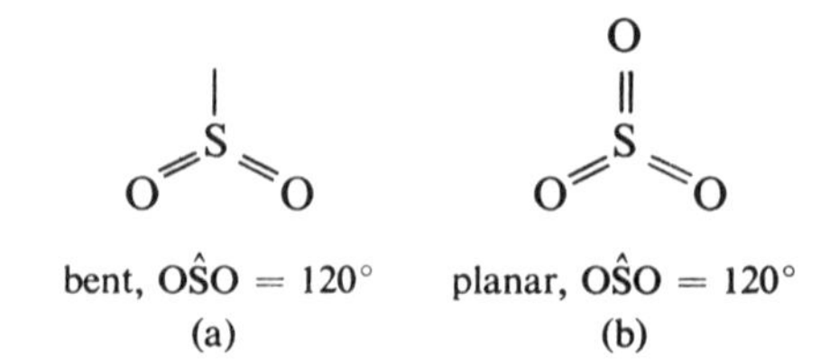

Fig 9.21 Valence structures for (a) SO_2, (b) SO_3.

9.9 RESONANCE

The valence structures written for some of the polyatomic ions in section

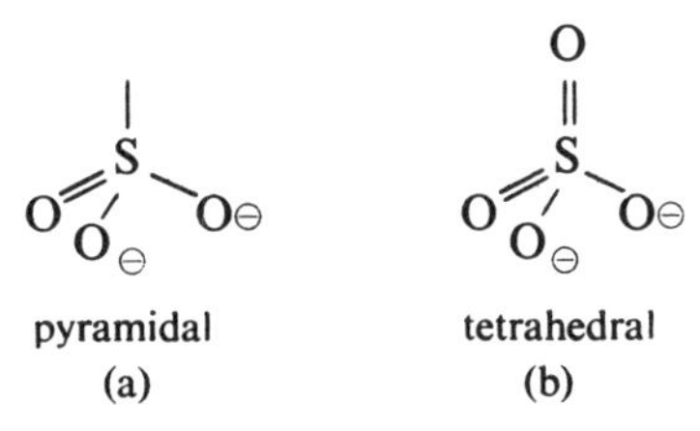

Fig 9.22 Valence structures for (a) SO_3^{2-}, (b) SO_4^{2-}.

9.8 are ambiguous. Thus, consider the carbonate anion:

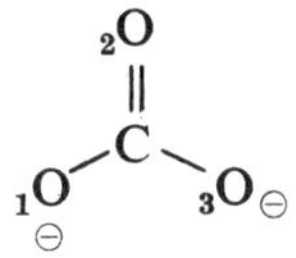

Fig 9.23 Valence structure for CO_3^{2-}.

where we have numbered the oxygen atoms. One of the oxygen atoms is involved in the double bond, while the other two have negative charges, and have formed only single bonds. Equally good structures for this ion would be:

Fig 9.24 Alternative valence structures for CO_3^{2-}.

In cases where such an ambiguity arises, the actual electronic structure of the ion (or molecule) turns out to be a combination, or 'average' of all such possible structures. Structural investigations of the carbonate ion have shown that all three carbon-oxygen bonds are identical, and the 'true' state of the ion is said to be a *resonance hybrid* of three equivalent structures:

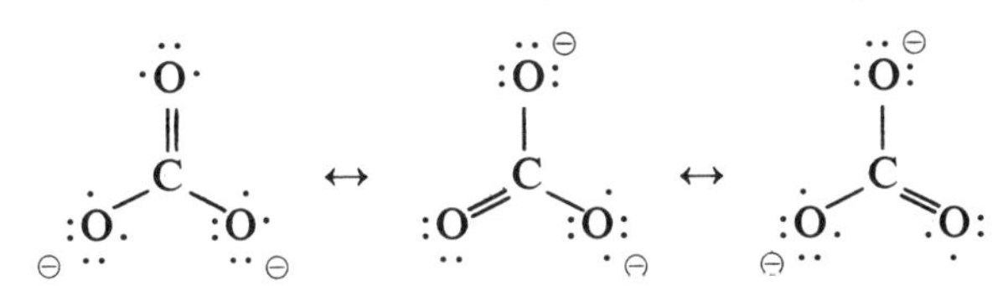

Fig 9.25 Resonance between equivalent canonical forms in CO_3^{2-}.

where any one of the three equivalent forms is called a *canonical form*, or *canonical structure*. It is most important to realize that these canonical forms *do not* exist in equilibrium with each other. A carbonate ion *never* has one double and two single bonds, but rather has three identical bonds which are intermediate in character between single and double. Furthermore, in this case, each oxygen will bear just two-thirds of a negative charge.

A similar situation exists for the planar nitrate anion, which has three equivalent canonical forms:

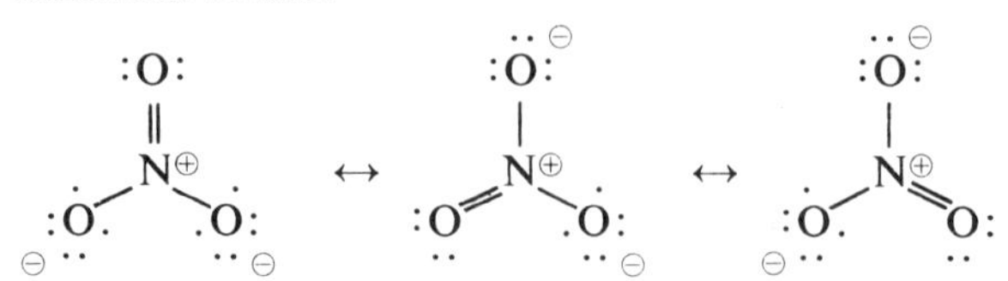

Fig 9.26 Resonance between equivalent canonical forms in NO_3^-.

The sulphate anion has six equivalent canonical forms:

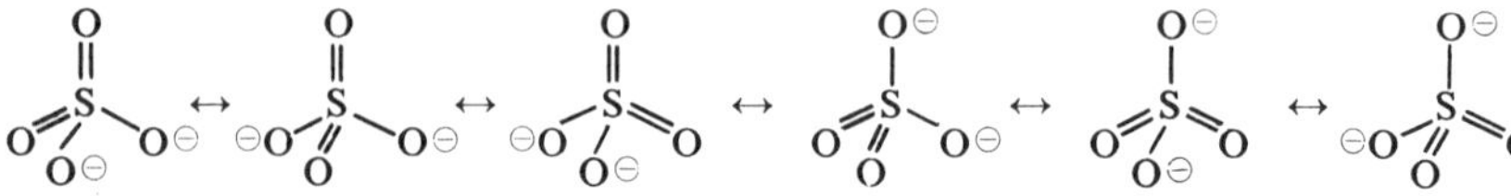

Fig 9.27 Resonance between equivalent canonical forms in SO_4^{2-}.

In accord with expectations, the sulphate anion is exactly tetrahedral and all the sulphur-oxygen bonds are identical.

Molecules, as well as ions, exhibit this phenomenon of resonance. Thus nitric acid can be written in two equivalent canonical forms:

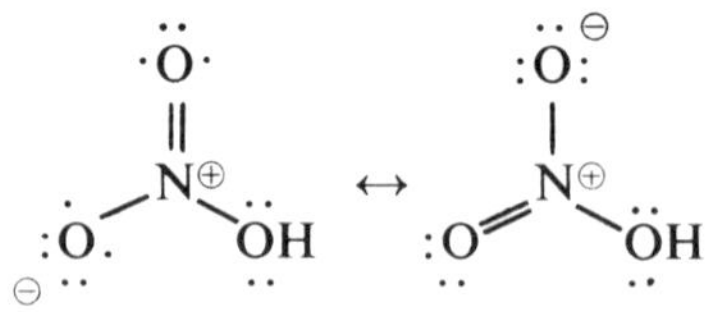

Fig 9.28 Resonance between equivalent canonical forms in HNO_3.

Ozone is of interest since, although structurally similar to sulphur dioxide, the central oxygen is required to obey an octet rule, hence we have two equivalent canonical forms:

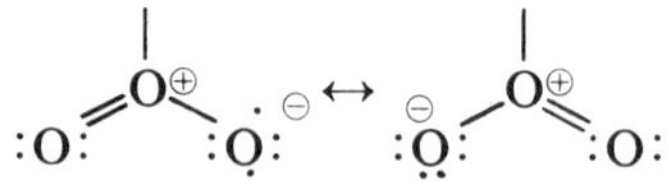

Fig 9.29 Resonance between equivalent canonical forms in O_3.

Each canonical form illustrates well the bent nature of the molecule, and the two oxygen-oxygen bonds are identical.

While it is helpful to recognize the possibility of resonance occurring, chemists still commonly write conventional valence structures for molecules and ions. Thus it is always 'correct' to write:

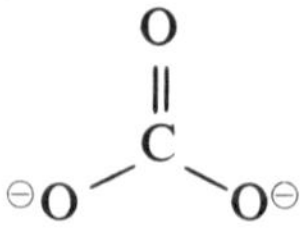

for the carbonate ion although the experienced chemist, on seeing such a structure, recognizes the possibility of resonance between equivalent canonical forms. Further development of this topic is beyond the scope of this book and in the text we will continue to write conventional valence structures.

QUESTIONS

1. The following sets of formulae represent structures which contain the same number of electrons: (a) BH_4^-; H_3O^+; (b) AlF_6^{3-}; SiF_6^{2-}; PF_6^-; SF_6. Show by writing valence structures that the entities in each set have a common stereochemistry.
2. Helium and neon do not form compounds with fluorine, but xenon, krypton and radon do. Suggest a reason for this difference in behaviour.
3. Phosphorus forms a compound phosphorus pentachloride (PCl_5), but nitrogen forms no corresponding compound. Account for this difference.
4. (a) Write down electronic structures for each of the atoms: H; Be; B; C; N; O; F; Mg; Al; Si; P; S; Cl.
   (b) Write valence structures for the following molecules: $BeCl_2$(g); MgF_2(g); NCl_3; BCl_3; H_2S; BrCl; AlH_3(g); HBr; $CHCl_3$; CH_2Cl_2; SiF_4.
5. (a) Write down electronic structures for each of the atoms: Se; As; I; Xe.
   (b) Write valence structures for the following molecules: H_2Se; AsH_3; HOCl; Cl_2O; CH_3Cl; BF_3; NF_3; SCl_4; ICl_3; XeF_4; ICl_2^+; ICl_2^-.
   In the cases when more than one geometrical arrangement is possible, give the alternative structures.
6. Write down the valence structures of gaseous boron trichloride (BCl_3) and of ammonia. Ammonia and boron trichloride react to form a compound BCl_3. NH_3. Suggest a valence structure for this entity.
7. Suggest valence structures for the following molecules and ions: CS_2; NCO^-; NOCl; $POCl_3$; NH_2^-; NO_2^+; SbF_5(g); $SOCl_2$; $COCl_2$.
8. Show by writing valence structures how it is that the carbonate and nitrate ions are planar, whilst the chlorate ion (ClO_3^-) is tetrahedral.
9. Show by writing valence structures that each of the following contains either double or triple bonds: CO_2; HCN; C_2N_2; N_2O_3.
10. In the phosphate ion (PO_4^{3-}) the phosphorus atom forms five covalent bonds. Write down one valence structure for the phosphate ion, showing its stereochemistry. From this structure suggest one possible valence structure for the phosphite ion PO_3^{3-}, and from this suggest its stereochemistry.
11. In selenic acid (H_2SeO_4) the selenium atom forms six covalent bonds. Draw one valence structure and suggest the stereochemistry.
12. Give a valence structure for the nitrite ion (NO_2^-).
13. The ozone molecule (O_3) contains three oxygen atoms arranged in a bent chain. The two bonds in the molecule are found to be equivalent. Describe in a short paragraph how the concept of resonance accounts for this observation.
14. Since chlorine may have more than four electron pairs in its outer shell we might represent the chlorine molecule as $\ddot{\underset{..}{Cl}}\equiv\ddot{\underset{..}{Cl}}$ What experimental evidence can be produced to favour the preferred structure $:\ddot{\underset{..}{Cl}}-\ddot{\underset{..}{Cl}}:$?

10

Bonding between Molecules

Structural investigations show that many liquids and solids consist of discrete molecules, held together by comparatively weak intermolecular forces. We have already stated that all forces between molecules are electrostatic in origin and we must ask how it is that molecules, which are electrically neutral, can attract one another. The answer lies in the properties of electric dipoles.

10.1 ELECTRIC DIPOLES

An electric dipole consists of two equal and opposite electric charges, separated by a definite distance, and may be represented:

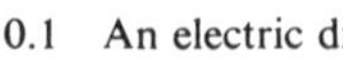

Fig 10.1 An electric dipole.

Forces exist between two electric dipoles which are due to the electrostatic attraction of oppositely-charged ends of the dipoles. Thus two dipoles attract or repel each other depending on the way they are mutually oriented as shown in Figure 10.2. Thus two dipoles will tend to arrange themselves in configurations like (c) and (d) in which nett attractive forces outweigh nett repulsive forces.

Dipoles may also be attracted by charged particles. In practice, dipoles in the vicinity of charged particles tend to orient themselves so that the end

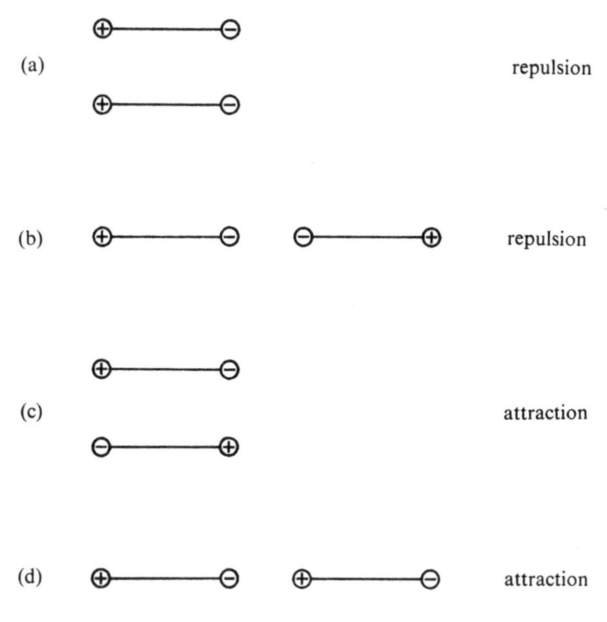

Fig 10.2 Various configurations for two dipoles.

of the dipole which has a charge of opposite sign from that of the particle comes close to the particle. This means, in effect, that charges first orient and then attract electric dipoles, as shown in Figure 10.3.

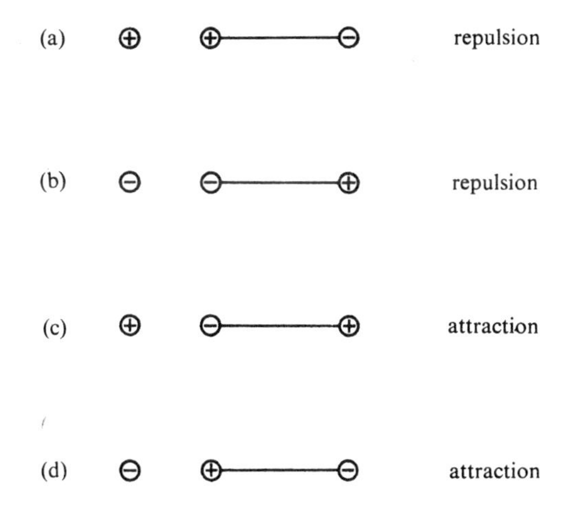

Fig 10.3 Various configurations for an electric charge and a dipole.

Thus the system (charge + dipole) will tend to arrange itself in configurations like (c) or (d), in which the opposite charges are adjacent, since these

configurations have a lower potential energy than (a) or (b). For configurations like (c) and (d) it can readily be shown that, for a given dipole, the potential energy of the system becomes lower as:
(i) the magnitude of the electric charge is increased;
(ii) the dipole is moved nearer the charge.
We noted in chapter 9 that many molecules were dipolar, e.g. HF and HCl could be represented as:

$$\overset{\delta+}{\mathrm{H}}\text{—}\overset{\delta-}{\mathrm{Cl}} \qquad \overset{\delta+}{\mathrm{H}}\text{—}\overset{\delta-}{\mathrm{F}}$$

In these molecules the electronegativity difference between the atoms has induced a permanent charge separation. Such molecules are said to be *permanent dipoles*, since they have a dipolar character which does not change with time. It is apparent that dipolar molecules will be able to attract each other, and will be attracted to ions.

10.2 DISPERSION FORCES

The electrons surrounding the nucleus of any free atom, on the time average, are distributed in a spherically symmetrical fashion, i.e. no free atom is a permanent dipole. Similarly, many simple symmetrical molecules such as oxygen (O_2) and methane (CH_4) have no permanent dipolar character; indeed no homonuclear diatomic molecule has any permanent dipolar character. Yet despite the lack of permanent dipolar character, atoms such as argon, and molecules such as oxygen and methane, form into solid lattices at low temperatures, indicating the presence of weak but definite forces of attraction between the atoms or molecules. It should be realized that there is no possibility of covalent bond formation between the entities just mentioned (argon atoms, oxygen molecules, methane molecules), so that a new form of bonding appears to operate in the solid lattices formed by these materials.

The forces involved in this weak bonding are called dispersion forces.

The simplest example of the operation of dispersion forces is given by the bonding of the noble gas atoms into crystal lattices. The strength of the bonding is indicated in Table 10.1.

Table 10.1

	M.P.(K)	*B.P.(K)*	ΔH_{sub} *(kJ mol⁻¹)*
He	—	4.21	0.082
Ne	24.5	27.17	1.75
Ar	83.8	87.92	6.64
Kr	105.9	120.9	8.92
Xe	161.3	165.1	12.64
Rn	202	208	—

where ΔH_{sub} is the heat content change in the sublimation process:

$$Y(s) \rightarrow Y(g)$$

where Y is a noble gas atom. It is useful to recognize that the boiling point of the liquid in these cases gives a good qualitative indication of the strength of the bonding between the inert gas atoms; the boiling point rises steadily as ΔH_{sub} rises. Notice that the strength of the bonding between the atoms increases with increasing atomic weight. As another example, consider the halogens:

Table 10.2

	M.P.(°C)	*B.P.(°C)*	ΔH_{fus} *(kJ mol⁻¹)*	ΔH_{sub} *(kJ mol⁻¹)*
F_2	−219.6	−187.9	0.51	7.05
Cl_2	−101	−34.05	6.43	26.8
Br_2	−7.2	58.2	10.5	40.5
I_2	113.6	184.5	15.7	57.5

where ΔH_{fus} and ΔH_{sub} are the heat content changes in the respective processes:

$$X_2(s) \rightarrow X_2(l) \qquad (\Delta H_{fus})$$

$$X_2(s) \rightarrow X_2(g) \qquad (\Delta H_{sub})$$

X_2 being a halogen molecule. Notice the increased strength of the bonding as the molecular weight of the halogen increases.

As implied in chapter 7, dispersion forces are fundamentally electrostatic in nature. They are directly due to the motions of electrons within atoms and molecules. At any instant of time an atom, e.g. a hydrogen atom, will have a particular configuration of nucleus and electron, say:

$$e^- \qquad \oplus$$

Fig 10.4 Instantaneous 'photograph' of a hydrogen atom.

Thus the atom will be instantaneously a dipole. Since the electron is continuously moving around the nucleus, the orientation of the dipole will be rapidly changing with time, and for all atoms there is no nett dipolar character. The fact remains that at any *instant* of time, any atom (or molecule) will be an *instantaneous dipole*, which can interact electrostatically with neighbouring instantaneous dipoles. In such interactions attractive forces slightly outweigh repulsive forces and there is a nett weak attractive force. Theoretical chemists have confirmed that this picture is reasonable, and have shown that:

The greater the number of electrons in an atom or molecule, the stronger the dispersion forces the atom or molecule will exert on another atom or molecule.

This conclusion is in line with the trends shown in Tables 10.1 and 10.2. It should be realized that dispersion forces are *always* present between *all* atoms and molecules, although they frequently pass unnoticed because of the presence of much stronger forms of bonding. All bonding due to the operation of dispersion forces can safely be labelled as 'weak bonding'.

10.3 FORCES INVOLVING PERMANENT DIPOLES —HYDROGEN BONDING

Molecules which are permanent dipoles will tend to pack together, aligned so that the positive and negative ends of successive molecules are adjacent. Figure 10.5 represents schematically the packing arrangement of dipolar molecules, such as HBr, HI, ICl, etc., into a crystal lattice. In such crystals it is difficult to decide how much of the weak intermolecular bonding is due to the attraction between permanent dipoles, and how much is due to dispersion forces. However, there is one manifestation of dipolar force known as *hydrogen bonding*, which is particularly strong. Consider the data in the following tables of physical properties of a number of molecular hydrides.

Table 10.3

	M.P.(°C)	*B.P.(°C)*	ΔH_{fus} *(kJ mol^{-1})*	ΔH_{sub} *(kJ mol^{-1})*
Group VII hydrides				
HF	−83.07	19.9	4.57	—
HCl	−114.19	−85.03	1.99	18.2
HBr	−86.86	−66.72	2.40	20.0
HI	−50.79	−35.35	2.86	22.56
Group VI hydrides				
H_2O	0.00	100.0	6.00	46.7
H_2S	−85.53	−60.31	2.38	21.08
H_2Se	−65.73	−41.3	2.51	21.81
H_2Te	−51	−2.3	—	25
Group V hydrides				
NH_3	−77.74	−33.40	5.65	29.05
PH_3	−133.75	−87.72	1.13	15.75
AsH_3	−116.3	−62.5	2.34	19.84
SbH_3	−88	−17	—	—
BiH_3	—	22	—	—
Group IV hydrides				
CH_4	−182.5	−161.5	0.94	9.0
SiH_4	−184.7	−111.4	0.66	12.76
GeH_4	−165.9	−88.4	0.92	14.9
SnH_4	−150	−51.8	—	—
PbH_4	—	−13	—	—

where ΔH_{fus} and ΔH_{sub} are the heat content changes for the respective processes:

$$\text{Hydride(s)} \rightarrow \text{Hydride(l)} \qquad (\Delta H_{fus})$$

$$\text{Hydride(s)} \rightarrow \text{Hydride(g)} \qquad (\Delta H_{sub})$$

If dispersion forces only were operating in bonding these hydride molecules together, we would expect a steady increase in ΔH_{sub} as the molecular weight of the hydride increased within a group, as indeed is evidenced by the Group IV hydrides. However, the lowest molecular weight hydride in each of Group V, VI and VII has an abnormally high ΔH_{sub} value, and we must conclude that the intermolecular bonding is abnormally strong for the three hydrides,

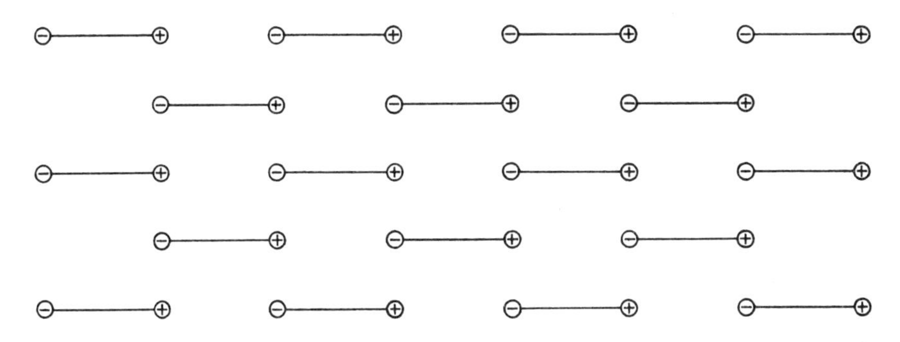

Fig 10.5 Packing of dipoles in a lattice.

H_3N, H_2O and HF. The strong intermolecular bonding between the molecules of these three hydrides is also well reflected in their melting and boiling points. Boiling points for the hydrides are shown graphically in Figure 10.6.

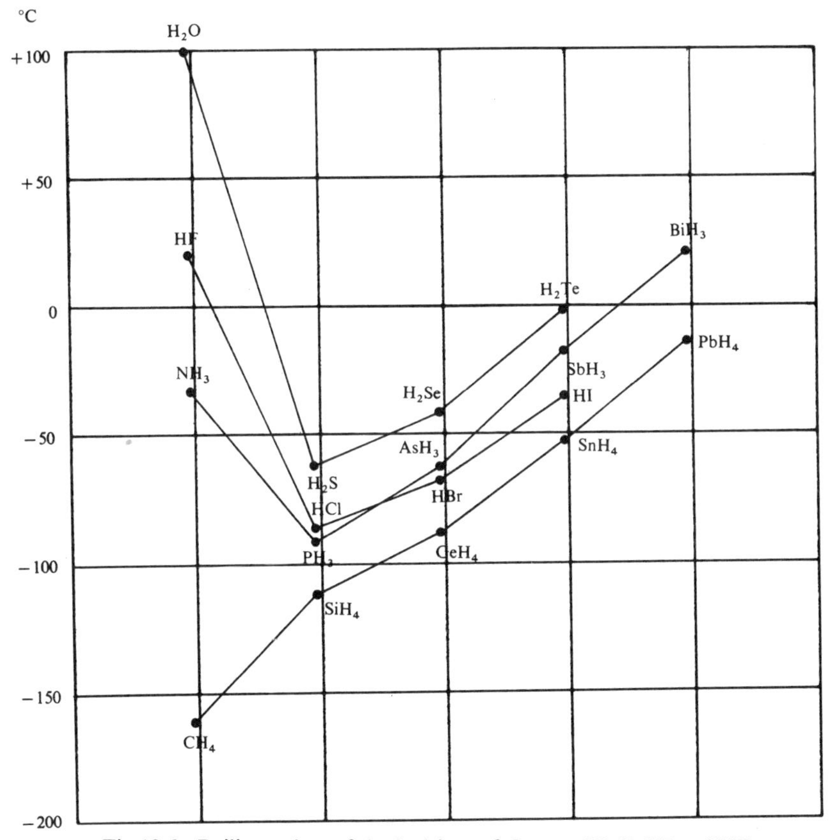

Fig 10.6 Boiling points of the hydrides of Groups IV, V, VI and VII.

The abnormal behaviour of the hydrides of nitrogen, oxygen and fluorine may be explained by assuming that hydrogen, when it is bonded to one of the three most electronegative atoms (fluorine, oxygen and nitrogen) has the ability to participate in a dipolar link of great strength with a molecule containing another of these three most electronegative atoms. A striking example of hydrogen bonding is provided by solid hydrogen fluoride, which consists of infinite planar zig-zag chains.

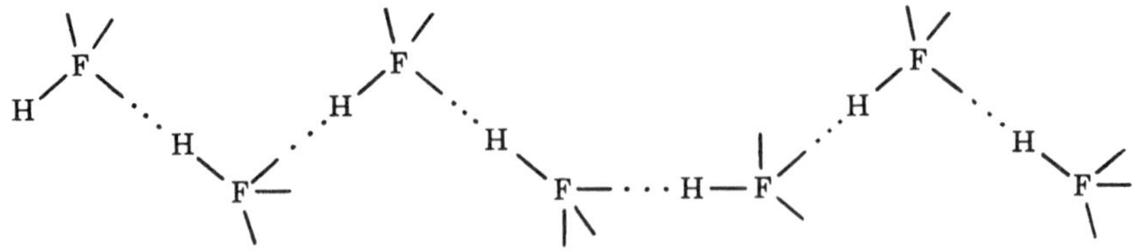

Fig 10.7 Chain of molecules in solid HF. The dots represent the hydrogen bonds.

In Figure 10.7 the dotted lines represent the 'hydrogen bonds'. It appears that the hydrogen of a hydrogen fluoride molecule, because of its partial positive charge, is attracted to the negative charge in one of the non-bonding pairs on the fluorine atom of an adjacent molecule. The amount of energy required to break one of the dipolar links in solid HF is not known, but is believed to be in the vicinity of 30-40 kJ mol^{-1}.

We may summarize the conditions for hydrogen bond formation:

(a) a hydrogen atom bonded to F, N or O, so that the hydrogen has an appreciable partial positive charge;

(b) an electronegative atom with a non-bonding pair which can attract the hydrogen atom described in (a).

Thus some typical hydrogen bonded links are shown in Figure 10.8.

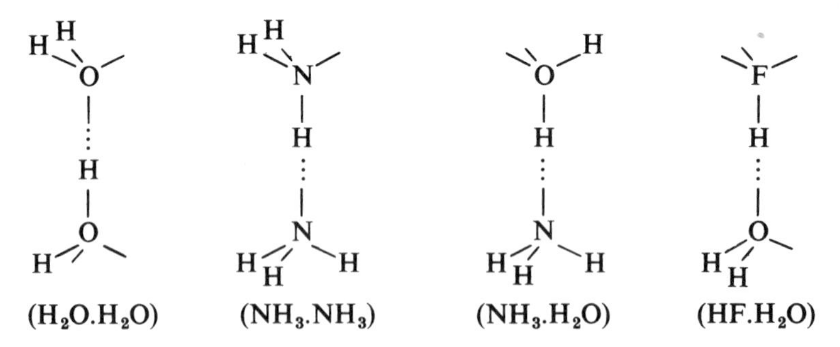

Fig 10.8 Some hydrogen-bonded entities. Dots represent the hydrogen bonds.

It would be generally true to say that hydrogen bonding can be classified as 'weak bonding'.

10.4 ICE

Hydrogen bonding plays an important role in determining the structures of ice and liquid water. The strength of the bonding between the water molecules in ice is fairly high:

$$H_2O(s) \rightarrow H_2O(g) \qquad \Delta H = +50.0 \text{ kJ}$$

A water molecule is known to be an essentially tetrahedral arrangement of four electron pairs around an oxygen atom, with protons embedded in the two bonding pairs. This shape, coupled with the ability to form strong dipolar links, determines the structures of ice and water. There are many slightly different structural forms of ice, but they all have the essential feature that every water molecule is surrounded by exactly four other water molecules, with each of which hydrogen bonds are formed. A single water molecule can form just four hydrogen bonds—two with its two hydrogens which may be attracted to the negative charge in the non-bonding pairs of other water molecules, and another two with its own two non-bonding pairs to which other hydrogen atoms may be attached. The result is the open network structure shown on Figure 10.9.

When ice melts, the rigid structure disrupts and partially collapses; the density of liquid water at 0°C is greater than that of ice at the same temperature. Thus, from having just four nearest neighbours in ice, each water molecule in liquid water at 0°C has, on the average, 4.4 nearest neighbours. We believe that the structure of liquid water is essentially 'ice-like' in that many water molecules will be hydrogen bonded to four others, even though the *long range order* of the ice structure has been lost.

10.5 ION-DIPOLE INTERACTION

An important form of bonding which lies on the borderline between weak and strong bonding exists between ions, and molecules which are permanent dipoles. We know that many substances, when dissolved in water, give solutions containing positive and negative ions, e.g. sodium chloride. In order to discuss the interaction that occurs between the individual ion and water, we would like to know the heat content change occurring when a free ion (say, $Na^+(g)$) is dissolved in water, i.e. the ΔH for processes of the type:

$$Na^+(g) \rightarrow Na^+(aq)$$

A process such as this cannot be carried out directly; we cannot take 'a bottle of sodium ions', dissolve some of the ions in water and measure the heat evolved, since sodium ions are not obtainable in any quantities except in combination with anions. However, the heat content changes represented in reactions like that shown above can be deduced from available thermochemical data using a suitable thermochemical cycle, and for sodium ions and fluoride ions it has been shown that:

$$Na^+(g) \rightarrow Na^+(aq) \qquad \Delta H = -398 \text{ kJ}$$

$$F^-(g) \rightarrow F^-(aq) \qquad \Delta H = -515 \text{ kJ}$$

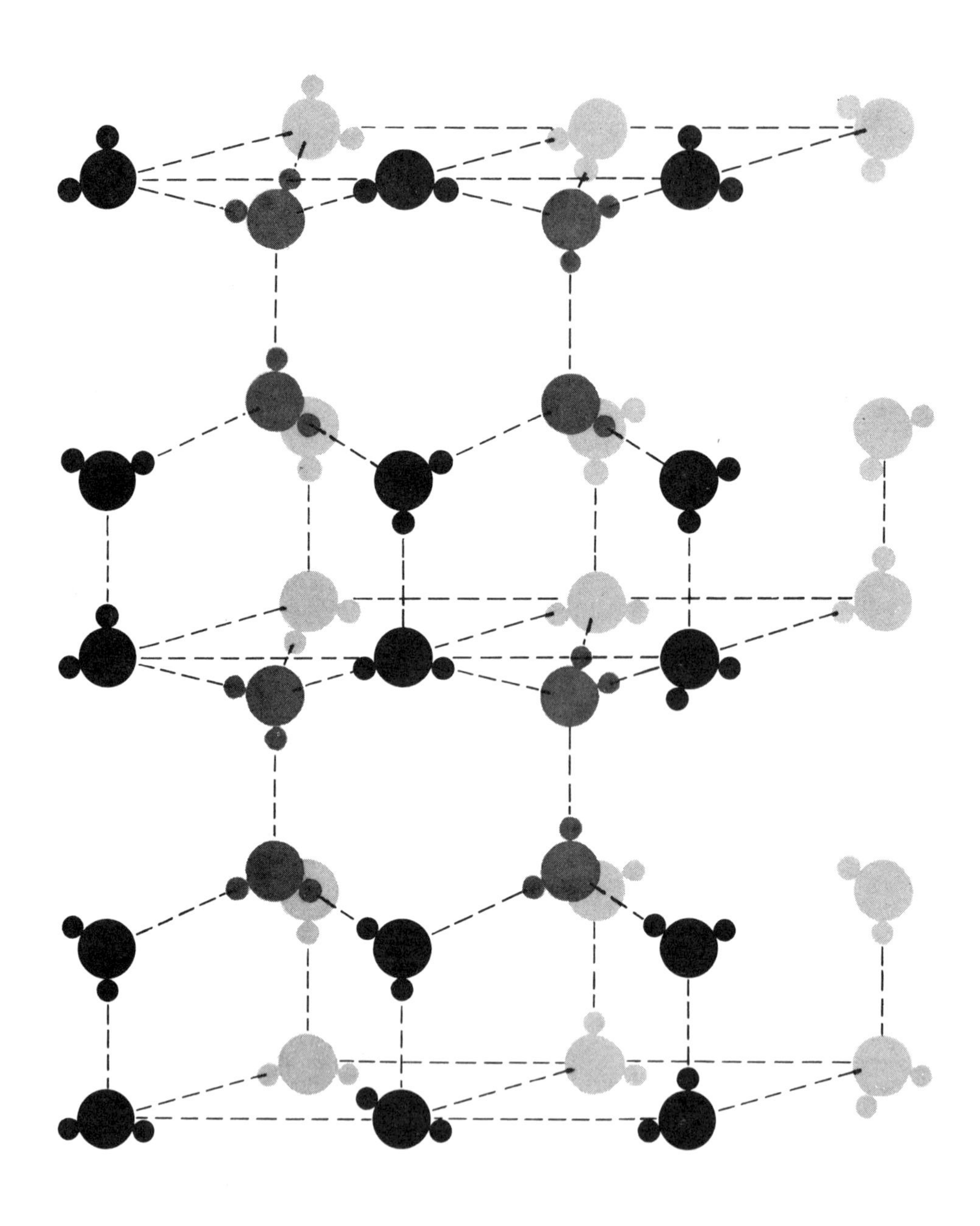

Fig 10.9 Structure of ice.

That is to say, both ions undergo exothermic reactions with water, and there is a marked drop in potential energy when free gaseous ions 'dissolve' in water. It seems that the drop in potential energy observed when a sodium ion is added to water is due to a specific bonding which occurs between the ion and water molecules, and it is likely that the bonding between ions and water molecules is due to the electrostatic attraction between the ion and the dipolar water molecule.

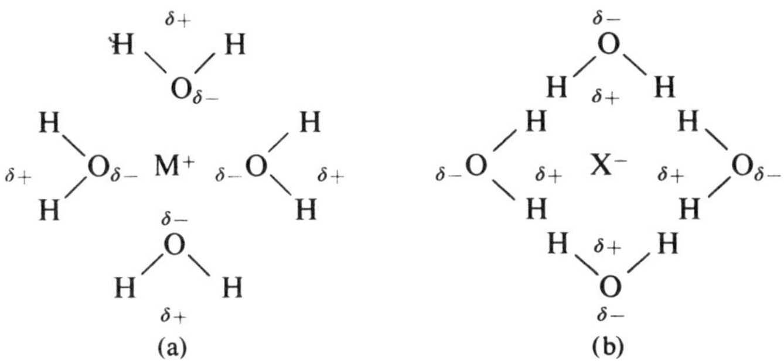

Fig 10.10 Orientation of water molecules around (a) positive ion, (b) negative ion (diagrammatic).

As shown in Figure 10.10, a positive ion would orient water molecules so that the hydrogens point away from the ion, while the negative ion would orient water molecules so that the non-bonding pairs point away from the ion. On the basis of our discussion in section 10.1 we might expect the strength of the bonding between an ion and a single dipolar molecule to:

(a) increase with the magnitude of the charge on the ion;
(b) increase as the radius of the ion decreases, thus allowing the dipole to move closer to the electric charge.

These suppositions are borne out by the table of data given below.

Table 10.4

Heats of hydration of ions

Ion	*Ionic radius* $\times 10^8$ *cm*	ΔH_{aq} *(kJ mol^{-1})*
Li^+	0.60	−507
Na^+	0.95	−398
K^+	1.33	−318
Rb^+	1.48	−289
Cs^+	1.69	−260
Ag^+	1.26	−464
Cd^{2+}	0.97	−1,760
F^-	1.36	−515
Cl^-	1.81	−373
Br^-	1.95	−339
I^-	2.16	−305

ΔH_{aq} is the heat content change in the process:

$$M^{n+}(g) \rightarrow M^{n+}(aq)$$

It is not possible to deduce from these data how strong is the bond between an ion and an individual water molecule. The heats of hydration quoted are for the complete hydration of the ion, and we do not know precisely how many water molecules are bound to a given ion in solution. It is believed that in aqueous solution the rather small lithium ion has four water molecules packed more or less tetrahedrally around itself. If this is so, we can estimate a rough value for the energy of the bond between a lithium ion and a water molecule as $507/4 \simeq 127$ kJ mol^{-1}. Most monatomic doubly- and triply-charged positive ions in solution are larger than Li^+ and it is generally believed that they can accommodate six water molecules around themselves in octahedral fashion, e.g. $Mg(H_2O)_6^{2+}$, $Cd(H_2O)_6^{2+}$, $Fe(H_2O)_6^{3+}$, $Al(H_2O)_6^{3+}$.

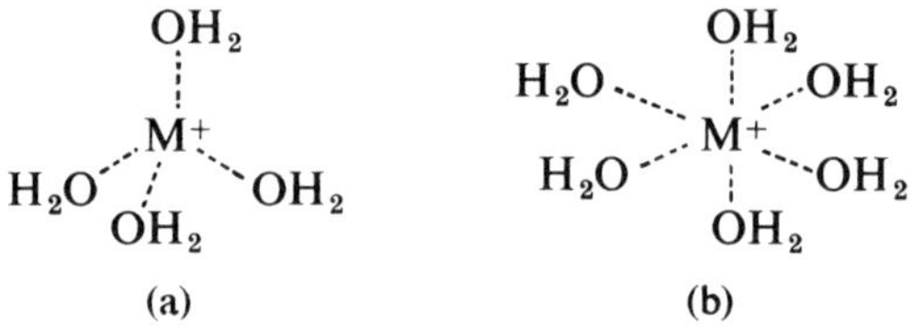

(a) OH_2, H_2O, OH_2, OH_2 around M^+

(b) OH_2, H_2O, OH_2, H_2O, OH_2, OH_2 around M^+

Fig 10.11 Packing of water molecules around positive ions (a) tetrahedral (b) octahedral.

Support for this contention comes from the structures of hydrated solids such as $MgCl_2.6H_2O$, $AlCl_3.6H_2O$ in which the 'waters of crystallization' are packed octahedrally around the cation (as shown in Figure 10.11) in an essentially ionic lattice (see chapter 11). Presumably these hydrated species persist in solution, although an ion such as $Al(H_2O)_6^{3+}$ would weakly attract and orient more water molecules around itself in aqueous solution, so that it is not strictly true to say that the hydrated aluminium cation *in water* has the formula $Al(H_2O)_6^{3+}$. Indeed, it is not possible to specify the exact 'formula' of *any* hydrated ion in solution.

For doubly- and triply-charged positive ions it is useful to distinguish between a primary *solvation sheath*, usually of six octahedrally-arranged water molecules next to the ion, and a *secondary solvation sheath* of more weakly-bonded water molecules, outside the primary sheath. Singly-charged ions are less strongly solvated, and the idea of a definite number of water molecules in a primary solvation sheath cannot be used.

QUESTIONS

1. Would one expect chlorine gas or fluorine gas to show the greater deviation from ideal gas behaviour? Give reasons for your answer.
2. What is the significance of the trends in the melting points and boiling points of the noble gases in terms of the forces between the atoms.

3. The boiling point of phosphine (PH_3) is -85 °C, whilst that of ammonia in the same group is -33.5 °C. How can we account for this difference in boiling point?
4. Which of the following molecules might have boiling points affected by hydrogen bonding: NH_3; CH_3—CH_3; $CH_3.OH$; $CH_3.NH_2$; HI.
5. How do you explain the fact that water expands when it freezes? Why are the values for the heat of fusion and heat of vaporization of water so very much greater than the corresponding values found for the hydrides of sulphur, selenium and tellurium in the same group of the periodic table?
6. In ionic compounds formed by the alkali metals, normally only Li^+ and Na^+ are hydrated, whereas at room temperature all the alkaline earth ions (Be^{2+}, Mg^{2+}, Ca^{2+}, Sr^{2+}, Ba^{2+}) are normally hydrated. Explain this difference in behaviour.
7. In each of the following pairs suggest in which hydrated ion the bonding between ion and water molecules is strongest: (a) $Fe(H_2O)_6^{3+}$, $Fe(H_2O)_6^{2+}$; (b) $Be(H_2O)_4^{2+}$, $Mg(H_2O)_6^{2+}$. Give reasons.
8. Ethane (C_2H_6) is only very slightly soluble in water, but ethanol ($CH_3.CH_2.OH$) is exceedingly soluble. Suggest a reason for this difference.
9. Sodium amide ($NaNH_2$) is an ionic solid containing sodium ions (Na^+) and amide ions (NH_2^-); it is soluble in liquid ammonia.
   (a) Suggest a valence structure for NH_2^-.
   (b) Show how you would expect ammonia molecules to be oriented about the sodium ion and the amide ion in solution.
10. Ammonium chloride dissolves in liquid ammonia yielding ammonium ions and chloride ions. How would you expect the ammonia molecules to be oriented around these ions?

11

Bonding into Infinite Arrays

Strong bonding into infinite arrays is exhibited by solids falling into one of the three classes already mentioned in section 7.4, viz.:
(a) network lattices; strong bonding in three dimensions;
(b) layer lattices; strong bonding in two dimensions;
(c) chain lattices; strong bonding in one dimension.
Network lattices are the most common of the above three classes, and a discussion of the bonding in network lattices will form a major subject in this chapter.

Apart from a few rare molecules such as the ion-pairs of the alkali metal halides, the bonding between the atoms in discrete molecules can be regarded as being essentially covalent. That is to say, definite electron pair bonds join together the atoms in almost all molecules. In network lattices, covalent bonding is less common, and two new types of bonding, metallic and ionic, make their appearance.

11.1 NETWORK LATTICES—BOND TYPES

It is a matter of general experience that most of the discrete molecules which normally exist in appreciable quantities at room temperatures consist of combinations of atoms of comparatively high electronegativity, e.g. H_2, F_2, Cl_2, Br_2, H_2O, CH_4, CCl_4, NH_3, SO_2. When molecules are formed between atoms with a large electronegativity difference, e.g. the NaCl, KCl, CsCl, KF diatomic molecules, they seem to exist in appreciable quantities only in the gas phase at high temperatures. This is because the most stable state of compounds such as those just listed is as an infinite array of atoms or ions *in which there are no discrete molecules.* It is only when sufficient energy is provided to disrupt the very stable lattices that discrete molecules can be obtained in the gas phase. A similar situation exists with combinations of atoms of low electronegativity. Molecules such as Li_2, Na_2, K_2 do exist, but only in appreciable numbers in the gas phase at high temperatures when we have disrupted the much more stable form of these elements, which is the metallic crystal.

The type of bonding observed in solids in which the atoms are strongly bound into network lattices takes on three extreme forms determined broadly by the electronegativities of the atoms involved. These are:

(a) metallic bonding; occurs when all atoms have *low* electronegativity;
(b) covalent bonding; occurs when all atoms have *high* electronegativity;
(c) ionic bonding; occurs when atoms have large electronegativity *differences.*

It must be emphasized that these three classifications represent three extremes of bonding type in the solid, and many lattices will be intermediate in character between any two of these three extremes.

11.2 NETWORK LATTICES—METALLIC BONDING

Metals are familiar substances in the modern world. The enormous technological importance of pure metals and alloys is due directly to their peculiar properties, particularly their low electrical resistivity, and their resistance to sudden stresses. Most modern building constructions depend on the strength and toughness of a steel framework. The physical properties regarded as being characteristic of metals are:

1 H																	2 He
3 Li	4 Be											5 B	6 C	7 N	8 O	9 F	10 Ne
11 Na	12 Mg											13 Al	14 Si	15 P	16 S	17 Cl	18 A
19 K	20 Ca	21 Sc	22 Ti	23 V	24 Cr	25 Mn	26 Fe	27 Co	28 Ni	29 Cu	30 Zn	31 Ga	32 Ge	33 As	34 Se	35 Br	36 Kr
37 Rb	38 Sr	39 Y	40 Zr	41 Nb	42 Mo	43 Tc	44 Ru	45 Rh	46 Pd	47 Ag	48 Cd	49 In	50 Sn	51 Sb	52 Te	53 I	54 Xe
55 Cs	56 Ba	57 La	72 Hf	73 Ta	74 W	75 Re	76 Os	77 Ir	78 Pt	79 Au	80 Hg	81 Tl	82 Pb	83 Bi	84 Po	85 At	86 Rn
87 Fr	88 Ra	89 Ac															

58 Ce	59 Pr	60 Nd	61 Pm	62 Sm	63 Eu	64 Gd	65 Tb	66 Dy	67 Ho	68 Er	69 Tm	70 Yb	71 Lu
90 Th	91 Pa	92 U	93 Np	94 Pu	95 Am	96 Cm	97 Bk	98 Cf	99 Es	100 Fm	101 Md	102 No	103 Lw

Fig. 11.1 The metallic elements.

(a) toughness, hardness;
(b) reflect light well when polished;
(c) can be bent and stretched;
(d) excellent conductors of heat and electricity.

The metals which have these properties *par excellence* are the transition elements, which include a high proportion of the technologically important metals. The elements which could reasonably be described as metallic at room temperatures are shown in Figure 11.1 and we note that 80 per cent of the elements in the periodic table rate as metals.

The one property that all these atoms have in common is a fairly low electronegativity; their comparatively low affinity for electrons gives us a clue as to the nature of the bonding in metallic solids.

Structural studies have shown that in all metallic crystals the atoms are close packed, or nearly so, having either eight or twelve nearest neighbours.

Fig. 11.2 A network lattice with metallic bonding—copper.

The most satisfactory picture of the bonding in these arrays is that of a collection of close-packed *positive ions*, embedded in a 'sea' of electrons.

The atoms have each given up custody of at least one electron, which then becomes the property of the whole metallic crystal. The force of attraction between the positive ions and the 'sea' of electrons then bonds the atoms strongly into the lattice. This picture explains particularly well the high electrical conductivity of metals, since the valence electrons, not being bound to any particular atom, can move more or less freely in an applied electric field.

When the valence electrons move in an applied electric field they do so at a speed which is determined partly by the intensity of the electric field. Thus in copper metal, in a potential gradient of 1 volt cm^{-1}, the electrons move at a speed of approximately 50 cm s^{-1}. Furthermore, the comparative ease with which metals can be bent and distorted is readily understood, since the positive ions can be moved relative to each other without seriously disturbing the bonding within the metal; the electron 'sea' simply flows to follow any change in shape of a crystal.

The strength of the bonding in the metallic state is best illustrated by the heat required for the sublimation process:

$$M(s) \rightarrow M(g) \qquad \Delta H_v$$

Some ΔH_v values for various metals are given in Table 11.1.

Table 11.1

Metal	ΔH_v *(kJ mol⁻¹)*	*B.P.(°C)*	*Metal*	ΔH_v *(kJ mol⁻¹)*	*B.P.(°C)*
Li	159	1,331	Fe	414	2,887
Na	109	890	Cr	393	2,642
Be	314	2,477	Ni	423	2,837
Mg	151	1,120	Pt	510	3,827
Ca	192	1,492	Au	351	2,707
Al	314	2,447	W	845	5,530

Notice that the boiling points of the metals are all fairly high, and give a good guide to the strength of the metallic bonding in the metallic crystal. All the metals exist largely as atoms in the gas phase, although there are some metals, such as the alkali metals, which on vaporization form small numbers of diatomic molecules.

Compounds formed between metals are called alloys, and do not usually obey the familiar rules of stoichiometry. Thus copper will dissolve zinc ranging from a trace to 38.4 atom per cent. Any alloy within this composition range is called 'brass'. Since brass has no one definite empirical formula it is often regarded as a simple mixture. Indeed, alloys in general are sometimes thought of as being mixtures of metals. We must be careful not to let the use of the word 'mixture' in this context obscure the fact that the *bonding* holding the atoms in alloys is strong metallic bonding, and in this sense, alloys are 'true' compounds.

11.3 NETWORK LATTICES—COVALENT BONDING

Atoms which have high electronegativity, together with the ability to form three or four electron pair bonds, are frequently involved in the formation of infinite arrays of covalently bonded atoms. The classical example of a covalent lattice is the diamond crystal, in which we have an array of carbon atoms, each covalently bonded to four other carbon atoms. Silicon exists in a similar crystalline form, and silicon carbide (SiC) consists of the same lattice with silicon atoms and carbon atoms on alternate lattice sites.

Another example of a lattice with covalent bonding is provided by solid silicon dioxide ('silica') SiO_2, in which each silicon atom is surrounded by four oxygen atoms, and each oxygen atom by two silicon atoms. We can therefore describe SiO_2 as a covalent lattice in which silicon forms four covalent bonds, and oxygen two covalent bonds. Obviously, since silicon and oxygen have different electronegativities, the silicon oxygen bonds will have some polar character.

Notice that the octet rule seems to be obeyed in each of the four lattices discussed. These covalently-bonded solids are all hard, brittle, and have a high electrical resistivity. The first two properties are expected, since to appreciably distort a covalently-bonded crystal, covalent bonds would have to be broken. The high electrical resistivity can be thought of as being due to the localization of the electrons in the covalent bonds, rendering them incapable of moving freely in an applied electric field.

As a measure of the bond strengths in a covalent lattice, we may quote the heat content changes for the process:

$$\text{C(s, diamond)} \rightarrow \text{C(g)} \qquad \Delta H = +718 \text{ kJ}$$

ΔH is the energy we must supply to the diamond lattice to separate the atoms completely; i.e. it is the 'heat of atomization' of diamond. Obviously, the covalent bonding in the diamond lattice is quite strong. Since melting or vaporizing these substances must necessarily involve the disruption of at least some of the covalent bonds in the lattices, the melting and boiling points of the solids might be expected to give a guide to the strength of the bonding.

Table 11.2

Solid	*M.P.(°C)*	*B.P.(°C)*
C (diamond)	—	3,800 (sublimes)
Si (silicon)	1,423	2,680
SiO_2	1,700	2,590

The high melting and boiling points of the solids suggest that the bonding in these lattices is quite strong.

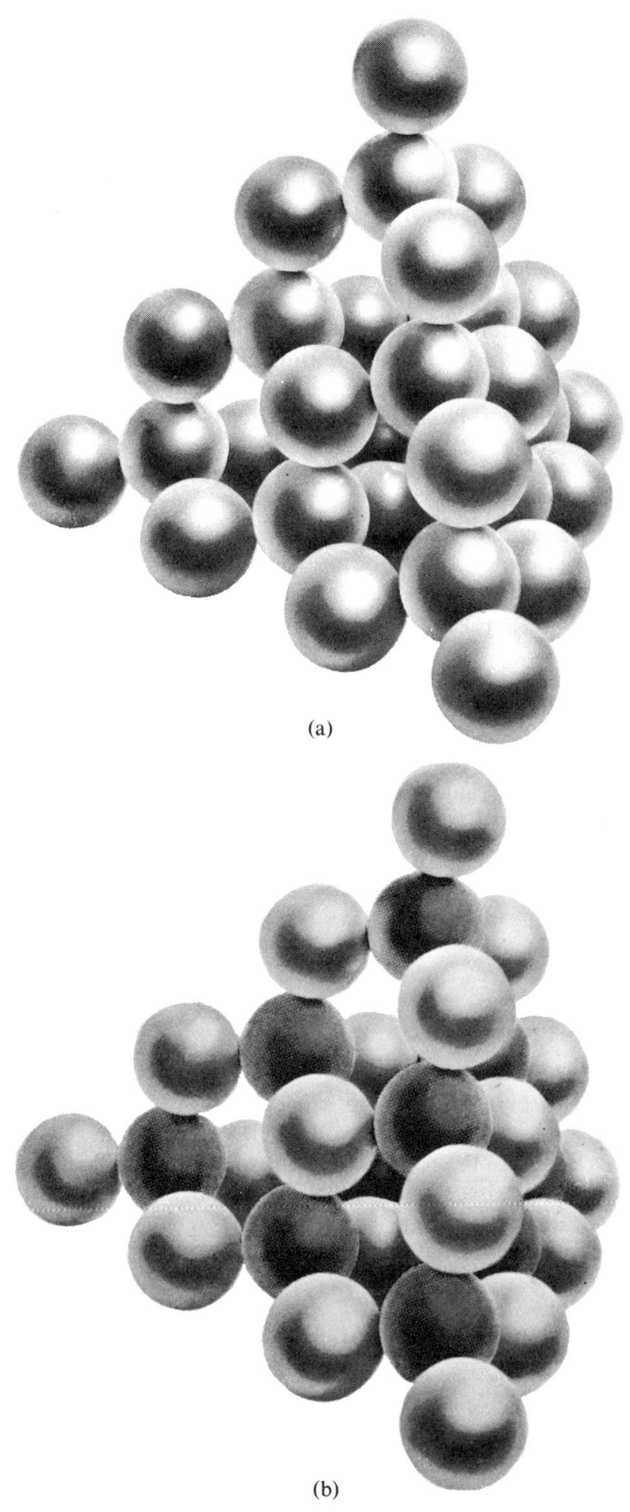

(a)

(b)

Fig 11.3 Some network lattices with covalent bonding (a) diamond, (b) silicon carbide.

11.4 NETWORK LATTICES—IONIC BONDING

(1) General survey

Ionic lattices are formed when atoms with large electronegativity differences form a compound. In these lattices, electrons are transferred from the atoms of low electronegativity to the atoms of high electronegativity, and the whole array is then bonded together by the attraction between the positive and negative ions. Ionic solids are characteristically formed between the elements of Group I and II of the periodic table (the alkali and alkaline earth metals) representing the atoms of low electronegativity; and oxygen, sulphur and the elements of Group VII representing the atoms of high electronegativity. In compounds formed between the two groups of atoms just mentioned, an octet rule is obeyed. The alkali metal atoms form singly-charged positive ions (Li^+, Na^+, K^+, Rb^+, Cs^+, Fr^+); while alkaline earth metal atoms form doubly-charged cations (Be^{2+}, Mg^{2+}, Ca^{2+}, Sr^{2+}, Ba^{2+}, Ra^{2+}). On the other side of the periodic table, the halogen atoms form singly-charged negative ions in their ionic compounds (F^-, Cl^-, Br^-, I^-) while oxygen and sulphur atoms exist as O^{2-} and S^{2-} in their ionic compounds. All these ions listed, with the exception of Li^+ and Be^{2+} which have a filled K-shell, have four electron pairs in their valence shells. Since we know the charges the various monatomic ions will possess, we can immediately write down the stoichiometric formula of a large number of ionic compounds by observing the requirement that a stable lattice must be electrically neutral.

The halides, oxides and sulphides of Group I are all network lattices, and the fluorides, oxides, sulphides and some chlorides of Group II are also network lattices. (As we shall see later, the bromides and iodides of Group II are layer lattices.)

EXERCISE

Assuming only monatomic ions are present, write down the stoichiometric formulae of the following compounds: radium fluoride, beryllium fluoride, calcium sulphide, rubidium sulphide, caesium iodide, magnesium fluoride, francium oxide.

(2) Evidence for the existence of ions in solids

The experimental evidence for the existence of ions in solids is mostly indirect. One of the most frequently quoted pieces of evidence is the fact that substances such as molten sodium chloride are excellent conductors of electricity. Furthermore, if molten NaCl is electrolysed by passing direct current through the melt, sodium is deposited at the negative electrode (suggesting Na^+) and chlorine evolved at the positive electrode (suggesting Cl^-). This evidence proves fairly conclusively that there are ions present in the liquid, but does not really prove that they ever existed in the solid, although it is certainly suggestive of this. All of the other evidence for the existence of ions in, say, the solid alkali halides, is equally circumstantial; however, it is not possible to explain their structures and stabilities with any other reasonable hypothesis.

(3) Structures of ionic solids

The structures of ionic compounds appear to be determined by the requirement that the ions be packed in the arrangement of lowest potential energy. To describe the structure of *any* solid is quite difficult, since there are so many subtly different ways of arranging atoms in three dimensions. However, one very important property of a structure, which is quite easy to discuss, is the co-ordination number of the atoms (or ions) in a solid. We may define in general:

The co-ordination number of an atom or ion in any environment is equal to the number of atoms it has for nearest neighbours.

Thus in sodium chloride each sodium ion is surrounded by six chloride ions, and hence sodium is said to have a co-ordination number of 6, or to be 6 co-ordinated. Similarly, each chloride ion is surrounded by six sodium ions, so that the chloride ion in sodium chloride has the co-ordination number of 6. In caesium chloride on the other hand, each caesium cation is surrounded by eight chloride ions, and each chloride ion is surrounded by eight caesium ions. Thus both caesium and chloride ions have co-ordination numbers of 8. We sometimes speak of a sodium chloride crystal as exhibiting '6 : 6 co-ordination' and a caesium chloride crystal as exhibiting '8 : 8 co-ordination'. We might ask why it is that sodium chloride shows 6 : 6 co-ordination, while caesium chloride shows 8 : 8 co-ordination? A caesium ion (radius 1.7×10^{-8} cm) is larger than a sodium ion (radius 0.95×10^{-8} cm) and is therefore able to fit more negative ions around itself. The relative sizes of the ions are such that a caesium ion can neatly fit eight chloride ions around itself, while a sodium ion can only fit six. Thus 6 : 6 co-ordination gives the lowest potential energy for sodium chloride, and 8 : 8 co-ordination gives the lowest potential energy for caesium chloride.

The strength of the forces holding the ions together in ionic solids is indicated by the heat content changes for a process such as:

$$NaCl(s) \rightarrow Na^+(g) + Cl^-(g) \qquad \Delta H = +753 \text{ kJ}$$

The heat content change for this process tells us the amount of energy we must supply in order to disrupt the ionic crystal and produce free separated ions. The amount of energy required to so disrupt one mole of a crystal is called the *lattice energy* of the crystal. The process of actually producing separated ions from an ionic crystal cannot be carried out in practice. When an ionic lattice is heated and vaporized, the vapour consists of molecules which are small fragments of the lattice. Thus sodium chloride vapour contains large numbers of NaCl diatomic molecules. There are certainly no free sodium or chloride ions in the vapour. The energy required to break up a crystal to give free ions can, however, be arrived at indirectly, by using appropriate thermochemical cycles. In Table 11.3 lattice energies are shown for a number of ionic solids. ΔH is the heat content change for the process:

$$MX(s) \rightarrow M^{n+}(g) + X^{n-}(g)$$

where M is a Group I or Group II metal, and X is a halogen or oxygen or sulphur.

Table 11.3

Lattice	*M.P.(°C)*	*B.P.(°C)*	ΔH *(kJ)*
NaCl	808	1,465	753
KF	856	1,502	797
CsF	682	—	723
CsCl	645	1,300	623
MgO	3,802	decomp.	3,933
CaO	2,587	decomp.	3,523
CuS	decomp. at ~220		3,020

The very high melting and boiling points of these ionic solids are an indication of the great stability of the lattices *relative* to the molecular forms

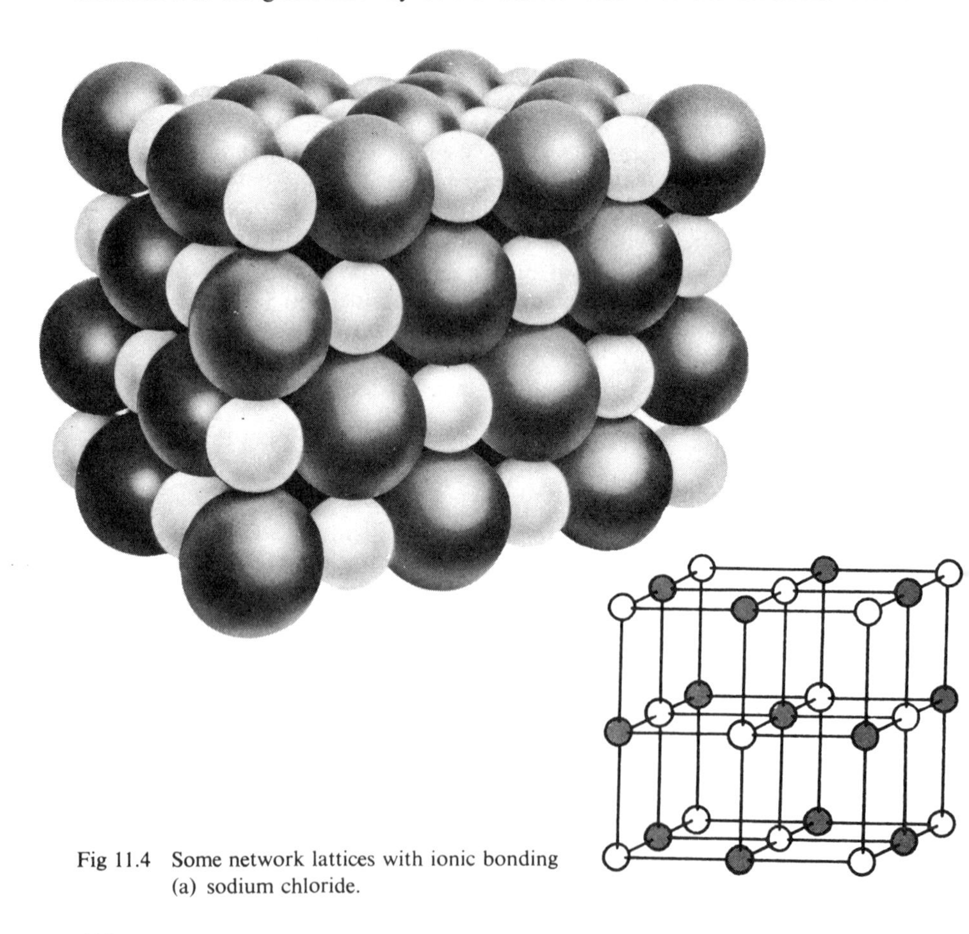

Fig 11.4 Some network lattices with ionic bonding (a) sodium chloride.

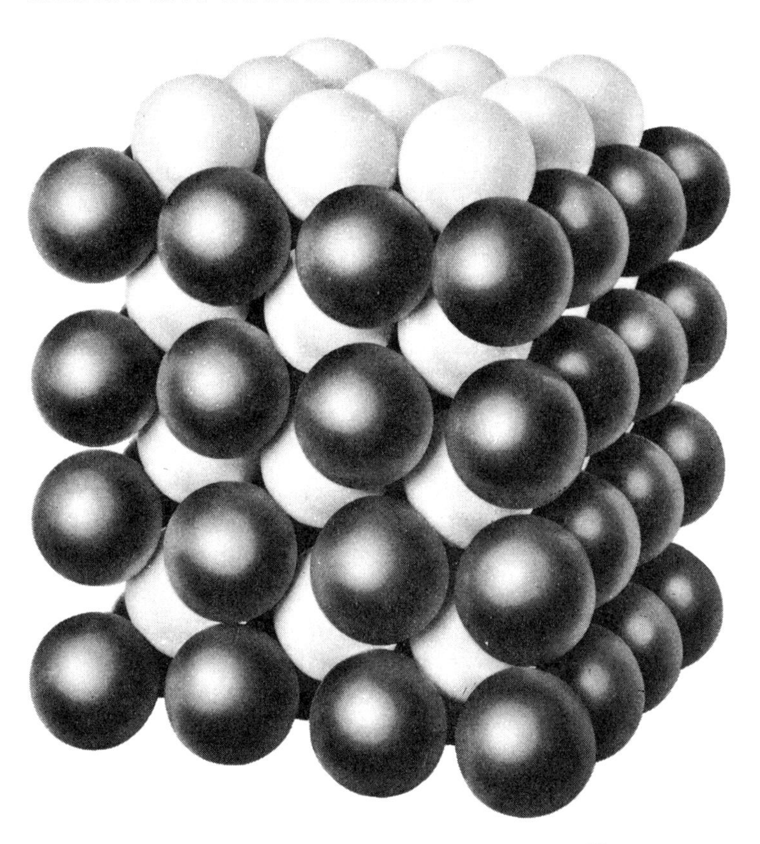

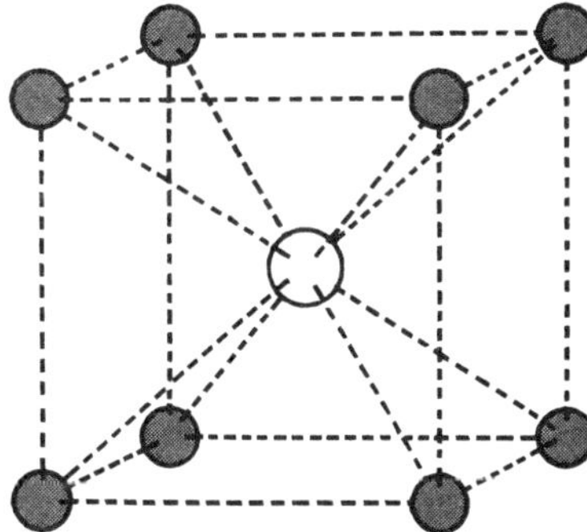

Fig 11.4 Some network lattices with ionic bonding
(b) caesium chloride.

of these compounds. Thus potassium chloride is much more stable as an ionic lattice than as KCl molecules. Indeed, the high boiling point of potassium chloride is an indication of how difficult it is to produce molecules in the gas phase from the very stable lattice.

Up to this point we have considered only those ionic compounds formed between the metals of Group I and II, and the non-metals, sulphur, oxygen and the halogens. These non-metals also form network lattice compounds with many transition metals, e.g. the silver halides, AgF, AgCl, AgBr, and AgI, which contain the Ag^+ ion. Other transition metals, such as copper, zinc and cadium, form network lattices with fluorine, oxygen and sulphur, e.g. CuF, CuO, ZnS and CdS. Although we would loosely call such compounds 'ionic', in fact the actual nature of the bonding in a solid such as ZnS has long been a matter of some debate. In zinc sulphide, each zinc atom (or ion) is surrounded by four sulphur atoms (or ions), and each sulphur atom by four zinc atoms, so that the structure is similar to that of silicon carbide. The electronegativity difference between zinc and sulphur is not too great, and many chemists, by invoking the concept of resonance (see section 9.9) would describe the bonding in the crystal as being 'partially covalent'. We shall not pursue this topic except to point out that at the moment there is no known way of satisfactorily deciding the exact nature of the bonding in such compounds. However, at this level, no harm is done by thinking of these solids as containing ions. Thus the compounds CuF, CuO, ZnS and CdS may be regarded as containing the ions Cu^+, Cu^{2+}, Zn^{2+} and Cd^{2+} respectively.

(4) Ionic lattices containing molecular ions

Many ionic lattices contain molecular ions, rather than monatomic ions. A few examples of these ions are given in Tables 11.4 and 11.5.

All molecular ions can be involved in lattice formation in which the ions are packed and held together just as are the monatomic ions in the alkali halides; thus $MgCl_2.6H_2O$ contains $Mg(H_2O)_6^{2+}$ ions and Cl^- ions packed together in a lattice. NH_4NO_3 consists of NH_4^+ and NO_3^- ions. The lattice energies of such compounds are of the same order of magnitude as those of simpler ionic compounds, thus:

$$NH_4Cl(s) \rightarrow NH_4^+(g) + Cl^-(g) \qquad \Delta H = +640 \text{ kJ}$$

The stoichiometry of compounds containing molecular ions may immediately be deduced from the charges on the ions, and the requirement that a stable lattice must be electrically neutral.

EXERCISE
Write down stoichiometric formulae for: ammonium perchlorate, lithium fluoroborate, hydronium perchlorate, rubidium carbonate, radium sulphate, francium peroxide, sodium persulphide, nitronium perchlorate.

11.5 LAYER LATTICES

Some elements exhibit covalent bonding into layer lattices. Black phosphorus, one of the allotropes of phosphorus, consists of layers of atoms, each forming three covalent bonds. The layers are held together by dispersion forces. Another example of a covalently-bonded layer lattice is graphite, the most stable allotrope of carbon. This has a structure in which the carbon atoms are bonded into infinite sheets of hexagons of carbon atoms. Dispersion forces then hold the sheets together.

Graphite is a reasonably good conductor of electricity, since the valence electrons are apparently able to move across the layers under the influence of an applied electric field.

Most layer lattices are compounds between two different atoms, with moderate electronegativity differences. Compounds of general formula MX_2 or MX_3 where M is a metal and X is chloride, bromide, iodide, or sulphide, very commonly form layer lattices, e.g. $MgCl_2$, $FeCl_3$, $CdBr_2$, CdI_2, SnS_2, $AlCl_3$. The bonding within the layers in these layer lattices is often loosely described as 'ionic'. Again, as with network lattices like zinc sulphide, many chemists prefer to judge individual cases on their merits, and would call the bonding in compounds such as cadmium iodide (CdI_2), for example, 'partially covalent'. However, as before, no harm can come from thinking of these lattices as containing ions. The structures of CdI_2 and $FeCl_3$

Table 11.4

Some molecular cations

Name	*Formula*	*Valence structure*	*Example of solid containing ion*
Hydronium	H_3O^+	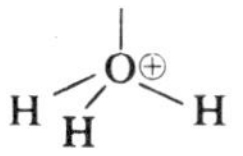 O⊕ bonded to three H, with one lone-pair line	$H_3O.ClO_4$ —hydronium perchlorate
Ammonium	NH_4^+	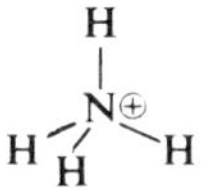 N⊕ bonded to four H	NH_4Cl —ammonium chloride
Nitronium	NO_2^+	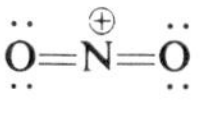 $\ddot{O}=\overset{\oplus}{N}=\ddot{O}$	$N_2O_5(NO_2.NO_3)$ —nitrogen pentoxide (nitronium nitrate)
Hexaquo magnesium	$Mg(H_2O)_6^{2+}$	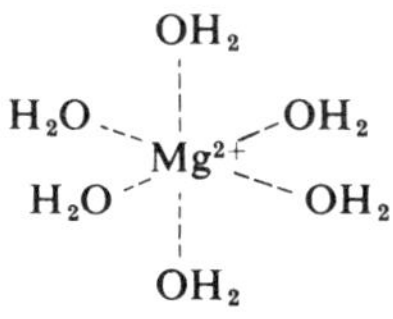 Mg^{2+} surrounded by six OH_2	$MgCl_2.6H_2O$ —hydrated magnesium chloride
Hexaquo aluminium	$Al(H_2O)_6^{3+}$	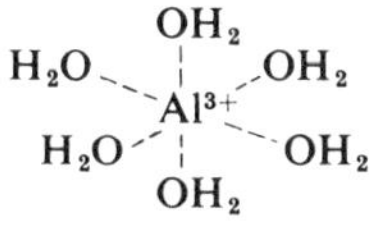 Al^{3+} surrounded by six OH_2	$AlCl_3.6H_2O$ —hydrated aluminium chloride
Cobalt(III) hexammine	$Co(NH_3)_6^{3+}$	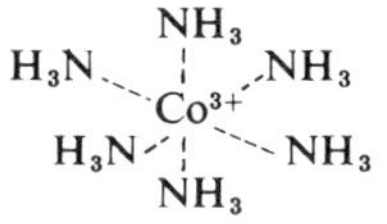 Co^{3+} surrounded by six NH_3	$Co(NH_3)_6.Cl_3$

are shown in Figure 11.6. We would describe the structure of an iron(III) chloride layer as consisting of two sheets of closely-packed chloride ions, held together by Fe^{3+} ions between the sheets, arranged so that each Fe^{3+} ion is surrounded by six chloride ions. In the crystal of $FeCl_3$, these composite layers are stacked one upon the other, and held together by dispersion forces.

Table 11.5

Name	*Formula*	*Valence structure*	*Example of compound*
Peroxide	O_2^{2-}	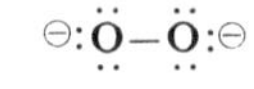 ⊖:Ö–Ö:⊖	Na_2O_2 —sodium peroxide
Persulphide	S_2^{2-}	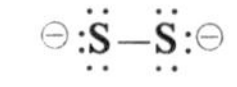 ⊖:S̈–S̈:⊖	FeS_2 —'iron pyrites' (iron(II) persulphide)
Acetylide	C_2^{2-}	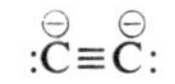 ⊖:C≡C:⊖	CaC_2 —'calcium carbide' (calcium acetylide)
Perchlorate	ClO_4^-	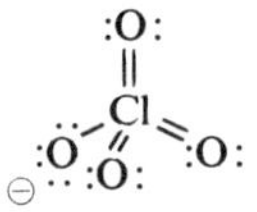 :O: ‖ Cl (=O, =O, –O:⊖)	$KClO_4$ —potassium perchlorate
Chlorate	ClO_3^-	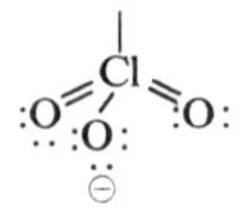 Cl (=O, =O, –O:⊖)	$KClO_3$ —potassium chlorate
Hydrogen sulphate	HSO_4^-	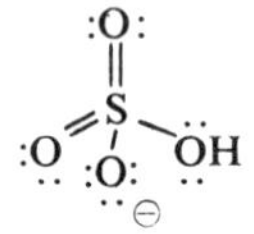 :O: ‖ S (=O, –OH, –O:⊖)	$NaHSO_4$ —sodium hydrogen sulphate
Fluoroborate	BF_4^-	 :F: – B⊖ (–F, –F, –F)	$NaBF_4$ —sodium fluoroborate
Hydrogen carbonate	HCO_3^-	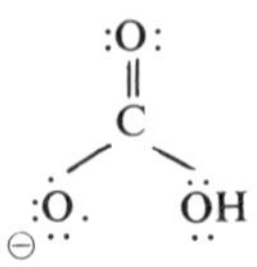 :O: ‖ C (–O:⊖, –OH)	$NaHCO_3$ —sodium hydrogen carbonate

Some layer lattice compounds have quite low boiling points, e.g. aluminium chloride sublimes at 183°C. We must be careful in drawing from this

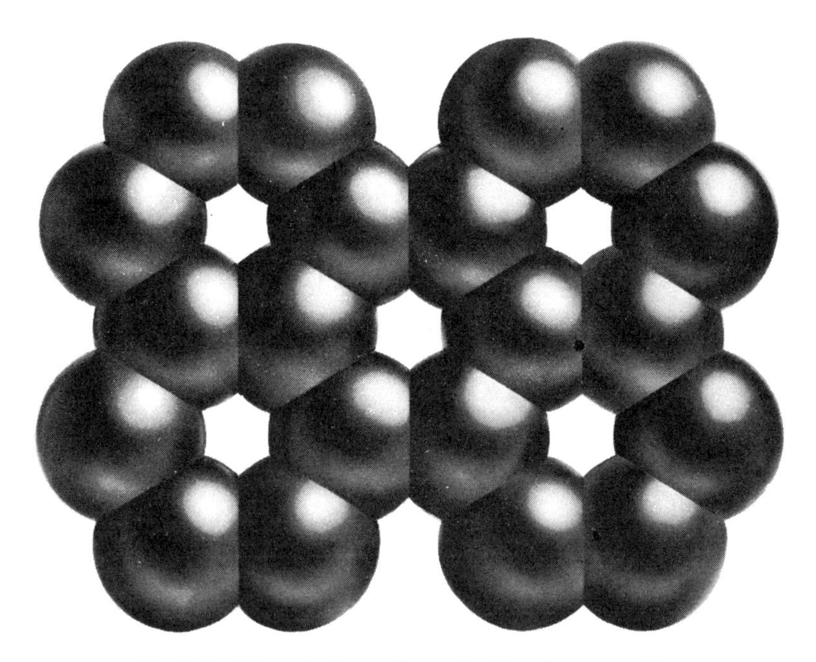

Fig 11.5 A layer lattice with covalent bonding—graphite.

sort of evidence any conclusions about the strength of the bonding in solid $AlCl_3$. In this case, it turns out that essentially ionic bonding in solid $AlCl_3$ is quite strong and the low boiling point arises because solid $AlCl_3$ produces quite stable Al_2Cl_6 molecules on melting and on vaporization. We will discuss the structure of this molecule in chapter 30.

It is worth noting that many of the layer lattice compounds, e.g. $FeCl_3$, $AlCl_3$, form related hydrated compounds ($FeCl_3.6H_2O$, $AlCl_3.6H_2O$) which are network lattices. In these lattices, hydrated cations, e.g. $Fe(H_2O)_6^{3+}$, $Al(H_2O)_6^{3+}$, can be readily distinguished.

QUESTIONS

1. On the basis of the electronegativities of the atoms involved, and your general chemical knowledge, place the following solid compounds into their correct structural classification, and then discuss the nature of the bonding in each case: $CaCl_2$; PCl_3; RbF; CCl_4; MgO; BrCl; H_2O; HBr; SiC.
2. Classify the following solid compounds into their correct structural family, and discuss the expected bonding in each case. Write valence structures for any discrete molecules or molecule-ions: Xe; CH_3Cl; SiO_2; CO_2; Cu; CaF_2; Na_2CO_3; KF; $HClO_4$; HF; NH_4NO_3; $AlCl_3$; $FeCl_3.6H_2O$; ZnS; $KHSO_4$.
3. Account for the fact that at room temperature metals are good electrical conductors, but ionic network solids are not.
4. Silica (SiO_2) is a high melting point solid whilst carbon dioxide (CO_2) at room temperature is a gas. Account for this difference in physical state in terms of the structure of these two substances.

(a)

(b)

Fig 11.6 Some layer lattices
(a) cadmium iodide, (b) iron(III) chloride.

5. Propose chemical experiments to show that diamond and graphite are different structural forms of the one element.
6. Write down the electron configuration of each *ion* involved in the substances KF, Na_2O, $MgBr_2$, LiH.
7. Explain what is meant when we say that the ions in calcium oxide solid show 6 : 6 co-ordination, whilst those in caesium chloride (CsCl) show 8 : 8 co-ordination.
8. Sodium fluoride and magnesium oxide have the same structure in the crystal and the interionic distances in the two compounds are not greatly different. Suggest why magnesium oxide has a higher melting point than sodium fluoride.
9. Why do network solids have higher heats of vaporization than solids, such as iodine, which are composed of discrete molecules?
10. A student interpreted the formula MgF_2 to mean that the crystal was made up of magnesium atoms and fluorine molecules. How would you convince him that his structure was an unlikely one?
11. Predict the order of increasing boiling point for the following substances containing iodine and explain the basis of your prediction: I_2, NaI, CI_4.
12. Consider each of the following in the solid phase: Na, Si, H_2S, He, KF, HF. Which would be an example of
    (a) a solid in which strong hydrogen bonds exist between molecules?
    (b) a solid which is a good electrical conductor?
    (c) a solid which is a poor electrical conductor, but conducts on melting?
    (d) a solid consisting of atoms held together only by weak dispersion forces?
    (e) a solid in which the atoms are covalently bonded together into a network?
    (f) a solid containing angular dipolar molecules?
13. Contrast the bonding between atoms in solid argon, copper, diamond, with respect to expected bond type and expected bond strength.
14. Write an essay explaining why it is often better to use a structural classification of solids rather than a classification based on bond type.

12

Rates of Chemical Reactions

The idea of the rate of a chemical reaction is used intuitively by many people. For instance, when we insist on an egg boiled for 4.1 minutes, we base this demand on our experience of how rapidly egg albumen hardens at 100°C. The effect of temperature on reaction rates is also encountered. Woe betide any budding cook who imagines that four minutes for an egg in tepid water is equivalent to the same time in boiling water. The 'setting' of concrete, the 'fading' of dyes, the 'drying' of ink and the 'rusting' of iron are but a few further examples of the many chemical reactions whose rates we are aware of in our everyday lives. Quite apart from the technological interest, it is important to appreciate some of the factors influencing reaction rates in order that we have a proper understanding of basic chemistry.

12.1 THE CONCEPT OF A REACTION RATE

The expression '*the rate of a chemical reaction*' is meant to convey 'how much' of a particular chemical is either consumed or produced in a stated time interval. For instance, when a dilute aqueous solution containing iodide ions is added to a dilute aqueous solution containing perdisulphate ions, the following reaction occurs:

$$2I^-(aq) + S_2O_8^{2-}(aq) \rightarrow I_2(aq) + 2SO_4^{2-}(aq)$$

If a little starch solution is also added to this reaction mixture, we then have a sensitive method for detecting the iodine formed in this reaction, as a blue coloration will become visible when the concentration of $I_2(aq)$ reaches the very low value of 10^{-5} M. In practice it is found that some seconds or even minutes must elapse before this blue coloration is visible. We may conclude that the I^--$S_2O_8^{2-}$ reaction needs a finite time to produce an iodine concentration of 10^{-5} M.

The rate of the I^--$S_2O_8^{2-}$ reaction may be estimated from the time in seconds required for the iodine concentration to rise from its initial zero value to 10^{-5} M. In a typical case, this time would be 80 seconds at 25°C and the rate of the reaction could be estimated as follows:

$$\text{rate of formation of } I_2(aq) = \frac{\text{increase in } [I_2]}{\text{elapsed time}}$$

$$= \frac{[I_2]_2 - [I_2]_1}{t_2 - t_1}$$

where $(t_2 - t_1)$ represents the elapsed time in seconds, $[I_2]$ represents the molar concentration of iodine, the subscripts 2 and 1 referring to final and initial measurements respectively. Thus $[I_2]_1 = 0$ and $[I_2]_2 = 10^{-5}$

$$\therefore \text{ rate of formation of } I_2(aq) = \frac{10^{-5} - 0}{80}$$

$$= 1.25 \times 10^{-7} \text{ M s}^{-1}$$

Since for every mole of I_2 formed, 1 mole of $S_2O_8^{2-}$ has been consumed, the rate of consumption of $S_2O_8^{2-}$ is also 1.25×10^{-7} M s^{-1}. Similarly, the stoichiometry of the reaction indicates that the rate of reaction of I^-, and the rate of formation of SO_4^{2-} are each twice this rate, namely 2.5×10^{-7} M s^{-1}. Notice that the rates of consumption of reactants and the rates of formation of products are simply related by whole numbers determined by the stoichiometry of the reaction.

Reactions such as the above, when all reactants are taken in the same physical state, either gaseous or as an aqueous solution, are termed *homogeneous reactions*. In each case the reaction rate may be expressed in units of M s^{-1}. A large number of chemical reactions cannot be classified as homogeneous. The rusting of iron involves the reaction of oxygen gas with the surface of solid iron. The dissolving of metallic zinc in hydrochloric acid is a reaction between a liquid and a solid. The reaction between metallic aluminium and iodine crystals involves two solids which have distinct, recognizable physical forms. All these reactions are examples of *heterogeneous* reactions wherein reaction occurs between reactants in distinctly different physical forms. A study of the rate of heterogeneous reactions is often quite complicated because factors such as the nature of the surfaces of solids are involved. This factor is readily seen if the same mass of zinc powder and granulated zinc is added to hydrochloric acid. Moreover, the rates of heterogeneous reactions cannot be always expressed in the simple units M s^{-1}. We shall restrict most of our subsequent discussions to the rates of homogeneous reactions.

12.2 VARIATION OF REACTANT CONCENTRATION WITH TIME

We shall now consider an example of a homogeneous reaction in more detail. A suitable reaction is the decomposition of dinitrogen pentoxide. Either in the gas phase, or when dissolved in a solvent such as carbon tetrachloride, N_2O_5 decomposes at a conveniently measurable rate according to the equation

$$2N_2O_5 \rightarrow 4NO_2 + O_2$$

The data in Table 12.1 give measured values of the concentration of N_2O_5 at various elapsed times of reaction.

Table 12.1

Decomposition of N_2O_5 in CCl_4 solution at 45°C

Elapsed time *(s)*	Δt *(s)*	$[N_2O_5]$ *(M)*	$-\Delta[N_2O_5]$ *(M)*	*Reaction rate* *($M\,s^{-1}$)*
0	0	2.10	—	—
100	100	1.95	0.15	1.5×10^{-3}
300	200	1.70	0.25	1.3×10^{-3}
700	400	1.31	0.39	0.99×10^{-3}
1000	300	1.08	0.23	0.77×10^{-3}
1700	700	0.76	0.32	0.45×10^{-3}
2100	400	0.56	0.14	0.35×10^{-3}
2800	700	0.37	0.19	0.27×10^{-3}

In this table,

$$\Delta N_2O_5 = [N_2O_5]_2 - [N_2O_5]_1$$

and

$$\Delta t = t_2 - t_1$$

where the subscripts 2 and 1 again refer to the final and initial states respectively. Since $[N_2O_5]$ decreases with time, $[N_2O_5]_2 < [N_2O_5]_1$. The rate of the reaction is calculated in the last column from the decrease in N_2O_5 concentration $-\Delta[N_2O_5]$, in an elapsed time interval Δt. Inspection of this last column shows that the *rate of reaction decreases as time proceeds.* As the concentration of N_2O_5 is decreasing during this time, it suggests that the rate of the reaction is in some way dependent upon the concentration of the reactant. The lower the reactant concentration, the slower does the reaction proceed.

This situation is represented graphically in Figures 12.1 and 12.2 which show the variation of concentration of N_2O_5 with time, and the variation of reaction rate with time.

In the next and subsequent section, we shall explore quantitatively the nature of this dependence of reaction rate upon concentration.

The rate of any homogeneous reaction, expressed in the units $M\,s^{-1}$, decreases as the time of reaction increases. The rate depends in some way on the concentration of the reactants.

12.3 RATE LAWS AND RATE CONSTANTS

We began this chapter using the term 'reaction rate' to mean the rate at

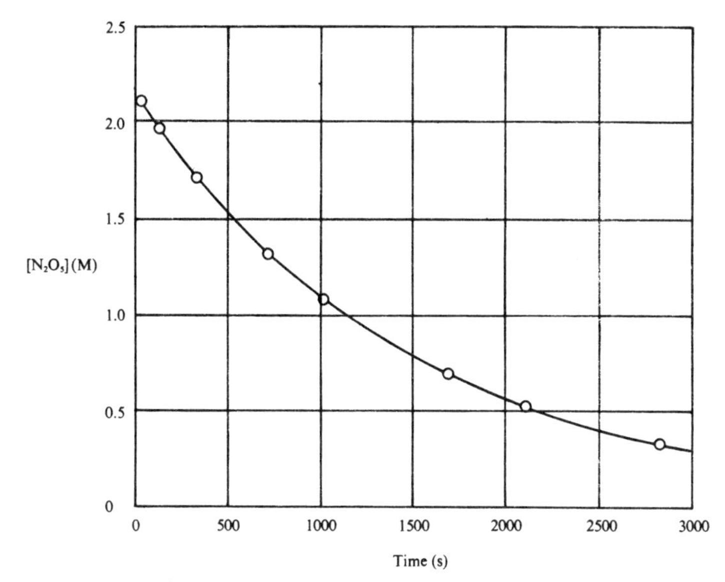

Fig 12.1 The variation of N_2O_5 concentration with time.

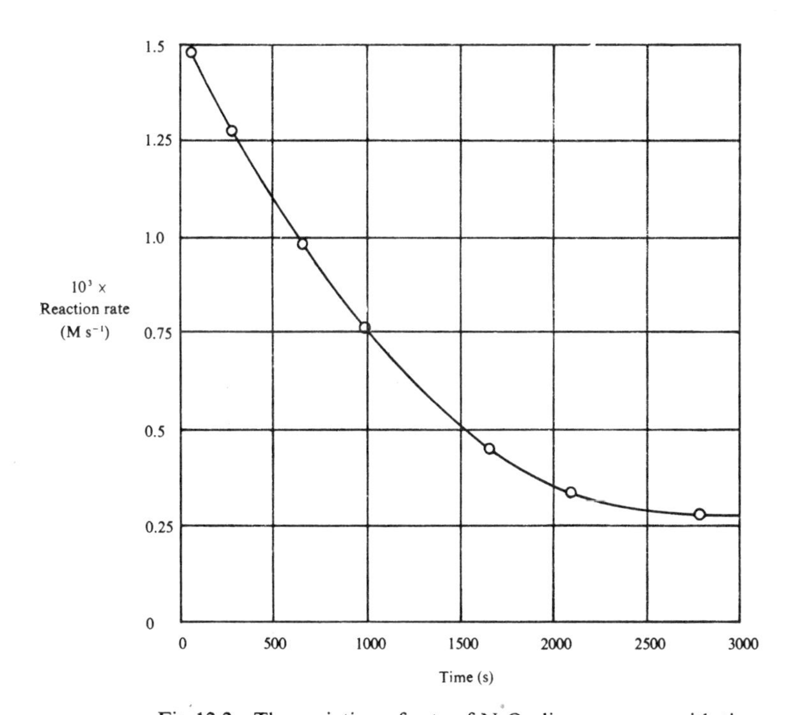

Fig 12.2 The variation of rate of N_2O_5 disappearance with time.

which a specified reactant is used up, or the rate at which a specified product was formed.

We have seen that the reaction rate (R) is dependent upon the concentration of reactant. In this section we shall illustrate one means of experimentally determining the nature of this dependence of rate on concentration. This will lead to an expression referred to as a *rate law*. In section 12.4 we shall consider a more general method for determining rate laws.

Although it is not a widely applicable procedure, the measurement of initial rates of reaction is sometimes a very convenient means for deducing the dependence of rate upon concentration. Consider, as an example, the gaseous reaction between hydrogen and nitric oxide

$$2H_2(g) + 2NO(g) \rightarrow 2H_2O(g) + N_2(g)$$

The rate of this reaction may be followed experimentally by measuring the decrease in pressure during the reaction as 4 mole of gaseous reactants gives rise to 3 mole of gaseous products. From this it is possible to calculate the rate of nitrogen production in M s^{-1}. The initial rates for a set of six experiments using different initial concentrations of reactants for this system are quoted in Table 12.2.

Table 12.2

Rate of reaction between H_2 and NO (800°C)

Experiment	*Initial concentration (M)*		*Initial rate of N_2 production ($M\ s^{-1}$)*
	[NO]	[H_2]	
1	6.00×10^{-3}	1.00×10^{-3}	3.19×10^{-3}
2	6.00×10^{-3}	2.00×10^{-3}	6.36×10^{-3}
3	6.00×10^{-3}	3.00×10^{-3}	9.56×10^{-3}
4	1.00×10^{-3}	6.00×10^{-3}	0.48×10^{-3}
5	2.00×10^{-3}	6.00×10^{-3}	1.92×10^{-3}
6	3.00×10^{-3}	6.00×10^{-3}	4.30×10^{-3}

Experiments 1 to 3 show how the rate varies with concentration of H_2, keeping the concentration of NO constant. Doubling the hydrogen concentration is seen to double the rate, and when the hydrogen concentration is increased threefold, so is the rate. This suggests the rate is directly proportional to the hydrogen concentration, thus

$$\text{rate} \propto [H_2] \tag{1}$$

Experiments 4 to 6 show how the rate varies with concentration of NO, keeping the concentration of H_2 constant. When the NO concentration is doubled, i.e. from [NO] = 0.001 M to [NO] = 0.002 M, the rate is found to increase four times, i.e. from 0.478×10^{-3} M s^{-1} to 1.92×10^{-3} M s^{-1}. When the NO concentration is tripled, i.e. from [NO] = 0.001 M to [NO] = 0.003 M, the rate is found to increase nine times, i.e. 0.478×10^{-3} M s^{-1} to

4.30×10^{-3} M s^{-1}. This suggests the rate is directly proportional to the square of the concentration of NO, thus

$$\text{rate} \propto [\text{NO}]^2 \tag{2}$$

Likewise, if the concentration of NO initially were (say) 0.005 M, the rate would be twenty-five times the initial rate of 0.478×10^{-3} M.

Combining (1) and (2)

the overall rate $\propto [H_2][NO]^2$

or rate $= k[H_2][NO]^2$

where k is a constant at this temperature, called the *rate constant.*

This expression is described as a *rate law* in that it describes the functional dependence of the reaction rate upon the concentration of the various reactants. **The form of the rate law for any particular homogeneous reaction can only be found by experiment.** *One cannot deduce the form of the rate law from a stoichiometric equation.*

The data in Table 12.2 may now be used to deduce the value of the constant of proportionality, k.

Writing that

$$R = k[H_2][NO]^2$$

and substituting typical values for the first experiment in Table 12.2

$$3.19 \times 10^{-3} = k(1.00 \times 10^{-3})(6.00 \times 10^{-3})^2$$

$$\therefore \quad k = 8.8 \times 10^4 \text{ M}^{-2}\text{ s}^{-1}$$

EXERCISE

Verify that the units of k are M^{-2} s^{-1}.

The rate constant k has a constant value regardless of the reactant concentration in a particular reaction provided the temperature is constant. The complete rate of expression for the H_2-NO reaction at 800°C may then be summarized as

$$R = 8.8 \times 10^4[H_2][NO]^2 \text{ M s}^{-1}$$

This rate expression enables one to predict the rate of this reaction for any chosen values of the reactant concentrations.

EXERCISE

Calculate the rate of the reaction of hydrogen and nitric oxide at 800°C, for $[H_2]$ = 0.002 M, and [NO] = 0.003 M.

Similar studies have been made on the rates of other homogeneous reactions and the following generalizations may be made:

(a) The rate law which reproduces the experimentally determined reaction rate is an algebraic expression and involves a rate constant, which is constant at constant temperature, and concentrations of reactants raised to specified powers. Most commonly, these powers are 0, 1, 2 and $\frac{1}{2}$.

(b) The rate law for any reaction may be deduced *only* from the experimental measurement of reaction rates.
(c) The rate law bears *no necessary resemblance* to the stoichiometric equation for a given reaction.

Rate law expressions where the concentration term is raised to the first power are said to be *first order* rate laws, and reactions described by such rate laws are referred to as first order reactions. As we shall see in the next section, it can be experimentally shown that the decomposition of dinitrogen pentoxide is a first order reaction. That is, the rate of the reaction

$$2N_2O_5 \rightarrow 4NO_2 + O_2$$

is given by the rate law

$$R = k[N_2O_5]$$

Where the concentration term appears in a rate law either as a square, or as a product of two different first power concentration terms, the reaction is described as *second order*. Thus the decomposition of hydrogen iodide

$$2HI(g) \rightarrow H_2(g) + I_2(g)$$

is second order, for its rate is given by the experimentally-found rate law

$$R = k[HI]^2$$

and for the reaction by which gaseous hydrogen and iodine combine

$$H_2(g) + I_2(g) \rightarrow 2HI(g)$$

the rate law is also second order, being

$$R = k[H_2][I_2]$$

This reaction is said to be first order in $[H_2]$ and first order in $[I_2]$, and it is thus overall a second order reaction. It is interesting to notice that for the analogous combination of hydrogen and bromine

$$H_2(g) + Br_2(g) \rightarrow 2HBr(g)$$

the experimentally found rate law in the initial stages of reaction is

$$R = k[H_2][Br_2]^{\frac{1}{2}}$$

The overall order of reaction would be described as 3/2. These two reactions therefore have different rate laws although they are stoichiometrically similar.

The reaction between hydrogen and nitric oxide described in this section is said to be *third order*. It is first order in $[H_2]$ and second order in $[NO]$.

Cases occur where increasing the concentration of a reactant decreases the rate of reaction. Thus the rate of atmospheric oxidation of solutions of Fe^{2+} according to the equation

$$O_2 + 4H^+ + 2Fe^{2+} \rightarrow 4Fe^{3+} + 2H_2O$$

is decreased by increasing $[H^+]$. Such solutions of iron(II) prepared for analytical use are made up in acid solution to reduce the rate of atmospheric oxidation.

These examples serve to illustrate that the stoichiometry of a reaction cannot be used to deduce a rate law. The rate law is derived from a series of experimental rate measurements.

12.4 EXPERIMENTAL DETERMINATION OF RATE LAWS AND RATE CONSTANTS

Here we shall describe a general method by which rate laws, and hence order of reactions, may be found experimentally. We shall take as our example the decomposition of N_2O_5 considered earlier. However, before proceeding, it is necessary to refine one description of the rate of the reaction. Previously we have used, symbolically,

$$R = -\Delta[N_2O_5]/\Delta t,$$

and the rate thus calculated is really an average rate, because the rate will not be exactly the same at the different times t_1 and t_2. To have a more precise identification of the rate with a particular time t, and a particular $[N_2O_5]$, it is necessary to make the time interval Δt as small as possible.

We are concerned then with the ratio $\Delta[N_2O_5]/\Delta t$ as Δt tends to an infinitesimally small value. From a study of the differential calculus, it is known that this ratio tends to a definite limiting value as Δt tends to zero. More generally, $\Delta c/\Delta t$ tends to a limiting value, where c is the concentration of the species which is decreasing with time. This limiting value is referred to as the differential coefficient of the concentration with respect to time, and is given the symbol $\mathrm{d}c/\mathrm{d}t$. If a negative sign is put in front of the symbol, a negative rate is indicated. In such a case, the concentration, and the rate, become smaller as time goes on.

The rate law for the N_2O_5 decomposition may now be written

$$-\frac{\mathrm{d}[N_2O_5]}{\mathrm{d}t} = k[N_2O_5]$$

or more generally

$$-\frac{\mathrm{d}c}{\mathrm{d}t} = kc$$

(Note that a second order reaction would have the form

$$-\frac{\mathrm{d}c}{\mathrm{d}t} = kc^2)$$

We now proceed to put the equation

$$-\frac{\mathrm{d}c}{\mathrm{d}t} = kc$$

into a form in which the concentration (c) is expressed as an explicit function of the time of reaction (t).

Rearranging,

$$\frac{1}{c}\frac{\mathrm{d}c}{\mathrm{d}t} = -k$$

Carrying out the process of integration with respect to t to both sides, we obtain the result

$$2.303 \log c = -kt + I$$

where I is a constant known as an integration constant. The value of this constant is found by setting the concentration to its initial value c_0, when $t = 0$,

$$\therefore I = 2.303 \log c_0$$

$$\therefore 2.303 \log c_0 = -kt + 2.303 \log c_0$$

or

$$\log c = -\frac{kt}{2.303} + \log c_0$$

This equation is of the general form y = ax + b. A plot of log c against t

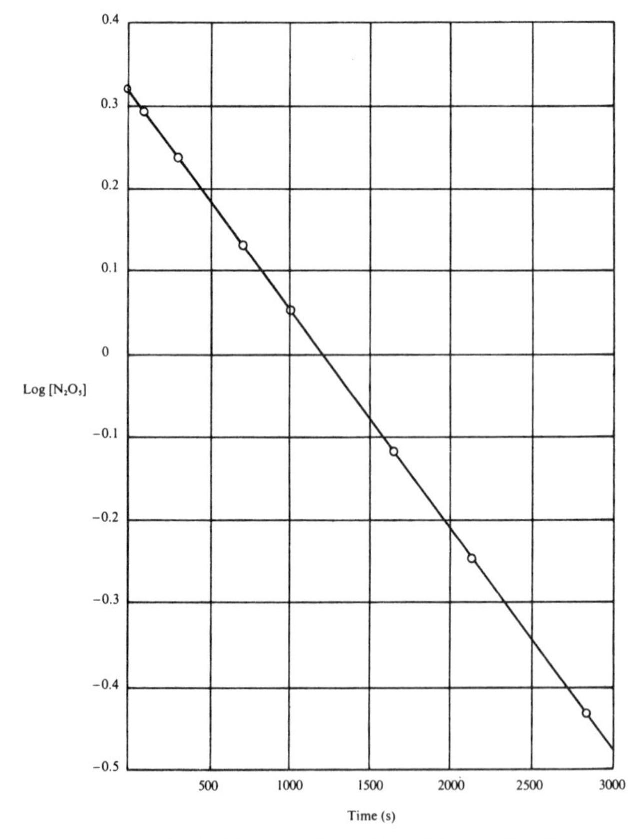

Fig 12.3 The variation of log $[N_2O_5]$ with time.

therefore yields a straight line, whose slope is $-k/2.303$, and whose intercept is $\log c_0$.

So to test if a reaction follows a first order rate law, we plot $\log c$ against t, and see if we obtain a straight line relation. If so, the reaction is first order. If we do not obtain a straight line from this plot, the reaction is not first order.

The plot of $\log [N_2O_5]$ against elapsed time is shown in Figure 12.3. That it is a straight line plot provides experimental evidence for the statement made previously, that this is a first order reaction.

Further, the gradient of the line corresponds to $-k/2.303$, and hence from the graph we can arrive at the value of the rate constant. In this case the rate constant has the value

$$k = 6.2 \times 10^{-4}\ s^{-1}$$

For second and higher order rate laws, it is also possible to produce integrated rate expressions which enable determination of the order of reaction and the rate constant from experimental data. The details of the procedures need not concern us here.

12.5 CATALYSTS

The rates of chemical reactions are influenced by a number of factors, one of which is the concentrations of those reactants involved in the rate law. A second factor which may markedly affect reaction rates is the presence of catalysts in the reacting system. Catalysts have traditionally been defined as substances which accelerate reactions without themselves being consumed in the reaction. In recent years studies of *how* catalysts function in accelerating reactions have led to a broader meaning for the term, and some substances which are used up in reactions are described as catalysts by practising chemists. Rather than offer a precise definition, we shall simply stress that catalysts do affect the rate of reactions, and *always* take part in the reaction they are catalysing.

Catalytic action which occurs in a single phase, such as an entirely gaseous reaction, or a reaction occurring completely in solution, is described as *homogeneous catalysis*. A readily observable example is the effect of manganous ions catalysing the reaction between permanganate and excess oxalate ions. The reaction

$$2MnO_4^- + 5C_2O_4^{2-} + 16H^+ \rightarrow 2Mn^{2+} + 10CO_2 + 8H_2O$$

is readily observed by change in colour of the solution from pink to colourless; the reaction is quite slow at room temperature, but the presence of a small amount of Mn^{2+}, say from $MnSO_4$ crystals added to the reacting solution, increases the rate of decoloration very considerably.

Likewise the presence of such metallic cations as Fe^{2+} and Cu^{2+} causes a marked increase in the rate of the reaction

$$2I^- + S_2O_8^{2-} \rightarrow 2SO_4^{2-} + I_2$$

This change is observable by the colour of the iodine formed.

An important example of solution catalysis is the study of reactions catalysed by acids and bases. Acid-base catalysis has been found very important in governing the rates of many organic reactions and many physiological processes. For example, the hydrolysis of an ester such as methyl acetate to give the alcohol, methanol, and the carboxylic acid, acetic acid, is catalysed by H^+. Thus for the reaction

$$\underset{\text{methyl acetate}}{CH_3CO.OCH_3} + H_2O \overset{H^+}{\rightleftharpoons} \underset{\text{methanol}}{CH_3OH} + \underset{\text{acetic acid}}{CH_3COOH}$$

the experimental rate law for the forward reaction is found to be

$$R = k[CH_3CO.OCH_3][H^+]$$

This shows that the rate is dependent upon not only a reactant but also on the concentration of the catalyst being used. It is interesting to note that the reverse reaction can occur and that this reaction of methanol and acetic acid to produce the ester is also catalysed by H^+. Studies of the catalysis of reactions by acids and bases have been very important in helping chemists to determine the *mechanisms* of reactions, i.e. the detailed sequences of molecular rearrangements which occur in the transformation of reactants into products.

An example of gaseous homogeneous catalysis is the use of nitric oxide to catalyse the oxidation of SO_2 to SO_3:

$$2SO_2 + O_2 \rightarrow 2SO_3$$

The presence of NO increases the rate at which SO_3 is produced.

The action of solid surfaces in catalysing reactions is the main type of *heterogeneous catalysis*. Heterogeneous catalysts are also used for the sulphur dioxide oxidation reaction. Thus finely divided solid platinum, or solid vanadium pentoxide, catalyse this reaction through the initial adsorption of the reactant molecules on their surface. This oxidation of sulphur dioxide is an important stage in the manufacture of sulphuric acid.

Another instance of the heterogeneous catalysis is the ammonia synthesis

$$N_2 + 3H_2 \rightarrow 2NH_3$$

The reacting gases are passed over a surface of finely-divided metallic iron. Here the adsorption of nitrogen molecules on the iron surface leads to a weakening of the bond between the atoms in the nitrogen molecule. Traces of other substances, promoters, are added to prevent the sintering together of the iron crystals at the high temperature of the reaction. Such sintering would reduce the effective surface area of the catalyst and thus reduce the adsorption of the nitrogen molecules on the surface. Thus catalysis in large-scale industrial reactions may be quite complex in practice.

It is interesting to note that, in biological systems, enzymes function as catalysts in an analogous manner to a solid surface. The enzyme is an extremely large molecular species and the molecule undergoing decomposition is adsorbed on to a specific site on the enzyme 'surface'. The efficiency of the enzyme 'solid catalysts' is usually much greater than catalysts as yet devised by chemists. In fact, if the human body had to rely on man-devised catalysts for its metabolic processes, then the body would often need to withstand reaction temperatures exceeding 100°C and the digestion of foods would often proceed for reaction times of some days.

12.6 EFFECT OF TEMPERATURE ON REACTION RATES

At higher temperatures, homogeneous reactions usually proceed at faster rates. Thus, hydrogen and iodine do not undergo reaction at an observable rate at 25°C, but at 300°C reaction occurs at a high rate

$$H_2(g) + I_2(g) \rightarrow 2HI(g)$$

As mentioned previously, the rate of this reaction is summarized by the rate law

$$R = k[H_2][I_2]$$

At constant concentrations of H_2 and I_2, an increased rate at higher temperatures must be due to an increased value of the rate constant, k. Some typical values of k are listed in Table 12.3, together with corresponding values of the reaction rate at initial hydrogen and iodine concentrations of 0.02 M.

In solution, the analytically important reaction between MnO_4^- and $C_2O_4^{2-}$ proceeds so slowly at room temperature in the absence of a catalyst that sharp endpoints cannot be obtained by titration. If the reaction mixture is heated to about 60°C before the total stoichiometric amount of titrant has been added, the reaction occurs rapidly and titration of the hot solution to the endpoint is quite feasible.

Table 12.3

Rate constant values for the $H_2 + I_2$ reaction

Reaction temperature (°C)	*Rate constant ($M^{-1}\,s^{-1}$)*	*Reaction rate $M\,s^{-1}$ (at $[H_2] = [I_2] = 0.02\,M$)*
360	1.31×10^{-3}	0.5×10^{-6}
394	6.68×10^{-3}	2.67×10^{-6}
437	46.9×10^{-3}	18.8×10^{-6}
465	138.7×10^{-3}	55.48×10^{-6}

The rate of the reaction between $S_2O_8^{2-}$ and I^- is increased with temperature, and this is shown by the shorter times at which I_2 is first visible.

The rates of most reactions increase with temperature, although cases are known where increasing temperatures reduce the rate of reaction.

QUESTIONS

1. The reaction between nitrite ion NO_2^- and iodide ion I^- in acid solution may be represented by the equation

$$2NO_2^- + 4H^+ + 2I^- \rightarrow I_2 + 2NO + 2H_2O$$

The experimental rate law for this reaction is

$$R = k[NO_2^-][I^-][H^+]^2$$

How would the rate of the reaction be altered if
(a) $[H^+]$ and $[I^-]$ were kept constant but $[NO_2^-]$ was doubled?
(b) $[I^-]$ and $[NO_2^-]$ were kept constant but $[H^+]$ was doubled?
(c) $[I^-]$ and $[NO_2^-]$ were kept constant but $[H^+]$ was halved?
(d) all concentrations were doubled?

2. For the gaseous oxidation reaction

$$2NO(g) + O_2(g) \rightarrow 2NO_2(g)$$

state the relationship between
(a) the rate of consumption of NO and
(b) the rate of consumption of O_2 and
(c) the rate of production of NO_2.

3. For the reaction

$$2A + B + C \rightarrow 3D + E$$

the rate of formation of E is found to double if [A] is doubled, keeping [B] and [C] constant, and is found to double if [B] is doubled keeping [A] and [C] constant. The rate is unaffected by [C].
Write the rate law for the reaction.

4. The rate of the reaction

$$H_2O_2 + 2I^- + 2H^+ \rightarrow I_2 + 2H_2O$$

may be calculated by measuring the time for the first appearance of I_2 in the solution, i.e. the time required for the $[I_2]$ to reach 10^{-5} M.
(a) For a particular experiment in which $[H_2O_2] = 0.010$, $[I^-] = 0.010$, and $[H^+] = 0.10$, calculate the reaction rate if I_2 first appears after 5.7 seconds.
(b) In a second experiment in which $[H_2O_2] = 0.005$, $[I^-] = 0.010$ and $[H^+] = 0.10$, calculate the reaction rate if I_2 first appears after 11.5 seconds.
(c) From these calculations show that the reaction rate depends upon $[H_2O_2]$ raised to the first power, i.e.

$$\text{that } R \propto [H_2O_2]$$

(d) Given the further information that the rate law is

$$R = k[H_2O_2][H^+][I^-]$$

calculate the rate constant k.
(e) Verify that the unit of k is $M^{-2}\,s^{-1}$.
(f) Predict the rate of reaction if

$$[H_2O_2] = 0.05,\ [H^+] = 0.10, \text{ and } [I^-] = 0.02\ M.$$

5. The rate of the reaction

$$H_3O^+ + NH_3 \rightarrow NH_4^+ + H_2O$$

was studied in 1960 and the rate expression shown to be

$$R = k[H_3O^+][NH_3]$$

where $k = 4.3 \times 10^{10}\ M^{-1}\,s^{-1}$ at 25°C. Calculate the rate of this acid-base reaction the instant 400 cm³ 0.10 M ammonia is mixed with 200 cm³ 0.05 M nitric acid. Calculate the initial rate of formation of ammonium ion, and using this initial rate as an estimate calculate the time necessary for nitric acid to react completely. Why would this calculated time be only an estimate?

6. Suggest experimental means by which the rates of the following reactions could be followed:
   (a) $CaCO_3(s) \rightarrow CaO(s) + CO_2(g)$
   (b) $2NO(g) + 2H_2(g) \rightarrow N_2(g) + 2H_2O(g)$
   (c) $Cl_2(g) + 2Br^-(aq) \rightarrow Br_2(l) + 2Cl^-(aq)$
   (d) $N_2O_4(g) \rightarrow 2NO_2(g)$
7. In the lungs, oxygen from the air is able to dissolve in blood and a constant concentration of 1.6×10^{-6} M oxygen is maintained by continuous dissolution of oxygen from respired air. The oxygen reacts in the blood with a compound of iron called haemoglobin (abbreviated here as Hb) to yield a bright red-compound, oxy-haemoglobin, HbO_2

$$Hb + O_2 \rightarrow HbO_2$$

Since the blood is recirculated by the heart the haemoglobin concentration is also maintained at a constant concentration in the lung capillaries of 8×10^{-6} M. The rate of formation of oxy-haemoglobin, R, is described by the rate law:

$$R = k[Hb][O_2]$$

where $k = 2.1 \times 10^6$ M^{-1} s^{-1} at 37°C (the body temperature).
   (a) For the constant concentrations of haemoglobin and oxygen given, calculate the rate of oxy-haemoglobin formation and the rate of oxygen consumption.
   (b) In a sample of 200 cm^3 blood, calculate the number of mole of oxygen consumed in this volume per second and hence per hour.
   (c) If the number of mole of oxygen consumed per hour were measured as a gas at 150 mmHg pressure and 37°C, what volume of oxygen would be consumed?
   (d) In some illnesses it is necessary that the rate of oxy-haeomglobin formation should be increased to 1.1×10^{-4} M s^{-1}. Since the haemoglobin concencentration is fixed at 8×10^{-6} M, to what value must the oxygen concentration be raised. Suggest a method whereby this oxygen concentration might be increased. (Hint: the concentration of oxygen dissolved in blood is directly proportional to the pressure of oxygen breathed into the lungs.)
8. The rate law for the reaction

$$2HI \rightarrow H_2 + I_2$$

is

$$R = k[HI]^2$$

   (a) How would the rate change if the concentration of HI was reduced to $\frac{1}{3}$ its original value?
   (b) In what way must the HI concentration be changed in order to double the rate?
   (c) State the unit of the rate constant k.
9. For the reaction

$$H_2O_2(aq) + 2H^+(aq) + 2I^-(aq) \rightarrow 2H_2O + I_2(aq)$$

the following data were obtained

Concentration of H_2O_2 (M)	*Concentration of I^- (M)*	*Concentration of H^+ (M)*	*Rate of formation of I_2 (M s^{-1})*
0.010	0.010	0.10	1.75×10^{-6}
0.030	0.010	0.10	5.25×10^{-6}
0.030	0.020	0.10	1.05×10^{-5}
0.030	0.020	0.20	1.05×10^{-5}

(a) Write down the rate law for the reaction.
(b) Calculate a value for the rate constant using the above data.
(c) Give the unit of the rate constant.

10. The compound nitramide, NH_2NO_2, is a white crystalline solid at room temperatures. It is readily soluble in water, in which it decomposes at a measurable rate according to:

$$NH_2NO_2(aq) \rightarrow N_2O(g) + H_2O(l)$$

Nitrous oxide (N_2O) is insoluble in water, so the rate of evolution of nitrous oxide may be used to measure the rate of decomposition of nitramide. Starting with 100 cm^3 of a 0.50 M solution of nitramide in water, the volume of N_2O evolved is measured at 0°C and 100 kN m^{-2} pressure.

Volume of N_2O evolved (cm^3)	*Time (s)*
0.00	0
5.60	500
11.20	1000

(a) What percentage of the nitramide has decomposed after 1000 seconds?
(b) Calculate the initial reaction rate, expressed as the rate of disappearance of nitramide.
(c) Given that the rate law for the decomposition of nitramide is

$$R = k[NH_2NO_2],$$

calculate the rate constant for the reaction.

11. In the presence of acid, sucrose is converted into a mixture of glucose and fructose;

$$\underset{\text{sucrose}}{C_{12}H_{22}O_{11}} + H_2O \rightarrow \underset{\text{glucose}}{C_6H_{12}O_6} + \underset{\text{fructose}}{C_6H_{12}O_6}$$

The following rate data were obtained at 25°C:

Time (s)	0	1800	3600	5400	7800	10,800
Concentration of sucrose (M)	2.005	1.804	1.615	1.451	1.259	1.069

Show that the conversion of sucrose follows a first order rate law and determine the value of the rate constant at 25°C.

12. The gas phase reaction

$$SO_2Cl_2 \rightarrow SO_2 + Cl_2$$

has a first order rate constant of $2.3 \times 10^{-5}\ s^{-1}$ at 320°C. What percentage of SO_2Cl_2 would remain if the temperature is held at 320°C for 200 minutes?

13

Homogeneous Equilibria

In carrying out calculations we have frequently assumed that chemical reactions proceed to completion. By this we mean that reactants which are not in excess are totally used up by conversion into products. For instance, in calculating the amount of hydrogen bromide obtainable by reacting 1 mole of hydrogen with 1 mole of bromine, we would assume (correctly) that 2 mole of hydrogen bromide would be formed according to

$$H_2(g) + Br_2(g) \rightarrow 2HBr(g)$$

If, however, we carried out the analogous reaction with 1 mole of hydrogen and 1 mole of gaseous *iodine*, we would find that however long we allowed the mixture of hydrogen and iodine to react, we could *never* obtain 2 mole of hydrogen iodide. The reaction of hydrogen with iodine at ordinary temperatures simply will not proceed to completion. The quantitative treatment of this and similar systems is taken up in this and the following four chapters.

13.1 EXTENT OF CHEMICAL REACTION

We shall consider first the results obtained from an experiment in which a mixture of gaseous hydrogen and gaseous iodine is reacted for some time in a closed vessel of volume 1000 cm^3 and at a fixed temperature of 458 °C. This leads to the formation of gaseous hydrogen iodide in accord with the equation

$$H_2(g) + I_2(g) \rightarrow 2HI(g)$$

Not all of the reactants are used up, however, and after a certain time the composition of the mixture in the vessel, comprising unreacted hydrogen, unreacted iodine and hydrogen iodide, remains unchanged with time. A definite amount of each reactant remains and a definite amount of product is formed. Some typical data for four different initial mixtures of H_2 and I_2 in a 1000 cm^3 vessel are set out in Table 13.1.

Table 13.1

Data on the reaction $H_2 + I_2 \rightarrow 2HI$ (458°C)

	Initial amounts of reactants (mole)		*Final amounts of reactants and product (mole)*		
	n(H_2)	n(I_2)	n(H_2)	n(I_2)	n(HI)
Run 1	1.1966×10^{-2}	0.6941×10^{-2}	0.5617×10^{-2}	0.0594×10^{-2}	1.2699×10^{-2}
2	1.2009×10^{-2}	0.8402×10^{-2}	0.4580×10^{-2}	0.0973×10^{-2}	1.4858×10^{-2}
3	1.2277×10^{-2}	0.9959×10^{-2}	0.3842×10^{-2}	0.1524×10^{-2}	1.6871×10^{-2}
4	1.2290×10^{-2}	0.8781×10^{-2}	0.4567×10^{-2}	0.1058×10^{-2}	1.5445×10^{-2}

Applying our knowledge of stoichiometry to the equation, we expect that for every mole of H_2 reacted 1 mole of I_2 will be used, and 2 mole of HI will form. That this is so may be seen by comparison of our calculated predictions with the experimental data. So considering run 1, number of mole H_2 used = $(1.1966 - 0.5617) \times 10^{-2} = 0.6349 \times 10^{-2}$.
$\therefore$ we expect that 0.6349×10^{-2} mole I_2 will be used.
Experimentally, $(0.6941 - 0.0594) \times 10^{-2} = 0.6347 \times 10^{-2}$ mole I_2 is used.
We expect, too, that $(2 \times 0.6349) \times 10^{-2}$ mole HI will form.
Experimentally, 1.2699×10^{-2} mole HI is formed.

EXERCISE
Check the agreement of the other sets of data.

If we consider the changes in composition in terms of reactant and product *concentrations* expressed in mol dm^{-3}, we note that during the course of reaction the H_2 and I_2 concentrations decrease and the concentration of HI increases. This occurs until, after a certain time, no further concentration changes occur. This may be represented graphically as in Figure 13.1. This reaction between H_2 and I_2 will be considered further in subsequent sections, but we shall now look at the decomposition reaction of hydrogen iodide, represented by the equation

$$2HI(g) \rightarrow H_2(g) + I_2(g)$$

In an experiment, a series of 1000 cm^3 flasks each containing different initial amounts of hydrogen iodide were heated to the same constant temperature, and hydrogen and iodine formed in definite amounts on decomposition. Again it is found that at a certain stage no further changes occurred in the composition of the mixtures and the corresponding data are set out in Table 13.2.

EXERCISE
Verify that the experimental results agree with the prediction that each mole of HI which decomposes produces half a mole of H_2, and half a mole of I_2.

Table 13.2

Data on the reaction $2HI \rightarrow H_2 + I_2$ (458°C)

	Initial amount of HI (mole)	*Final amounts of reactant and products (mole)*		
		n(*HI*)	n(*H_2*)	n(*I_2*)
Run 5	1.5200×10^{-2}	1.1807×10^{-2}	0.1696×10^{-2}	0.1696×10^{-2}
6	1.2866×10^{-2}	0.9999×10^{-2}	0.1433×10^{-2}	0.1433×10^{-2}
7	3.7863×10^{-2}	2.9435×10^{-2}	0.4213×10^{-2}	0.4213×10^{-2}

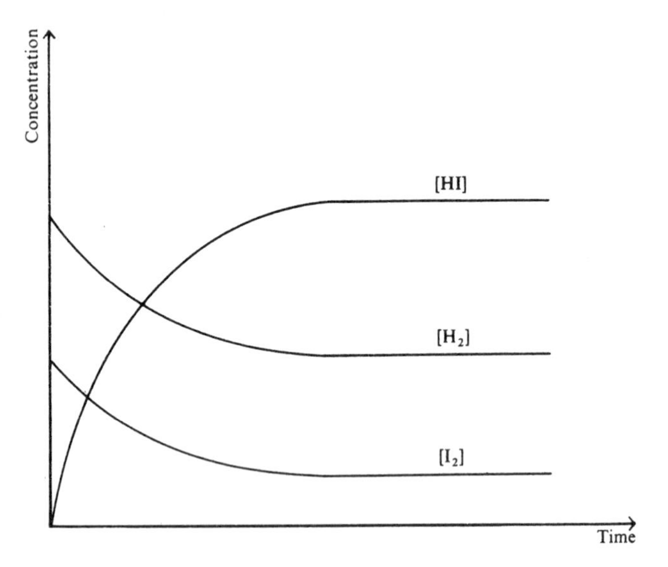

Fig. 13.1 Concentration–time graph for $H_2 + I_2 \rightarrow 2HI$.

During the course of this reaction the concentration of HI decreases and the concentrations of H_2 and I_2 increase correspondingly. After a certain time each concentration remains constant. This is represented graphically in Figure 13.2.

13.2 DYNAMIC NATURE OF THE EQUILIBRIUM STATE

In the previous chapter we saw that the rate of reactions depends, among other things, on the concentration of the reacting species. For both the reaction between H_2 and I_2 to form HI, and for the decomposition of HI into H_2 and I_2, the rate of reaction is found to decrease with decreasing concentration of reactants.

Considering the reaction between H_2 and I_2, as the reaction proceeds the concentration of each reactant decreases, so the rate of the reaction therefore decreases. Paralleling this decrease in concentration of H_2 and I_2 is an

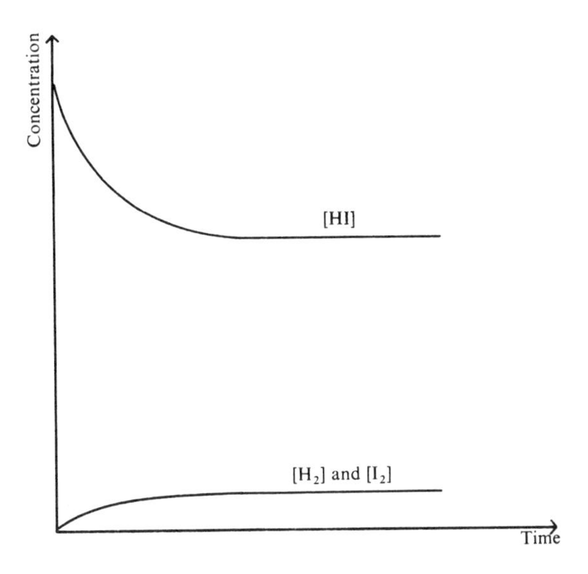

Fig 13.2 Concentration–time graph for $2HI \rightarrow H_2 + I_2$.

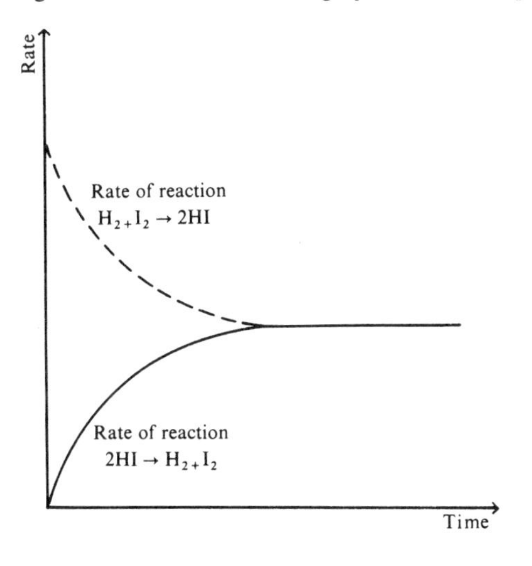

Fig 13.3 Rate–time graph for $H_2 + I_2 \rightarrow 2HI$.

increase in concentration of HI, and this causes an increase in the rate of the decomposition reaction to re-form H_2 and I_2. Thus the rate of this reverse reaction rises from initially zero while the rate of the forward reaction decreases. When the stage is reached when concentrations remain constant with time, the rate of the forward reaction must be precisely equal to the rate of the reverse reaction. The rate–time graph analogous to the concentration–time graph of Figure 13.1 is shown in Figure 13.3.

A system in which both forward and reverse reactions proceed at identical rates is described as existing in a state of dynamic equilibrium. It is important to appreciate that the equilibrium state is not one where reactions have 'stopped', and nothing is happening; it is the state where there is *no observable change* simply because, in this case, the rate at which hydrogen and iodine give rise to hydrogen iodide is the same as the rate at which hydrogen iodide decomposes back to hydrogen and iodine.

This equilibrium state is represented by the equation

$$H_2 + I_2 \rightleftharpoons 2HI$$

The reaction $H_2 + I_2 \rightarrow 2HI$ is called the *forward* reaction, while the reaction $2HI \rightarrow H_2 + I_2$ is called the *reverse* reaction, or back reaction.

The continual reaction between the species in a system at equilibrium may be detected experimentally. If a very small trace of deuterium, D_2, an isotope of hydrogen, is added to an equilibrium mixture of H_2, I_2 and HI, in a short time DI may be detected in the mixture. This shows that the forward reaction is still proceeding. Likewise, addition of DI to the original system could be used to demonstrate that the reverse reaction is still proceeding. However, the addition of more than trace amounts of a species would itself upset the equilibrium.

13.3 CONSTANCY OF THE EQUILIBRIUM CONCENTRATION FRACTION

Let us now take the data in Tables 13.1 and 13.2, express the reactant and product amount as molar concentrations, and see if these sets of concentrations are related. If we evaluate the fraction

$$\frac{[HI]}{[H_2][I_2]}$$

for each set of equilibrium concentrations, there is no apparent relation between the sets of data. If we next evaluate the fraction

$$\frac{[HI]^2}{[H_2][I_2]}$$

we find this fraction to have a *constant value*, within experimental limits, for each set of equilibrium concentrations at the given temperature. This is illustrated in Table 13.3.

Notice that the fraction which gives a constant is the one which takes account of the stoichiometry of the reaction

$$H_2 + I_2 \rightleftharpoons 2HI$$

by raising the [HI] to the second power in the equilibrium expression, and the $[H_2]$ and $[I_2]$ to the first power. The equilibrium value of this fraction is called an equilibrium constant, designated K_c, for the reaction written above, i.e.,

$$K_c = \frac{[HI]^2}{[H_2][I_2]}$$

Table 13.3

The equilibrium system $H_2 + I_2 \rightleftharpoons 2HI$ (458°C)

	Equilibrium concentrations (mol dm⁻³)			$\frac{[HI]}{[H_2][I_2]}$	$\frac{[HI]^2}{[H_2][I_2]}$
	$[H_2]$	$[I_2]$	$[HI]$		
Run 1	0.5617×10^{-2}	0.0594×10^{-2}	1.2699×10^{-2}	0.264	48.4
2	0.4580×10^{-2}	0.0973×10^{-2}	1.4858×10^{-2}	0.300	48.6
3	0.3842×10^{-2}	0.1524×10^{-2}	1.6871×10^{-2}	0.347	48.6
4	0.4567×10^{-2}	0.1058×10^{-2}	1.5445×10^{-2}	0.313	49.3
5	0.1696×10^{-2}	0.1696×10^{-2}	1.1807×10^{-2}	0.410	48.5
6	0.1433×10^{-2}	0.1433×10^{-2}	0.9999×10^{-2}	0.489	48.7
7	0.4213×10^{-2}	0.4213×10^{-2}	2.9435×10^{-2}	0.166	48.8

We would find by considering many more sets of equilibrium concentrations of H_2, I_2 and HI that, at this temperature, this fraction always had the value of 48.7 ± 0.5.

Let us now consider another example of a chemical equilibrium, this time involving molecules in solution. The dissociation of nitrogen tetroxide to nitrogen dioxide proceeds in chloroform solution according to:

$$N_2O_4 \rightleftharpoons 2NO_2$$

In Table 13.4 we quote data obtained experimentally by dissolving nitrogen tetroxide in chloroform and noting by precise measurements of the colour of the resulting solution, the amounts of nitrogen dioxide produced. For this reaction we find that the concentration fraction

$$\frac{[NO_2]^2}{[N_2O_4]}$$

remains constant at constant temperature whatever the initial concentration of nitrogen tetroxide.

The two systems discussed above are described as *homogeneous*, for in each, both reactant and product species co-exist in a single phase; in the first case all species are in the gas phase, in the second case both species are in solution. They provide examples which illustrate a general relation found to exist between the concentrations of reactants and products in a homogeneous chemical system which is at equilibrium. The generalization may be stated as follows:

Table 13.4

The equilibrium system $N_2O_4 \rightleftharpoons 2NO_2$ in chloroform (25°C)

Equilibrium concentrations (M)		$\frac{[NO_2]^2}{[N_2O_4]}$
$[NO_2]$	$[N_2O_4]$	
0.0010	0.0016	6.2×10^{-4}
0.0031	0.0155	6.1×10^{-4}
0.0060	0.057	6.3×10^{-4}
0.0110	0.195	6.2×10^{-4}

If an equilibrium mixture contains substances A, B, . . . X, Y . . . related according to the equation

$$aA + bB + \ldots \rightleftharpoons xX + yY + \ldots$$

then, at any fixed temperature, it is found experimentally that

$$\frac{[X]^x[Y]^y \ldots}{[A]^a[B]^b \ldots} = K_c$$

where K_c is a constant at constant temperature, and [X], etc., represented the equilibrium concentration of a gas X, or a dissolved species X in solution, in mol dm^{-3}. The constant K_c is called an equilibrium constant for the reaction in question.

It must be emphasized that the equation just given is applicable only to systems in which equilibrium has been attained. We have expressed A, B . . . as reactants, X, Y . . . as products, and expressed the concentration fraction in terms of

$$\frac{[\text{products}]}{[\text{reactants}]}$$

For each particular equilibrium equation, the concentrations of the species on the right-hand side are conventionally placed in the numerator. Thus for the H_2/I_2 reaction discussed earlier, the equilibrium constant

$$K_c = \frac{[HI]^2}{[H_2][I_2]} = 48.7 \text{ at } 458°C$$

refers to the equation

$$H_2 + I_2 \rightleftharpoons 2HI$$

on the other hand, the equilibrium constant

$$K'_c = \frac{[H_2][I_2]}{[HI]^2}$$

refers to the equation

$$2HI \rightleftharpoons H_2 + I_2$$

In fact, this equilibrium constant is the reciprocal of the previous constant, and has the value $1/48.7 = 0.0205$, $K_c = 1/K'_c$. Likewise, the equilibrium constant

$$K''_c = \frac{[HI]}{[H_2]^{\frac{1}{2}}[I_2]^{\frac{1}{2}}}$$

refers to the equation

$$\tfrac{1}{2}H_2 + \tfrac{1}{2}I_2 \rightleftharpoons HI$$

and has a value $\sqrt{48.7} = 6.98$, i.e. $K''_c = \sqrt{K_c}$.

Thus it is necessary to refer the actual numerical value of an equilibrium constant to a specific equation.

EXERCISE
Calculate the equilibrium constant, K_c, at 25°C, for the reaction

$$NO_2 = \frac{1}{2}N_2O_4$$

using the data of Table 13.4.

EXAMPLE 1
'The equilibrium constant, K_c, for the reaction

$$N_2(g) + 3H_2(g) \rightleftharpoons 2NH_3(g)$$

at 350°C is 1.85.' Explain the meaning of this statement.

SOLUTION
At 350°C, when N_2, H_2 and NH_3 are present together in *any equilibrium mixture* the equilibrium concentration fraction

$$\frac{[NH_3]^2}{[N_2][H_2]^3} = 1.85 = K_c$$

irrespective of the individual concentrations of each individual gas.

EXERCISE
Write down an expression for the equilibrium constant for the reaction

$$Fe^{3+}(aq) + 2Cl^-(aq) \rightleftharpoons FeCl_2^+(aq)$$

EXAMPLE 2
The equilibrium constant, K_c, for the reaction

$$2SO_2(g) + O_2(g) \rightleftharpoons 2SO_3(g) \quad \text{(a)}$$

is 800 at 527°C. Find the equilibrium constant for the equilibrium:

$$SO_3(g) \rightleftharpoons SO_2(g) + \frac{1}{2}O_2(g) \quad \text{(b)}$$

at this temperature.

SOLUTION
The equilibrium constant referring to equation (a) is

$$\frac{[SO_3]^2}{[SO_2]^2[O_2]} = K_c = 800$$

The equilibrium constant referring to equation (b) is

$$\frac{[SO_2][O_2]^{\frac{1}{2}}}{[SO_3]} = K'_c$$

Now

$$\frac{[SO_3]^2}{[SO_2]^2[O_2]} = 800$$

Taking square roots of each side,

$$\therefore \quad \frac{[SO_3]}{[SO_2][O_2]^{\frac{1}{2}}} = \sqrt{800} = 28.2$$

Taking the reciprocal of each side,

$$\therefore \quad \frac{[SO_2][O_2]^{\frac{1}{2}}}{[SO_3]} = 1/28.2 = 0.0354 = K'_c$$

The required equilibrium constant is 0.0354.
Notice that a quoted equilibrium constant always refers to a specifically written chemical equation.

EXAMPLE 3

Consider the equilibrium established between triiodide ions, iodine and iodide ions in aqueous solution

$$I_3^-(aq) \rightleftharpoons I_2(aq) + I^-(aq)$$

(a) In one particular equilibrium mixture, the concentrations of I_3^-, I_2 and I^- are respectively 0.0067 M, 0.001 M and 0.01 M. Determine the equiii-brium constant for the reaction.
(b) In another equilibrium mixture at the same temperature, the I_2 and I^- concentrations are respectively 0.001 M and 0.20 M. Calculate the concentration of I_3^-.

SOLUTION

(a) The equilibrium constant is given by

$$K = \frac{[I_2][I^-]}{[I_3^-]} = \frac{0.001 \times 0.01}{0.0067}$$

$$= 0.0015$$

(b) Since

$$\frac{[I_2][I^-]}{[I_3^-]} = 0.0015$$

$$\therefore \quad [I_3^-] = \frac{[I_2][I^-]}{0.0015}$$

and thus in this case

$$[I_3^-] = \frac{0.001 \times 0.20}{0.0015}$$

$$= 0.13 \text{ M}$$

EXERCISE

The following equilibrium is established in solution:

$$Ag^+(aq) + 2NH_3(aq) \rightleftharpoons Ag(NH_3)_2^+(aq)$$

In a particular equilibrium mixture, the concentrations of Ag^+, NH_3 and $Ag(NH_3)_2^+$ are respectively 0.0067 M, 10^{-3} M and 0.1 M. Calculate the equilibrium constant for the reaction.

In another equilibrium mixture, the concentrations of $Ag(NH_3)_2^+$ and NH_3 are respectively 0.01 M and 0.10 M. Calculate the concentration of Ag^+.

13.4 MAGNITUDE OF EQUILIBRIUM CONSTANTS

In equilibrium mixtures which have been formed from approximately the appropriate stoichiometric amounts of each initial reactant, i.e., when there is no considerable excess of any starting reactant, and in the absence of side reactions, the magnitude of the equilibrium constant normally serves as a guide to the relative proportions of products to reactants in the equilibrium mixture. This implies that the equilibrium constant may be used as a measure of the extent to which a reaction proceeds, since a high relative proportion of products to reactants would indicate a considerable extent of reaction while a very low ratio of products to reactants would indicate reaction to only a slight extent.

Thus if K_c is very large, say of the order of 10^5 or higher, when equilibrium is reached from initial reactants, there would be almost complete conversion of reactants to products, whereas if K_c is of the order of 10^{-5} or lower, when equilibrium is reached at a given temperature, only a negligible fraction of reactants will have been converted to products. For values of K_c between these extremes particularly if $10^{-2} < K_c < 10^2$, the reactants and products at equilibrium are present in comparable proportions.

So considering the reaction by which hydrogen halides may be prepared by direct union of the elements,

$$H_2(g) + X_2(g) \rightleftharpoons 2HX(g)$$

(where X = F, Cl, Br or I) at the same temperature we find

$$H_2(g) + F_2(g) \rightleftharpoons 2HF(g) \qquad K = 10^{47}$$

$$H_2(g) + Cl_2(g) \rightleftharpoons 2HCl(g) \qquad K = 10^{17}$$

$$H_2(g) + Br_2(g) \rightleftharpoons 2HBr(g) \qquad K = 10^{9}$$

$$H_2(g) + I_2(g) \rightleftharpoons 2HI(g) \qquad K = 1$$

This series of equilibrium constants indicates two things. First, the extent of the forward reaction decreases down the series, showing that at equilibrium the proportion of hydrogen halide formed to unreacted elements decreases down the series. Second, the magnitudes indicate that the first three reactions all go virtually to completion whereas, as we have already seen, the formation of hydrogen iodide proceeds only to a partial extent, and when equilibrium is established there are quite considerable concentrations of hydrogen and iodine still present.

It is important to realize that the equilibrium constant in measuring the extent of a reaction, i.e. how *far* a reaction proceeds, does not give any information at all as to the rate of the reaction, i.e., how *fast* it proceeds. These two aspects of a reaction should be regarded as quite independent. For example, when iron is burned in oxygen, the oxidation proceeds rapidly, whereas the rusting of iron in a moist atmosphere is quite slow. But the extent of both these reactions is comparable—they both proceed virtually to completion. Many reactions encountered in elementary chemistry do proceed both rapidly and almost to completion. Some, however, such as the oxidation of sulphur dioxide to sulphur trioxide in oxygen at 400°C, proceed very slowly but almost completely, while others, such as the oxidation of nitric oxide to nitrogen dioxide in oxygen at 400°C, proceeds rapidly but only to a moderate extent.

13.5 EQUILIBRIUM CONSTANTS IN GASEOUS SYSTEMS

In arriving at a general relation between the amounts of reactants and products present in a chemical system at equilibrium (see section 13.3), we chose to express these amounts in terms of the molar concentrations of the reactants and products present. This is a convenient practice which is often used by chemists. However in the case of equilibria in gaseous systems, it is much more convenient to express the amount of a gas present in terms of its partial pressure, rather than its molar concentration. Since, for any gas

$$pV = nRT$$

$$p = (n/V)RT$$

$$= 10^3[\text{gas}]\,RT$$

where p is the pressure in N m^{-2}, V the volume in m^3 and [gas] is the concentration of the gas in mol dm^{-3}. Thus, for any gas,

$$p \propto [\text{gas}]$$

It follows therefore, that for the equilibrium

$$H_2(g) + I_2(g) \rightleftharpoons 2HI$$

we can write either

$$K_c = \frac{[HI]^2}{[H_2][I_2]}, \quad \text{or} \quad K_p = \frac{p^2_{HI}}{p_{H_2}p_{I_2}}$$

where, if K_c is a constant, K_p must also be constant. Also for the equilibrium

$$N_2O_4(g) \rightleftharpoons 2NO_2(g)$$

we can write either

$$K_c = \frac{[NO_2]^2}{[N_2O_4]} \quad \text{or} \quad K_p = \frac{p^2_{NO_2}}{p_{N_2O_4}}$$

The numerical value of K_p will not necessarily be the same as K_c for any gaseous system. Indeed the numerical value of K_p will sometimes depend on the pressure units chosen to express the partial pressures of gases involved in the equilibrium. While it might seem natural to use N m^{-2} as the appropriate pressure unit, there is a long-standing convention in chemistry that the *atmosphere* should be the pressure unit used for expressing K_p values. Note that

$$p\,(\text{gas})\text{ in atmosphere} = \frac{p\,(\text{gas})\text{ in N m}^{-2}}{101{,}325}$$

The reasons for continuing this practice, at least for the present, are partly explained in the next section. In order then to arrive at a relation between K_p and K_c for any particular case we have

$$p = 10^3\,[\text{gas}]\,RT$$

$$\therefore \quad \frac{p}{101{,}325} = \frac{10^3\,[\text{gas}]\,RT}{101{,}325}$$

$$\therefore p\,(\text{in atm}) = \frac{10^3 \times 8.31}{101{,}325}[\text{gas}]\,T$$

$$= 0.082\,[\text{gas}]\,T$$

Thus, for the equilibrium

$$H_2(g) + I_2(g) \rightleftharpoons 2HI(g)$$

$$K_p = \frac{p^2_{HI}}{p_{H_2}p_{I_2}} = \frac{[HI]^2(0.082)^2T^2}{[H_2][I_2](0.082)^2T^2}$$

$$= \frac{[HI]^2}{[H_2][I_2]} = K_c$$

i.e. K_p and K_c are numerically identical.
However for

$$N_2O_4(g) \rightleftharpoons 2NO_2(g)$$

$$K_p = \frac{p^2_{NO_2}}{p_{N_2O_4}} = \frac{[NO_2]^2(0.082)^2T^2}{[N_2O_4]0.082T}$$

$$= \frac{[NO_2]^2}{[N_2O_4]}0.082T = 0.082T.K_c$$

at 298 K, $K_c = 5.4 \times 10^{-3}$, so

$$K_p = 5.4 \times 10^{-3} \times 0.082 \times 298 = 0.132$$

remembering that pressures must be expressed in the otherwise obsolescent pressure unit of atmosphere. It turns out that the numerical value of K_p will be the same as that of K_c only when the stoichiometry is such that we have the same number of mole on each side of the equilibrium. Otherwise, K_p and K_c will be numerically different.

We conclude this section by stressing that, for gaseous equilibria, it is standard practice among chemists to use K_p as the equilibrium constant, with pressures expressed in the units of atmosphere.

13.6 THE ACTIVITY CONCEPT

The concept of activity is basic to any *advanced* treatment of chemical equilibrium. We shall briefly introduce it here because, although *all* the material in this text can be handled adequately without any reference to 'activity', some familiarity with the concept may prove helpful in understanding the relation between the homogeneous equilibria discussed to date, and 'mixed phase' equilibria, such as gas-solid and solubility equilibria (see chapter 16). Every species taking part in a chemical equilibrium, whether it be solid, liquid, gas or a species in solution, may be assigned a definite activity. The activity of any substance which takes part in a chemical equilibrium is a measure of its 'effective concentration' in the equilibrium system. We shall consider three separate cases:

(1) For an ideal gas, the activity is defined to be exactly proportional to its partial pressure, i.e.,

$$a\,(\text{gas}) \propto p\,(\text{gas})$$

While a real gas will always deviate from ideal behaviour, it is none the less a good approximation at all ordinary pressures to say, for any gas

$$a\,(\text{gas}) \propto p\,(\text{gas})$$

(2) For a species in solution, we can write for an ideally-behaving solute, that its activity is proportional to its concentration, i.e.,

$$a\,(\text{solute}) \propto c\,(\text{solute})$$

While it is true, as we shall see in later chapters, that solutions containing ions are often *not* 'ideal', it is frequently a useful approximation to assume that the activity of a solute is proportional to its concentration, i.e., for any solute

$$a\,(\text{solute}) \propto c\,(\text{solute})$$

(3) For any pure solid participating in an equilibrium it happens that its activity is constant, provided only that *some* of the solid is present. The *amount* of solid present is unimportant. Direct evidence for this assertion is presented in chapter 16. For a solid then

$$a\,(\text{solid}) = \text{constant}.$$

Now for reasons of mathematical convenience we wish to define activity so that it has no dimensions; i.e. activity is to be a pure number. This is achieved by defining the activity of a substance as a ratio.

(1) For an ideal gas, we have

$$a\,(\text{gas}) = \frac{p\,(\text{gas})}{p^\circ\,(\text{gas})}$$

where the activity has been equated to the ratio of the pressure of the gas (in *any* chosen units) to the pressure of the gas in a defined *standard state*. For gases, the universally accepted convention is to take the standard state pressure, p°, as being equal to the standard atmosphere, so for an ideal gas:

$$a\,(\text{gas}) = \frac{p\,(\text{in N m}^{-2})}{101{,}325} = \frac{p\,(\text{in atm})}{1}$$

The activity of a gas is thus numerically equal to the partial pressure of the gas expressed in the obsolescent pressure unit, the atmosphere.

From the point of view of the modern trend towards the use of SI units, it would undoubtedly be more convenient to use a standard state pressure of 1 N m^{-2} rather than 101,325 N m^{-2} (1 atmosphere). If 1 N m^{-2} were chosen as the standard state pressure, ideal gas activities would then be numerically equal to the pressure in N m^{-2}. However, such a change at this point in time would be extremely confusing, since there are already in existence vast tabulations of data used in advanced physical chemistry which are based on a standard state pressure of 101,325 N m^{-2}; these tabulations would have to be substantially recalculated if the universally-accepted standard state of 1 atmosphere were altered.

(2) For an ideal solute, we have

$$a\,(\text{solute}) = \frac{c\,(\text{solute})}{c^{\circ}\,(\text{solute})}$$

where c is the concentration of the solute, and c° is the standard state concentration of the solute in the same units. When molar concentration is chosen as the appropriate unit for c, the universally accepted convention is to take the standard state concentration to be 1 mol dm^{-3}, so for an ideal solute

$$a\,(\text{solute}) = c\,(\text{in mol dm}^{-3})$$

We can summarize the above discussion by noting

(a) **the activity of an ideal gas is a pure number, numerically equal to the pressure of the gas, expressed in atmosphere.**

(b) **the activity of an ideal solute in solution is a pure number, numerically equal to the concentration of the solute expressed in mol dm^{-3}.**

The generalization we made in section 13.3 can now be restated, such that for the reaction

$$a\text{A} + b\text{B} + \ldots \rightleftharpoons x\text{X} + y\text{Y} + \ldots$$

$$K = \frac{a_{\text{X}}^{x} a_{\text{Y}}^{y} \ldots}{a_{\text{A}}^{a} a_{\text{B}}^{b} \ldots}$$

where the equilibrium constant K is dimensionless, and a will be numerically equal to the partial pressure in atmosphere of an ideal gas, and the molar concentration of an ideal solute. Thus K, as defined above, will be numerically equal to the K_{p} defined in section 13.5 for gaseous equilibria, and K_{c} when used for solution equilibria in section 13.3 It is because equilibrium constants are most conveniently defined in terms of activities, and are hence dimensionless, that we have not assigned specific dimensions to them in this text. Thus, unless there are explicit indications to the contrary, we have always expressed a gas activity as being numerically equal to its partial pressure in atmosphere, and solute activity as being numerically equal to solute molarity.

Again, unless there is explicit indication to the contrary, we have always assumed that both gases and solutes behave ideally.

13.7 RELATION BETWEEN K AND INDIVIDUAL GAS PRESSURES

A fundamental consequence of the constancy of K in any particular gaseous system at equilibrium is that while the partial pressures of *individual* gases involved in the equilibrium may vary over a wide range, the equilibrium partial pressure ratio is always the same at constant temperature, *irrespective of the value of any of the individual partial pressures.* Let us consider some examples:

EXAMPLE 1

In a particular gaseous mixture at equilibrium,

$$COCl_2 \rightleftharpoons CO + Cl_2$$

the partial pressures are found to be

$$p_{CO} = 16{,}000 \text{ N m}^{-2}, \quad p_{Cl_2} = 26{,}200 \text{ N m}^{-2} \quad \text{and} \quad p_{COCl_2} = 20{,}700 \text{ N m}^{-2}.$$

Calculate K, the equilibrium constant for the reaction.

SOLUTION

$$K = \frac{p_{CO}p_{Cl_2}}{p_{COCl_2}} \quad \text{where pressures are expressed in atmosphere}$$

$$\therefore K = \frac{(16{,}000/101{,}300)(26{,}200/101{,}300)}{(20{,}700/101{,}300)}$$

$$= \frac{0.158 \times 0.258}{0.204} = 0.20$$

EXERCISE

K for the dissociation of $COCl_2$ is 0.2 (see Example 1); calculate the partial pressure of Cl_2 in a mixture in which the partial pressures of $COCl_2$ and CO are 10,000 N m^{-2} and 50,000 N m^{-2} respectively.

A major reason for the use of partial pressure as the measure of gas 'activity' in gaseous equilibria is the fact that actual experimental results are very often obtained by direct pressure measurement. The next example shows how pressure measurement may be used to obtain a K value.

EXAMPLE 2

0.05 mole of N_2O_4 is introduced into a vessel of volume 1000 cm^3 at 25°C (298 K). The total gas pressure at equilibrium is measured to be 142,500 N m^{-2}. Calculate K for the dissociation of N_2O_4 to NO_2.

SOLUTION

First calculate the initial pressure (p_i) of the N_2O_4 before any dissociation occurs.

From the ideal gas equation, we obtain

$$p_i = \frac{nRT}{V} = \frac{0.05 \times 8.31 \times 298}{10^{-3}}$$

$$= 112{,}400 \text{ N m}^{-2} = \frac{112{,}400}{101{,}325} \text{ atm}$$

$$= 1.22 \text{ atm}$$

Now at equilibrium, the total pressure (p_{total}) will be greater than 1.22 atm, because for each mole of N_2O_4 used up by dissociation, 2 mole of NO_2 forms. Remember that the pressure depends only on the number of molecules present at a given volume and temperature, and is independent of the type of molecule.

$$p_{total} = \frac{142{,}500}{101{,}325} = 1.40 \text{ atm}$$

$$= p_{N_2O_4} + p_{NO_2}$$

If the change in partial pressure of N_2O_4 when equilibrium is reached is x,

$$p_{N_2O_4} = p_i - x$$

and

$$p_{NO_2} = 2x$$

$$\therefore\ p_{total} = (p_i - x) + 2x = p_i + x$$

$$\therefore\ 1.40 = 1.22 + x$$

$$\therefore\ x = 0.18$$

Now

$$K = \frac{p^2_{NO_2}}{p_{N_2O_4}} = \frac{(2x)^2}{p_i - x}$$

$$= \frac{(0.36)^2}{1.04} = 0.125$$

The method outlined above may be quite generally used. The procedure just described may be reversed, and a given K value can be used to calculate actual equilibrium partial pressures.

EXERCISE

Using the data in Example 1, calculate the partial pressure of NO_2 in the equilibrium mixture formed when 0.01 mole of N_2O_4 is admitted to a 1000 cm³ vessel at 25°C.

A further complication arises in an equilibrium system when extra reactant or product is added, e.g., in the gaseous dissociation

$$PCl_5(g) \rightleftharpoons PCl_3(g) + Cl_2(g)$$

1 mole each of PCl_3 and Cl_2 is provided for each mole of PCl_5 decomposed. However, extra chlorine can be independently added to the system, and

provided the system comes to equilibrium, the ratio

$$\frac{p_{Cl_2}p_{PCl_3}}{p_{PCl_5}}$$

will still be a constant at equilibrium.

EXAMPLE 3

0.03 mole of PCl_5 is added to a 1000 cm^3 vessel at 250°C. 0.02 mole of Cl_2 is also added. The equilibrium constant for the reaction is given to be

$$\frac{p_{Cl_2}p_{PCl_3}}{p_{PCl_5}} = 1.71$$

Calculate (a) the partial pressures of Cl_2, PCl_3 and PCl_5 when the system attains equilibrium, and (b) the final total pressure attained.

SOLUTION

(a) Since some PCl_3 must be formed for the system to achieve equilibrium, the reaction

$$PCl_5 \rightleftharpoons PCl_3 + Cl_2$$

must proceed to some extent in the forward direction.
Let p_i be the initial partial pressure of PCl_5, before any dissociation has occurred, and p_j be the initial partial pressure of Cl_2 before any dissociation has occurred. From the general gas equation:

$$pV = nRT$$

$$\therefore p = \frac{nRT}{V}$$

$$\therefore p_i = \frac{0.03 \times 8.31 \times 523}{1000 \times 10^{-6}} = 131{,}000 \text{ N m}^{-2}$$

$$= 1.29 \text{ atm}$$

and

$$p_j = \frac{0.02 \times 8.31 \times 523}{1000 \times 10^{-6}} = 87{,}000 \text{ N m}^{-2}$$

$$= 0.86 \text{ atm}$$

Let x be the change in the partial pressure of PCl_5 due to dissociation. Since 1 mole of PCl_3 is formed for each mole of PCl_5 decomposed

$$p_{PCl_3} = x$$

$$p_{PCl_5} = p_i - x = 1.29 - x$$

and since, for each mole of PCl_3 produced, 1 mole of Cl_2 is also produced

$$p_{Cl_2} = p_j + x = 0.86 + x$$

From the equilibrium law

$$K = 1.71 = \frac{p_{Cl_2} p_{PCl_3}}{p_{PCl_5}} = \frac{(0.86 + x)x}{1.29 - x}$$

$$\therefore\ x^2 + 0.85x - 2.21 = 0$$

Solving,
$$x = 0.68$$

$$\therefore\ p_{PCl_3} = 0.68 \text{ atm}$$

$$p_{PCl_5} = 0.61 \text{ atm}$$

$$p_{Cl_2} = 1.54 \text{ atm}$$

We check the correctness of this result by noting that

$$\frac{0.68 \times 1.54}{0.61} = 1.71$$

(b) total pressure, $p_{total} = p_{PCl_5} + p_{PCl_3} + p_{Cl_2}$

$$= 0.61 + 0.68 + 1.54$$

$$= 2.83 \text{ atm} = 287 \text{ kN m}^{-2}$$

EXERCISE

Using the data of Example 2, calculate the total pressure at equilibrium when 0.01 mole of PCl_5 and 0.005 mole of PCl_3 are added to a vessel of volume 1500 cm^3.

13.8 ALTERING THE STATE OF AN EQUILIBRIUM IN THE GAS PHASE

To illustrate further some of the implications of the equilibrium law for gaseous systems, we shall here consider the effects of both adding one of the reacting species to a system at equilibrium, and changing the total volume of a system at equilibrium. We shall discuss this in terms of the equilibrium

$$CO(g) + Cl_2(g) \rightleftharpoons COCl_2(g)$$

for which

$$K = \frac{p_{COCl_2}}{p_{CO} p_{Cl_2}} = 5 \text{ at } 600°C$$

Suppose the equilibrium partial pressures in one specific equilibrium mixture to be $p_{COCl_2}(1)$, $p_{CO}(1)$ and $p_{Cl_2}(1)$.

(a) *Addition of a reacting species to an equilibrium mixture at constant volume and constant temperature.*

If some further Cl_2 is added to the system, then instantly on addition

$$p_{Cl_2} > p_{Cl_2}(1),\ p_{CO} = p_{CO}(1) \text{ and } p_{COCl_2} = p_{COCl_2}(1)$$

and

$$\frac{p_{COCl_2}}{p_{CO} p_{Cl_2}} < K$$

The system is returned to an equilibrium position by the reaction

$$CO + Cl_2 \rightarrow COCl_2$$

proceeding until equilibrium is re-established, thus increasing the numerator term in the fraction, and decreasing the denominator term. When equilibrium is restored (remembering constant volume)

$$p_{COCl_2}(2) > p_{COCl_2}(1) \text{ and } p_{CO}(2) < p_{CO}(1)$$

Moreover,

$$\frac{p_{COCl_2}(2)}{p_{CO}(2)p_{Cl_2}(2)} = K = \frac{p_{COCl_2}(1)}{p_{CO}(1)p_{Cl_2}(1)}$$

so

$$p_{Cl_2}(2) > p_{Cl_2}(1)$$

showing that not all the added chlorine has been used up. Quantitatively, consider the data in Figure 13.4.

EXERCISE

Show that if some Cl_2 were removed from the initial equilibrium mixture (e.g. by addition of metallic mercury which would react to form $HgCl_2$, but would not react with other species present) when equilibrium is re-established

$$p_{COCl_2}(2) < p_{COCl_2}(1) \qquad p_{CO}(2) > p_{CO}(1)$$

and that this requires that

$$p_{Cl_2}(2) < p_{Cl_2}(1)$$

despite the fact that chlorine is formed by the further decomposition of phosgene.

We may state as a general conclusion that, *in any equilibrium system, addition of extra reactant will cause the reaction to proceed in the forward direction so as to re-establish equilibrium, while addition of a product will cause 'back reaction' to occur, again to re-establish equilibrium.*

(b) *Changing the total volume of the system at constant temperature.* Suppose the volume of the system is doubled by allowing the gas mixture to expand into an evacuated vessel of the same size connected to the original vessel or by pressure change in a system with a movable piston.

At the instant of doubling the volume, each partial pressure is halved:

$$p_{COCl_2} = \tfrac{1}{2}p_{COCl_2}(1) \qquad p_{CO} = \tfrac{1}{2}p_{CO}(1) \qquad p_{Cl_2} = \tfrac{1}{2}p_{Cl_2}(1)$$

Thus

$$\frac{p_{COCl_2}}{p_{CO}p_{Cl_2}} = 2K$$

and since this fraction exceeds K the system is not in equilibrium. Equilibrium is reattained by some $COCl_2$ decomposing to produce CO and Cl_2, thus effecting a decrease in the concentration fraction, i.e., by the reaction

$$COCl_2 \rightarrow CO + Cl_2$$

At the second equilibrium position $p_{COCl_2}(2) < p_{COCl_2}(1)$ since there is less of it *and* it is present in a greater volume.

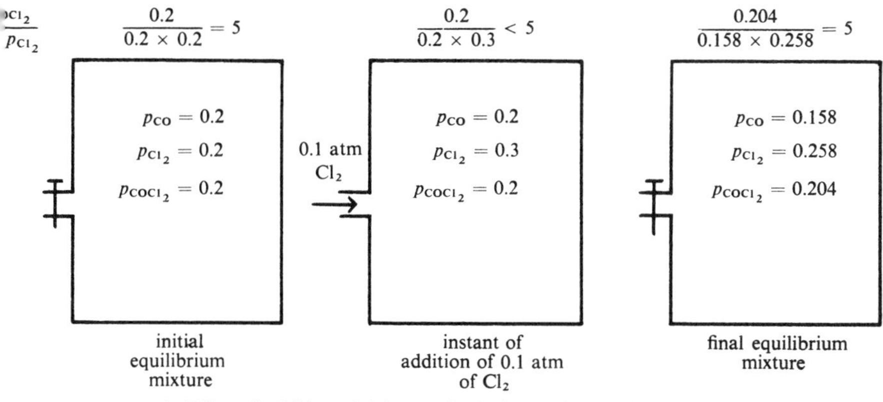

1. Effect of addition of 0.1 atm of chlorine to the equilibrium system.

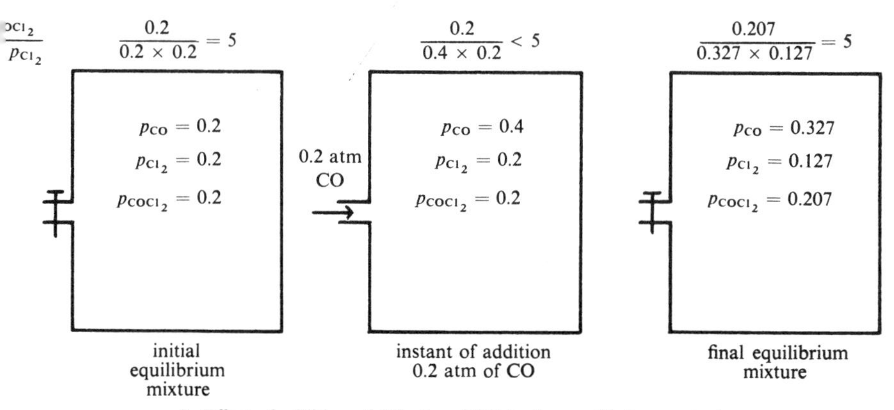

2. Effect of addition of 0.2 atm of CO to the equilibrium system.

Fig 13.4 Equilibrium data for the system $CO + Cl_2 \rightleftharpoons COCl_2$.

Since
$$\frac{p_{COCl_2}(2)}{p_{CO}(2)p_{Cl_2}(2)} = K = \frac{p_{COCl_2}(1)}{p_{CO}(1)p_{Cl_2}(1)}$$
it follows that $p_{CO}(2)$ will be less than $p_{CO}(1)$ and $p_{Cl_2}(2)$ will be less than $p_{Cl_2}(1)$, *despite the fact that there are greater masses of these species at the second equilibrium*, due to the reaction written above. The equilibrium law, remember, illustrates partial pressure changes, not mass changes.

The data in Figure 13.5 show pressure changes for an increase, and then a decrease in volume, carried out separately on an initial equilibrium mixture.

EXERCISE

Show that doubling the volume in an equilibrium system

$$H_2(g) + I_2(g) \rightleftharpoons 2HI(g)$$

does *not* result in a change in the relative *proportions* of the species present, even though each individual partial pressure will be reduced.

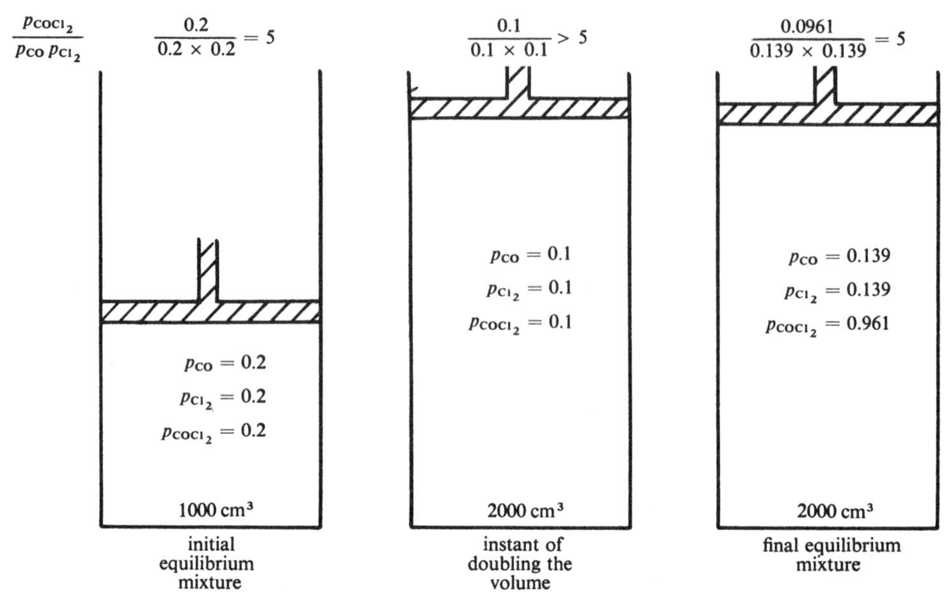

1. Effect of doubling the volume of the equilibrium system

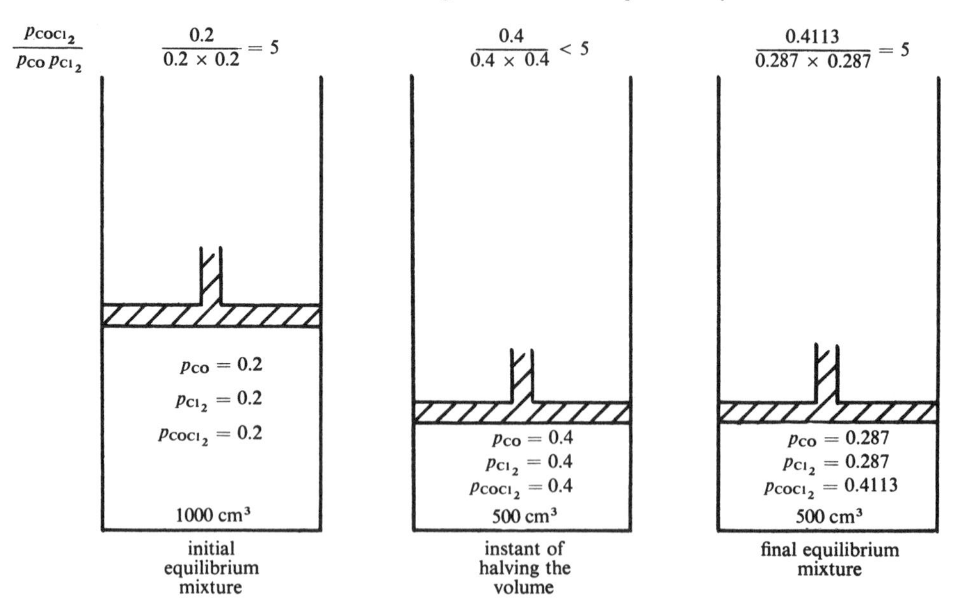

2. Effect of halving the volume of the equilibrium system.

Fig 13.5 Further data for the equilibrium $CO + Cl_2 \rightleftharpoons COCl_2$.

EXAMPLE

Changing the total volume of a system is one means of changing the pressure exerted by a system. Thus reduction of volume leads to an increase in pressure. Increases in total pressure arise, however, in other ways. One such is the addition of a reacting gaseous species to the system at constant volume and constant temperature. Another is the addition of a non-reacting gaseous species to the system at constant volume and constant temperature. By considering the addition of gaseous N_2 to the phosgene equilibrium at constant volume, show that the position of equilibrium remains unchanged, and that the equilibrium state of the system is not altered by the addition.

SOLUTION

For this equilibrium system

$$COCl_2 \rightleftharpoons CO + Cl_2$$

at any given temperature

$$\frac{p_{CO}p_{Cl_2}}{p_{COCl_2}} = \text{a constant}$$

Suppose that to this system some nitrogen is added at constant volume. Since the nitrogen will not react with CO, the mass (and number of mole) of CO will be unchanged by the addition. Further, since there is no volume change during the addition, the partial pressure of CO will not be changed. Similarly, as the N_2 will not react either with Cl_2 or $COCl_2$, their partial pressures stay unchanged.

Despite the increase in total pressure, the equilibrium partial pressures of each gas stay unchanged; the partial pressure ratio remains unaltered, and the system remains in a state of equilibrium. Note that if the N_2 was added at constant total pressure, thus allowing an increase in the volume of the system, the partial pressures of the various species would have altered with a consequent shift in the position of equilibrium.

13.9 RELATION BETWEEN K AND INDIVIDUAL CONCENTRATIONS FOR SOLUTION EQUILIBRIA

A fundamental consequence of the constancy of K in any particular equilibrium in solution is that while the concentration of *individual* solutes involved in the equilibrium may vary over a wide range, the equilibrium concentration fraction is always the same at constant temperature, *irrespective of the values of any of the individual concentrations.* Let us illustrate with some examples.

EXAMPLE 1

0.10 mole of N_2O_4 is added to 500 cm^3 of chloroform. At equilibrium, the concentration of NO_2 is measured, by noting the intensity of the brown colour of NO_2 in the solution, to be 0.011 M at 25°C, calculate K for the equilibrium.

SOLUTION

The concentration of N_2O_4, assuming no dissociation, would be 0.20 M. Since for every mole of NO_2 formed 0.5 mole of N_2O_4 must have dissociated, the change in $[N_2O_4]$ must be $0.5 \times 0.011 = 0.0055$. Thus, in the equilibrium mixture

$$[N_2O_4] = 0.20 - 0.0055 = 0.1945 \text{ M}$$

$$\text{and } [NO_2] = 0.011$$

$$K = \frac{[NO_2]^2}{[N_2O_4]} = \frac{(0.011)^2}{0.1945} = 6.2 \times 10^{-4}$$

The reverse of the procedure described above may be carried out by using a known K to calculate equilibrium concentrations.

EXAMPLE 2

Tribromide ions in aqueous solution are in equilibrium with bromide ions and molecular bromine, according to

$$Br_3^- \rightleftharpoons Br_2 + Br^-$$

where
$$K = \frac{[Br_2][Br^-]}{[Br_3^-]} = 0.05 \text{ at } 25°\text{C}$$

Calculate the concentrations of Br_2, Br^- and Br_3^- at equilibrium in a solution formed by dissolving 0.10 mole of $NaBr_3$ in 1000 cm^3 of water at 25°C.

SOLUTION

The initial concentration of Br_3^-, before any dissociation occurs, is 0.10 M. Suppose that at equilibrium, when Br_3^- has partially dissociated, x mol dm^{-3} of Br_3^- has been used up. Then at equilibrium,

$$[Br_3^-] = 0.10 - x$$

Since for every mole of Br_3^- dissociated we get 1 mole of Br_2 and 1 mole of Br^-, we have

$$[Br^-] = [Br_2] = x$$

Now
$$K = \frac{[Br_2][Br^-]}{[Br_3^-]} = \frac{x^2}{0.10 - x} = 0.05$$

$$\therefore x^2 + 0.05x - 0.005 = 0$$

Solving the quadratic,

$$x = 0.05$$

$$\therefore [Br_3^-] = 0.05 \text{ M and also}$$

$$[Br_2] = [Br^-] = 0.05 \text{ M}$$

A more complex situation arises when extra amounts of *one* reactant or product are added to a reaction mixture. This is illustrated in the following example.

EXAMPLE 3

0.10 mole of Br^- and 0.02 mole of Br_2 are added to 500 cm³ of water. Calculate the concentrations of Br_3^-, Br_2 and Br^- present at equilibrium, given that

$$\frac{[Br_2][Br^-]}{[Br_3^-]} = 0.05$$

SOLUTION

$[Br^-]$ before any Br_3^- is formed will be 0.20 M. $[Br_2]$ before any Br_3^- is formed will be 0.04 M. Let x be the concentration of Br_3^- present at equilibrium. Since for each mole of Br_3^- formed, 1 mole each of Br_2 and Br^- must be consumed, we have

$$[Br^-] = 0.20 - x$$

$$[Br_2] = 0.04 - x$$

Thus

$$\frac{(0.04 - x)(0.20 - x)}{x} = 0.05$$

This simplifies to the quadratic

$$x^2 - 0.29x + 0.008 = 0$$

So

$$x = 0.031, \quad \text{and}$$

$$[Br^-] = 0.169 \text{ M}$$

$$[Br_2] = 0.009 \text{ M}$$

$$[Br_3^-] = 0.031 \text{ M}$$

EXERCISE

Using the data of Example 1 calculate the equilibrium concentration of N_2O_4 and NO_2 formed when 0.10 mole of N_2O_4 is added to 1 dm³ of chloroform at 25°C. ($K = 6.2 \times 10^{-4}$.)

13.10 ALTERING THE STATE OF AN EQUILIBRIUM MIXTURE IN SOLUTION

We have emphasized the constancy of the equilibrium concentration fraction at constant temperature, despite the fact that the concentration of *individual* chemical species may vary quite markedly. This, of course, is the crux of the equilibrium law. The point is further illustrated by considering the effect of adding extra reactant to an equilibrium system in solution. Consider the equilibrium reaction between the tribromide ion, bromine and bromide ion:

$$Br^-(aq) + Br_2(aq) \rightleftharpoons Br_3^-(aq)$$

$$\frac{[Br_3^-]}{[Br^-][Br_2]} = 20 = K$$

Suppose one specific equilibrium mixture of the solutes to be denoted by

$[Br_3^-]_1$, $[Br^-]_1$ and $[Br_2]_1$. If some further bromide ion is added to the system, then instantly on addition

$$[Br^-] > [Br^-]_1, \quad [Br_2] = [Br_2]_1 \quad \text{and} \quad [Br_3^-] = [Br_3^-]_1$$

and
$$\frac{[Br_3^-]}{[Br^-][Br_2]} < K$$

indicating that the system is not in an equilibrium state. To restore equilibrium, the concentration fraction must increase, and this is achieved by the occurrence of the reaction

$$Br^-(aq) + Br_2(aq) \rightarrow Br_3^-(aq)$$

until equilibrium is re-established, by increasing the numerator term in the concentration fraction expression, and decreasing the denominator term. When equilibrium is restored

$$[Br_3^-]_2 > [Br_3^-]_1 \quad \text{and} \quad [Br_2]_2 < [Br_2]_1$$

and naturally

$$\frac{[Br_3^-]_2}{[Br^-]_2[Br_2]_2} = K = \frac{[Br_3^-]_1}{[Br^-]_1[Br_2]_1}$$

Furthermore $[Br^-]_2 > [Br^-]_1$, showing that all the extra Br^- has not been used up.
Quantitatively, the point is illustrated in Figure 13.6.

EXERCISE
(a) Describe qualitatively the effect of halving the bromine concentration in the initial equilibrium mixture shown in Figure 13.6 on the concentrations of bromide and tribromide ions.
(b) Calculate, by the method shown in 13.9, the new equilibrium concentrations of bromide, tribromide and bromine when the bromine concentration is halved as described in (a).

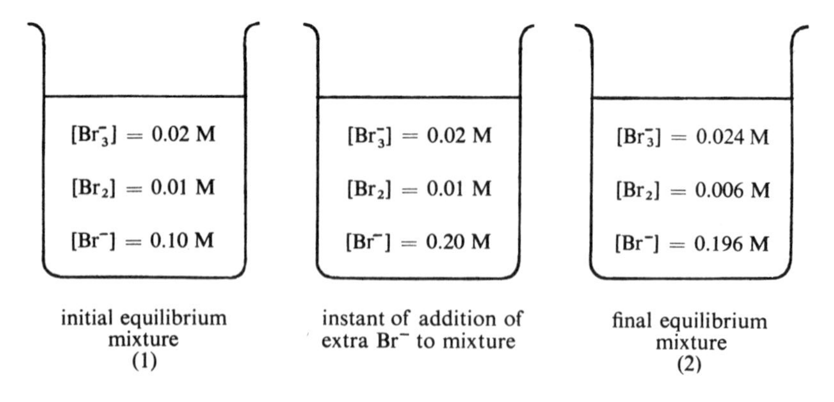

Fig 13.6 Effect of addition of Br^- to the equilibrium mixture $Br^- + Br_2 \rightleftharpoons Br_3^-$.

We may again state as a general conclusion that, in any equilibrium system, addition of extra reactant will cause the reaction to proceed in the forward direction, while addition of extra product will cause 'back reaction' to occur. In each case the reaction occurs so as to re-establish an equilibrium solution.

13.11 COMPETING EQUILIBRIA

The usefulness of the equilibrium constant for calculating reactant and product concentrations in a system at equilibrium is further illustrated in systems where there is more than one established equilibrium. Examples are available for both gaseous and solution equilibria. We shall consider only gaseous equilibria in this section; the same principles may however be applied unchanged to solution equilibria, and indeed examples of competing equilibria in solution are discussed in chapters 14, 15 and 16. Consider, for example, the gaseous system

$$PCl_5 \rightleftharpoons PCl_3 + Cl_2 \tag{1}$$

where

$$K_1 = \frac{p_{PCl_3}p_{Cl_2}}{p_{PCl_5}} \tag{1'}$$

If some carbon monoxide is added, some of the chlorine will react according to

$$CO + Cl_2 \rightleftharpoons COCl_2 \tag{2}$$

and

$$K_2 = \frac{p_{COCl_2}}{p_{CO}p_{Cl_2}} \tag{2'}$$

When a final state of equilibrium is established, we will have present in the system PCl_5, PCl_3, CO, $COCl_2$ and Cl_2, and their concentrations will be related by the expressions given for K_1 and K_2. Thus the two equilibria will exist simultaneously in the vessel.

The vessel contains specific equilibrium partial pressures of five species—PCl_5, PCl_3, Cl_2, CO and $COCl_2$. Four of these, however, are described by the equation

$$CO(g) + PCl_5(g) \rightleftharpoons COCl_2(g) + PCl_3(g) \tag{3}$$

Thus a third, but not 'independent', equilibrium must exist in addition to the other two. By definition

$$K_3 = \frac{p_{COCl_2}p_{PCl_3}}{p_{CO}p_{PCl_5}} \tag{3'}$$

Since this third equilibrium is not independent of the other two (it exists as a consequence of the other two), K_3 is not independent of K_1 and K_2.

In fact, from (1′),

$$p_{Cl_2} = K_1 \times \frac{p_{PCl_5}}{p_{PCl_3}}$$

Substituting into (2′)

$$K_2 = \frac{p_{COCl_2}}{p_{CO}} \times \frac{p_{PCl_3}}{K_1 p_{PCl_5}}$$

$$= \frac{K_3}{K_1}$$

$$\text{or } K_3 = K_1 \times K_2$$

Thus the three equilibrium constants are related, and since these equilibrium concentration fractions are constant irrespective of individual concentrations, it follows that, if any two of the equilibrium constants are known, the third can be calculated.

Equation (3) may be regarded as the sum of equations (1) and (2), and its equilibrium constant is the product of those of the other two reactions, i.e.:

$$PCl_5 \rightleftharpoons PCl_3 + Cl_2 \qquad K_1$$

$$CO + Cl_2 \rightleftharpoons COCl_2 \qquad K_2$$

$$(1) + (2) \qquad PCl_5 + CO \rightleftharpoons PCl_3 + COCl_2 \qquad K_3 = K_1 \times K_2$$

EXAMPLE 1

At 1565 K, for the water gas reaction

$$CO(g) + H_2O(g) \rightleftharpoons CO_2(g) + H_2(g) \tag{1}$$

and

$$K_1 = 0.36$$

and for the oxidation of carbon monoxide

$$CO(g) + \tfrac{1}{2}O_2(g) \rightleftharpoons CO_2(g) \tag{2}$$

and

$$K_2 = 9.8 \times 10^5$$

Use these data to find the equilibrium constant K_3 for the decomposition of steam.

$$H_2O(g) \rightleftharpoons H_2(g) + \tfrac{1}{2}O_2(g) \tag{3}$$

SOLUTION

$$K_3 = \frac{p_{H_2} p_{O_2}^{\frac{1}{2}}}{p_{H_2O}}$$

Now from (2), since

$$K_2 = \frac{p_{CO_2}}{p_{CO} p_{O_2}^{\frac{1}{2}}}$$

$$\therefore p_{O_2}^{\frac{1}{2}} = \frac{p_{CO_2}}{K_2 \times p_{CO}}$$

Substituting into the expression above

$$K_3 = \frac{p_{H_2}}{p_{H_2O}} \times \frac{p_{CO_2}}{K_2 p_{CO}}$$

$$= \frac{K_1}{K_2}$$

$$= \frac{0.36}{9.8 \times 10^5}$$

$$= 3.7 \times 10^{-7}$$

Alternatively, by subtraction

	$CO + H_2O \rightleftharpoons CO_2 + H_2$	K_1
	$CO + \frac{1}{2}O_2 \rightleftharpoons CO_2$	K_2
(1) − (2)	$H_2O \rightleftharpoons H_2 + \frac{1}{2}O_2$	$K_3 = K_1/K_2$

EXAMPLE 2

Show how the equilibrium constants K_1 and K_2 for the respective reactions

$$H_2(g) + Cl_2(g) \rightleftharpoons 2HCl(g) \quad (1)$$

$$2H_2(g) + O_2(g) \rightleftharpoons 2H_2O(g) \quad (2)$$

are related to the equilibrium constant K_3 for the reaction

$$4HCl(g) + O_2(g) \rightleftharpoons 2H_2O(g) + 2Cl_2(g) \quad (3)$$

(One could envisage these three equilibria arising from the addition of O_2 to an equilibrium mixture of H_2, Cl_2 and HCl.)

Equation (3) is obtained by doubling equation (1), and subtracting this from equation (2)

	$2H_2 + O_2 \rightleftharpoons 2H_2O$	K_2
	$2H_2 + 2Cl_2 \rightleftharpoons 4HCl$	K_1^2
(2)−2×(1)	$O_2 - 2Cl_2 \rightleftharpoons 2H_2O - 4HCl$	
i.e.	$O_2 + 4HCl \rightleftharpoons 2H_2O + 2Cl_2$	$K_3 = K_2/K_1^2$

3.12 CATALYSTS AND EQUILIBRIUM

Catalysts, as we have seen, alter the rate of reactions. It has been found experimentally however that the *presence of a catalyst alters the rate of both forward and reverse reactions to precisely the same extent.* There are no 'one-way' catalysts. So the presence of a catalyst does not alter the *position* of an equilibrium; not only does K remain constant, but the individual equilibrium concentrations remain unchanged.

In the industrially important equilibrium reaction

$$2SO_2 + O_2 \rightleftharpoons 2SO_3$$

SO_3 may be formed from SO_2 and O_2 more rapidly in the presence of suitable catalysts, but as seen in Table 13.5 the proportion of SO_3 formed from given initial starting materials is unaffected by the catalysts.

Table 13.5

Catalysis of the $2SO_2 + O_2 \rightleftharpoons 2SO_3$ reaction

T 527°C	p_{SO_2}*(atm)*	p_{O_2}*(atm)*	p_{SO_3}*(atm)*	K	*Time taken*
(a) uncatalysed	0.2	0.4	0.192	12.2	days
(b) catalysed by gaseous NO	0.2	0.4	0.192	12.2	minutes
(c) catalysed by solid V_2O_5 supported on asbestos	0.2	0.4	0.192	12.2	minutes
(d) catalysed by finely-divided Pt supported on silica-gel	0.2	0.4	0.192	12.2	minutes

13.13 RELATION OF TEMPERATURE TO EQUILIBRIUM CONSTANT

In all the situations discussed so far, we have specified a fixed, unchanging temperature. We found that the percentage of a specific product or reactant may be increased or decreased at equilibrium by altering conditions such as the relative proportions of initial reactants or by imposing changes of condition upon an equilibrium mixture. In each case we stressed that the value of the equilibrium concentration fraction, i.e. the equilibrium constant, remained unchanged. The equilibrium constant does vary with temperature, however, and values of K at different temperatures are known for many reactions.

Consider, for instance, the data in Table 13.6 for the well-studied ammonia synthesis equilibrium

$$N_2 + 3H_2 \rightleftharpoons 2NH_3$$

for which

$$K = \frac{p^2_{NH_3}}{p_{N_2}p^3_{H_2}}$$

This table records the experimentally-determined equilibrium constants for the reaction written above at a series of temperatures. There is a continuous decrease in the equilibrium constant with increase in temperature.

In the case of the decomposition of N_2O_4, there is a continuous increase in the equilibrium constant with increasing temperature. We find

T(°C)	$K = p^2_{NO_2}/p_{N_2O_4}$
25	0.141
35	0.317
45	0.677

Table 13.6

Equilibrium constants for the reaction
$N_2 + 3H_2 \rightleftharpoons 2NH_3$

T (°C)	K
350	0.0266
400	0.0129
450	0.00659
500	0.00381

In the reaction

$$N_2 + 3H_2 \rightleftharpoons 2NH_3$$

it is found that the forward reaction of this equilibrium system is exothermic, while for the N_2O_4 decomposition

$$N_2O_4 \rightleftharpoons 2NO_2$$

it is found that the forward reaction of this equilibrium system is endothermic.

It is a matter of general experience that,

(a) **for those equilibria in which the forward reaction is exothermic, i.e. ΔH negative, the equilibrium constant decreases with rise in temperature;**

(b) **for those equilibria in which the forward reaction is endothermic, i.e. ΔH positive, the equilibrium constant increases with rise in temperature.**

EXAMPLE 1

For the forward reaction in the equilibrium system

$$SO_2 + \tfrac{1}{2}O_2 \rightleftharpoons SO_3 \qquad \Delta H = -93.6\ \text{kJ}$$

Predict the effect of increasing temperature on K, and thus on the proportion of SO_3 in a specific equilibrium mixture at constant pressure. (By convention, ΔH refers to the forward reaction.)

SOLUTION

Since the forward reaction is exothermic, K decreases with rising temperature, and so the proportion of SO_3 in a particular equilibrium mixture decreases as the temperature rises.

EXAMPLE 2

For the forward reaction

$$\tfrac{1}{2}H_2(g) + \tfrac{1}{2}Br_2(g) \rightleftharpoons HBr(g) \qquad \Delta H = -35.1\ \text{kJ}$$

How will a decrease in temperature affect the extent of dissociation of gaseous hydrogen bromide?

SOLUTION

The dissociation reaction is endothermic so K decreases with decrease in temperature. Thus the proportion of products H_2 and Br_2 is less at lower temperatures. The extent of dissociation is therefore reduced at lower temperatures.

When data from a large number of equilibria are studied, it is found that a simple relation exists between K, the equilibrium constant, ΔH, the enthalpy change for the reaction, and the absolute temperature. This relation is

$$\log K = -\frac{\Delta H}{2.303RT} + C$$

where C is a constant. ΔH is the heat content change for the forward reaction, R is the gas constant and T the absolute temperature. This relation holds for all equilibria, both in the gas phase and in solution. This relation can be used to quantitatively predict the way in which K will vary with temperature, or alternatively it can be used to obtain ΔH for a reaction by measuring K as a function of temperature. Plots of log K against $1/T$ are shown in Figure 13.7. They are straight lines of slope $-(\Delta H/2.303R)$.

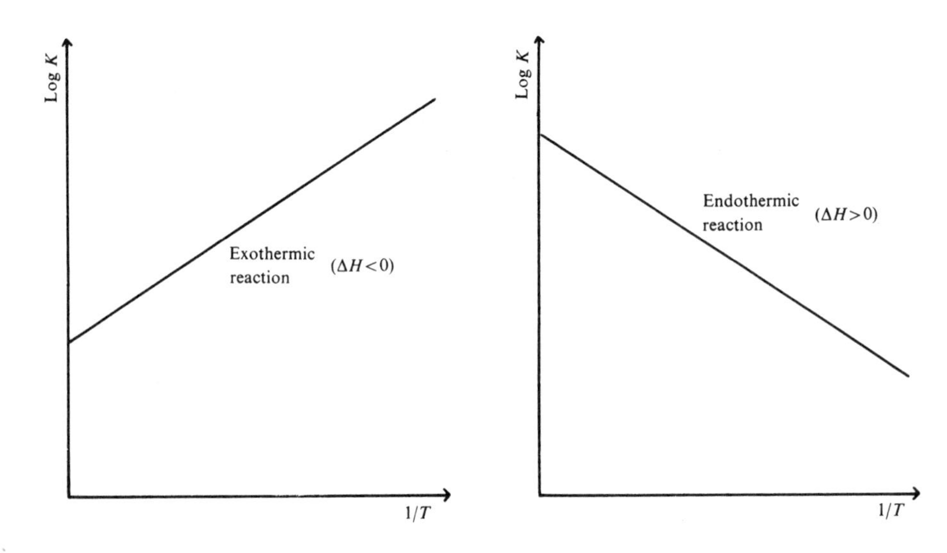

Fig 13.7 Graph of log K against $1/T$.

QUESTIONS

1. Write expressions for the equilibrium constant for each of the following homogeneous reactions:
   (a) $N_2(g) + O_2(g) \rightleftharpoons 2NO(g)$
   (b) $Cl_2(g) + 3F_2(g) \rightleftharpoons 2ClF_3(g)$
   (c) $CO(g) + \frac{1}{2}O_2(g) \rightleftharpoons CO_2(g)$
   (d) $3O_2(g) \rightleftharpoons 2O_3(g)$

2. For the equilibrium system $A + B \rightleftharpoons C$, the equilibrium concentrations in solution in mole dm^{-3} are:

$$[A] = 0.1 \quad [B] = 0.2 \quad [C] = 0.3$$

Calculate the equilibrium constant for the reaction.

3. For the equilibrium system $3A + B \rightleftharpoons C + 2D$, the equilibrium concentrations in solution in mole dm^{-3} are:

$$[A] = \tfrac{1}{2} \quad [B] = \tfrac{1}{3} \quad [C] = \tfrac{1}{4} \quad [D] = \tfrac{1}{5}$$

Calculate the equilibrium constant for the reaction.

4. The equilibrium constant for the reaction $A + B \rightleftharpoons C + D$ is 4. If in a gaseous equilibrium mixture $[A] = 5$, $[B] = 2$ and $[D] = 4$, find $[C]$.
5. The equilibrium constant for the equilibrium system $2X \rightleftharpoons 3Y$ is $\tfrac{1}{8}$. If $p_X = 8$, find p_Y.
6. For the equilibrium $X \rightleftharpoons Y + Z$, a 500 cm^3 flask contains 1 mole X, 1.5 mole Z and 2.5 mole Y in equilibrium together at 0°C. Find the equilibrium constant for the reaction. If a second equilibrium mixture has $p_Y = 10$ atm and $p_X = 6$ atm, find p_Z.
7. The following data represent equilibrium concentrations of dissolved N_2O_4 and dissolved NO_2 in chloroform solution, where N_2O_4 dissociates according to $N_2O_4 \rightleftharpoons 2NO_2$. In each case evaluate the fractions $[NO_2]/[N_2O_4]$, and $[NO_2]^2/[N_2O_4]$, and explain the significance of your calculations.

$[N_2O_4]$	$[NO_2]$
0.129	1.17×10^{-3}
0.227	1.61×10^{-3}
0.324	1.85×10^{-3}
0.405	2.13×10^{-3}

All concentrations are expressed in mole dm^{-3}.

8. (a) An equilibrium mixture contains 2.0 mole bromine, 1.25 mole hydrogen and 0.5 mole hydrogen bromide, at a fixed temperature, and is present in a 4000 cm^3 vessel. Determine the equilibrium constant for the reactions represented by
   (i) $2HBr(g) \rightleftharpoons H_2(g) + Br_2(g)$
   (ii) $\tfrac{1}{2}H_2(g) + \tfrac{1}{2}Br_2(g) \rightleftharpoons HBr(g)$
   (b) In another experiment conducted at this temperature, some HBr was admitted into an evacuated 2000 cm^3 vessel, and when equilibrium was attained, some had decomposed, yielding 6.32 mole bromine as one product.
   (i) What was the concentration of each species present at equilibrium?
   (ii) What mass of HBr was originally let into the vessel?
9. Suppose 3 mole HCl and 2 mole O_2 are introduced into a 5000 cm^3 vessel and the temperature held constant at 450°C until equilibrium is attained according to the reaction:

$$4HCl(g) + O_2(g) \rightleftharpoons 2H_2O(g) + 2Cl_2(g)$$

$$\Delta H = -113 \text{ kJ}$$

   (a) From this data, could the equilibrium constant be calculated? If so, find its value. If not, what further data would be needed?
   (b) How will the value of K for the system at 450°C and 1 atm compare with that at 550°C and 1 atm?

(c) If the temperature is maintained at 450°C, but the system permitted to expand, i.e. the volume increases until the pressure is reduced to 0.5 atm, how will the relative concentrations at equilibrium compare with those at 450°C and 1 atm pressure?
(d) How will the value of K for the system at 450°C and 1 atm compare with that at 450°C and 0.5 atm?
(e) Show how the equilibrium constant expression for this reaction is related to those for the reactions:
(i) $2H_2O(g) \rightleftharpoons 2H_2(g) + O_2(g)$
(ii) $2HCl(g) \rightleftharpoons H_2(g) + Cl_2(g)$

10. Considering the hypothetical equilibrium system in the gas phase

$$A + B \rightleftharpoons C + D$$

for which ΔH is negative, how will the equilibrium partial pressure of D be altered by:
(a) halving the volume of the system at constant temperatures;
(b) removing A by chemical reaction, at constant volume and temperature;
(c) adding C at constant volume and temperature;
(d) lowering the temperature keeping the volume constant;
(e) adding a catalyst to the equilibrium mixture at constant volume and temperature?

11. Complete the following table of data for the equilibrium

$$H_2(g) + CO_2(g) \rightleftharpoons H_2O(g) + CO(g) \text{ at } 986°C$$

Experiment	*Initial concentrations*				*Equilibrium concentrations*				K
	$[H_2]$	$[CO_2]$	$[H_2O]$	$[CO]$	$[H_2]_e$	$[CO_2]_e$	$[H_2O]_e$	$[CO]_e$	
1	1.0	1.0	0	0	0.44	0.44	0.56	0.56	—
2	0	0	1.0	1.0	—	—	—	—	1.60
3	1.0	—	0	0	0.27	1.27	0.73	0.73	1.61
4	2.0	2.0	0	0	0.88	—	—	—	1.60
5	0	0	—	—	0.88	0.88	1.12	1.12	—
6	1.0	—	—	1.0	0.883	0.883	1.117	1.117	1.59
7	0.2	0.4	0.6	0.8	—	—	0.456	—	1.60

12. In the following questions, consider the equilibrium:

$$PCl_5(g) \rightleftharpoons PCl_3(g) + Cl_2(g)$$

for which $K = 1.70$ at 250°C.
(a) If PCl_5 is 48.5% dissociated at 200°C, 1 atm, and 97% dissociated at 300°C, 1 atm, state whether the decomposition reaction is exothermic, or endothermic.
(b) If 1 mole $PCl_5(g)$ in a 1000 cm^3 flask at 250°C is allowed to come to equilibrium, find the equilibrium partial pressure of each species present.
(c) If 5 mole $PCl_5(g)$ were initially present as in (b) calculate the equilibrium concentrations of each species.
(d) Proceed as in (b) and (c) for the following initial mixtures of species:
(i) 1 mole of PCl_3 and 1 mole Cl_2,
(ii) 1 mole of PCl_5 and 1 mole Cl_2.

14

Acids and Bases

In the last chapter we dealt with the general principles of chemical equilibria for reactions both in the gas phase and in solution. In this chapter we shall concern ourselves specifically with equilibria involving acids and bases, mainly in aqueous solution. We shall precede the specific consideration of acids and bases with a brief, general discussion of aqueous solutions.

4.1 SOLUTIONS

A solution of sodium chloride in water differs from a solution of sugar in water in an important respect. The former conducts electricity while the latter does not. Solutions which conduct electric current are described as *electrolytes*, and those which do not conduct current are termed *non-electrolytes.*

Conductivity in solution is attributed to the presence of ions which are free to move independently. Substances such as NaCl and $BaSO_4$, which are essentially ionic in the pure state, give rise to free ions when in solution. Saturated solutions of these two compounds would differ markedly in their conductivity because of the far greater solubility of NaCl. In each case, however, all of the solid which has dissolved is present in the solution as free ions, viz. Na^+, Cl^-, and Ba^{2+}, SO_4^{2-}.

Not only substances which exist as ions in the pure state form ionic solutions; some substances which do not consist of ions form electrolyte solutions because of reaction with water (hydrolysis). Thus pure hydrogen chloride, and pure ammonia hydrolyse when dissolved in water and the following equilibria are set up:

$$HCl(aq) + H_2O(l) \rightleftharpoons H_3O^+(aq) + Cl^-(aq) \quad (1)$$

$$NH_3(aq) + H_2O(l) \rightleftharpoons NH_4^+(aq) + OH^-(aq) \quad (2)$$

The relative extents to which these substances hydrolyse may be estimated by the conductivities of the aqueous solutions, provided solutions of the same analytical concentration are considered. Thus the conductivity of 0.1 M HCl solution is found to be far greater than that of 0.1 M NH_3 solution. In the solution of HCl, virtually all HCl molecules may be regarded as hydrolysed to ions—the forward reaction in (1) proceeds to completion. In the ammonia solution only a small fraction of the NH_3 molecules hydrolyse to NH_4^+ and OH^-—the forward reaction in (2) proceeds only to a slight extent (approximately 1 per cent for a 0.1 M solution).

Electrolyte solutions such as HCl solution, in which hydrolysis is almost complete, are frequently described as *strong electrolytes*. Electrolyte solutions such as NH_3 solution, in which hydrolysis is slight, are described as *weak electrolytes*.

Not all electrolytes fall readily into these two categories: as well as those which are virtually completely hydrolysed (strong electrolytes), and those which are only slightly hydrolysed (weak electrolytes), there are those in which hydrolysis occurs to some intermediate extent, say 50 per cent. This means that the terms strong and weak are of limited significance as a method of classification, and we shall see that a more quantitative scheme for comparing extents of hydrolysis is necessary.

Solutions of ionic crystalline substances, such as NaCl and $BaSO_4$ mentioned above, could be definitely described as strong electrolytes. This is because the solutions comprise ions, and ions only, there being no dissolved 'molecules' at all in the solution, just as there are none in the solid ionic lattice.

14.2 THE PROTON IN WATER

One particular cation, the proton H^+, deserves special mention since, being a unique ion in having no electrons, its effective radius is far smaller than the radii of all other ions. A charge spread over a relatively minute volume gives rise to a very high surface-charge density, so that in solution the proton exerts a considerable attractive force towards the negative oxygen end of dipolar water molecules. One thus expects strong bonding between protons and water molecules and an aquated species in solution, $H^+(aq)$, may be postulated. Thermochemical evidence certainly supports this contention; there is a considerable quantity of energy liberated on aquation of a proton, far in excess of that for other ions in aqueous solutions.

$$H^+(g) + aq \rightarrow H^+(aq) \qquad \Delta H = -1180 \text{ kJ}$$

$$Na^+(g) + aq \rightarrow Na^+(aq) \qquad \Delta H = -396 \text{ kJ}$$

It has been suggested that one water molecule in particular is closely bound to each proton, and the entity H_3O^+ has been postulated to represent the aquated proton in solution. A principal justification for this assertion is the proved existence of this ion H_3O^+ in certain solid hydrate crystals. Solid perchloric acid hydrate, at one time represented as $HClO_4.H_2O$, has been demonstrated to exist as H_3O^+ cations and ClO_4^- anions. The three hydrogen atoms are equivalent, and each is bonded to the same oxygen. The substance is a stable ionic crystal, with the same structure as ammonium perchlorate, $NH_4^+ClO_4^-$.

Further evidence for the nature of the hydrated proton is obtained by considering the heat of hydration of the proton (−1180 kJ) as made up of two parts:

$$H^+(g) + H_2O(g) \rightarrow H_3O^+(g) \qquad \Delta H = -762\ \text{kJ}$$

$$H_3O^+(g) + aq \rightarrow H_3O^+(aq) \qquad \Delta H = -418\ \text{kJ}$$

The heats of these individual reactions cannot be obtained by direct experiment, but can be calculated from the appropriate thermochemical cycles. Thus the heat of hydration is considered to arise from the formation of the species H_3O^+, and its subsequent hydration. This close similarity between the heats of hydration of Na^+ (−396 kJ) and H_3O^+ (−418 kJ) suggest both are hydrated similarly. Just as Na^+ is hydrated, so H_3O^+ itself is further hydrated, and there is evidence to suggest one important hydrated species in solution to be $H_3O^+(H_2O)_3$, i.e. $H_9O_4^+$. This species is illustrated in Figure 14.1.

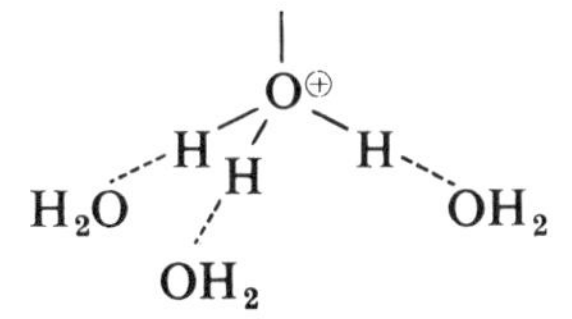

Fig 14.1 The $H_9O_4^+$ ion.

Generally there is little to recommend the writing of $H_9O_4^+$ to represent the hydrated proton, partly because it would lead to unnecessarily clumsy equations, and partly because this entity itself is loosely attracted to more distant water molecules. Provided that the hydration is recognized, H^+ is an adequate representation, although for emphasis we shall sometimes write $H^+(aq)$. The notation H_3O^+ is useful when considering specific proton transfers, and will be used where convenient.

14.3 CHANGING CONCEPTS OF ACIDS AND BASES

A discussion of the current usage of the terms *acid* and *base* is preceded by a brief consideration of aspects of earlier ideas of the nature of acids and bases. These earlier ideas, developed as they were within the framework of available knowledge at particular times, were useful in correlating chemical information and in showing some system and order in a range of reactions. By attempting to understand something of their usefulness and their inadequacies, we gain insight into the way in which concepts change, and the way terms acquire new meanings without entirely losing the old.

An emphasis in an early criterion of acid-base behaviour was placed on the formation of 'salts' with properties different from those of the acid and base from which they could be made. Sodium nitrate, for instance, was viewed as the product of sodium hydroxide and nitric acid. Today we recognize as far more fundamental the nature of sodium nitrate as an ionic crystalline aggregate of Na^+ cations and NO_3^- anions, and we tend to think of 'salts' in this sense. So it would not be illogical to regard Na^+Cl^-, $Na^+CH_3CO.O^-$ and Na^+OH^- as 'salts', although the last would certainly not have been so considered at the turn of this century. A further problem arises since ionic compounds represent an ideal extreme to which many substances approximate, more or less closely. So the term 'salt' is no longer rigorously defined. We shall rarely use the term, and only in contexts in which possible ambiguities of meaning do not arise.

The idea that acids are compounds containing hydrogen replaceable by metals had predictive value in the period in which it was invoked (Liebig, 1838). Thus substances showing acidic properties could be expected to act as sources of hydrogen. Also, the means of preparing certain 'salts' by reacting the appropriate acid with a particular metal were clarified. This definition, however, was not suggestive of *why* this combined hydrogen produced acidic properties, and it offered no explanation of the relative strengths of different acids. So with the development of an understanding of electrolytic dissociation, more useful definitions were put forward and accepted.

The criteria attributed by Arrhenius were that acids could produce H^+ in solution, and bases could produce OH^- in solution. This emphasized the similarity of all neutralization processes, which were represented by the simple general equation:

$$H^+ + OH^- \rightarrow H_2O$$

Measurements of the heat changes associated with this process using different acids and bases were made in 1922 and the results are tabulated in Table 14.1.

Table 14.1

Heats of neutralization (ΔH) at 20°C

Acid + Base	ΔH *(kJ)*
HCl + NaOH	−57.25
HNO_3 + KOH	−57.27
HCl + NaOH	−57.10
HNO_3 + NaOH	−57.29
HCl + LiOH	−57.20
HNO_3 + LiOH	−57.33
	Mean −57.24

These results confirmed that the same heat energy was evolved irrespective of the acid and base used in neutralization, and it was concluded that the same reaction was always involved in neutralization:

$$H^+(aq) + OH^-(aq) \rightarrow H_2O(l) \qquad \Delta H = -57.24 \text{ kJ}$$

Comparing acids and bases of the same analytical concentration, the extent to which H^+ and OH^- were produced in solution provided a basis for contrasting strengths. The distinction between strong and weak electrolytes arose in this context. (The lack of precision in the terms strong and weak has already been mentioned.) Two major drawbacks soon emerged. First, there was no ready way of comparing acid and base strengths *quantitatively*. Second, the criteria confined acid-base reactions to systems in which water is the solvent, and since that time the development of studies in non-aqueous solvents has progressed significantly.

The value of the newer concepts we shall develop lies in the way they answer the limitations just noted—their essential feature is an inherent simplicity and generality.

14.4 THE BRONSTED-LOWRY CONCEPT OF ACIDS AND BASES

A definition of the terms acid and base, applicable to other solvents as well as water, and from which a clear-cut scheme for comparing acid and base strengths can be made, was devised by Bronsted, and independently by Lowry, in 1923.

A substance functions as an acid when it donates a proton to a base: a substance functions as a base when it accepts a proton from an acid.

Often this is put more succinctly:

Acids are proton donors.
Bases are proton acceptors.

Notice, however, that protons do not exist in isolation. An acid must donate a proton to something, and the entity which accepts this proton is a base. Thus the definitions imply acids and bases always react together.

The acid which has exactly one proton more than a particular base is called the *conjugate acid* of that base. Likewise, the base which has exactly one proton less than a particular acid is called the *conjugate base* of that acid. Such an acid-base pair is called a *conjugate pair*.

Some examples of acid-base reactions

(a) In an aqueous solution of hydrochloric acid HCl functions as an acid, and water functions as a base:

$$\underset{acid}{HCl} + \underset{base}{H_2O} \rightleftharpoons \underset{acid}{H_3O^+} + \underset{base}{Cl^-}$$

The conjugate acid-base pairs are HCl/Cl^- and H_3O^+/H_2O.

(b) In an ammonia solution NH_3 functions as a base, and water functions as an acid:

$$\underset{base}{NH_3} + \underset{acid}{H_2O} \rightleftharpoons \underset{acid}{NH_4^+} + \underset{base}{OH^-}$$

The conjugate acid-base pairs are NH_4^+/NH_3 and H_2O/OH^-

(c) From reactions (a) and (b) we notice that water can function both as an acid and as a base. Pure water itself is slightly ionized:

$$\underset{acid}{H_2O} + \underset{base}{H_2O} \rightleftharpoons \underset{acid}{H_3O^+} + \underset{base}{OH^-}$$

Such substances are described as amphoteric, or said to be ampholytes.

(d) When a solution of NH_4NO_3 (principally NH_4^+ and NO_3^-) is added to a solution of Na_2CO_3 (principally Na^+ and CO_3^{2-}), the NH_4^+ acts as an acid, and CO_3^{2-} acts as a base:

$$\underset{acid}{NH_4^+} + \underset{base}{CO_3^{2-}} \rightleftharpoons \underset{acid}{HCO_3^-} + \underset{base}{NH_3}$$

The acid-base conjugate pairs are NH_4^+/NH_3 and HCO_3^-/CO_3^{2-}. One would not describe the sodium carbonate as a base; it is the *source* of the base CO_3^{2-}; likewise, the ammonium nitrate is the source of the acid NH_4^+. A substance such as sodium hydroxide is also to be regarded, not as itself a base, but as a source of the base OH^-. Note also that the ion HCO_3^- is amphoteric. In this equation it is an acid whose conjugate base is CO_3^{2-}. When it functions as a base, its conjugate acid is H_2CO_3.

(e) When solid sodium hydride, NaH, is added to water, hydrogen is evolved and the solution becomes alkaline:

$$\underset{base}{H^-} + \underset{acid}{H_2O} \rightleftharpoons \underset{acid}{H_2} + \underset{base}{OH^-}$$

The conjugate acid of H^- is seen to be H_2. Normally we would not think of H_2 as displaying either acid or base properties. In this reaction, however, within the terms of our definition, it is an acid.

(f) When gaseous hydrogen chloride is allowed to contact gaseous ammonia, fumes of solid white ammonium chloride form:

$$\underset{base}{NH_3(g)} + \underset{acid}{HCl(g)} \rightleftharpoons \underset{acid}{NH_4^+} + \underset{base}{Cl^-}$$

This is an example of an acid-base reaction occurring in the gas phase.

14.5 ROLE OF WATER

Most of the acid-base reactions we consider occur in aqueous solution. As we have seen, water is itself ionized to a very slight extent:

$$2H_2O \rightleftharpoons H_3O^+ + OH^-$$

This equilibrium exists not only in pure water, but in every aqueous solution, irrespective of any other substances present. In such cases, H_3O^+ in the equation above represents the total hydrogen ion concentration in the solution from all sources, and OH^- represents the total hydroxide ion concentration. Applying the equilibrium law to this equilibrium at constant temperature:

$$\frac{[H_3O^+][OH^-]}{[H_2O]^2} = \text{a constant } (K')$$

In moderately dilute aqueous solutions the concentration of water $[H_2O]$ is considerably greater than the concentration of any other species. Because of its presence in considerable excess over all other reagents in a solution, its concentration is substantially constant (around 55 mole dm^{-3}), irrespective of the nature and concentration of other species present. This may be seen by examination of the data in Table 14.2, showing $[H_2O]$ in a range of aqueous solutions.

Table 14.2

Concentration of water in various solutions

Solution	*Density (g dm^{-3})*	*Reagent concentration (g dm^{-3})*	*Water concentration*	
			(g dm^{-3})	*(M)*
Pure water	1000	—	1000	55.5
0.1 M HCl	1001	3.6	997	55.5
0.5 M HCl	1008	18.0	990	55.0
1.0 M HCl	1018	36.5	981	54.5
0.5 M H_2SO_4	1032	49.0	984	54.6
0.5 M NaOH	1021	20.0	1001	55.5
1.0 M NaOH	1042	40.0	1002	55.6
0.1 M $Ba(OH)_2$	1015	17.0	998	55.5

So
$$[H_3O^+][OH^-] = K' [H_2O]^2 = K_w$$

where K_w is an equilibrium constant termed the *self-ionization constant of water*. It has the value of 1×10^{-14} at 25°C. This product of hydrogen ion concentration and hydroxide ion concentration is constant in *all aqueous solutions* at constant temperature.

In pure water and in neutral solutions such as sodium chloride, the H_3O^+ and OH^- arise only from water ionization. In acidic and basic solutions, H_3O^+ and OH^- arise from various sources, but the product $[H_3O^+][OH^-]$ is the same for *all* these solutions at the same temperature.

Neutral, acidic and basic solutions may now be precisely defined.

(a) A solution is **neutral** if

$$[H_3O^+] = [OH^-] = 10^{-7}$$

(b) A solution is **acidic** if

$$[H_3O^+] > [OH^-], \text{ i.e. if } [H_3O^+] > 10^{-7}, [OH^-] < 10^{-7}$$

(c) A solution is **basic** if

$$[H_3O^+] < [OH^-], \text{ ie. if } [H_3O^+] < 10^{-7}, [OH^-] > 10^{-7}$$

Various acid-base indicators change colour over regions of $[H_3O^+]$ around 10^{-7}. While the colour of an indicator in solution is frequently used to decide whether it is acidic or alkaline, the acidity of a solution is fundamentally defined by comparing $[H_3O^+]$ with that in pure water.

14.6 THE pH NOTATION

The $[H_3O^+]$ in solutions of acids and bases varies quite widely. Thus in 0.1 M HCl solution $[H_3O^+] = 0.1 = 10^{-1}$, in pure water $[H_3O^+] = 10^{-7}$, while in 0.1 M NaOH solution, $[H_3O^+] = 10^{-13}$. (A full explanation of these assertions is deferred until section 14.9.) It is convenient to introduce an alternative unit to measure such a wide range of $[H_3O^+]$, and this is termed the pH of a solution. It is defined such that

if $[H_3O^+] = 10^{-x}$, then $pH = x$.
So in 0.1 M HCl, since $[H_3O^+] = 10^{-1}$, then $pH = 1$.
In pure water, since $[H_3O^+] = 10^{-7}$, then $pH = 7$.
In 0.1 M NaOH, since $[H_3O^+] = 10^{-13}$, then $pH = 13$.

Formally pH is related to $[H_3O^+]$ by the equation

$$pH = -\log_{10}[H_3O^+]$$

In most aqueous solutions $[H_3O^+] < 1$, so by taking the *negative* logarithm of $[H_3O^+]$, a positive number is obtained for the pH.

EXAMPLE 1

A particular solution has $[H_3O^+] = 2 \times 10^{-4}$. Calculate its pH.

SOLUTION

Since $[H_3O^+] = 2 \times 10^{-4} = 10^{0.3} \times 10^{-4} = 10^{-3.7}$

$\therefore pH = 3.7$

EXAMPLE 2

A solution has a pH of 5.2. Calculate its $[H_3O^+]$ and $[OH^-]$.

SOLUTION

Since $pH = 5.2$

$\therefore [H_3O^+] = 10^{-5.2} = 10^{0.8} \times 10^{-6} = 6.3 \times 10^{-6}$

Now as $[H_3O^+][OH^-] = 10^{-14}$ in every aqueous solution,

$[OH^-] = 10^{-14}/6.3 \times 10^{-6} = 1.6 \times 10^{-9}$

Notice from these two examples, that *increasing* pH corresponds to a *decreasing* $[H_3O^+]$.

14.7 STRENGTHS OF ACIDS AND BASES

Different acids hydrolyse in water to differing extents. The greater the extent of hydrolysis, the stronger is the acid. Qualitatively we may describe (a) HCl as a strong acid—

$$HCl + H_2O \rightleftharpoons H_3O^+ + Cl^- \quad \textit{(virtually complete)}$$

(b) HSO_4^- as a moderately weak acid—

$$HSO_4^- + H_2O \rightleftharpoons H_3O^+ + SO_4^{2-} \quad \textit{(partial)}$$

(c) acetic acid $CH_3CO.OH$, as a weak acid—

$$H_3C{-}C(=O)OH + H_2O \rightleftharpoons H_3O^+ + H_3C{-}C(=O)O^- \qquad (\textit{slight})$$

i.e. $$CH_3CO.OH + H_2O \rightleftharpoons H_3O^+ + CH_3CO.O^-$$

(d) NH_4^+ as a very weak acid—

$$NH_4^+ + H_2O \rightleftharpoons H_3O^+ + NH_3 \qquad (\textit{very slight})$$

The use of descriptive terms such as strong and weak has limited value because of the *range* of acid strength. It is necessary therefore that we compare acid strengths quantitatively.

For the general reaction of the acid HA with water,

$$HA + H_2O \rightleftharpoons H_3O^+ + A^-$$

Applying the equilibrium law at constant temperature

$$\frac{[H_3O^+][A^-]}{[HA][H_2O]} = K, \text{ where } K \text{ is a constant}$$

As discussed in a previous section, $[H_2O]$ is effectively constant,

$$\therefore \frac{[H_3O^+][A^-]}{[HA]} = K[H_2O] = K_a$$

This equilibrium concentration fraction is a particular case of an equilibrium constant and is represented by the symbol K_a, which is called the *acidity constant* of the acid. A numerical value can be assigned as the acidity constant for each acid. Generally acidity constants for acids are quoted at 25°C. The magnitude of this constant is a measure of the extent to which the acid is hydrolysed in water. The greater the extent of hydrolysis, the larger the equilibrium constant for the reaction as defined by the equations above.

At 25°C, the acidity constants for the acids above are

$$K_a(HCl) = 10^7 \qquad K_a(HSO_4^-) = 10^{-2}$$

$$K_a(CH_3CO.OH) = 1.75 \times 10^{-5} \qquad K_a(NH_4^+) = 6.3 \times 10^{-10}$$

showing that the greater the extent of the hydrolysis reaction, the larger is K_a, i.e. the stronger is the acid.

Just as we found it convenient to introduce pH as a means of expressing a wide range of $[H_3O^+]$, so here it is convenient to define a quantity pK_a to express the wide range of values of K_a. We define

$$pK_a = -\log_{10}K_a$$

Hence if $K_a(HCl) = 10^7$, $pK_a = -7$

$K_a(HSO_4^-) = 10^{-2}$, $pK_a = 2$

$$K_a(CH_3CO.OH) = 1.75 \times 10^{-5} = 10^{0.24} \times 10^{-5} = 10^{-4.76}, \qquad pK_a = 4.8$$

$$K_a(NH_4^+) = 6.3 \times 10^{-10} = 10^{0.8} \times 10^{-10} = 10^{-9.2}, \qquad pK_a = 9.2$$

Just as acidity constants may be defined for acids, and may be used to compare acid strengths, so also one may define a basicity constant for a base.

For the general reaction of the base B with water,

$$B + H_2O \rightleftharpoons BH^+ + OH^-$$

Applying the equilibrium law at constant temperature,

$$\frac{[BH^+][OH^-]}{[B][H_2O]} = K, \text{ where } K \text{ is a constant}$$

As discussed previously, $[H_2O]$ is effectively constant,

$$\therefore \quad \frac{[BH^+][OH^-]}{[B]} = K[H_2O] = K_b$$

K_b is defined as the basicity constant of a base, and

$$pK_b = -\log_{10} K_b$$

The relative magnitudes of K_b and pK_b values for different bases give a measure of the strengths of the bases.

14.8 RELATION BETWEEN K_a AND K_b

Consider the hydrolysis reactions for the acid HA, and its conjugate base A^-:

$$HA + H_2O \rightleftharpoons H_3O^+ + A^- \qquad (1)$$

$$A^- + H_2O \rightleftharpoons HA + OH^- \qquad (2)$$

The acidity and basicity constants are defined as:

$$K_a(HA) = \frac{[H_3O^+][A^-]}{[HA]}$$

$$K_b(A^-) = \frac{[HA][OH^-]}{[A^-]}$$

Now the product

$$K_a \times K_b = \frac{[H_3O^+][A^-]}{[HA]} \times \frac{[HA][OH^-]}{[A^-]}$$

$$= [H_3O^+][OH^-]$$

which has been previously defined as the self-ionization constant for water, K_w.

Thus $$K_a \times K_b = K_w$$

Since K_w is a constant for *all* aqueous solutions at any given temperature (at 25°C, $K_w = 1 \times 10^{-14}$), there is no need to tabulate values of K_b since these may be readily obtained from the K_a values of the conjugate acids. Moreover, the one K_a expression, for say an acid HA, is applicable not only to a solution of HA in water, but also to a solution of the base A^- in water, since such a solution has a specific $[A^-]$, [HA] and $[H_3O^+]$. Likewise, this same K_a expression is equally applicable to such solutions as a mixture of HA and Na^+A^-, or to a solution of HA in HCl—in fact to any solution for which there exists a [HA], $[A^-]$ and $[H_3O^+]$.

EXAMPLE 1

Find the $K_b(NH_3)$ at 25°C, given $K_a(NH_4^+) = 6.3 \times 10^{-10}$.

SOLUTION

Since $$K_b = K_w/K_a$$

$$K_b = 10^{-14}/6.3 \times 10^{-10} = 1.6 \times 10^{-5}.$$

EXAMPLE 2

Find the pK_b (HS^-) at 25°C, given pK_a $(H_2S) = 7.2$.

SOLUTION

Since $$K_a \times K_b = K_w$$

$$\therefore\ pK_a + pK_b = pK_w = 14$$

$$\therefore\ pK_b(HS^-) = 14 - 7.2 = 6.8.$$

In Table 14.3, the K_a values and pK_a values of a number of acids are listed, and from this the relative strengths of their conjugate bases may be deduced.

Notice also that the stronger is an acid, the weaker is its conjugate base. Thus HCN is a stronger acid than HCO_3^-, and CN^- is therefore a weaker base than CO_3^{2-}. However, while it is true that the strong acids have weak conjugate bases, it does *not* follow that weak acids necessarily have strong conjugate bases. The NH_4^+ is a weak acid, about 0.01% hydrolysed in a 0.1 M solution, and ammonia is a weak base, about 1% hydrolysed in a 0.1 M solution.

14.9 POLYPROTIC ACIDS

A number of common acids can donate either two or more protons, and are called *polyprotic acids*. Sulphuric acid, H_2SO_4, is in particular a diprotic (or dibasic) acid, since it hydrolyses in two stages:

$$H_2SO_4 + H_2O \rightleftharpoons HSO_4^- + H_3O^+ \qquad (1)$$

$$HSO_4^- + H_2O \rightleftharpoons SO_4^{2-} + H_3O^+ \qquad (2)$$

The relative extents of these two stages of hydrolysis give a guide to the actual entities present in a sulphuric acid solution. Thus in 0.1 M H_2SO_4, it is found that nearly all H_2SO_4 molecules are hydrolysed to HSO_4^-, but only

Table 14.3
The relative strengths of acids and their conjugate bases

Increasing acid strength ↑	*Acid*	Ka	*p*Ka	*Conjugate base*	*Decreasing base strength* ↑
	$HClO_4$	*ca.* 10^{+10}	−10	ClO_4^-	
	H_2SO_4	*ca.* 10^{+9}	−9	HSO_4^-	
Strong acids	HCl	*ca.* 10^{+7}	−7	Cl^-	Extremely
	H_3O^+	0.5×10^2	−1.7	H_2O	weak
	HSO_4^-	1×10^{-2}	2.0	SO_4^{2-}	bases
Mod. weak	H_3PO_4	8.0×10^{-3}	2.1	$H_2PO_4^-$	
acids	$Fe(H_2O)_6^{3+}$	8.0×10^{-4}	3.1	$Fe(H_2O)_5OH^{2+}$	
Weak acids	$CH_3CO.OH$	1.8×10^{-5}	4.8	$CH_3CO.O^-$	
	$H_2PO_4^-$	6.3×10^{-8}	7.2	HPO_4^{2-}	
	HCN	8.0×10^{-10}	9.1	CN^-	Weak bases
	NH_4^+	6.3×10^{-10}	9.2	NH_3	
	HCO_3^-	5.0×10^{-11}	10.3	CO_3^{2-}	Mod. weak
	HPO_4^{2-}	5.0×10^{-13}	12.3	PO_4^{3-}	bases
	$Na(H_2O)_x^+$	2.0×10^{-15}	14.7	$Na(H_2O)_{x-1}(OH)$	
Extremely	H_2O	2.0×10^{-16}	15.7	OH^-	
weak acids	CH_3CH_2OH	*ca.* 10^{-16}	16	$CH_3CH_2O^-$	Strong
	NH_3	*ca.* 10^{-30}	30	NH_2^-	bases

a small fraction of these ions are further hydrolysed to SO_4^{2-}. Notice that in equation (1), HSO_4^- functions as a Bronsted base, the base conjugate to the acid H_2SO_4, whereas in equation (2), the same entity HSO_4^- functions as a Bronsted acid, the acid conjugate to the base SO_4^{2-}. Thus HSO_4^- is amphoteric. Other common examples of diprotic acids are H_2S, H_2CO_3, and oxalic acid, $H_2C_2O_4$.

EXERCISE
Classify the following entities as Bronsted acids, or bases, or ampholytes, or neither: H_2S, HS^-, S^{2-}, S, H_2CO_3, HCO_3^-, CO_3^{2-}, CO_2.

Phosphoric acid, H_3PO_4, is an example of a triprotic (or tribasic) acid, since it undergoes three successive hydrolyses. Just as in the case of the diprotic acids above, one can assign acidity constants, designated K_{a1}, K_{a2}, etc., to each stage, to indicate the extent of hydrolysis. Thus

$$H_3PO_4 + H_2O \rightleftharpoons H_2PO_4^- + H_3O^+ \qquad K_{a1} = 8. \times 10^{-3}$$
$$H_2PO_4^- + H_2O \rightleftharpoons HPO_4^{2-} + H_3O^+ \qquad K_{a2} = 6.3 \times 10^{-8}$$
$$HPO_4^{2-} + H_2O \rightleftharpoons PO_4^{3-} + H_3O^+ \qquad K_{a3} = 5 \times 10^{-13}$$

Thus the representations $K_a(H_2PO_4^-)$ and K_{a2} for H_3PO_4 are synonymous. Notice that the decreasing magnitude of successive acidity constants indicates that the extent of hydrolysis decreases progressively. This may be correlated with the expected increasing difficulty of removing further protons from species becoming progressively more negatively charged. The sequence here indicates that H_3PO_4 is an acid, both $H_2PO_4^-$ and HPO_4^{2-} are ampholytes, while PO_4^{3-} is a base.

In oxyacids the 'acidic' hydrogens are always attached to oxygen atoms. Thus the monoprotic acids shown in Figure 14.2 have the valence structures indicated.

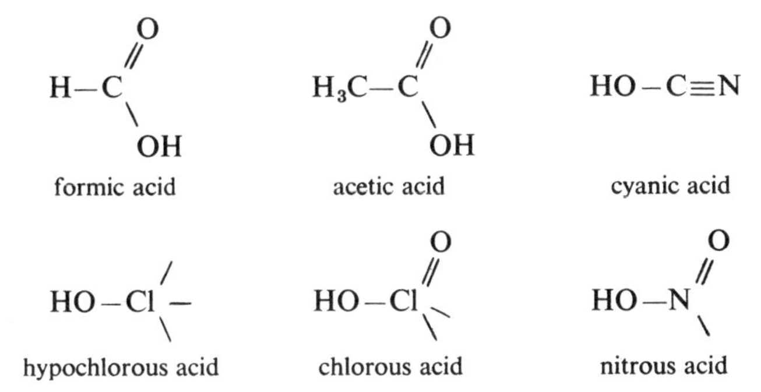

Fig 14.2 Valence structures of some monoprotic acids.

In the acids H_3PO_4, H_3PO_3 and H_3PO_2, it is found experimentally that these acids are triprotic, diprotic and monoprotic, respectively, although each has three hydrogen atoms in the molecule. This experimental behaviour is in accord with the known structures of these acids represented in Figure 14.3. It is only the hydrogen atoms in hydroxy groups, and not those bonded directly to phosphorus, which are 'acidic'.

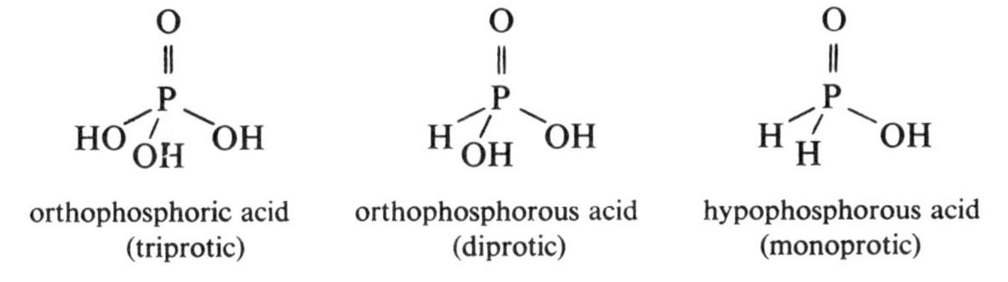

Fig 14.3 Valence structures of H_3PO_4, H_3PO_3 and H_3PO_2.

14.10 SOME NUMERICAL CALCULATIONS

In these numerical examples, attention is focused upon the concentrations of various entities present in some acidic and basic solutions. In particular we calculate the $[H_3O^+]$ and the pH of a number of selected solutions.

EXAMPLE 1

Estimate $[H_3O^+]$ and pH in each of the following solutions:
(a) 0.1 M HNO_3, (b) 0.2 M NaOH, (c) 0.02 M $Ba(OH)_2$, (d) 0.5 M H_2SO_4.

SOLUTION

(a) As nitric acid is virtually completely hydrolysed:

$$HNO_3 + H_2O \rightarrow H_3O^+ + NO_3^-$$

every mole of HNO_3 gives rise to 1 mole of H_3O^+ and 1 mole of NO_3^-. Thus

$$[H_3O^+] = 0.1\text{ M and pH} = 1$$

Note that in every aqueous solution there is an additional H_3O^+ contribution from water dissociation. In this case the $[H_3O^+]$ from water dissociation is *less* than 10^{-7} M, and may be neglected.

(b) Every mole of NaOH yields 1 mole of OH^- in solution.

$$NaOH \rightarrow Na^+ + OH^-$$

$$\therefore\ [OH^-] = 0.2$$

$$\therefore\ [H_3O^+] = 10^{-14}/2 \times 10^{-1} = 5 \times 10^{-14}$$

since in every aqueous solution, $[H_3O^+][OH^-] = 10^{-14}$

Now $$5 \times 10^{-14} = 10^{0.7} \times 10^{-14} = 10^{-13.3}$$

$$\therefore\ \text{pH} = 13.3$$

(c) Every mole of $Ba(OH)_2$ yields 2 mole of OH^- in solution.

$$Ba(OH)_2 \rightarrow Ba^{2+} + 2OH^-$$

$$\therefore\ [OH^-] = 2 \times 0.02 = 0.04$$

$$\therefore\ [H_3O^+] = 10^{-14}/4 \times 10^{-2} = 2.5 \times 10^{-13}$$

If $[H_3O^+] = 2.5 \times 10^{-13} = 10^{0.7} \times 10^{-13} = 10^{-12.6}$

$$\therefore\ \text{pH} = 12.6$$

(d) H_2SO_4 is a dibasic acid, and hydrolyses in two stages

$$H_2SO_4 + H_2O \rightleftharpoons HSO_4^- + H_3O^+ \qquad (\textit{virtually complete})$$

$$HSO_4^- + H_2O \rightleftharpoons SO_4^{2-} + H_3O^+ \qquad (\textit{partial})$$

Every mole of H_2SO_4 is virtually completely hydrolysed to 1 mole of H_3O^+ and 1 mole of HSO_4^-, and the HSO_4^- is partially hydrolysed further. If HSO_4^- were completely undissociated

$$[H_3O^+]\text{ would be }0.5\text{ M} = 5 \times 10^{-1} = 10^{-0.3}$$

$$\therefore\ \text{pH would be }0.3$$

If HSO_4^- were completely dissociated

$$[H_3O^+]\text{ would be }2 \times 0.5\text{ M} = 1\text{ M}$$

$$\therefore\ \text{pH would be }0$$

To determine the actual value between these extremes, one needs K_a (HSO_4^-), a measure of the extent of this second dissociation. In fact, it can be shown that

$$[H_3O^+] = 0.51\text{ M} = 5.1 \times 10^{-1} = 10^{0.71} \times 10^{-1}$$

$$\therefore\ \text{pH} = 0.29$$

EXAMPLE 2

Given $K_a(H_2S) = 10^{-7.2}$ and $K_a(HS^-) = 10^{-14.2}$, calculate the equilibrium constant for the reaction

$$H_2S + 2H_2O \rightleftharpoons 2H_3O^+ + S^{2-}$$

and show how $[S^{2-}]$ in a saturated solution of H_2S depends upon $[H_3O^+]$. The saturation solubility of H_2S in water is 0.1 M.

SOLUTION

For

$$H_2S + H_2O \rightleftharpoons H_3O^+ + HS^- \qquad (1)$$

$$\frac{[H_3O^+][HS^-]}{[H_2S]} = 10^{-7.2}$$

and

$$HS^- + H_2O \rightleftharpoons H_3O^+ + S^{2-} \qquad (2)$$

$$\frac{[H_3O^+][S^{2-}]}{[HS^-]} = 10^{-14.2}$$

Adding (1) and (2) gives the desired equation, the equilibrium constant for which is given by

$$\frac{[H_3O^+][HS^-]}{[H_2S]} \times \frac{[H_3O^+][S^{2-}]}{[HS^-]} = 10^{-7.2} \times 10^{-14.2} = 10^{-21.4}$$

$$\therefore \frac{[H_3O^+]^2 \times [S^{2-}]}{[H_2S]} = 10^{-21.4}$$

In 0.1 M H_2S, it is precisely true that

$$[H_2S] + [HS^-] + [S^{2-}] = 0.1$$

but it is a good approximation to write

$$[H_2S] = 0.1$$

as the extent of hydrolysis is so small.

$$\therefore [H_3O^+]^2[S^{2-}] = 10^{-22.4}$$

$$\therefore [S^{2-}] = 10^{-22.4}/[H_3O^+]^2$$

indicating that $[S^{2-}]$ increases with decreasing $[H_3O^+]$, i.e. with increasing alkalinity of the solution. For example, in a neutral solution, $[H_3O^+] = 10^{-7}$ then $[S^{2-}] = 10^{-22.4}/10^{-14} = 10^{-8.4}$ whereas, in a solution in which $[H_3O^+] = 10^{-1}$, such as a solution 0.1 M with respect to HCl, then $[S^{2-}] = 10^{-22.4}/10^{-2} = 10^{-20.4}$.

EXAMPLE 3

Calculate the pH of a solution of 0.1 M ammonia in which has been dissolved 4.0 g of solid ammonium nitrate in 250 cm³ of solution.

SOLUTION

$$4.0 \text{ g } NH_4NO_3 = 4/80 \text{ mole} = 0.05 \text{ mole}$$

$\therefore$ the concentration of NH_4NO_3 in solution is 0.05 mole in 250 cm^3 or 0.2 mole dm^{-3}, i.e. $[NH_4^+] = 0.2$, $[NO_3^-] = 0.2$. The principal entities in the solution are NH_3 molecules, NH_4^+ and NO_3^- ions and, of course, a large excess of H_2O molecules.

$$\text{Since } [NH_3] = 0.1 \text{ and } [NH_4^+] = 0.2$$

we shall neglect the very slight hydrolysis of each of these ions in water. As the equilibrium

$$NH_4^+ + H_2O \rightleftharpoons NH_3 + H_3O^+$$

must exist in this solution, we may use the fact that the concentration fraction

$$\frac{[NH_3][H_3O^+]}{[NH_4^+]}$$

has a constant value in *any* solution containing these ions present in equilibrium. Now $pK_a(NH_4^+) = 9.2$,

$$\therefore \frac{[NH_3][H_3O^+]}{[NH_4^+]} = 10^{-9.2} = 10^{0.8} \times 10^{-10} = 6.3 \times 10^{-10}$$

$$\therefore \frac{0.1 \times [H_3O^+]}{0.2} = 6.3 \times 10^{-10}$$

$$\therefore [H_3O^+] = 1.26 \times 10^{-9} = 10^{0.10} \times 10^{-9}$$

$$\therefore pH = 8.9$$

Note that, although the concentration of the acid NH_4^+ is twice that of the base NH_3, the solution is basic. This reflects the fact that NH_3 is a stronger base than is NH_4^+ an acid. Alternatively one could solve the problem using the equation

$$NH_3 + H_2O \rightleftharpoons NH_4^+ + OH^-$$

to find $[OH^-]$ first, and then use

$$K_w = [H_3O^+][OH^-]$$

to find the $[H_3O^+]$.

EXAMPLE 4

Calculate and compare the pH and the percentage hydrolysis in solutions of (a) 1 M acetic acid, (b) 0.1 M acetic acid, and (c) 0.1 M sodium acetate.

SOLUTION

(a) Since $K_a(CH_3CO.OH) = 1.8 \times 10^{-5}$, for the hydrolysis

$$CH_3CO.OH + H_2O \rightleftharpoons H_3O^+ + CH_3CO.O^-$$

$$\frac{[H_3O^+][CH_3CO.O^-]}{[CH_3CO.OH]} = 1.8 \times 10^{-5} = 10^{0.2} \times 10^{-5} = 10^{-4.8} \qquad (1)$$

We make two assumptions:

(i) As K_a is very small, we assume that the extent of hydrolysis is sufficiently slight to write

$$[CH_3CO.OH] = 1 \quad (2)$$

as an approximation to the precise statement that

$$[CH_3CO.OH] + [CH_3CO.O^-] = 1 \quad (3)$$

(ii) As every mole of acetic acid which reacts produces 1 mole of H_3O^+ and 1 mole of $CH_3CO.O^-$, we write

$$[H_3O^+] = [CH_3CO.O^-] \quad (4)$$

This neglects the contribution of H_3O^+ from water dissociation, which will be *less* than 10^{-7} M.

Making the approximations in (2) and (4) and substituting into (1) gives

$$\frac{[H_3O^+]^2}{1} = 10^{-4.8}$$

$$\therefore [H_3O^+] = 10^{-2.4}$$

$$\therefore pH = 2.4$$

Also since $[CH_3CO.O^-] = 10^{-2.4}$ and $[CH_3CO.OH] = 1$,

$$\therefore \frac{[CH_3CO.O^-]}{[CH_3CO.OH]} = 10^{-2.4} = 10^{0.6} \times 10^{-3} = 4 \times 10^{-3}$$

thus if the fraction of molecules hydrolysed is 0.004, then the extent of hydrolysis is 0.4 per cent.

Note that this very small fraction of molecules hydrolysed is consistent with assumption made in (i). Also, since $[H_3O^+]$ from the acid dissociation is as high as $10^{-2.4}$, the assumption in (ii), that we neglect the H_3O^+ from water dissociation, is quite reasonable. The $[H_3O^+]$ from the acid is seen to be more than 10,000 times that from the water dissociation.

If the pH is obtained by exact calculation, i.e. without making the simplifying approximations indicated here, the value obtained is 2.41.

EXERCISE

In this particular example, the approximation (i) is more serious than approximation (ii). By writing that

$$[CH_3CO.OH] = 1 - [CH_3CO.O^-] = 1 - [H_3O^+],$$

solve the quadratic equation

$$\frac{[H_3O^+]^2}{1 - [H_3O^+]} = 10^{-4.8} = 1.8 \times 10^{-5}$$

to obtain a more accurate value of the pH.

(b) Similarly, invoking these approximations for 0.1 M acetic acid,

$$\frac{[H_3O^+]^2}{0.1} = 10^{-4.8}$$

$$\therefore [H_3O^+]^2 = 10^{-4.8} \times 10^{-1} = 10^{-5.8}$$

$$\therefore [H_3O^+] = 10^{-2.9}$$

$$\therefore pH = 2.9$$

Also, $[CH_3CO.O^-] = 10^{-2.9}$ and $[CH_3CO.OH] = 10^{-1}$

$$\therefore \frac{[CH_3CO.O^-]}{[CH_3CO.OH]} = 10^{-1.9} = 10^{0.1} \times 10^{-2} = 1.2 \times 10^{-2}$$

indicating some 1.2 per cent hydrolysis.

Notice that as the acid becomes more dilute the $[H_3O^+]$ decreases (the pH increases), but the extent of ionization increases.

(c) Sodium acetate is completely dissociated in solution:

$$NaCH_3CO.O(s) \rightarrow Na^+(aq) + CH_3CO.O^-(aq)$$

The acetate is slightly hydrolysed:

$$CH_3CO.O^- + H_2O \rightleftharpoons CH_3CO.OH + OH^-$$

Since $K_a(CH_3CO.OH) = 10^{-4.8}$, $K_b(CH_3CO.O^-) = 10^{-14}/10^{-4.8} = 10^{-9.2}$.

$$\therefore \frac{[CH_3CO.OH][OH^-]}{[CH_3CO.O^-]} = 10^{-9.2}$$

Making assumptions analogously as before

(a) assume $[CH_3.CO.O^-] = 1$, since hydrolysis is very slight;

(b) assume $[CH_3CO.OH] = [OH^-]$, neglecting $[OH^-]$ from water dissociation

$$\therefore \frac{[OH^-]^2}{0.1} = 10^{-9.2}$$

$$\therefore [OH^-] = 10^{-5.1}$$

$$\therefore [H_3O^+] = 10^{-14}/10^{-5.1} = 10^{-8.9}$$

$$\therefore pH = 8.9$$

$$\text{Also } \frac{[CH_3CO.OH]}{[CH_3CO.O^-]} = \frac{10^{-5.1}}{10^{-1}} = 10^{-4.1} = 10^{0.9} \times 10^{-5} = 8 \times 10^{-5}$$

thus the percentage hydrolysis is 0.008, i.e. about 0.01 per cent.

14.11 ACIDITY IN SELECTED SOLUTIONS

The acidity or alkalinity of solutions of various compounds may be interpreted in terms of complete dissociation to free ions with possible hydrolysis of the dissolved ions. We consider some specific examples.

(1) In NaCl solution

$$NaCl(s) \rightarrow Na^+ + Cl^- \qquad (\textit{complete})$$

Neither Na^+ nor Cl^- hydrolyses to any significant extent.

From the water dissociation

$$[H_3O^+] = [OH^-] = 10^{-7}$$

and the solution is neutral.
(2) In NH_4Cl solution

$$NH_4Cl(s) \rightarrow NH_4^+ + Cl^- \qquad (complete)$$

The NH_4^+ hydrolyses to a slight extent

$$NH_4^+ + H_2O \rightleftharpoons NH_3 + H_3O^+ \qquad (slight)$$

so $$[H_3O^+] > [OH]^-, \text{ i.e. } [H_3O^+] > 10^{-7}$$

and the solution is acidic.

(3) In Na_2CO_3 solution

$$Na_2CO_3(s) \rightarrow 2Na^+ + CO_3^{2-} \qquad (complete)$$

the CO_3^{2-} is hydrolysed to a slight extent

$$CO_3^{2-} + H_2O \rightleftharpoons HCO_3^- + OH^- \qquad (slight)$$

so $$[H_3O^+] < [OH^-], \text{ i.e. } [H_3O^+] < 10^{-7}$$

and the solution is alkaline.
(4) In $NaHCO_3$ solution

$$NaHCO_3(s) \rightarrow Na^+ + HCO_3^- \qquad (complete)$$

The HCO_3^- may hydrolyse both as an acid and base

$$HCO_3^- + H_2O \rightleftharpoons CO_3^{2-} + H_3O^+ \qquad pK_a(HCO_3^-) = 10.3$$
acid

$$HCO_3^- + H_2O \rightleftharpoons H_2CO_3 + OH^- \qquad pK_b(HCO_3^-) = 7.6$$
base

Whether the solution is acidic or basic depends upon the relative extents of these two possible reactions. Since $pK_b(HCO_3^-) < pK_a(HCO_3^-)$ and hence $K_b(HCO_3^-) > K_a(HCO_3^-)$ the last reaction proceeds to the greater extent, and the solution is alkaline.
(5) In $(NH_4)_2S$ solution

$$(NH_4)_2S(s) \rightarrow 2NH_4^+ + S^{2-} \qquad (complete)$$

Both cation and anion will hydrolyse

$$NH_4^+ + H_2O \rightleftharpoons NH_3 + H_3O^+ \qquad (1)$$

$$S^{2-} + H_2O \rightleftharpoons HS^- + OH^- \qquad (2)$$

The cation hydrolysis by itself would increase $[H_3O^+]$ in the solution, whereas the anion hydrolysis by itself would increase $[OH^-]$ in the solution. Whether the solution is acidic or basic depends upon the relative extents of these two reactions. Since $pK_a(NH_4^+) = 9.2$ and $pK_b(S^{2-}) = 0.2$, the S^{2-} is a much stronger base than is NH_4^+ an acid, and the solution is alkaline. The reaction could alternatively be directly represented by the equation

$$\underset{acid}{NH_4^+} + \underset{base}{S^{2-}} \rightleftharpoons \underset{base}{NH_3} + \underset{acid}{HS^-}$$

but this by itself would not give an indication of acidity or alkalinity.

(6) That some ions are quite strongly hydrated has already been established, and will be further discussed in chapter 15. Here we notice one consequence of hydration—the capacity of a hydrated ion to function as a Bronsted acid. Thus in a solution of $FeCl_3$

$$FeCl_3(s) \rightarrow Fe^{3+}(aq) + 3Cl^-(aq) \qquad (\textit{complete})$$

The entity $Fe^{3+}(aq)$ may be represented as $Fe(H_2O)_6^{3+}$ to show the proton transfer reaction it undergoes in aqueous solution

$$Fe(H_2O)_6^{3+} + H_2O \rightleftharpoons Fe(H_2O)_5OH^{2+} + H_3O^+$$

$$pK_a = 3.1$$

indicating that the hydrated iron(III) ion is a stronger acid than, say, acetic acid. The acidity of such an aquo-cation is considered due to a drift toward the metal ion of bonding electrons in the O—H bond of the aquated water molecule. This drift of electronic charge increases the ease with which the O—H bond may be broken and a proton transferred to an adjacent water molecule. The relative strengths of some other hydrated cations is shown in Table 14.4

Table 14.4

Acid strengths of hydrated cations

Cation	pK_a	*Cation*	pK_a	*Cation*	pK_a
$Na^+(aq)$	14.7	$Mg^{2+}(aq)$	11.4	$Al^{3+}(aq)$	4.9
$Ca^{2+}(aq)$	12.7	$Zn^{2+}(aq)$	9.6	$Cr^{3+}(aq)$	3.9
$Ag^+(aq)$	11.7	$Pb^{2+}(aq)$	8.8	$Fe^{3+}(aq)$	3.1

14.12 NEUTRALIZATION

In chapter 4 a number of calculations of the stoichiometric aspects of neutralization reactions have been presented. We see now the way in which the pH changes during the neutralization of an acid with a base.

Consider the titration of 50 cm^3 of 1 M HCl to which we progressively add small volumes of 2 M NaOH. To simplify calculations we suppose the 50 cm^3 of acid is diluted with water to a volume of 1000 cm^3. The volume change on addition of the base will be neglected for the purpose of approximate calculation.

Table 14.5

Progressive neutralization of HCl with NaOH

Mole HCl initially	*Volume 2 M NaOH added (cm³)*	*Mole NaOH added*	*Mole of excess reagent*	$[H_3O^+]$	$[OH^-]$	*pH*
0.050	nil	nil	0.050(HCl)	5×10^{-2}	2×10^{-13}	1.3
0.050	10	0.020	0.030(HCl)	3×10^{-2}	3.3×10^{-13}	1.5
0.050	20	0.040	0.010(HCl)	1×10^{-2}	1×10^{-12}	2.0
0.050	22.5	0.045	0.005(HCl)	5×10^{-3}	2×10^{-12}	2.3
0.050	24.5	0.049	0.001(HCl)	1×10^{-3}	1×10^{-11}	3.0
0.050	25.0	0.050	equivalence	1×10^{-7}	1×10^{-7}	7.0
0.050	25.5	0.051	0.001(NaOH)	1×10^{-11}	1×10^{-3}	11.0
0.050	27.5	0.055	0.005(NaOH)	2×10^{-12}	5×10^{-3}	11.7
0.050	30	0.060	0.010(NaOH)	1×10^{-12}	1×10^{-2}	12.0
0.050	40	0.080	0.030(NaOH)	3.3×10^{-13}	3×10^{-2}	12.5

50 cm³ of 1 M HCl represents 0.050 mole, irrespective of the volume in which it exists. The equivalence point in the titration will be reached when 0.050 mole of NaOH is added, i.e. 25 cm³ of 2 M solution. When the volume of NaOH added is less than 25 cm³ we shall have excess acid, when it exceeds

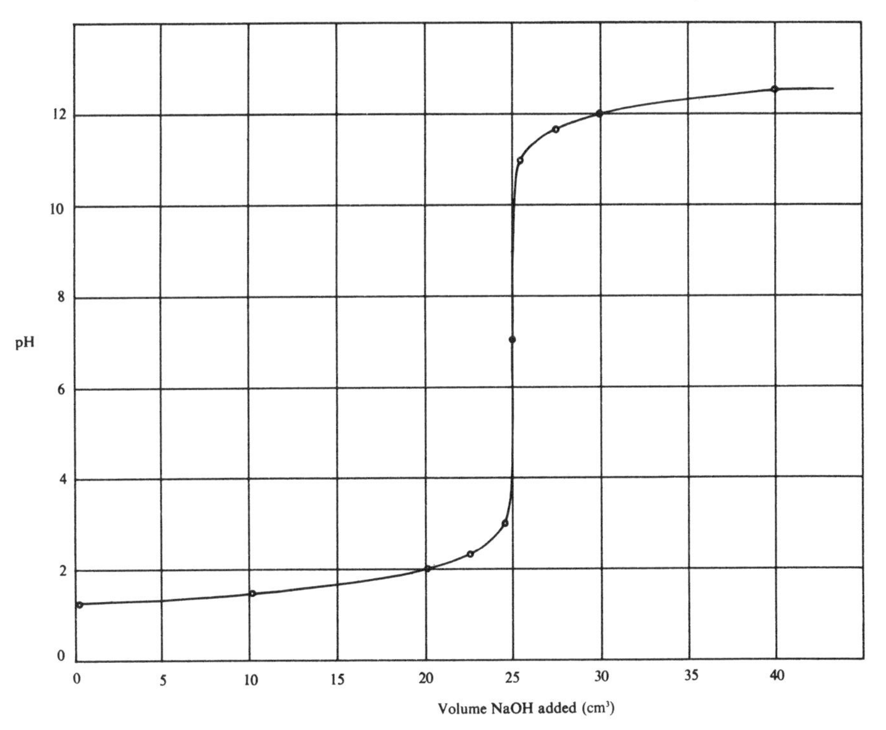

Fig 14.4 Strong acid/strong base titration curve.

25 cm^3 the base will be in excess. In Table 14.5 we have calculated the changing $[H_3O^+]$, $[OH^-]$ and pH of the solution as the volume of NaOH solution is progressively increased.

It is convenient to represent such a titration graphically, showing the way the pH changes as more alkali is added. Such a *titration curve* is shown in Figure 14.4.

From this curve we notice two types of pH change with addition of alkali:

(a) in the regions where one reagent is present in considerable excess, there is very little change in pH for the addition of quite considerable amounts of the other reagent;

(b) near the equivalence point, a very marked pH change occurs for the addition of very slight amounts of added reagent.

EXERCISE

Draw an analogous titration curve to show the pH change when a solution of 1 M NaOH is gradually neutralized by one of 1 M HCl.

The use of acid-base indicators for detecting the equivalence point in titrations arises from the fact that these indicators are themselves weak acids, which are converted to their conjugate base of a different colour at a particular pH. Thus methyl red changes colour around pH = 5, and phenolphthalein changes colour around pH = 9. For the titration just considered, either of these indicators would be suitable for, at the equivalence point, the addition of one drop of alkali would cause a pH change from say, 4 to 10.

14.13 EFFECT OF SOLVENT ON ACID STRENGTH

It was indicated in section 14.7 that, although the strengths of acids such as $HClO_4$, H_2SO_4 and HCl in aqueous solution were listed, their relative strengths could not be distinguished in dilute aqueous solution. Water, although a very weak base, is sufficiently strong virtually to completely hydrolyse each of these acids. In a solvent less basic than water, however, each acid is but partially solvolysed, and the relative extents of solvolysis may be obtained. Thus we may quote pK_a values for these acids in glacial acetic acid as solvent

$$HClO_4 + CH_3CO.OH \rightleftharpoons CH_3C(OH)_2^+ + ClO_4^- \quad pK_a = 4.87$$

$$H_2SO_4 + CH_3CO.OH \rightleftharpoons CH_3C(OH)_2^+ + HSO_4^- \quad pK_a = 7.24$$

$$HCl + CH_3CO.OH \rightleftharpoons CH_3C(OH)_2^+ + Cl^- \quad pK_a = 8.55$$

Notice here acetic acid functions as a base; its conjugate acid is the $CH_3C(OH)_2^+$ cation. In acetic acid as solvent, perchloric acid is a fairly weak acid and hydrogen chloride is a very weak acid. The Cl^- behaves as a much stronger base than it does in water.

In a sufficiently acidic solvent even a substance such as HNO_3 may become a base. Thus in liquid HF as solvent

$$\underset{base}{HNO_3} + \underset{acid}{HF} \rightleftharpoons \underset{acid}{H_2NO_3^+} + \underset{base}{F^-}$$

Analogously, in more basic solvents than water, the strength of acids increases by comparison with water, i.e. the extent of dissociation increases. So in liquid ammonia as solvent, acetic acid becomes a strong acid, as the reaction

$$\underset{acid}{CH_3CO.OH} + \underset{base}{NH_3} \rightleftharpoons \underset{acid}{NH_4^+} + \underset{base}{CH_3CO.O^-}$$

proceeds virtually to completion.

In general we may define K_a for any solvent S by the equation

$$HA + S \rightleftharpoons SH^+ + A^-$$

$$K_a = \frac{[SH^+][A^-]}{[HA]}$$

EXERCISE

Write defining equations for the K_a of the acid NH_4^+ in (a) water, (b) liquid HF, (c) liquid ammonia.

In which solvent will NH_4^+ be a strong acid, and in which will it be weakest?

Thus the principles of proton transfer enunciated by Bronsted and Lowry to redefine acids and bases are readily applicable to non-aqueous solvents, and the equilibrium constants which measure the extent of solvolysis provide a measure of acid-base strengths in such solvents.

QUESTIONS

1. 2.00 g of solid pellets of sodium hydroxide are dissolved in water and made up to 500 cm^3 of solution. State
   (a) the entities present in the solution,
   (b) the $[OH^-]$,
   (c) the $[H_3O^+]$,
   (d) the pH of the solution.
2. 2.0 cm^3 of 14 M HNO_3 is made up to 400 cm^3 of solution with water. Calculate
   (a) the molarity (the analytical concentration) of the resulting solution,
   (b) the $[H_3O^+]$ of the solution,
   (c) the pH of the solution,
   (d) the mass of solid KOH which would have to be added to the 400 cm^3 of this solution to increase the pH to 7.0.
3. Calculate the pH of solutions whose $[H_3O^+]$ are as follows:
   (a) 10^{-4} M
   (b) 0.03 M
   (c) 5.3×10^{-5} M.
4. Calculate the $[H_3O^+]$ and $[OH^-]$ in solutions with the following pH:
   (a) 8.0,
   (b) 0.2,
   (c) 12.3.
5. Calculate the pK_a of the following acids, whose K_as are as follows:
   (a) $K_a(HSO_3^-) = 1 \times 10^{-7}$
   (b) $K_a(HCO.OH) = 2 \times 10^{-4}$
   (c) $K_{a1}(H_3PO_4) = 7.5 \times 10^{-3}$

6. Calculate the K_a of the following acids, whose pK_as are given:
   (a) pK_a(HN_3) = 4.72
   (b) pK_a(HCN) = 9.32
   (c) pK_a(H_3AsO_4) = 2.22
7. State the approximate pH of each of the following solutions of hydrochloric acid:
   (a) a 0.1 M solution, i.e. 10^{-1} M,
   (b) a 0.01 M solution, i.e. 10^{-2} M,
   (c) a 0.001 M solution, i.e. 10^{-3} M,
   (d) a 0.00000001 M solution, i.e. 10^{-8} M.
   Comment briefly on your answers, especially in (d).
8. Decide which of the following solutions could be prepared from the reagents specified. Give a short explanation of each decision:
   (a) a solution of pH = − 1 using nitric acid and water;
   (b) a solution of pH = 14 using sodium hydroxide and water;
   (c) a solution of pH = 16 using sodium hydroxide and water.
9. Given $K_a(HCO_3^-) = 6 \times 10^{-11}$ and $K_b(HCO_3^-) = 3.33 \times 10^{-8}$, calculate the equilibrium constants for the reactions:
   (a) $2HCO_3^- \rightleftharpoons H_2CO_3 + CO_3^{2-}$
   (b) $H_2CO_3 + 2H_2O \rightleftharpoons 2H_3O^+ + CO_3^{2-}$
10. Formic acid, HCO.OH, and ammonia are both weak electrolytes since both hydrolyse to only a slight extent in water, as indicated by K_a(HCO.OH) = 2×10^{-4} and $K_b(NH_3) = 1.8 \times 10^{-5}$.
   (a) Calculate the equilibrium constant for the reaction

   $$HCO.OH + NH_3 \rightleftharpoons NH_4^+ + HCO.O^-$$

   remembering that $K_w = 10^{-14}$.
   If equal volumes of 0.1 M HCO.OH and 0.1 M NH_3 are mixed:
   (b) decide whether the solution formed would be a good electrical conductor.
   (c) state qualitatively the approximate concentrations of each of the following entities which exist in the resulting solution:

   $$HCO.OH,\ HCO.O^-,\ NH_3,\ NH_4^+,\ H_3O^+,\ OH^-$$

11. Hydrazoic acid, HN_3, is a weak monobasic acid which hydrolyses in water according to the equation:

   $$HN_3 + H_2O \rightleftharpoons H_3O^+ + N_3^-$$

   The pK_a of this acid is 4.72.
   (a) Calculate the $[H_3O^+]$ in a 0.1 M solution of the acid, and hence determine the approximate concentrations of each of the following species in solution:

   $$HN_3,\ N_3^- \text{ and } OH^-;$$

   (b) estimate the percentage hydrolysis of the acid;
   (c) state the pH of the solution.
12. Find the $[H_3O^+]$ and the pH in:
   (a) a solution of 0.2 M NaOH,
   (b) a solution of 0.5 M HCl,
   (c) a mixture of 40 cm^3 of (a) and 25 cm^3 of (b),
   (d) the solution formed when 0.5 g solid NaOH is added to solution (c).

13. Find the $[H_3O^+]$ and the pH in:
    (a) a solution of 0.5 M NH_4Cl, given $pK_a(NH_4^+) = 9.2$;
    (b) a mixture of 20 cm^3 1 M NH_3 and 20 cm^3 1 M HCl;
    (c) a mixture of 40 cm^3 2 M HCl, 20 cm^3 2 M NH_3 and 20 cm^3 2 M NaOH.
14. What mass of solid ammonium bromide must be added to 1000 cm^3 of 0.1 M ammonia solution to obtain a solution of pH = 9.0?
15. Discuss qualitatively the nature of the entities present in 0.1 M solutions of each of:
    (a) NH_4NO_3; (b) $(NH_4)_2CO_3$; (c) NaHS; (d) $Al_2(SO_4)_3$.
    State explicitly what numerical constants you would need to consider to predict whether the solutions of (b) and (c) were acidic, basic or neutral.
16. (a) The acidity constant of a certain weak monobasic acid is 1×10^{-8}.
        (i) What is the pH of an 0.05 M solution of this acid?
        (ii) Find the percentage hydrolysis of the acid at this concentration.
    (b) The pH of a solution of a second weak monobasic acid is found experimentally to be 4.0 at 0.01 M.
        (i) Find the acidity constant of this acid.
        (ii) Determine the percentage hydrolysis of the acid at this concentration.
    (c) A third weak monobasic acid is found to be 0.7% hydrolysed at 0.02 M concentration.
        (i) Calculate the pH of the solution.
        (ii) Find the acidity constant of the acid.
17. (a) How many mole of solid NaOH must be added to 200 cm^3 of 0.1 M HCl in order that the resulting solution have a pH of 7?
    (b) How many mole of solid NH_4Cl must be added to 200 cm^3 of 0.1 M NH_3 in order that the resulting solution have a pH of 9?
18. (a) A solution containing 2.34 g NH_4Cl in 250 cm^3 of solution is found to have a pH of 5.0. Use this information to calculate $K_a(NH_4^+)$ and hence $K_b(NH_3)$.
    (b) A solution containing 0.49 g NaCN in 1000 cm^3 of solution has a pH of 10.7. Use this information to calculate K_a(HCN).
19. For the weak dibasic acid hydrogen selenide H_2Se, $K_{a1} = 1.7 \times 10^{-4}$ and $K_{a2} = 1 \times 10^{-10}$:
    (a) calculate the equilibrium constant for the reaction:

$$H_2Se + 2H_2O \rightleftharpoons 2H_3O^+ + Se^{2-}$$

    (b) Name the entities present in 0.05 M solution of H_2Se, and indicate their approximate concentrations qualitatively, e.g. moderately high, low, or very low.
    (c) Estimate by approximate calculation the concentrations of these entities referred to in (b).
    (d) By analogy with the ion HSe^-, comment on the assertion that the hydroxide ion, OH^- is actually amphoteric.
20. Orthoarsenic acid H_3AsO_4 is a triprotic acid with $pK_{a1} = 2.22$, $pK_{a2} = 6.98$ and $pK_{a3} = 11.53$.
    (a) By writing appropriate equations indicate concisely what is meant by this statement.
    (b) Discuss qualitatively the entities present in each of the following solutions, and give some indication of the relative concentrations of them.

        (a) 0.2 M H_3AsO_4
        (b) 0.2 M Na_2HAsO_4

21. The K_a of hypobromous acid HOBr is 2×10^{-9}:
    (a) calculate the pH of an 0.05 M solution of HOBr;
    (b) calculate the pH of an 0.05 M solution of NaOBr;
    (c) calculate the pH of a mixture of 50 cm^3 of (a) and 30 cm^3 of (b).
22. Benzoic acid is a weak monobasic acid which hydrolyses thus:

$$C_6H_5CO.OH + H_2O \rightleftharpoons C_6H_5CO.O^- + H_3O^+$$

$$K_a(C_6H_5CO.OH) = 6.4 \times 10^{-5}$$

    (a) calculate the pH of a saturated solution of benzoic acid in water, given the solubility is 2.06 g dm^{-3};
    (b) find the approximate pH of an 0.01 M aqueous solution of sodium benzoate $C_6H_5CO.ONa$;
    (c) when 1.22 g of solid benzoic acid is neutralized with 0.4 M NaOH solution, 25.0 cm^3 of the latter is required. Show how, from these data, the molecular weight of benzoic acid can be obtained experimentally;
    (d) explain why, in the solution formed from (c) at the equivalence point, the pH is greater than 7.

15

Complex Ion Equilibria

15.1 THE NATURE OF COMPLEX IONS IN SOLUTION

We have seen that dissolved ions in an environment of water molecules exist as hydrated ions. Although the number of water molecules closely associated with a given ion is in most cases ill defined or unknown, it is nevertheless convenient more or less arbitrarily to assign specific numbers of water molecules, *hydration numbers*, to certain ions in water. Thus the ion $Mg^{2+}(aq)$ is represented $Mg(H_2O)_6^{2+}$, in agreement with the stoichiometry of hydrated salts such as $Mg(H_2O)_6(ClO_4)_2$ and $Mg(H_2O)_6Cl_2$. Likewise, the ion $Ag^+(aq)$ is represented $Ag(H_2O)_2^+$, and the ion $Zn^{2+}(aq)$ is represented $Zn(H_2O)_4^{2+}$, although the experimental evidence for such hydration numbers is partially indirect, and cannot be discussed here. That the $Cr^{3+}(aq)$ ion may be represented as $Cr(H_2O)_6^{3+}$ has been quite firmly established, but for many other hydrated ions the evidence is inconclusive.

Molecules firmly bonded to an ion in solution are termed *ligands*. In an environment where other polar molecules are available, either instead of water, as in liquid ammonia for example, or as well as water, as in an aqueous ammonia solution, one may expect these other molecules to function also as ligands. Thus in 0.001 M NH_3 solution, silver ions form both $Ag(NH_3)(H_2O)^+$ and $Ag(NH_3)_2^+$ in comparable amounts, zinc ions remain chiefly as $Zn(NH_3)(H_2O)_3^{2+}$, while copper(II) ions form principally $Cu(NH_3)_2(H_2O)_4^{2+}$ and $Cu(NH_3)_3(H_2O)_3^{2+}$. At higher concentrations of ammonia, say 0.1 M NH_3, these cations exist principally as $Ag(NH_3)_2^+$, $Zn(NH_3)_4^{2+}$ and $Cu(NH_3)_4(H_2O)_2^{2+}$, respectively. These ions are known as *complex ions*.

Not only polar molecules, but ions, too, may replace ligand water molecules. So if to a small volume of dilute copper(II) sulphate solution some concentrated hydrochloric acid is slowly added, the colour of the solution changes from pale blue through to a deep yellow-green. This is associated with progressive replacement of water ligands about $Cu(H_2O)_6^{2+}$ by Cl^- to produce $Cu(H_2O)_5Cl^+$, $Cu(H_2O)_4Cl_2$, $Cu(H_2O)_3Cl_3^-$ and $Cu(H_2O)_2Cl_2^-$ in solution. The common anionic ligands include the halide ions, cyanide ion CN^-, thiocyanate SCN^-, and hydroxide ion OH^-.

Just as many hydrated ions in solution form characteristic hydrated salts, e.g. $Al(H_2O)_6Cl_3$, $Be(H_2O)_4SO_4$, so many complex ions form solid complex compounds, e.g. $Cu(NH_3)_4SO_4.2H_2O$ and $Zn(NH_3)_4Cl_2$. In such solids the arrangement of the ligands around the central cation follows a specific geometrical orientation, and the shapes of a number of complex ions are shown in Figure 15.1. The shapes of these ions in aqueous solutions are believed to be closely similar to the shapes established in solid compounds.

Some ligands occupy more than one co-ordination position around a central ion. The oxalate anion, $^-O.OC–CO.O^-$, for instance, co-ordinates to two positions in such complex anions as $Fe(C_2O_4)_3^{3-}$. This ion exists both in aqueous solution and in the green complex solid $K_3Fe(C_2O_4)_3$. Compounds containing such ligands are termed *chelate compounds*, and the oxalate ion in such a compound would be called a chelating agent.

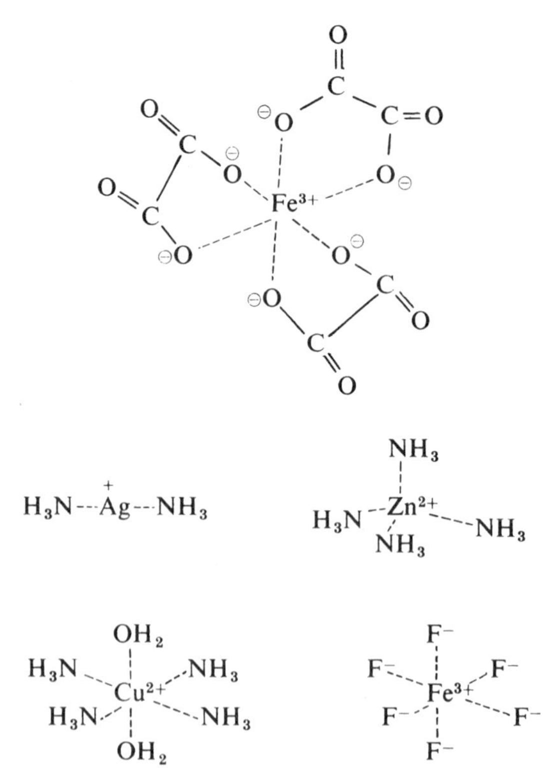

Fig 15.1 The shapes of some complex ions.

It may be noticed in passing that, although our concern will be with a few simple complex compounds and their behaviour in aqueous solution, the study of vast numbers of complex compounds has led chemists to a much deeper understanding of the nature of the bonding between metal cations and complexing agents. Furthermore, some very important naturally-occurring substances consist of complexes, e.g. the iron complex, haem, a part of haemoglobin, is the red pigment in the red blood corpuscles, and the magnesium complex, chlorophyll, the green colouring matter in plants.

15.2 COMPLEX ION EQUILIBRIA

In many cases, values of the equilibrium constants for the stepwise formation of complex ions have been measured. Thus the formation of $Cu(NH_3)_4(H_2O)_2^{2+}$ takes place in four stages, each of which represents the replacement of one water ligand by an ammonia molecule. Each stage is characterized by an individual equilibrium constant. Without loss of rigour, and for simplicity, we shall omit the water molecules from the following equations. Equilibrium constants quoted are at room temperature.

$$Cu^{2+} + NH_3 \rightleftharpoons Cu(NH_3)^{2+} \qquad (1)$$

$$K_1 = \frac{[Cu(NH_3)^{2+}]}{[Cu^{2+}][NH_3]} = 1.41 \times 10^4$$

$$Cu(NH_3)^{2+} + NH_3 \rightleftharpoons Cu(NH_3)_2^{2+} \qquad (2)$$

$$K_2 = \frac{[Cu(NH_3)_2^{2+}]}{[Cu(NH_3)^{2+}][NH_3]} = 3.16 \times 10^3$$

$$Cu(NH_3)_2^{2+} + NH_3 \rightleftharpoons Cu(NH_3)_3^{2+} \qquad (3)$$

$$K_3 = \frac{[Cu(NH_3)_3^{2+}]}{[Cu(NH_3)_2^{2+}][NH_3]} = 7.77 \times 10^2$$

$$Cu(NH_3)_3^{2+} + NH_3 = Cu(NH_3)_4^{2+} \qquad (4)$$

$$K_4 = \frac{[Cu(NH_3)_4^{2+}]}{[Cu(NH_3)_3^{2+}][NH_3]} = 1.35 \times 10^2$$

In an ammoniacal solution of copper(II) ions, these four equilibria are all established. The actual concentrations of the various copper–ammonia complexes will be determined by the concentration of free ammonia. The 'free' ammonia concentration, written as $[NH_3]$ in (1) to (4), is the concentration of ammonia remaining in solution after some of the initially-added ammonia has been complexed in the various copper(II) complexes.

Figure 15.2 shows the relative distribution of Cu^{2+} among the copper(II)–ammonia complexes as a function of the free ammonia concentration as calculated from the four equilibrium constants K_1 to K_4. At $[NH_3] = 10^{-3}$ M, 1.0 per cent of the total copper(II) ion exists as $Cu(H_2O)_6^{2+}$, 14.2 per cent as $Cu(NH_3)(H_2O)_5^{2+}$, 45.0 per cent as $Cu(NH_3)_2(H_2O)_4^{2+}$, 35.0 per cent as

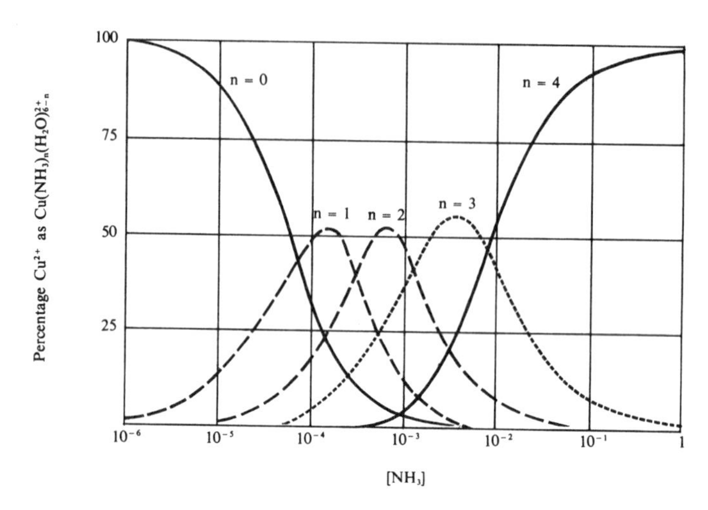

Fig 15.2 Variation of concentration of copper(II)–ammonia complexes as a function of ammonia concentration.

$Cu(NH_3)_3(H_2O)_3^{2+}$ and only 4.8 per cent as $Cu(NH_3)_4(H_2O)_2^{2+}$. As the free ammonia concentration is increased, the percentage of $Cu(NH_3)_4(H_2O)_2^{2+}$ increases sharply and the percentage of the lower ammonia complexes falls. At $[NH_3] = 10^{-1}$ M, 93.2 per cent of all the copper(II) ion exists as $Cu(NH_3)_4(H_2O)_2^{2+}$, 6.7 per cent as $Cu(NH_3)_3(H_2O)_3^{2+}$ and only 0.1 per cent as $Cu(NH_3)_2(H_2O)_4^{2+}$. In this system of stepwise equilibria, there are several co-existing complexes. Thus equilibrium (2) becomes significant before complete conversion to 100 per cent $Cu(NH_3)(H_2O)_5^{2+}$ occurs via equilibrium (1) (e.g. $[NH_3] = 10^{-4}$ M in Figure 15.2). This is a common situation in any system of successive equilibria.

The equilibrium for formation of any of the ammonia complexes can be expressed in terms of $Cu(H_2O)_6^{2+}$ and NH_3. By adding equations (1) to (4) we obtain:

$$Cu^{2+} + 4NH_3 \rightleftharpoons Cu(NH_3)_4^{2+}$$

for which

$$\frac{[Cu(NH_3)_4^{2+}]}{[Cu^{2+}][NH_3]^4} = K_1 \times K_2 \times K_3 \times K_4 = 4.7 \times 10^{12}$$

This particular equilibrium constant is referred to as the *stability constant* K_{st} of the complex ion $Cu(NH_3)_4^{2+}$, i.e.

$$K_{st}\{Cu(NH_3)_4^{2+}\} = \frac{[Cu(NH_3)_4^{2+}]}{[Cu^{2+}][NH_3]^4} = 4.7 \times 10^{12}$$

Likewise the stability constant of the complex ion $Cu(NH_3)_3^{2+}$ is the equilibrium constant referring to the equation

$$Cu^{2+} + 3NH_3 \rightleftharpoons Cu(NH_3)_3^{2+}$$

15.3 STABILITY CONSTANTS

The significance of the stability constants of complex ions, such as that for $Cu(NH_3)_4^{2+}$ to which we have referred, lies in the fact that such equilibrium constants give a measure of the extent to which complex ion formation occurs from the central ion and ligands, and so gives a measure of the 'stability' of the complex in solution.

In general terms, for the reaction

$$M(aq) + nL \rightleftharpoons ML_n(aq)$$

$$K_{st} = \frac{[ML_n]}{[M][L]^n} \text{ at constant temperature.}$$

For example, consider the stability constants for the following silver complexes:

(a) $Ag^+ + 2Cl^- \rightleftharpoons AgCl_2^-$ $\quad K_{st} = \frac{[AgCl_2^-]}{[Ag^+][Cl^-]^2} = 1.1 \times 10^5$

(b) $Ag^+ + 2NH_3 \rightleftharpoons Ag(NH_3)_2^+$ $\quad K_{st} = \frac{[Ag(NH_3)_2^+]}{[Ag^+][NH_3]^2} = 1.6 \times 10^7$

(c) $Ag^+ + 2S_2O_3^{2-} \rightleftharpoons Ag(S_2O_3)_2^{3-}$ $\quad K_{st} = \frac{[Ag(S_2O_3)_2^{3-}]}{[Ag^+][S_2O_3^{2-}]^2} = 1.0 \times 10^{13}$

(d) $Ag^+ + 2CN^- \rightleftharpoons Ag(CN)_2^-$ $\quad K_{st} = \frac{[Ag(CN)_2^-]}{[Ag^+][CN^-]^2} = 1.0 \times 10^{21}$

The larger the magnitude of the stability constant, the greater is the extent to which the forward reaction proceeds. Thus the proportion of complexed Ag^+ to uncomplexed Ag^+, for given quantities of initial reagents, will be highest for CN^- as complexing ligand, and lowest for Cl^-.

An important consequence follows. If to a solution containing Ag^+, say $AgNO_3$ solution, one added both an ammonia solution and a solution containing $S_2O_3^{2-}$, say from $Na_2S_2O_3$ solution, there would be a competition between NH_3 molecules and $S_2O_3^{2-}$ for the co-ordination positions around Ag^+. As the stability constants indicate, $S_2O_3^{2-}$ complexes to a more considerable extent than does NH_3, and this would result in the complex ion $Ag(S_2O_3)_2^{3-}$ existing as the predominant complexed species in the solution. Alternatively, the addition of a solution of $S_2O_3^{2-}$ to a solution already containing $Ag(NH_3)_2^+$ would result in the replacement of some NH_3 molecules by $S_2O_3^{2-}$. Likewise, addition of CN^- to either a solution of $Ag(NH_3)_2^+$ or $Ag(S_2O_3)_2^{3-}$ would result in the formation of $Ag(CN)_2^-$. On the other hand, the addition of ammonia solution to one containing $Ag(CN)_2^-$ would not result in any significant ligand replacement.

In general, then, one can successively replace the ligands around a particular complex ion by adding a stronger complexing agent.

Where different complex ions are coloured, such changes can be readily observed. For example, the successive additions to a solution containing

$Fe(H_2O)_6^{3+}$ of solutions of potassium chloride (saturated), potassium thiocyanate and sodium fluoride result in a series of colour changes. The initial pale yellow iron(III) solution changes first to yellow, then blood red, and finally colourless. This is correlated with the successive formation from $Fe(H_2O)_6^{3+}$ of $FeCl_4^-$, $Fe(NCS)_6^{3-}$ and FeF_6^{3-} as the principal entities at each successive stage. Each successive ligand becomes more firmly bound to the central ion.

Many complexing ligands, such as NH_3, CN^- and F^-, act as bases in water and in acidic solutions these ligands will be protonated. Thus the pyramidal NH_3 ligand, with an available lone pair of electrons, may be converted by acid to the tetrahedral NH_4^+ ion which does not complex with a metal cation like Ag^+. For the $Ag(NH_3)_2^+$ complex, when $[NH_3] = 10^{-2}$ M then the value of K_{st} indicates that $[Ag(NH_3)_2^+]/[Ag^+] = 1.6 \times 10^3$; the silver cation is virtually complexed completely. However, if the same solution is acidified to pH 4, then the acidity constant K_a of the NH_4^+ cation allows us to predict that the free NH_3 concentration would fall from 10^{-2} to 6×10^{-8}. The ratio $[Ag(NH_3)_2^+]/[Ag^+]$ would therefore fall from 1.6×10^3 in a basic ammonia solution to 5.8×10^{-56} in a solution at pH = 4 and the $Ag(NH_3)_2^+$ complex is said to be 'destroyed' by acid. Many metal complexes can be prevented from forming when acidic solutions of basic ligands are used.

QUESTIONS

1. A solution which is 0.2 M in $Ag(NH_3)_2^+$ is made 1 M in NH_3. What will be the concentration of Ag^+ in the solution? $K_{st}[Ag(NH_3)_2^+] = 1.6 \times 10^7$.
2. What concentration of NH_3 will reduce $[Zn^{2+}]$ to 1×10^{-6} in a solution which is 0.1 M with respect to $Zn(NH_3)_4^{2+}$? $K_{st}[Zn(NH_3)_4^{2+}] = 1 \times 10^9$.
3. What equilibrium concentration of free $S_2O_3^{2-}$ would be required in 1 dm^3 of 0.001 M $AgNO_3$ to reduce $[Ag^+]$ to 10^{-14}? $K_{st}[Ag(S_2O_3)_2^{3-}] = 1 \times 10^{13}$.
4. A mass of 45.50 g of the complex salt $Cu(NH_3)_4SO_4$ is dissolved in water and made up to a volume of 2 dm^3.
   (a) Calculate the analytical concentration of the solution.
   (b) Assuming that no dissociation of the complex cation to free copper ions(II) and ammonia molecules occurs, state the concentration of $Cu(NH_3)_4^{2+}$ in the solution.
   (c) Assuming that dissociation of the complex cation does occur, and that $K_{st}[Cu(NH_3)_4^{2+}] = 2 \times 10^{13}$, find the actual concentration of each of the entities $Cu(NH_3)_4^{2+}$, NH_3 and Cu^{2+} in the solution. (Hint: Suppose that x mole dm^{-3} Cu^{2+} form from the dissociation of the complex ion.)
   (d) State the percentage dissociation which has occurred.

16

Heterogeneous Equilibria

In the three preceding chapters we have discussed various examples of chemical equilibria when all the reacting species were in the same phase; they were all either gaseous species, or all species in solution. In this chapter we will deal with mixed phase or heterogeneous equilibria. Two aspects of this topic will be dealt with: solid–gas equilibria, and solid–solution equilibria.

16.1 THE ROLE OF THE SOLID IN SOLID–GAS EQUILIBRIA

The simplest case of a solid–gas equilibrium occurs when we consider the equilibrium between a solid and its vapour. Thus, at room temperature, solid iodine is in equilibrium with iodine vapour,

$$I_2(s) \rightleftharpoons I_2(g)$$

Some data are given in Table 16.1.

Table 16.1
Vapour pressure of solid iodine

Mass of iodine (g)	*Pressure in atmosphere*			
	38.7°C	*73.2°C*	*97.5°C*	*116.5°C*
1	0.00132	0.0132	0.0526	0.132
5	0.00132	0.0132	0.0526	0.132
10	0.00132	0.0132	0.0526	0.132
50	0.00132	0.0132	0.0526	0.132

The important feature of these results is that the pressure exerted by the solid iodine at a definite temperature is *independent of the amount of iodine present*. That is, in the equilibrium,

$$I_2(s) \rightleftharpoons I_2(g)$$

the 'effective concentration' of solid iodine is a constant, provided *some* solid iodine is present. Chemists express this observation by saying that the *activity* of solid iodine is a constant. So we can write for the equilibrium constant:

$$K' = \frac{p(I_2, g)}{a(I_2, s)}$$

where $a(I_2, s)$ is a constant.
Clearly, K' will be constant whatever value $a(I_2, s)$ has, so we can also write

$$K'a(I_2, s) = K = p(I_2, g)/1$$

This simply expresses the obvious experimental fact that $p(I_2, g)$ is a constant at constant temperature. That is, the vapour pressure of iodine is the equilibrium constant, for the equilibrium system shown above.

Results analogous to the above are found whenever solids are involved in any chemical or physical equilibrium, viz. the 'effective concentration' or 'activity' of a solid is independent of *how much* solid is present, so that the activities of solids are always constant. Obviously, the numerical value of an equilibrium constant involving a solid will depend on the actual numerical value of the activity of the solid. In practice, the value chosen for the activity is always unity: i.e. at any temperature.

$$\text{activity of any pure solid} = 1$$

This convention is analogous to the conventions listed earlier where we equated the activity of a gas to its pressure (in atmosphere), and the activity of a solute to its molar concentration.

16.2 SOME EXAMPLES OF SOLID—GAS EQUILIBRIA

A striking example of the decomposition of a solid is provided by ammonium hydrosulphide NH_4HS, which dissociates at 25°C according to

$$NH_4HS(s) \rightleftharpoons NH_3(g) + H_2S(g)$$

$$K = p_{NH_3} \cdot p_{H_2S}$$

where the pressures are in atmosphere.

EXAMPLE

A sample of solid ammonium hydrosulphide is introduced into an evacuated container. Dissociation occurs, and the total pressure reached at equilibrium is 62.8 kN m^{-2}. Calculate the equilibrium constant for the dissociation.

SOLUTION

The pressure in atmosphere is

$$\frac{62,800}{101,325} = 0.62 \text{ atm}$$

$$\therefore p_{NH_3} + p_{H_2S} = 0.62$$

Since for every mole of NH_4HS decomposed we get equimolar amounts of NH_3 and H_2S

$$p_{NH_3} = p_{H_2S} = \tfrac{1}{2} \times 0.62 = 0.31$$

$$\therefore K = p_{NH_3} \times p_{H_2S}$$

$$= (0.31)^2 = 0.095$$

EXAMPLE

Given that the equilibrium constant for the dissociation of $NH_4HS(s)$ is 0.095 at 25°C, calculate the individual partial pressures of ammonia and hydrogen sulphide when solid NH_4HS is added to a vessel containing a partial pressure of ammonia of 50.66 kN m^{-2}.

SOLUTION

Pressure of ammonia before dissociation of NH_4HS has occurred is

$$\frac{50.66}{101.325} = 0.5 \text{ atm}$$

Let x be final partial pressure of H_2S. Since an equal amount of ammonia must also be found by dissociation of NH_4HS

$$p_{NH_3} = 0.5 + x$$

$$\therefore 0.095 = x(0.5 + x)$$

or

$$x^2 + 0.50x - 0.095 = 0$$

$$\therefore x = 0.147$$

i.e.,

$$p_{H_2S} = 0.147$$

$$p_{NH_3} = 0.5 + 0.147 = 0.647$$

and, verifying the result, we note that

$$K = 0.147 \times 0.647 = 0.095$$

Another common type of solid-gas equilibrium is that in which there are two non-miscible solids, as in

$$CaCO_3(s) \rightleftharpoons CaO(s) + CO_2(g)$$

$$K = p_{CO_2}$$

At 800°C, the pressure of CO_2 in equilibrium with $CaCO_3$(s) and CaO(s) is 22,300 N m^{-2}.
Thus the equilibrium constant is

$$K = \frac{22{,}300}{101{,}325} = 0.22$$

EXERCISE
Sodium bicarbonate decomposes according to

$$2NaHCO_3(s) \rightleftharpoons Na_2CO_3(s) + H_2O(g) + CO_2(g)$$

(a) Write down an expression for the equilibrium constant.
(b) At a particular temperature, the total gas pressure in equilibrium with $NaHCO_3$(s) is 33.8 kN m^{-2}. Calculate the value of the equilibrium constant at this temperature.

16.3 SOLUBILITY EQUILIBRIA

It is a common experience that a substance may apparently dissolve in one solvent but not in another. Grease, for example, seems insoluble in water, but dissolves in carbon tetrachloride. Nevertheless, all solids dissolve in all solvents to some extent, so our description of substances as either soluble or insoluble in a given solvent is rather imprecise. More usefully we may describe a substance as soluble if a mass of the order of one gram or more (say) dissolves in 1 dm^3 of solvent, whereas insoluble substances are those in which (say) less than 0.001 gram dissolves in this volume of solvent. Obviously there is no line of demarcation between soluble and insoluble substances, there is only a difference in degree of solubility.

In this chapter, we shall show that the solubility process can be described by the principles of chemical equilibrium. After a brief general consideration of solubility, we concentrate attention upon the solubility of those sparingly soluble substances which give rise to ions in solution. When a molecular solid, such as glucose $C_6H_{12}O_6$, or iodine I_2, dissolves in water, individual molecules break away from their fixed lattice positions, and move more or less independently in the environment of water molecules. The solution process could be represented by

$$C_6H_{12}O_6(s) \rightarrow C_6H_{12}O_6(aq)$$

$$I_2(s) \rightarrow I_2(aq)$$

In the case of compounds such as NaCl or AgCl, which give rise to ions when they dissolve in water, the corresponding equations for the solution process would be

$$NaCl(s) \rightarrow Na^+(aq) + Cl^-(aq)$$

$$AgCl(s) \rightarrow Ag^+(aq) + Cl^-(aq)$$

It is an experimental observation that at a given temperature particular compounds each have a definite solubility in water. If, as with glucose and sodium chloride, a considerable amount of solid dissolves in a definite amount (e.g. 1000 cm^3) of solution, the solids are described as soluble. Only very small amounts of I_2 and AgCl dissolve per dm^3 of solution at room temperature, and these solids are loosely described as insoluble, or more precisely as *sparingly soluble*.

The number of moles of solute which dissolve per unit volume of solution is the solubility of that substance. A saturated solution represents an equilibrium condition, since no further amount of solid dissolves in a given volume of saturated solution at a fixed temperature, whatever the amount of solid added.

Consider the general case of the dissolution in pure water of a solid MX, which produces ions $M^+(aq)$ and $X^-(aq)$ only

$$MX(s) \rightleftharpoons M^+(aq) + X^-(aq)$$

In a saturated solution, i.e. at equilibrium, the concentrations of $M^+(aq)$ and $X^-(aq)$ will be fixed. If the equilibrium lies well to the right (e.g. if MX represents NaCl) the substance will be 'soluble'; if it lies well to the left (e.g. if MX represents AgCl) the substance would ordinarily be termed 'insoluble'. The actual position of equilibrium, as expressed by the solubility of MX in water, will be essentially determined by the balance between the two factors mentioned in chapter 7, viz.:

(a) the tendency of the system to a potential energy minimum;

(b) the disruptive effects of kinetic energy, causing the system to tend to a more disordered state.

We shall consider these factors independently. Thermochemical data show that for some ionic solids, the ions in the solid lattice have a lower potential energy than the solvated ions in water, while for other ionic solids this situation is reversed. Some data are set out in Table 16.2, where ΔH

Table 16.2

Heats of solution and solubility of some electrolytes

System	ΔH *(kJ)*	*Solubility at 25°C (g dm⁻³)*
$KCl(s) \rightarrow K^+(aq) + Cl^-(aq)$	+17.3	420
$KBr(s) \rightarrow K^+(aq) + Br^-(aq)$	+20.0	600
$NaCl(s) \rightarrow Na^+(aq) + Cl^-(aq)$	+3.9	360
$NaBr(s) \rightarrow Na^+(aq) + Br^-(aq)$	−0.18	900
$KOH(s) \rightarrow K^+(aq) + OH^-(aq)$	−55.6	1200
$PbCl_2(s) \rightarrow Pb^{2+}(aq) + 2Cl^-(aq)$	+22.1	11
$BaCl_2(s) \rightarrow Ba^{2+}(aq) + 2Cl^-(aq)$	+8.8	40
$AgF(s) \rightarrow Ag^+(aq) + F^-(aq)$	−50.1	1950

represents the heat of solution of the solid, i.e. the heat evolved per mole of solid dissolved in water, as defined in section 5.6. When ΔH is positive, the ions in the solid lattice have a lower potential energy than the solvated ions in water. When ΔH is negative, the solvated ions in solution have the lower potential energy.

It has been seen in earlier chapters that both the lattice energies of ionic solids and the heats of solvation of ions in water are large quantities; the ΔH values quoted in the table represent the rather small differences between these two comparatively large terms.

Inspection of Table 16.2 shows that in some cases the position of equilibrium favours the state of lower potential energy indicating that factor (a) is of paramount importance. In other soluble substances the higher potential energy state is favoured, and in these cases the operation of factor (b) is of overriding importance.

When we come to consider the likely effects of the kinetic energy possessed by ions, it turns out that the disordering effect accompanying the break-up of the highly ordered ionic lattice to form ions in solution is nicely balanced by the ordering of solvent molecules, bound in the solvation sheaths around these dissolved ions. In some cases the ions in the solid lattice represent a more ordered state than the ions in water, while in other cases the situation is reversed. As mentioned earlier, this degree of order in a particular substance can be measured quantitatively, although it is beyond the scope of this book to discuss such measurements.

We must conclude with the sobering observation that there is no simple way of theoretically predicting the solubilities of ionic solids in water. This is not surprising since apparently either of the two factors (a) or (b) may dominate in deciding the extent of solubility, and each of these factors may only be quantitatively evaluated by direct experimental measurement.

16.4 SOLUBILITY EQUILIBRIUM AND SOLUBILITY PRODUCT

Consider again the dissolution of a solid MX in water, to form ions $M^+(aq)$ and $X^-(aq)$:

$$MX(s) \rightleftharpoons M^+(aq) + X^-(aq)$$

When a saturated solution is formed in the presence of excess solid, a dynamic equilibrium exists when the rate of dissolution of solid into dissolved ions is equal to, and balanced by, the rate at which ions in solution precipitate back on the solid lattice.

A few simple experiments show that the amount of MX(s) which is in contact with the aqueous solution does not alter the solubility. In other words, by finding the concentrations of the ions $M^+(aq)$ and $X^-(aq)$ in solutions containing different masses of MX(s), it is found that the position of equilibrium is absolutely fixed provided that *some* MX(s) is present in contact with the solution; the amount of MX(s) present is completely irrelevant. This means that, as with solid–gas equilibria, provided *some* solid MX(s) is present, its 'effective concentration' or 'activity' remains constant in the system of solid + solution, and the 'activity' of MX(s) is thus taken, by common agreement, to have the constant value of unity. So, for the equilibrium above we can write directly

$$[M^+][X^-] = \textit{a constant at constant temperature}$$

and this constant is designated K_s, termed the solubility product constant.

More generally, and omitting charges, for

$$M_aX_b \rightleftharpoons aM + bX$$

$$K_s = [M]^a[X]^b$$

where M and X may be ions.

The solubility product is the equilibrium constant for an equilibrium between a solid and its saturation solution, e.g.

(i) $$BaSO_4(s) \rightleftharpoons Ba^{2+} + SO_4^{2-}$$

$$K_s = [Ba^{2+}][SO_4^{2-}]$$

Experimentally, it is known that

$$K_s(BaSO_4) = 1 \times 10^{-10} \text{ at } 25°C$$

(ii) $$Ag_2CO_3(s) \rightleftharpoons 2Ag^+ + CO_3^{2-}$$

$$K_s = [Ag^+]^2[CO_3^{2-}] = 8 \times 10^{-12} \text{ at } 25°C$$

(iii) $$Al(OH)_3(s) \rightleftharpoons Al^{3+} + 3OH^-$$

$$K_s = [Al^{3+}][OH^-]^3 = 1 \times 10^{-33} \text{ at } 25°C$$

(iv) $$C_6H_{12}O_6(s) \rightleftharpoons C_6H_{12}O_6(aq)$$

$$K_s = [C_6H_{12}O_6]$$

The 'solubility product' of glucose is seen to be the same quantity as the solubility.

EXERCISE

Write down the solubility product expression for the following substances: CuI, Ag_2CrO_4, $Fe(OH)_3$, $Ca_3(PO_4)_2$.

Just as we found it convenient to define an alternative way of expressing acidity constants K_a, in terms of pK_a, so analogously one could define a quantity pK_s, such that

$$pK_s = -\log_{10} K_s$$

Thus for $BaSO_4$,

$$K_s = 1 \times 10^{-10}$$

$$\therefore \quad pK_s = 10$$

and for Ag_2CO_3,

$$K_s = 8.2 \times 10^{-12} = 10^{0.91} \times 10^{-12} = 10^{-11.09}$$

$$\therefore pK_s = 11.09$$

The approximate solubility products of a range of substances are listed in Table 16.3.

Table 16.3
Solubility products of selected compounds (25°C)

Compound	Ks	*Compound*	Ks	*Compound*	Ks
AgCl	1.5×10^{-10}	$CaCO_3$	4.8×10^{-9}	$Fe(OH)_2$	1×10^{-15}
AgBr	4×10^{-13}	$SrCO_3$	1×10^{-9}	$Fe(OH)_3$	1×10^{-38}
AgI	1×10^{-16}	$BaCO_3$	5×10^{-9}	Ag_2SO_4	1.2×10^{-5}
CaF_2	4×10^{-11}	Ag_2CrO_4	1×10^{-12}	$BaSO_4$	1×10^{-10}
$PbCl_2$	1.7×10^{-5}	$PbCrO_4$	2×10^{-14}	$PbSO_4$	2×10^{-8}
PbI_2	8×10^{-9}	$BaCrO_4$	2×10^{-10}	PbS	10^{-28}
Ag_2CO_3	8×10^{-12}	$Al(OH)_3$	1×10^{-33}	ZnS	10^{-23}
$MgCO_3$	1×10^{-5}	$Cu(OH)_2$	6×10^{-20}	CuS	10^{-40}

16.5 A LIMITATION OF THE SOLUBILITY PRODUCT PRINCIPLE

In considering the equilibrium

$$MX(s) \rightleftharpoons M^+ + X^-$$

for which

$$K_s(MX) = [M^+][X^-]$$

it has been tacitly assumed that, as defined here, K_s will be independent of the total concentration of ions of the solution. It has been found experimentally that this is not so. For example, the solubility of AgCl when dissolved in a solution of 1 M KNO_3 is some 10 per cent higher than its solubility in water, although neither ion K^+ nor NO_3^- reacts with the AgCl. In general terms, the validity of the principle as we have stated it is restricted to solutions in which the total concentration of ions is low, say around 0.01 M or less.

Two important consequences follow from this. First, the solubility product relation is quite inaccurate and inapplicable for compounds which are fairly soluble (say for those for which concentrations of the order of 0.05 M or higher are attainable). One would not define a solubility product for soluble substances such as NaCl and K_2CrO_4. Second, for compounds of low solubility, the principle as we have stated it holds well when considering solubility in pure water. In cases where there are moderate concentrations of dissolved ions from other sources in solutions, it can only be applied very approximately.

16.6 RELATION BETWEEN SOLUBILITY AND SOLUBILITY PRODUCT

The solubility product relation implies a specific interdependence of solubility and solubility product, for pure compounds dissolved in water. If either of these quantities, the solubility s or the solubility product K_s, is known, then the other quantity may be calculated.

EXAMPLE 1

Calculate the solubility of $BaSO_4$ in water, given K_s $(BaSO_4) = 1 \times 10^{-10}$.

SOLUTION

Suppose the solubility be s M. Since

$$BaSO_4(s) \rightleftharpoons Ba^{2+} + SO_4^{2-}$$

we see that s mole $BaSO_4$ gives rise to s mole Ba^{2+}, and to s mole SO_4^{2-}

$$\therefore\ [Ba^{2+}] = s \text{ and } [SO_4^{2-}] = s$$

whence $$K_s = [Ba^{2+}][SO_4^{2-}] = s \times s = s^2 = 1 \times 10^{-10}$$

$$\therefore\ s = 1 \times 10^{-5}\ \text{M}$$

$$= 1 \times 10^{-5} \times 233.3\ \text{g dm}^{-3}$$

$$= 2.33 \times 10^{-3}\ \text{g dm}^{-3}$$

i.e. a saturated solution of $BaSO_4$ in water contains 0.0023 gram of dissolved solid per dm^3 of solution.

EXAMPLE 2

Calculate the solubility of Ag_2CO_3 in water, given $K_s\,(Ag_2CO_3) = 8.2 \times 10^{-12}$ at 25°C.

SOLUTION

Suppose the solubility be s M. Since

$$Ag_2CO_3(s) \rightleftharpoons 2Ag^+ + CO_3^{2-}$$

we see that s mole Ag_2CO_3 gives rise to $2s$ mole Ag^+, and to s mole CO_3^{2-}

$$\therefore\ [Ag^+] = 2s \text{ and } [CO_3^{2-}] = s$$

whence $$K_s = [Ag^+]^2[CO_3^{2-}] = (2s)^2(s) = 4s^3$$

$$\therefore\ 4s^3 = 8.2 \times 10^{-12}$$

$$\therefore\ s^3 = 2.05 \times 10^{-12}$$

$$\therefore\ s = 1.27 \times 10^{-4}$$

i.e. the solubility of $Ag_2CO_3 = 1.27 \times 10^{-4}$ M

$$= 1.27 \times 10^{-4} \times 275.8\ \text{g dm}^{-3}$$

$$= 3.5 \times 10^{-2}\ \text{g dm}^{-3}$$

Notice from these two examples, that although the solubility product of silver carbonate is *less* than that of barium sulphate, the solubility of silver carbonate (1.3×10^{-4} M) *exceeds* that of barium sulphate (1.0×10^{-5} M). This shows that when substances of different formula type are considered, the solubility products do not directly indicate relative solubilities. When substances are of the same formula type, the solubilities, measured in M, are directly related to the solubility products.

EXERCISE

Arrange the following substances, whose solubility products are quoted, in order of increasing solubility.

$$K_s(PbSO_4) = 2 \times 10^{-8} \qquad K_s(ZnS) = 10^{-23}$$
$$K_s(AgBr) = 4 \times 10^{-13} \qquad K_s(CuCO_3) = 1 \times 10^{-10}$$
$$K_s(BaC_2O_4) = 1 \times 10^{-7}$$

If the solubilities are measured in g dm^{-3}, this order may not be preserved. Why?

EXAMPLE 3

The solubility of PbF_2 is 0.52 g dm^{-3}. Calculate K_s (PbF_2).

SOLUTION

$$0.52 \text{ g dm}^{-3} = 0.52/245 \text{ M} = 2.12 \times 10^{-3} \text{ M}$$

The equation

$$PbF_2(s) \rightleftharpoons Pb^{2+} + 2F^-$$

indicates that for each mole of PbF_2 dissolved, 1 mole of Pb^{2+} and 2 mole of F^- form. So

$$[Pb^{2+}] = 2.12 \times 10^{-3} \text{ and } [F^-] = 2 \times 2.12 \times 10^{-3}$$

$$\begin{aligned} K_s(PbF_2) &= [Pb^{2+}][F^-]^2 \\ &= (2.12 \times 10^{-3})(4.24 \times 10^{-3})^2 \\ &= 3.8 \times 10^{-8} \end{aligned}$$

EXERCISE

The solubility of calcium sulphate is 1.06 g dm^{-3} at 25°C. Show that $K_s(CaSO_4) = 6.1 \times 10^{-5}$.

16.7 ASSUMPTIONS UNDERLYING THESE CALCULATIONS

In the preceding and subsequent calculations, it is necessary to make certain assumptions, one of which has already been mentioned in section 16.5. Two others, of fundamental importance, are considered here.

(a) We assume that all of the dissolved compound is present in solution entirely as ions. Thus the relation

$$MX(s) \rightleftharpoons M^+(aq) + X^-(aq)$$

is a reduction from the general case

$$MX(s) \rightleftharpoons MX(aq) \rightleftharpoons M^+(aq) + X^-(aq)$$

where it is assumed

$$MX(aq) = 0$$

It is thus postulated that such substances behave as strong electrolytes, in that complete dissociation into ions has occurred. Note that this is not to suggest that the bonding in the *solid* is necessarily ionic; substances such as ZnS fall into this category, where a considerable degree of covalent character exists in the bonds.

(b) We assume that the ions derived from the solid remain as such in solution. We neglect to take account of any interaction between these ions and solvent water molecules other than simple solvation. For instance, the solubility product for calcium carbonate may be determined experimentally, and it is found that

$$K_s(CaCO_3) = 4.8 \times 10^{-9} \qquad (25°C)$$

Computing the solubility of $CaCO_3$ from these data leads to a solubility of 6.9×10^{-5} M. Experimentally, however, a direct determination of solubility gives close to twice this value; the solubility is actually 1.3×10^{-4} M. The discrepancy arises because CO_3^{2-}, formed in solution from the process

$$CaCO_3(s) \rightleftharpoons Ca^{2+} + CO_3^{2-}$$

hydrolyses

$$CO_3^{2-} + H_2O \rightleftharpoons HCO_3^- + OH^-$$

To achieve the expected concentration of free CO_3^{2-} as determined from the experimental solubility product ($[CO_3^{2-}] = 6.9 \times 10^{-5}$ M), a *greater* amount of $CaCO_3$ must dissolve than that calculated by assuming no hydrolysis of the CO_3^{2-} in solution.

While for convenience we shall make the assumptions referred to, it is important that we recognize that they are being made.

16.8 SOLUBILITY OF MX IN THE PRESENCE OF X

The general solubility product expression for a salt M_aX_b

$$K_s = [M]^a[X]^b$$

indicates that for the solid M_aX_b in equilibrium with a solution containing M and X, the fraction

$$[M]^a[X]^b$$

is always constant at a given temperature, irrespective of the individual concentrations of M and X. We now consider the solubility of M_aX_b, not only in water, but in solutions which contain other sources of the ions M or X. (In view of the comments in section 16.5 the solubilities obtained here will be only approximate.)

Suppose $BaSO_4$ is added to a solution of sodium sulphate. In such a case, provided some solid $BaSO_4$ is present, it is still true that

$$K_s(BaSO_4) - [Ba^{2+}][SO_4^{2-}] = 1 \times 10^{-10}$$

where $[SO_4^{2-}]$ arises from SO_4^{2-} ions due to dissolved Na_2SO_4 and from SO_4^{2-} ions due to dissolved $BaSO_4$. If the Na_2SO_4 solution is 0.1 M, say, then $[SO_4^{2-}]$ in the solution far exceeds that obtained when $BaSO_4$ dissolves in water. It follows then, that the $[Ba^{2+}]$ in the sodium sulphate solution must be markedly less than that when $BaSO_4$ dissolves in water. The solubility of the compound is measured by the concentration of the ion present in lower concentration, i.e. by the $[Ba^{2+}]$, since the $[Ba^{2+}]$ measures the number of mole dm^{-3} of $BaSO_4$ dissolved. So $BaSO_4$ is considerably less soluble in Na_2SO_4 solution than it is in water.

EXAMPLE 1

Calculate the solubility of $BaSO_4$ in 0.1 M Na_2SO_4 solution. Compare Example 1, section 16.4.

SOLUTION

If the solubility of $BaSO_4$ is s M, each mole of dissolved solid gives rise to s M of Ba^{2+}, and s M of SO_4^{2-}. In the solution thus formed

$$[Ba^{2+}] = s, \text{ and } [SO_4^{2-}] = s + 0.1$$

Since
$$[Ba^{2+}][SO_4^{2-}] = 1 \times 10^{-10}$$

$$\therefore \; s(s + 0.1) = 1 \times 10^{-10}$$

and the quadratic equation may be solved for s. An approximate solution is obtained by noticing that in a saturated solution of $BaSO_4$ in water $[SO_4^{2-}] = 1 \times 10^{-5}$. Here, $[SO_4^{2-}]$ from $BaSO_4$ dissolving will be *less* than this.

∴ it is a very good approximation to write

$$(s + 0.1) \sim 0.1$$

$$\therefore \quad s(0.1) = 1 \times 10^{-10}$$

$$\therefore \quad s = 1 \times 10^{-9}$$

Thus the solubility of $BaSO_4$ in the 0.1 M Na_2SO_4 solution is only 1×10^{-9} M, whereas in water it is 1×10^{-5} M.

EXERCISE

Find the solubility of $BaSO_4$ in a 0.01 M $BaCl_2$ solution.

EXAMPLE 2

Calculate the solubility of CaF_2 in

(a) water
(b) 0.2 M $Ca(NO_3)_2$ solution
(c) 0.2 M NaF solution

$$K_s(CaF_2) = 4 \times 10^{-11} \qquad (25°C)$$

SOLUTION

(a) If the solubility is s M, as shown in Example 2, section 16.6,

$$4s^3 = 4 \times 10^{-11}$$

$$\therefore \; s^3 = 1 \times 10^{-11} = 10 \times 10^{-12}$$

$$\therefore \; s = 2.2 \times 10^{-4} = 0.00022 \text{ M.}$$

(b) If s mole CaF_2 dissolve in 1 dm^3 of 0.2 M Ca^{2+} solution, then a further s mole Ca^{2+}, and $2s$ mole F^-, form

$$\therefore\ [Ca^{2+}] = 0.2 + s, \text{ and } [F^-] = 2s$$

$$\therefore\ (0.2 + s)(2s)^2 = 4 \times 10^{-11}$$

where s has a different value from the previous case, and in fact here s will be less than 0.0002 M,

$\therefore$ a valid approximation is

$$(0.2)(2s)^2 = 4 \times 10^{-11}$$

$$\therefore\ 0.8\,s^2 = 4 \times 10^{-11}$$

$$\therefore\quad s^2 = 5 \times 10^{-11} = 50 \times 10^{-12}$$

$$\therefore\quad s = 7.1 \times 10^{-6}\ M$$

(c) Arguing in an analogous manner, if the solubility again be s M, (where s is again less than 0.0002 M)

$$\therefore\ [Ca^{2+}] = s \text{ and } [F^-] = (2s + 0.2)$$

$$\therefore\ s(2s + 0.2)^2 = 4 \times 10^{-11}$$

Again, since s and $2s$ are both very small in comparison with 0.2, we approximate to

$$s(0.2)^2 = 4 \times 10^{-11}$$

$$\therefore\ s = 4 \times 10^{-11}/4 \times 10^{-2}$$

$$= 1 \times 10^{-9}\ M$$

These calculations involve an assumption similar to those in section 16.7. We assume that the slightly soluble solids do not react with the ions in the solutions. It would not be possible to use this procedure to determine the solubility of $Zn(OH)_2$ in a solution of NaOH. This is because $Zn(OH)_2$ reacts with OH^- if the $[OH^-]$ is sufficiently high to form $Zn(OH)_4^{2-}$

$$Zn(OH)_2 + 2OH^- \rightleftharpoons Zn(OH)_4^{2-}$$

thereby *increasing* considerably the solubility of $Zn(OH)_2$ above its value in water. Such cases will be considered more fully in section 16.11.

16.9 PREDICTING PRECIPITATION: IONIC PRODUCT AND SOLUBILITY PRODUCT

Suppose that a solution which is 0.5 M in Ca^{2+} is mixed with an equal volume of a solution 0.2 M in SO_4^{2-}. This may be achieved, say, by mixing 0.5 M $Ca(NO_3)_2$ and 0.2 M Na_2SO_4 solutions.

Instantly on mixing, since the volume of the solution is doubled,

$$[Ca^{2+}] = 0.25 \quad \text{and} \quad [SO_4^{2-}] = 0.1$$

so the expression

$$[Ca^{2+}][SO_4^{2-}] = 0.25 \times 0.1$$
$$= 2.5 \times 10^{-2}$$

Note immediately that this product is *not* the solubility product of $CaSO_4$, since $K_s(CaSO_4)$ is an equilibrium constant which implies a solid phase present and in equilibrium with its saturated solution. In fact

$$K_s(CaSO_4) = 6.1 \times 10^{-5}$$

The product we have calculated is termed the ionic product, and it is merely a concentration product applicable in any situation—it is *not* an equilibrium constant.

This ionic product is over a hundred times greater than the solubility product, implying that the solution formed by mixing Ca^{2+} and SO_4^{2-} is supersaturated with these ions. The concentrations of each of these ions are lowered by the reaction

$$Ca^{2+} + SO_4^{2-} \rightarrow CaSO_4(s)$$

which proceeds until $[Ca^{2+}][SO_4^{2-}]$ is reduced from 2.5×10^{-2} to the equilibrium value 6.1×10^{-5}. When this stage is reached, no further $CaSO_4$ precipitates.

As a general rule, note that **precipitation occurs whenever the ionic product exceeds the solubility product** and will continue until the ionic product equals the solubility product.

Suppose, however, that equal volumes of a solution 0.001 M in Ca^{2+} and of a solution 0.002 M in SO_4^{2-} are mixed. In this case, the ion concentrations on mixing are

$$[Ca^{2+}] = 0.0005 \quad \text{and} \quad [SO_4^{2-}] = 0.001$$

The ionic product

$$[Ca^{2+}][SO_4^{2-}] = 5 \times 10^{-4} \times 10^{-3}$$
$$= 5 \times 10^{-7}$$

is *less* than $K_s(CaSO_4)$; the solution formed is unsaturated in Ca^{2+} and SO_4^{2-}, and no precipitation would occur.

EXAMPLE

20 cm³ of 0.01 M $AgNO_3$ solution is added to 80 cm³ of 0.05 M K_2CrO_4 solution. Determine whether a precipitate would form on mixing the solutions if $K_s(Ag_2CrO_4) = 1.7 \times 10^{-12}$.

SOLUTION

In 0.01 M $AgNO_3$ solution, $[Ag^+] = 0.01$
If 20 cm³ of such solution is diluted to 100 cm³,

$$[Ag^+] = 0.01 \times 20/100 = 0.002$$

In 0.05 M K_2CrO_4 solution, $[CrO_4^{2-}] = 0.05$
If 80 cm^3 of such solution is diluted to 100 cm^3,

$$[CrO_4^{2-}] = 0.05 \times \frac{80}{100} = 0.04$$

Thus the ionic product

$$[Ag^+]^2[CrO_4^{2-}] = (2 \times 10^{-3})^2(4 \times 10^{-2}) = 1.6 \times 10^{-7}$$

is far greater than the solubility product. On mixing, then, the solution becomes supersaturated with Ag^+ and CrO_4^{2-}, and the reaction

$$2Ag + CrO_4^{2-} \rightarrow Ag_2CrO_4(s)$$

proceeds, precipitating solid Ag_2CrO_4, until the ionic product is reduced to 1.7×10^{-12}.

EXERCISE
Decide whether a precipitate may be expected to form if 0.001 g $BaCl_2$ is added to 1 dm^3 of 0.001 M Na_2CO_3 solution.

It is important to realize that these considerations give no information on the *rate* at which a precipitate forms, and thus on the rate of obtaining equilibrium. When solutions of moderately high concentration are used, precipitations of the type discussed in this chapter are usually quite rapid. When very dilute solutions are used, precipitation may be very slow, sometimes continuing over many hours.

The term 'ionic product' should not be confused with the equilibrium constant for the self-ionization of water, designated K_w. This is sometimes referred to as the ionic product of water. It is, however, an *equilibrium* constant, which is not the case when the term 'ionic product' is generally used.

16.10 DISSOLVING OF PRECIPITATES

For the equilibrium system

$$MX(s) \rightleftharpoons M^+(aq) + X^-(aq)$$

we have seen that $K_s = [M^+][X^-]$

If this equilibrium can be upset to allow the forward reaction

$$MX(s) \rightarrow M^+(aq) + X^-(aq)$$

to proceed, then the solid MX(s) will dissolve. This will happen only if the ionic product $[M^+][X^-]$ is no longer equal to the solubility product K_s, and in fact for the forward reaction to take place the ionic product must be *reduced* below the value of K_s.

This may be achieved by the addition of a reagent to the initial equilibrium system which reacts with one (or both) of the ions in solution, thus reducing the ionic product. If sufficient of the reagent is added, continually removing, by reaction, more and more of the reacting ion, equilibrium will not be re-attained and the solid will completely dissolve. If insufficient reagent is added, some solid dissolves, but some remains undissolved; equilibrium would be re-established, and the ionic product would again equal K_s.

For example, precipitates of metallic hydroxides such as magnesium hydroxide, copper(II) hydroxide and iron(III) hydroxide can be dissolved in acids. Representing iron(III) hydroxide by $Fe(OH)_3$ we would have

$$Fe(OH)_3 \rightleftharpoons Fe^{3+} + 3OH^-$$

$$K_s = [Fe^{3+}][OH^-]^3$$

If to this system some dilute HCl were added, H_3O^+ from the acid would react with OH^- from the solution

$$H_3O^+ + OH^- \rightarrow 2H_2O$$

This will occur to a very considerable extent, since H_2O is only slightly dissociated. So the $[OH^-]$ will be reduced below its equilibrium value, and

$$[Fe^{3+}][OH^-]^3 \text{ becomes } \textit{less} \text{ than } K_s$$

The system is no longer in equilibrium and to restore equilibrium the forward reaction

$$Fe(OH)_3 \rightarrow Fe^{3+} + 3OH^-$$

proceeds, and some $Fe(OH)_3$ dissolves.

Again $CaCO_3$ dissolves in acid by a similar process

$$CaCO_3(s) \rightleftharpoons Ca^{2+} + CO_3^{2-}$$

$$K_s = [Ca^{2+}][CO_3^{2-}]$$

In the presence of H_3O^+,

$$H_3O^+ + CO_3^{2-} \rightarrow HCO_3^- + H_2O$$

or

$$2H_3O^+ + CO_3^{2-} \rightarrow CO_2 + 3H_2O$$

reducing $[CO_3^{2-}]$ below the equilibrium value. Thus, the ionic product is reduced below the solubility product and the $CaCO_3$ dissolves.

In the next section we shall consider another case of dissolution of precipitates by removal of one of the ions in solution by complex formation.

16.11 DISSOLVING PRECIPITATES BY COMPLEX FORMATION

We have seen in section 16.10 that the necessary condition for dissolving a compound is that its ionic product be reduced below its solubility product. This can be achieved by addition of a reagent which reacts with one of the ions in the system. In particular, the concentration of certain ions in solution may be markedly reduced by addition of a reagent with which the ion forms a complex. For example, if to a precipitate of copper(II) hydroxide

$$Cu(OH)_2(s) \rightleftharpoons Cu^{2+} + 2OH^-$$

some ammonia solution is added, complex ion formation

$$Cu^{2+} + 4NH_3 \rightleftharpoons Cu(NH_3)_4^{2+}$$

considerably reduces $[Cu^{2+}]$, and the ionic product $[Cu^{2+}][OH^-]^2$ is lowered below $K_s\{Cu(OH)_2\}$.

The amount of precipitate dissolved by a given amount of an added reagent depends upon the magnitude of the solubility product of the compound and the stability constant of the complex.

When a 1 M NH_3 solution is added to a precipitate of AgCl, the precipitate is observed to dissolve. We may represent the reactions:

$$AgCl(s) \rightleftharpoons Ag^+ + Cl^- \quad (1)$$

$$Ag^+ + 2NH_3 \rightarrow Ag(NH_3)_2^+ \quad (2)$$

The formation of the complex ion reduces the concentration of free uncomplexed Ag^+, and thus reduces the ionic product $[Ag^+][Cl^-]$ below its equilibrium value K_s. Thus the system is no longer in equilibrium, and more AgCl dissolves in an attempt to increase the ionic product back to K_s. Whether this second equilibrium state can be attained depends upon the extent to which complex ion formation can occur, thus depending upon the amount of NH_3 added, the stability constant K_{st}, and also upon the solubility product K_s.

Thus when 1 M NH_3 solutions are added to precipitates of (say) 0.5 gram AgCl, AgBr and AgI, respectively, it is observed that the AgCl dissolves completely, AgBr dissolves partially and AgI appears insoluble. In fact, the solubility of AgI in the ammonia is appreciably greater than in water, but even in aqueous ammonia its solubility is still quite small.

EXAMPLE

Calculate and compare the masses of AgCl and of AgI which could dissolve in 1 dm^3 of 1.0 M ammonia solution.

$$K_s(AgCl) = 1.7 \times 10^{-10} \quad K_s(AgI) = 1 \times 10^{-16}$$

$$K_{st}\{Ag(NH_3)_2^+\} = 1.6 \times 10^7$$

SOLUTION

Suppose x mole AgCl dissolves per dm^3 of 1.0 M NH_3 solution.
Then when x mole AgCl has dissolved in solution:

$$[Cl^-] = x \quad (1)$$

$$[Ag^+] + [Ag(NH_3)_2^+] = x \quad (2)$$

Now provided some solid AgCl is present, irrespective of other entities in solution, such as ammonia, it is still true that

$$[Ag^+][Cl^-] = 1.7 \times 10^{-10} \quad (3)$$

Also, in the presence of some dissolved AgCl, furnishing Ag^+ and some ammonia, irrespective of the presence or absence of solid AgCl, it is still true that

$$[Ag(NH_3)_2^+]/[Ag^+][NH_3]^2 = 1.6 \times 10^7 \tag{4}$$

In a total of 1 mole of ammonia per dm^3,

$$[NH_3] + 2[Ag(NH_3)_2^+] = 1.0 \tag{5}$$

Two approximations may now be made:
Since K_{st} is large,

$$[Ag^+] \ll [Ag(NH_3)_2^+]$$

and we approximate (2) to give $[Ag(NH_3)_2^+] = x$

Since $$[Ag(NH_3)_2^+] \ll [NH_3]$$

we shall for simplicity approximate (5) to give $[NH_3] = 1.0$
Now combining (1) and (3) gives $[Ag^+] = 1.7 \times 10^{-10}/x$
Substituting these values for $[Ag(NH_3)_2^+]$, $[NH_3]$, and $[Ag^+]$ in (4)

$$\therefore \frac{x}{(1.7 \times 10^{-10}/x)(1.0)^2} = 1.6 \times 10^7 \tag{6}$$

whence $$x^2 = 1.6 \times 10^7 \times 1.7 \times 10^{-10} = 2.72 \times 10^{-3}$$

$$\therefore x = 5.2 \times 10^{-2}\,M$$

$$\therefore \text{solubility} = 5.2 \times 10^{-2} \times 143\ g\ dm^{-3}$$

$$= 7.4\ g\ dm^{-3}$$

Thus 7.4 gram AgCl dissolve per dm^3 of 1.0 M ammonia.

For AgI dissolving in ammonia, the approximations above are again valid, and carrying out the calculation we arrive at an equation analogous to (6) above, viz.

$$\frac{y}{(1 \times 10^{-16}/y)(1.0)^2} = 1.6 \times 10^7$$

from which we can show

$$y = 4 \times 10^{-5}\,M$$

$$\therefore \text{solubility} = 0.0094\ g\ dm^{-3}$$

Thus only 0.0094 gram AgI dissolves per dm^3 of 1.0 M ammonia.

While AgI then, dissolves to only a small extent in ammonia solution, a complexing solution which reacts to a more considerable extent with Ag^+ should dissolve a greater amount. Thus when 1.0 M $Na_2S_2O_3$ solution is used, it can be shown that over 7 g of AgI dissolve in 1 dm^3 of solution.

EXERCISE

Verify, using the approximations and procedure of the previous example, that the solubility of AgI in 1 dm^3 of 1.0 M $Na_2S_2O_3$ solution is 7.42 g dm^{-3}.

16.12 TEMPERATURE-DEPENDENCE OF SOLUBILITY PRODUCT

It has been emphasized that the essential condition for the constancy of an equilibrium constant is constant temperature. As expected, it is found experimentally that the solubility products of sparingly soluble compounds do vary with temperature. Strictly then, the temperature at which a K_s value is quoted should always be specified. Solubility products are generally referred to 25°C, and in this book this is so unless otherwise stated.

For many compounds, the solution process is endothermic. We would predict that for such substances the solubility products and the solubilities would rise with increasing temperature. That this is experimentally so may be seen by considering the increase in solubility and solubility product from AgCl and $BaSO_4$ with increasing temperature in Table 16.4. For both these compounds the heat of solution is positive, indicating an endothermic process.

Table 16.4

Variation of solubility and solubility product with temperature

AgCl	*0°C*	*10°C*	*25°C*	*50°C*	*100°C*
K_s		3.9×10^{-11}	1.4×10^{-10}	1.4×10^{-9}	2.2×10^{-8}
s(mg dm^{-3})	—	0.89	1.72	5.23	21.1
$BaSO_4$					
K_s	6.7×10^{-11}	8.9×10^{-11}	1.4×10^{-10}	2.1×10^{-10}	2.8×10^{-10}
s(mg dm^{-3})	1.9	2.2	2.8	3.36	3.9

Note the considerably greater increase in solubility of AgCl over this temperature range in contrast to that of $BaSO_4$.

EXERCISE

Present the data of Table 16.4 graphically, to show clearly the relative increases in solubility over this temperature range.

This variation in solubility of sparingly soluble compounds with temperature has an important consequence in gravimetric analysis. One technique by which purity of a precipitate is enhanced is by washing, preferably with a hot solution. Provided the extent of increase in solubility with temperature is slight, this procedure is quite satisfactory, but in those instances where the solubility is markedly temperature-dependent, washing and filtration must be carried out at room temperature to avoid appreciable solubility losses.

QUESTIONS

1. Calculate the molar solubility in each of the following compounds whose solubility products are given:
   (a) copper(I) iodide $K_s(CuI) = 5.0 \times 10^{-12}$
   (b) magnesium fluoride $K_s(MgF_2) = 7.0 \times 10^{-9}$
   (c) silver oxalate $K_s(Ag_2C_2O_4) = 5.0 \times 10^{-12}$

Comment on the fact that the solubility products of (a) and (c) are numerically equal, but the solubilities are not.

2. Calculate the solubility products of each of the following compounds whose solubilities are given:
   (a) silver bromide $s(AgBr) = 5.5 \times 10^{-7}$ M,
   (b) lead(II) iodate $s\{Pb(IO_3)_2\} = 4.2 \times 10^{-5}$ M,
   (c) silver phosphate $s(Ag_3PO_4) = 1.5 \times 10^{-5}$ M.
3. Calculate the solubility products of each of the following compounds whose solubilities are given:
   (a) magnesium carbonate $s(MgCO_3) = 0.43$ g dm^{-3},
   (b) lead(II) bromide $s(PbBr_2) = 4.28$ g dm^{-3}.
4. Arrange the following lead compounds in order of increasing solubility
   (a) when solubility is expressed in M,
   (b) when solubility is expressed in g dm^{-3}:

   PbC_2O_4, $PbCrO_4$, $PbSO_4$, $PbCO_3$

   $K_s(PbC_2O_4) = 3.4 \times 10^{-11}$ $\quad$ $K_s(PbCrO_4) = 1.8 \times 10^{-14}$

   $K_s(PbSO_4) = 1.0 \times 10^{-8}$ $\quad$ $K_s(PbCO_3) = 3.3 \times 10^{-14}$

5. Arrange the following silver compounds in order of increasing solubility, when solubility is expressed in M.

   $AgBrO_3$, Ag_3AsO_4, Ag_2CrO_4, AgN_3, Ag_3PO_4

   $K_s(AgBrO_3) = 5 \times 10^{-5}$ $\quad$ $K_s(Ag_3AsO_4) = 1 \times 10^{-23}$

   $K_s(Ag_2CrO_4) = 1 \times 10^{-12}$ $\quad$ $K_s(AgN_3) = 2.9 \times 10^{-9}$

   $K_s(Ag_3PO_4) = 1.6 \times 10^{-18}$

   (Note: detailed calculation is unnecessary in this question; rough estimates of each solubility are all that are required.)
6. Arrange the following metal sulphides in order of increasing solubility, by making rough estimates of each solubility in M.

   Ag_2S, CuS, Bi_2S_3, HgS

   $K_s(Ag_2S) = 1 \times 10^{-51}$ $\quad$ $K_s(CuS) = 4 \times 10^{-38}$

   $K_s(Bi_2S_3) = 1 \times 10^{-72}$ $\quad$ $K_s(HgS) = 1 \times 10^{-54}$

7. The solubility product of $PbSO_4$ is 1.0×10^{-8}. Find the solubility of this salt in M in:
   (a) pure water;
   (b) 0.1 M $Pb(NO_3)_2$ solution;
   (c) 0.1 M Na_2SO_4 solution.
8. The solubility product of $Ba(IO_3)_2$ is 6×10^{-10}. Find the solubility of this salt in g dm^{-3} in
   (a) pure water;
   (b) 0.05 M $BaCl_2$ solution;
   (c) 0.05 M KIO_3 solution.
9. In each of the following, decide whether a precipitate could be expected to form:
   (a) 0.005 g $AgNO_3$ solid is added to 2 dm^3 of 0.001 M NaCl solution;
   (b) 2 mg of solid $Ca(NO_3)_2$, and 2 mg of solid NaF are put into 500 cm^3 of water;

(c) 50 cm³ of 0.001 M $Mg(NO_3)_2$ solution is added to 200 cm³ of 0.001 M NaOH solution.

Relevant solubility products:

$K_s(AgCl) = 2 \times 10^{-10}$ $\quad K_s(CaF_2) = 4 \times 10^{-11}$

$K_s\{Mg(OH)_2\} = 1.2 \times 10^{-11}$

10. In each of the following, calculate the minimum concentration of lead ions needed to permit precipitation of the stated compounds in the solutions listed:
    (a) PbS, in 0.01 M Na_2S solution;
    (b) $PbCl_2$, in 0.01 M NaCl solution;
    (c) PbF_2, in 0.01 M NaF solution.

    $K_s(PbS) = 7 \times 10^{-28}$ $\quad K_s(PbCl_2) = 1.7 \times 10^{-5}$ $\quad K_s(PbF_2) = 7 \times 10^{-9}$

11. By referring to the equilibrium constants listed below, find the equilibrium constants for each of the following:
    (a) $Fe(OH)_3(s) + 3H_3O^+ \rightleftharpoons Fe^{3+} + 6H_2O$
    (b) $PbSO_4(s) + CrO_4^{2-} \rightleftharpoons PbCrO_4(s) + SO_4^{2-}$
    (c) $Mg(OH)_2(s) + 2CH_3CO.OH \rightleftharpoons Mg^{2+} + 2CH_3CO.O^- + 2H_2O$

    $K_s\{Fe(OH)_3\} = 4 \times 10^{-38}$ $\quad K_s(PbSO_4) = 2 \times 10^{-8}$

    $K_s(PbCrO_4) = 2 \times 10^{-14}$ $\quad K_a(CH_3CO.OH) = 1.7 \times 10^{-5}$

    $K_w = 10^{-14}$ $\quad K_s\{Mg(OH)_2\} = 1.2 \times 10^{-11}$

12. A solution contains 0.1 M concentrations of Al^{3+}, Fe^{3+} and Mg^{2+} as nitrates. To this solution a solution of sodium hydroxide is slowly added. Which metallic hydroxide precipitates first?

    $K_s\{Al(OH)_3\} = 2 \times 10^{-33}$ $\quad K_s\{Fe(OH)_3\} = 4 \times 10^{-38}$

    $K_s\{Mg(OH)_2\} = 1.2 \times 10^{-11}$

13. What range of hydroxide ion concentration would be suitable for precipitating 0.05 M Fe^{3+} alone in the presence of 0.05 M Fe^{2+} in a solution containing both cations?

    $K_s\{Fe(OH)_2\} = 1.6 \times 10^{-14}$ $\quad K_s\{Fe(OH)_3\} = 4 \times 10^{-38}$

14. In a solution which is 4 M with respect to NH_4Cl, and 0.2 M with respect to NH_3, calculate
    (a) $[OH^-]$;
    (b) the pH of the solution;
    (c) the lowest concentration of Fe^{2+} which must be present in the solution to lead to precipitation of $Fe(OH)_2$;
    (d) the lowest concentration of Fe^{3+} which must be present in the solution to lead to precipitation of $Fe(OH)_3$.

    In the light of your answers, suggest why it is advisable to oxidize iron (II) in a mixture containing both Fe^{2+} and Fe^{3+} if a quantitative precipitation of all metallic iron is to be effected. (Use data in Q. 13.)

15. The first and second acidity constants of H_2S are, respectively, approximately 10^{-7} and 10^{-15}.
    (a) Calculate the equilibrium constant for the reaction

    $$H_2S + 2H_2O \rightleftharpoons 2H_3O^+ + S^{2-}$$

    (b) Calculate the $[S^{2-}]$ in a solution of 0.1 M H_2S, if the pH of the solution is 2.
    (c) Decide whether ZnS could be precipitated from such a solution if the solution was also 0.1 M with respect to Zn^{2+}.

(d) State the restriction upon $[H_3O^+]$ in a 0.1 M H_2S solution if the precipitation of ZnS is to be prevented. $K_s(ZnS) = 1 \times 10^{-24}$.

16. In a solution 0.01 M with respect to each of Pb^{2+} and Mg^{2+}, separation is to be effected by precipitation of the less soluble carbonate.
   (a) Which carbonate will precipitate first?
   (b) What will be the concentration of the cation of this less soluble carbonate in the solution when the second carbonate begins to precipitate?

$$K_s(PbCO_3) = 1 \times 10^{-13} \qquad K_s(MgCO_3) = 1 \times 10^{-5}$$

17. Find the molar concentration of bromide ions in a solution of 0.1 M K_2CrO_4, which is in equilibrium with both solid AgBr and solid Ag_2CrO_4.

$$K_s(Ag_2CrO_4) = 1.6 \times 10^{-12} \qquad K_s(AgBr) = 4 \times 10^{-13}$$

18. A precipitate of AgI may dissolve completely in the presence of excess KI solution, due to formation of AgI_2^- and AgI_3^{2-} complex ions. When a precipitate of AgI has dissolved completely in excess of I^-, which of the following equilibrium concentrations would remain constant if a *further* excess of KI was added?

(a) $[Ag^+][I^-]$

(b) $\dfrac{[AgI_3^{2-}]}{[Ag^+][I^-]^2}$

(c) $\dfrac{[AgI_2^-]}{[Ag^+][I^-]^2}$

(d) $\dfrac{[AgI_3^{2-}]}{[AgI_2^-][I^-]}$

19. To a precipitate of silver bromide, a little ammonia solution is added, causing the precipitate to partially dissolve. If to this mixture some further ammonia is added, such that some precipitate still remains after this second addition, which of the following concentration fractions would have remained constant?

(a) $[Ag^+][Br^-]$

(b) $\dfrac{[Ag(NH_3)_2^+]}{[Ag^+][NH_3]}$

(c) $\dfrac{[NH_3]^2}{[Ag(NH_3)_2^+][Br^-]^2}$

(d) $\dfrac{[Ag^+][NH_3]^2}{[Ag(NH_3)_2^+]}$

17

Oxidation and Reduction

The terms 'oxidation' and 'reduction' have been used by chemists for almost a century. Oxidation reactions include such diverse processes as the combustion of fuels, the rusting and corrosion of metals, and respiration in living organisms. 'Reduction' has been used to describe the smelting of ore minerals to their metals. In chapter 14 we saw that earlier ideas of acid-base reactions could be generalized and extended by regarding these reactions as proton transfer processes. Similarly here we shall find that earlier ideas of oxidation and reduction reactions can be brought together, systematized and extended by formally regarding these reactions as electron transfer processes.

A striking feature of many reactions which involve transfer of electrons is that the energy of the chemical reaction may be directly converted to electrical energy. So in a flashlight dry-cell battery, and in the accumulator in a motor car, chemical reactions are directly harnessed to provide electrical power.

17.1 IDEAS OF OXIDATION AND REDUCTION

The elementary definition of oxidation and reduction was that oxidation involved addition of oxygen to a substance, and reduction the removal of oxygen from a substance. So the reaction

$$PbO(s) + CO(g) \rightarrow Pb(s) + CO_2(g)$$

may be regarded as an oxidation-reduction (redox) reaction, in which lead(II) oxide is *reduced* to lead, and carbon monoxide is *oxidized* to carbon dioxide.

In order to be able to regard the reaction

$$H_2S(g) + O_2(g) \rightarrow H_2O(l) + S(s)$$

as a redox reaction, the definitions above were extended. Here the oxidation of H_2S to sulphur involved the removal of hydrogen from the sulphide, and reduction of oxygen to water involved the addition of hydrogen.

The terms 'oxidation' and 'reduction' now have meanings far wider than these elementary definitions. Two important and related definitions of redox reactions in use today are considered in this and the next section. Just as with the terms 'acid' and 'base', the definitions we choose to adopt are those which are the most convenient in classifying and systematizing a wide variety of reactions. In the Bronsted-Lowry sense, acid-base reactions are proton transfer reactions, where an acid is defined as a proton donor and a base as a proton acceptor. Similarly redox reactions are defined as electron transfer reactions, where

an oxidant is an electron acceptor

a reductant is an electron donor.

In an oxidation-reduction reaction, the reductant donates electrons and is itself oxidized, while the oxidant accepts these electrons and is itself reduced. So

oxidation involves a loss of electrons

reduction involves a gain of electrons.

Consider the following examples:

(a) When metallic sodium reacts with gaseous chlorine, a white crystalline solid, sodium chloride, is formed.

$$2Na(s) + Cl_2(g) \rightarrow 2NaCl(s)$$

NaCl comprises Na^+ cations and Cl^- anions in a network lattice. So the reaction may be thought of as involving the transfer of an electron from each Na atom to each Cl atom. The reductant, sodium, is itself oxidized to Na^+, while the oxidant, chlorine, is itself reduced to Cl^-.

(b) If a solution of an iron(III) compound is added to a solution of a tin(II) compound, the Fe^{3+} cation is reduced to Fe^{2+}, and Sn^{2+} cation is oxidized to Sn^{4+}

$$2Fe^{3+}(aq) + Sn^{2+}(aq) \rightarrow 2Fe^{2+}(aq) + Sn^{4+}(aq)$$

Fe^{3+} is the oxidant, and Sn^{2+} is the reductant, since each of two Fe^{3+} accepts one electron from a Sn^{2+}.

(c) If an iron nail is placed in a solution of copper(II) sulphate, the nail becomes coated with metallic copper.

$$Fe(s) + Cu^{2+}(aq) \rightarrow Cu(s) + Fe^{2+}(aq)$$

Cu^{2+} is the oxidant, some Cu^{2+} cations being reduced to Cu metal, while Fe is the reductant. Some of the metallic Fe is oxidized to Fe^{2+}.

Notice that oxidation and reduction occur together—the reductant must donate its electrons to another species, and this species which accepts the electrons functions as an oxidant. To each oxidant then, there is a corresponding reduced form, and the two species may be called (analogously with acids and bases) a conjugate pair. Thus the conjugate pairs referred to in the above examples are (a) Na^+, Na and Cl_2, Cl^-; (b) Fe^{3+}, Fe^{2+} and Sn^{4+}, Sn^{2+}; (c) Fe^{2+}, Fe and Cu^{2+}, Cu. In each case the oxidized form has been written on the left-hand side.

The reaction

$$2Mg(s) + O_2(g) \rightarrow 2MgO(s)$$

is clearly one of oxidation when viewed from the elementary standpoint of oxygen addition. That it also involves a substance being reduced is not readily seen in this context. If, however, we adopt the electron transfer criterion, and remember that MgO is a network lattice of Mg^{2+} cations and O^{2-} anions, we see that the reaction may be regarded as oxidation of Mg to Mg^{2+}, and reduction of oxygen to oxide ion, O^{2-}. Mg is the reductant, and O_2 is the oxidant.

17.2 OXIDATION NUMBER

A formally similar reaction to

$$2Mg + O_2 \rightarrow 2MgO$$

is the burning of sulphur in oxygen

$$S + O_2 \rightarrow SO_2$$

Remembering that sulphur dioxide consists of discrete covalently-bonded molecules, we cannot readily regard this as oxidation-reduction in the sense of complete electron transfer from one species to another. Nevertheless, it is desirable that this reaction should be viewed as a redox reaction, and this is achieved through framing an alternative definition in terms of directed numbers assigned to the atoms, called *oxidation numbers*. The use of such numbers provides a means of deciding whether oxidation-reduction is involved in a particular reaction under consideration. Oxidation numbers may be determined as follows:

(1) The oxidation numbers of monatomic ions are simply the charges of the ions. Thus the oxidation numbers of the ions Na^+, Cl^-, Fe^{3+} are $+1$, -1 and $+3$, respectively.

(2) The oxidation numbers of atoms not in the form of ions, e.g. for C and O in CO_2, are established from the following formal rules:

(a) Atoms of elementary substances are given an oxidation number of zero (e.g. the oxidation number of Cl in elementary chlorine, Cl_2, is 0).

(b) Hydrogen is generally given an oxidation number $+1$ in its compounds (except when bonded to a metal as in sodium hydride NaH, where it has an oxidation number -1).

(c) Oxygen is generally given an oxidation number -2 in its compounds (except in peroxides, e.g. H_2O_2, where it is -1, and in F_2O, where it is $+2$).

The oxidation numbers of all other species are deduced from the further principles:

(d) that in neutral molecules the algebraic sum of the individual oxidation numbers is zero, and in ions the algebraic sum of the oxidation numbers is the charge on the ion;

(e) the more electronegative atom in the species has the negative oxidation number, the less electronegative atom has the positive number.

Some examples illustrate the procedure:

(i) NaCl : $\overset{+1}{\mathrm{Na}}\,\overset{-1}{\mathrm{Cl}}$ (ii) HCl : $\overset{+1}{\mathrm{H}}\,\overset{-1}{\mathrm{Cl}}$ (iii) $CaCl_2$: $\overset{+2}{\mathrm{Ca}}(\overset{-1}{\mathrm{Cl}})_2$

(iv) Al_2O_3 : $(\overset{+3}{\mathrm{Al}})_2(\overset{-2}{\mathrm{O}})_3$ (v) CO : $\overset{+2}{\mathrm{C}}\,\overset{-2}{\mathrm{O}}$ (vi) CO_2 : $\overset{+4}{\mathrm{C}}(\overset{-2}{\mathrm{O}})_2$

(vii) NaOH : $\overset{+1}{\mathrm{Na}}\,\overset{-2}{\mathrm{O}}\,\overset{+1}{\mathrm{H}}$ (viii) H_2SO_4 : $(\overset{+1}{\mathrm{H}})_2\overset{+6}{\mathrm{S}}(\overset{-2}{\mathrm{O}})_4$

(ix) $KMnO_4$: $\overset{+1}{\mathrm{K}}\,\overset{+7}{\mathrm{Mn}}(\overset{-2}{\mathrm{O}})_4$ (x) OH^- : $[\overset{-2}{\mathrm{O}}\,\overset{+1}{\mathrm{H}}]^-$ (xi) CO_3^{2-} : $[\overset{+4}{\mathrm{C}}(\overset{-2}{\mathrm{O}})_3]^{2-}$

(xii) $Cr_2O_7^{2-}$: $[(\overset{+6}{\mathrm{Cr}})_2(\overset{-2}{\mathrm{O}})_7]^{2-}$

We can now propose an alternative definition of oxidation-reduction to that given in section 17.1. The two definitions are quite closely related:

An atom undergoes oxidation when its oxidation number increases, and undergoes reduction when its oxidation number decreases.

Consider, for example, the following reactions:

(a) $$S(s) + O_2(g) \rightarrow SO_2(g)$$

i.e. $$\overset{0}{\mathrm{S}} + (\overset{0}{\mathrm{O}})_2 \rightarrow \overset{+4}{\mathrm{S}}(\overset{-2}{\mathrm{O}})_2$$

The oxidation number of sulphur has increased from 0 to +4. It has been oxidized. The oxidation number of oxygen has been decreased from 0 to −2. It has been reduced. Note carefully that *no physical significance* can be attached to these arbitrarily defined numbers. There is, for instance, no suggestion of the existence of a tetrapositive cation S^{4+} in SO_2.

(b) $$SO_2(g) + \tfrac{1}{2}O_2(g) \rightarrow SO_3(g)$$

i.e. $$\overset{+4}{\mathrm{S}}\,(\overset{-2}{\mathrm{O}})_2 + \tfrac{1}{2}(\overset{0}{\mathrm{O}})_2 \rightarrow \overset{+6}{\mathrm{S}}\,(\overset{-2}{\mathrm{O}})_3$$

The increase in oxidation number of sulphur shows this to be an oxidation reaction. Elemental oxygen has been reduced, since its oxidation state decreased from 0 to −2.

(c) $$SO_2(aq) + OH^-(aq) \rightarrow HSO_3^-(aq)$$

i.e. $$\overset{+4}{\mathrm{S}}\,(\overset{-2}{\mathrm{O}})_2 + [\overset{-2}{\mathrm{O}}\,\overset{+1}{\mathrm{H}}]^- \rightarrow [\overset{+1}{\mathrm{H}}\,\overset{+4}{\mathrm{S}}\,(\overset{-2}{\mathrm{O}})_3]^-$$

Here the oxidation state of each element remains unaltered, and so this reaction, by definition, is not a redox reaction.

(d) When a metal hydride such as NaH is added to water, hydrogen is evolved and the solution becomes alkaline.

$$H^-(aq) + H_2O \rightarrow H_2(g) + OH^-(aq)$$

i.e.
$$[\overset{-1}{(H)}]^- + \overset{+1}{(H)}_2\overset{-2}{O} \rightarrow \overset{0}{(H_2)} + [\overset{-2}{O}\,\overset{+1}{H}]^-$$

We may regard the oxidation number of H in the hydride anion to have increased from -1 to $+1$ in OH^-, while the oxidation number of H in the water to have decreased from $+1$ to 0 in H_2. The reaction is therefore one of oxidation-reduction.

Note that this last reaction also fits our definition and classification as an acid-base reaction. The fact that this reaction may be equally well classified under two headings highlights the fact that such classification schemes are not always mutually exclusive. Both classifications are useful, and we may regard this reaction both as an acid-base reaction, and as a redox reaction.

In a few instances ambiguities arise in assigning oxidation numbers. KI_3, for example, consists of the K^+ cations, and triodide anion I_3^-. Since K must therefore be assigned an oxidation number of $+1$, and I_3^- an overall oxidation number of -1, the nett oxidation number of the three iodines in I_3^- must also be -1. If these atoms were to all have the same oxidation number then this would be $-\frac{1}{3}$. Oxidation numbers, however, are almost invariably whole numbers, and in this case, we can assign them on the basis of the known valence structure of I_3^- (see Figure 25.3), wherein the three iodine atoms are arranged in a line, with a formal charge of -1 on the central atom. Thus the central atom has an oxidation number of -1, while the other two have oxidation numbers of zero. Again, with the thiosulphate anion $S_2O_3^{2-}$ direct application of the formal rules lead to assigning each sulphur atom an oxidation number $+2$. While this is adequate in many considerations of this ion, the structure of $S_2O_3^{2-}$ shows that the two sulphur atoms are not equivalent. One is at the centre of a tetrahedron bonded to the other four atoms, quite analogously with the structure of the sulphate ion, SO_4^{2-}. It is preferable, therefore, to assign the oxidation number $+6$ to the central sulphur atom, and an oxidation number of -2 to each surrounding atom, including the other sulphur atom in $S_2O_3^{2-}$.

Despite these difficulties the idea of oxidation numbers has proved sufficiently useful to warrant its retention.

17.3 STOICHIOMETRY OF REDOX REACTIONS

Many species in aqueous solutions generally act as oxidants (MnO_4^-, $Cr_2O_7^{2-}$, Cl_2, Fe^{3+}) whilst others generally act as reductants (e.g. Zn, SO_3^{2-}, Sn^{2+}, Fe^{2+}). A large number of possible redox reactions arise from various combinations of pairs of oxidants and reductants. To obtain the stoichiometry of some of these reactions, it is advisable to formulate a working procedure which simplifies the problem of balancing equations. We shall illustrate the procedure with both simple and more complex examples.

For the reaction between Fe^{3+} and I^- in solution to produce Fe^{2+} and I_2, one could probably write the equation directly. However, let us first write what are called partial equations for each redox pair separately:

for the oxidant $\qquad Fe^{3+} + e \rightarrow Fe^{2+} \qquad (1)$

for the reductant $\qquad 2I^- \rightarrow I_2 + 2e \qquad (2)$

The overall equation for the reaction is obtained by summing these partial equations, taking care to balance electrons. Note that the electrons 'lost' by the I^- anions are those which are 'gained' by Fe^{3+}. Thus the same electrons are regarded as transferred from one species (the I^-) to another (the Fe^{3+}). This is the basis of 'cancelling' the electrons, and is achieved by multiplying equation (1) by 2 and adding:

$$2Fe^{3+} + 2I^- \rightarrow 2Fe^{2+} + I_2 \qquad (3)$$

Equation (3) is the final stoichiometric equation.

This simple example illustrates one of the simplest methods of balancing redox equations. First consider the partial reactions of each redox pair separately; e.g. in the oxidation of copper metal to copper(II) ions with bromine in aqueous solution, during which the bromine is reduced to bromide, we would arrive at a balanced equation for this reaction by first writing down partial reactions. Thus

$$Br_2 + 2e \rightarrow 2Br^-$$

$$\text{and } Cu \rightarrow Cu^{2+} + 2e$$

Reaction is effected when bromine accepts electrons from copper metal so we add the two partial reactions and we get

$$Cu + Br_2 \rightarrow Cu^{2+} + 2Br^-$$

In a large variety of cases, partial reactions can be easily written down by inspection, since the change in oxidation number corresponds to simple changes in the charge on a single atom. Thus in the above example, the charge per bromine atom obviously changed from 0 to -1, whilst the charge per copper atom changed from 0 to $+2$.

Sometimes, however, it is not so easy to write down a partial reaction when polyatomic ions are involved, e.g. it is not obvious what the partial reactions are for the dichromate-chromium(III) and permanganate-manganese(II) redox pairs in acid solution. To arrive at partial reactions for these systems, we must use formal oxidati on numbers to obtain the desired equation.

EXAMPLE

Obtain the partial equation for the redox pair dichromate-chromium(III) in acid solution.

SOLUTION

(i) Write down the oxidant and its conjugate reductant

$$Cr_2O_7^{2-} \rightarrow Cr^{3+}$$

(ii) Balance with respect to the atom which is undergoing a change in oxidation number

$$Cr_2O_7^{2-} \rightarrow 2Cr^{3+}$$

(iii) Assign oxidation numbers to the atom undergoing a change in oxidation number.

$$\overset{(2\times+6)}{Cr_2O_7^{2-}} \rightarrow \overset{(2\times+3)}{2Cr^{3+}}$$

(iv) Balance oxidation numbers by adding the required number of electrons to the oxidant side of the equation.

$$\overset{(2\times+6)}{Cr_2O_7^{2-}} + 6e \rightarrow \overset{(2\times+3)}{2Cr^{3+}}$$

(v) In acid solution, balance oxygens by adding water.

$$Cr_2O_7^{2-} + 6e \rightarrow 2Cr^{3+} + 7H_2O$$

(vi) Balance hydrogens by adding H^+

$$Cr_2O_7^{2-} + 14H^+ + 6e \rightarrow 2Cr^{3+} + 7H_2O$$

Notice that total charge is now automatically balanced, and is +6 on either side.

EXERCISE

Determine the partial reaction for the redox pair permanganate (MnO_4^-) and manganese(II) in acid solution.

EXAMPLE

Write a balanced equation for the oxidation of sulphite to sulphate by dichromate ions, in which the dichromate ions are reduced to chromium(III) ions.

SOLUTION

The partial reaction for dichromate-chromium(III) has been determined in the previous example. We require the half-reaction for the oxidation of sulphite to sulphate. Proceeding in a stepwise fashion as before:

(i) and (ii) $$SO_3^{2-} \rightarrow SO_4^{2-}$$

(iii) $$\overset{(+4)}{SO_3^{2-}} \rightarrow \overset{(+6)}{SO_4^{2-}}$$

(iv) $$\overset{(+4)}{SO_3^{2-}} \rightarrow \overset{(+6)}{SO_4^{2-}} + 2e$$

(v) $$H_2O + SO_3^{2-} \rightarrow SO_4^{2-} + 2e$$

(vi) $$H_2O + SO_3^{2-} \rightarrow SO_4^{2-} + 2H^+ + 2e$$

This last must be combined with

$$Cr_2O_7^{2-} + 14H^+ + 6e \rightarrow 2Cr^{3+} + 7H_2O$$

Multiplying the sulphite $\rightarrow$ sulphate equation by three and adding, we obtain

$$Cr_2O_7^{2+} + 3SO_3^{2-} + 3H_2O + 14H^+ + 6e \rightarrow$$
$$6H^+ + 3SO_4^{2-} + 2Cr^{3+} + 7H_2O + 6e$$

and simplifying

$$Cr_2O_7^{2-} + 3SO_3^{2-} + 8H^+ \rightarrow 2Cr^{3+} + 3SO_4^{2-} + 4H_2O$$

Hydrogen ions are here represented by H^+ rather than H_3O^+ for simplicity, but this ion, as with all other ions in solution, must be understood to be hydrated.

The final equation is balanced both with respect to numbers of atoms and with respect to charges. No electrons appear in the final equation since they were introduced only as a useful working procedure. This method of obtaining balanced redox equations is but one of a number of procedures.

Another common reductant is ethanol, CH_3CH_2OH, which is known generally as 'alcohol'. In many countries today, a citizen driving with a suspected high concentration of this reductant in his bloodstream is liable to undergo 'breath analysis'. The usual analytical technique is for the subject to blow a definite volume of air from his lungs through an acidified dichromate solution. The orange colour of the dichromate is replaced by the green of Cr^{3+} if the subject has introduced alcohol into the oxidizing solution. The partial equations, and final equations are:

$$\times 2 \qquad Cr_2O_7^{2-} + 14H^+ + 6e \rightarrow 2Cr^{3+} + 7H_2O$$
$$\times 3 \qquad CH_3CH_2OH + H_2O \rightarrow CH_3CO.OH + 4H^+ + 4e$$

$$2Cr_2O_7^{2-} + 3CH_3CH_2OH + 16H^+ \rightarrow$$
$$4Cr^{3+} + 3CH_3CO.OH + 11H_2O$$

The number of mole of $Cr_2O_7^{2-}$, and hence the reduction in intensity of the orange colour, is directly related to the number of mole of alcohol injected. Justice afterwards is also, we are told, related to this number of mole of alcohol.

17.4 HARNESSING REDOX REACTIONS

In the previous section we found it convenient to 'split up' a redox reaction into two partial equations showing electron transfer. This procedure is quite reasonable since not only can many redox reactions in solution be carried out in a single vessel, they can also occur in two separate but connected vessels, and one can detect electrons transferred between them. For example, the reaction between Fe^{3+} and I^- to form Fe^{2+} and I_2 may be performed by placing the Fe^{3+} solution in one beaker, the I^- solution in a second, and dipping platinum wires connected to a galvanometer into each. To complete the electrical circuit a plug of filter paper soaked in KCl connects the two beakers. This arrangement is called a galvanic cell, or simply a cell.

The two arrangements of solution plus electrode of which the cell is constructed are each termed half-cells. That is, the solution containing

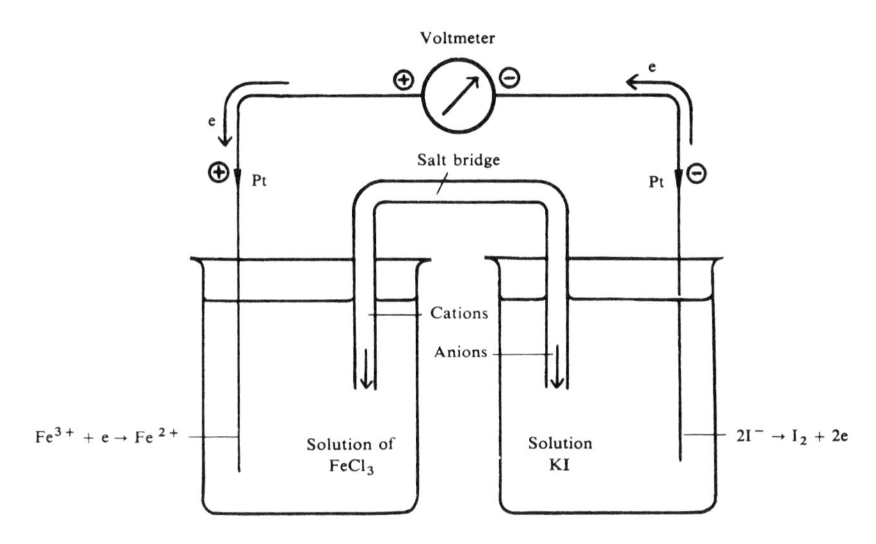

Fig. 17.1 Occurrence of a redox reaction in two separate compartments of a cell.

iron(III) and iron(II), into which dips a platinum wire, is called the iron(II)-iron(III) half-cell; the solution containing iodide and iodine, into which dips a platinum wire is called the iodine-iodide half-cell.

This particular cell is shown in Figure 17.1, and when the electrical circuit is completed as indicated, the galvanometer needle is deflected indicating electron flow, and in such a direction as to indicate that the electron flow along the wire is from the I^- to the Fe^{3+}. This observed electron flow is accompanied by observable chemical reaction, since brown I_2 forms around the wire in the I^- solution. Further, addition of a suitable reagent, such as ferricyanide ion $Fe(CN)_6^{3-}$, to the other electrode could show the formation of Fe^{2+} around the wire by observing the blue coloration obtained when Fe^{2+} reacts with $Fe(CN)_6^{3-}$.

Such observations are consistent with a reaction in which electrons are lost to the wire in the electrode compartment containing I^-. This is the oxidation reaction

$$2I^- \rightarrow I_2 + 2e$$

This electrode at which electrons are produced is the negative electrode of the cell. Similarly, electrons are consumed at the electrode compartment containing Fe^{3+} by the reduction reaction

$$Fe^{3+} + e \rightarrow Fe^{2+}$$

This is the positive electrode of the cell.

The cell provides an example of a redox reaction taking place without direct physical contact of oxidant with reductant. The experiment shows that free electron transfer can occur in a redox reaction, at least in the presence of the unreactive platinum electrodes. Similar tests with a variety of oxidants and reductants show that *most redox reactions in solution may be carried out in suitable cells*. Electron flow occurs between the reactants in the two separate compartments, and the same reaction occurs as in cases where reactants are mixed.

The filter paper plug, called a *salt bridge*, serves two functions: (a) it completes the electrical circuit. The circuit comprises a flow of electrons from the electrode dipping into I^-, through the external wire to the electrode dipping into Fe^{3+}, and a flow of ions between the two compartments through the salt bridge; (b) it provides cations and anions to complement the ions consumed and formed at the two electrodes. At the negative electrode, if 2 mole of I^- have been removed by the redox process, 2 mole of another singly-charged anion must replace this. These come from the chloride ions in the salt bridge. Similarly the reduction of each Fe^{3+} to a Fe^{2+} creates a surplus of a Cl^-. This surplus is balanced by a flow of K^+ cations from the salt bridge. So K^+ and Cl^- ions, in the same stoichiometric quantities, flow from the salt bridge into the two electrode compartments.

The fact that redox reactions can be carried out in cells, and electrons from these cells made to flow through an external circuit, opens up the possibility of harnessing chemical reactions to produce electrical energy directly. An early example, the so-called Daniell cell, has a zinc electrode dipping into a solution of zinc sulphate, and a copper electrode dipping into a solution of copper sulphate. A suitable arrangement for a laboratory experiment is shown in Figure 17.2.

As indicated by the galvanometer, electron flow is from the zinc electrode to the copper electrode. The reactions occurring at the two electrodes are:

$$Zn(s) \rightarrow Zn^{2+}(aq) + 2e$$

$$Cu^{2+}(aq) + 2e \rightarrow Cu(s)$$

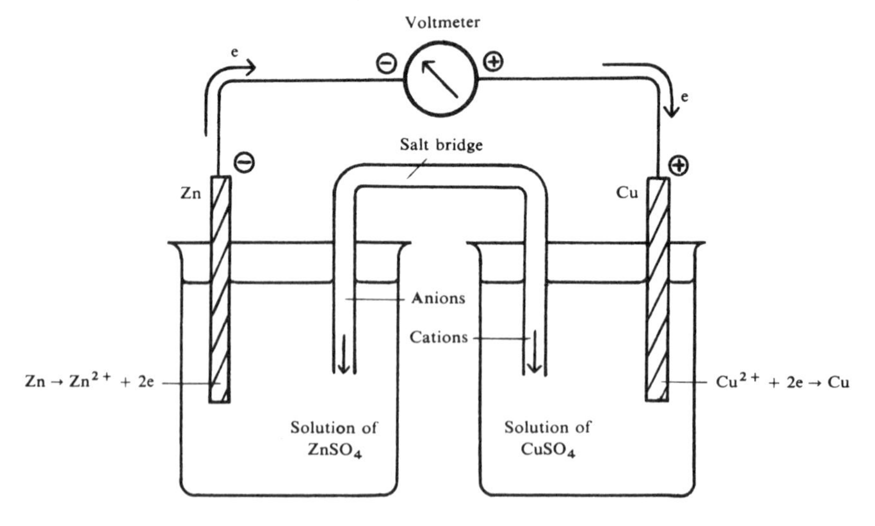

Fig. 17.2 A simple cell for energy production.

The oxidation reaction, producing electrons, occurs at the zinc electrode, which is the negative electrode. Reduction, where electrons are consumed, occurs at the copper electrode, which is the positive electrode.

The overall reaction is

$$Zn(s) + Cu^{2+}(aq) \rightarrow Zn^{2+}(aq) + Cu(s)$$

The stoichiometry of the reaction in the cell is identical with that occurring when zinc metal is added to an aqueous solution of copper sulphate.
The electrons produced in this cell may be used to do electrical work in a variety of ways, by flowing through a resistance in a heating applicance, by flowing through a lamp filament, or by flowing through the windings of an electrical motor. So chemical energy may be converted through electrical energy to heat, light and mechanical work. Such a method of generating electricity directly from suitable chemical reactions is often a more convenient and efficient method for energy production than the thermal methods described in chapter 5.

From this brief discussion of cells we have seen that:

(a) cells are constructed from two half-cells. Electrical connection between the solutions in the half-cells is made by direct contact between the solutions which will permit ions to flow between the half-cells. Electrical connection between the *electrodes* (solid conductors immersed in the solutions) by a conducting wire. Electrons may then flow between the electrodes.

(b) a half-cell consists of a solid (usually a metal) conductor immersed in a solution. Present in each half-cell is a conjugate redox pair. The half-cells mentioned to date have contained the conjugate pairs iron(II)-iron(III), iodine-iodide, copper(II)-copper and zinc(II)-zinc.

17.5 STANDARD REDOX POTENTIALS

The manner in which a conjugate redox pair is incorporated in a half-cell depends upon the physical form of the species involved. In the Fe^{3+}, Fe^{2+} redox pair, both species are available in aqueous solution and a platinum wire may be dipped into such a solution to conduct electrons to and from the solution. If, as with Zn^{2+}, Zn redox pair, the reduced form is a metal, the platinum wire may be dispensed with, as the metal itself can participate in chemical reactions and serve as an electron conductor. Finally, if one member of the redox pair is gaseous, as with Cl_2, Cl^- and H^+, H_2, a platinum wire dips into the solution of anion or cation, and the gas is bubbled through the solution and is adsorbed on the surface of the conducting metal. The arrangements used in these three cases are illustrated in Figure 17.3.
The half-cells illustrated in this figure are the Fe^{3+}, Fe^{2+}; the Zn^{2+}, Zn; and the H^+, H_2 half-cells respectively. In principle, a half-cell can be devised for every redox conjugate pair. Cells, capable of producing a flow of electrons, may then be built up by combining two separate half-cells.

The measurement of the *e.m.f.*, or potential, of a cell is in fact the measurement of *the electrical potential difference between the electrodes of two half-cells.* Since no one has been able to measure the absolute electrical potential

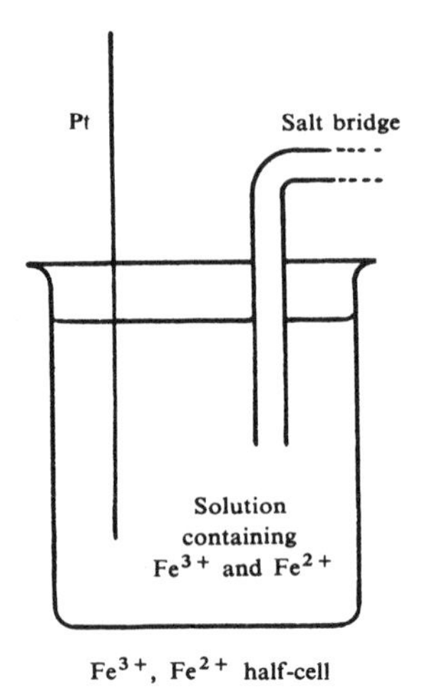

Fe^{3+}, Fe^{2+} half-cell

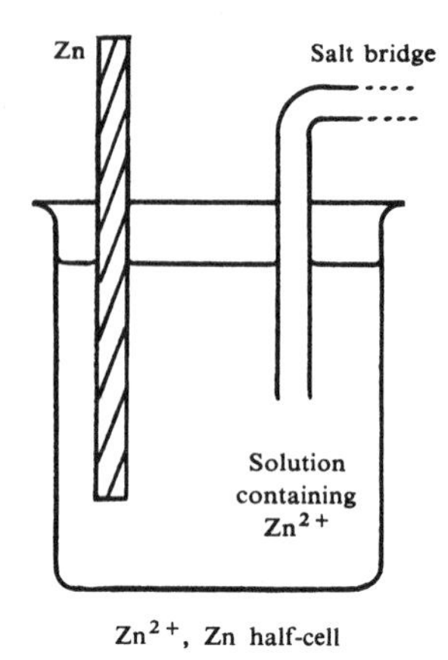

Zn^{2+}, Zn half-cell

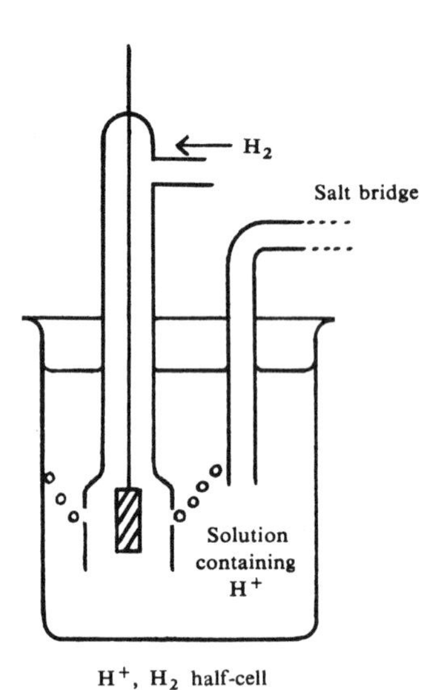

H^+, H_2 half-cell

Fig 17.3 Typical half-cell arrangements for various redox pairs.

of a single electrode, we must choose some arbitrary zero of electrical potential if we wish to assign numerical values to the potentials of individual half-cells. Chemists have chosen the so-called standard hydrogen half-cell (sometimes loosely referred to as the standard hydrogen electrode) as the arbitrary zero of electrical potential. The hydrogen half-cell shown in Figure 17.3 shows hydrogen gas bubbling around a platinum electrode which is dipping into a solution containing hydrogen ions. The potential taken up by the platinum electrode depends on both the hydrogen gas pressure, and the hydrogen ion concentration (or activity). The *standard* hydrogen half-cell is defined as being the one in which the hydrogen gas pressure is 1 standard atmosphere, and the hydrogen ion molarity is unity. (Put more strictly, the *standard* hydrogen half-cell is the one in which both hydrogen gas and hydrogen ion are in their *standard states* of unit activity.)

The potentials of all other half-cells are referred to that of the standard hydrogen half-cell. The potential of the Fe^{3+}, Fe^{2+} half-cell is thus the potential difference between the electrodes of a cell consisting of the standard hydrogen half-cell and the Fe^{3+}, Fe^{2+} half-cell. Now the potential of the Fe^{3+}, Fe^{2+} half-cell measured in this way is found to depend on the concentrations of Fe^{3+} and Fe^{2+} in the Fe^{3+}, Fe^{2+} half-cell (see section 17.6). So we define the *standard* potential of the Fe^{3+}, Fe^{2+} half-cell as being the one for the half-cell in which the Fe^{3+} and Fe^{2+} concentrations are 1 M. (Strictly, the half-cell in which the Fe^{3+} and Fe^{2+} are in their *standard* states of unit activity.) The potential of the cell set up in this latter way is found to be 0.77 volt at 25°C, with the electrode in the Fe^{3+}, Fe^{2+} half-cell being positive (Figure 17.4).

The complete cell reaction is obtained by adding the individual electrode reactions, thus,

$$H_2 \rightarrow 2H^+ + 2e \qquad \text{(negative electrode)}$$

$$Fe^{3+} + e \rightarrow Fe^{2+} \qquad \text{(positive electrode)}$$

$$H_2 + 2Fe^{3+} \rightarrow 2H^+ + 2Fe^{2+} \qquad \text{(cell reaction)}$$

Notice that electrons are produced at the hydrogen electrode (the negative

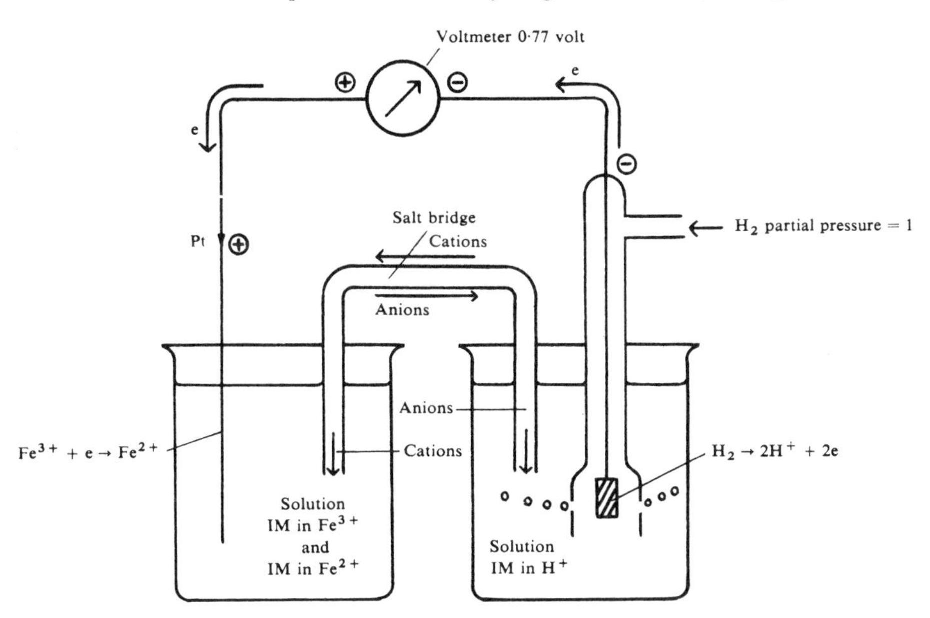

Fig 17.4 Measurement of the potential of the Fe^{3+}, Fe^{2+} half-cell.

electrode), and electrons are used up at the Fe^{3+}, Fe^{2+} electrode (the positive electrode).

We may summarize by defining the standard half-cell potential, or standard redox potential, E°, of any half-cell as being the *potential of the half-cell in question with all ionic species at a concentration of* 1 *M*, *relative to the standard hydrogen half-cell.* The standard potential, E°, has the same sign as the particular half-cell assumes relative to the standard hydrogen half-cell.

In the example quoted above, the standard redox potential of the Fe^{3+}, Fe^{2+} electrode is quoted as follows:

reaction	E° (volt)
$Fe^{3+} + e \rightarrow Fe^{2+}$	$+0.77$

Further examples are,

$Cu^{2+} + 2e \rightarrow Cu$	$+0.34$
$Zn^{2+} + 2e \rightarrow Zn$	-0.76

In the last two cases the cell reactions are, respectively,

$$H_2 + Cu^{2+} \rightarrow 2H^+ + Cu$$

and

$$Zn + 2H^+ \rightarrow H_2 + Zn^{2+}$$

Tables of standard redox potentials set up in this way have the advantage of permitting comparisons of oxidizing and reducing powers of the various systems included in the tabulation. We see that both Fe^{3+} and Cu^{2+} are stronger oxidants than H^+, since both metallic cations oxidize H_2 to H^+. On the other hand, Zn^{2+} is a weaker oxidant than H^+, for H^+ oxidizes metallic zinc, forming H_2 and Zn^{2+}. Furthermore, the magnitude of the standard potentials indicates that Fe^{3+} is a stronger oxidant than Cu^{2+}, and one would predict that if a cell comprising the Fe^{3+}, Fe^{2+} half-cell and the Cu^{2+}, Cu half-cell was formed, the spontaneous cell reaction would be

$$2Fe^{3+} + Cu \rightarrow 2Fe^{2+} + Cu^{2+}$$

17.6 STRENGTHS OF OXIDANTS AND REDUCTANTS

The four redox pairs just considered, with their standard redox potentials, are listed below.

	oxidized form	E°	reduced form	
increasing strength of oxidant ↑	Fe^{3+}	$+0.77$	Fe^{2+}	↓ increasing strength of reductant
	Cu^{2+}	$+0.34$	Cu	
	H^+	0.00	H_2	
	Zn^{2+}	-0.76	Zn	

Notice that the relative magnitude of E° gives a measure of the strengths both of the oxidants and the reductants. Of the four oxidants listed, Fe^{3+} is the strongest and Zn^{2+} is the weakest. Of the four reductants, Zn is the strongest and Fe^{2+} is the weakest.

The standard redox potentials for a large number of redox pairs are tabulated in Table 17.1. These E° values are measured in principle from the voltages of cells of the particular redox half-cell coupled with the standard hydrogen half-cell. All concentrations of solutions are 1 M, and the partial pressure of all gases is 1 atmosphere.

Table 17.1

Standard redox potentials

	Redox reaction			E° *(volt)*
Strongest oxidant	$F_2 + 2e$	$\rightarrow 2F^-$	Weakest reductant	+2.85
↑	$H_2O_2 + 2H^+ + 2e$	$\rightarrow 2H_2O$		+1.77
	$Ce^{4+} + e$	$\rightarrow Ce^{3+}$		+1.61
increasing strength of oxidant	$MnO_4^- + 8H^+ + 5e$	$\rightarrow Mn^{2+} + 4H_2O$		+1.52
	$PbO_2 + 4H^+ + 2e$	$\rightarrow Pb^{2+} + 2H_2O$		+1.46
	$Cr_2O_7^{2-} + 14H^+ + 6e$	$\rightarrow 2Cr^{3+} + 7H_2O$		+1.36
	$Cl_2 + 2e$	$\rightarrow 2Cl^-$		+1.36
	$Br_2 + 2e$	$\rightarrow 2Br^-$		+1.09
	$NO_3^- + 4H^+ + 3e$	$\rightarrow NO + 2H_2O$		+0.96
	$Hg^{2+} + 2e$	$\rightarrow Hg$	increasing strength of reductant	+0.85
	$NO_3^- + 2H^+ + e$	$\rightarrow NO_2 + H_2O$		+0.81
	$Ag^+ + e$	$\rightarrow Ag$		+0.80
	$Fe^{3+} + e$	$\rightarrow Fe^{2+}$		+0.77
	$O_2 + 2H^+ + 2e$	$\rightarrow H_2O_2$		+0.68
	$I_2 + 2e$	$\rightarrow 2I^-$		+0.54
	$O_2 + 2H_2O + 4e$	$\rightarrow 4OH^-$		+0.40
	$Cu^{2+} + 2e$	$\rightarrow Cu$		+0.34
	$Sn^{4+} + 2e$	$\rightarrow Sn^{2+}$		+0.15
	$S + 2H^+ + 2e$	$\rightarrow H_2S$		+0.14
Reference zero of potential	$2H^+ + 2e$	$\rightarrow H_2$		0.00
	$Pb^{2+} + 2e$	$\rightarrow Pb$		−0.13
	$Sn^{2+} + 2e$	$\rightarrow Sn$		−0.14
	$Fe^{2+} + 2e$	$\rightarrow Fe$		−0.44
	$Zn^{2+} + 2e$	$\rightarrow Zn$		−0.76
	$Al^{3+} + 3e$	$\rightarrow Al$		−1.67
	$Mg^{2+} + 2e$	$\rightarrow Mg$		−2.34
	$Na^+ + e$	$\rightarrow Na$		−2.71
	$Ca^{2+} + 2e$	$\rightarrow Ca$		−2.87
	$Ba^{2+} + 2e$	$\rightarrow Ba$		−2.90
	$K^+ + e$	$\rightarrow K$	↓	−2.93
Weakest oxidant	$Li^+ + e$	$\rightarrow Li$	Strongest reductant	−3.02

The table has been arranged with the strongest oxidant F_2 at the top, and the weakest oxidant Li^+ at the bottom. The strength of an oxidant depends on the ease with which the compound will accept electron(s). Thus F_2 has a greater tendency than any other oxidant listed to accept electrons from a reductant, and Li^+ has the least tendency. Conversely the F^- shows the least tendency of all the reductants to donate an electron, while Li shows the greatest tendency. Li is the strongest reductant, and F^- is the weakest. Notice this reciprocal or conjugate character of oxidants and reductants.

The stronger the oxidant of a given redox pair, the weaker is the conjugate reductant.

The standard redox potential of a redox pair, designated E°, is a measure of the strength of the oxidant and its conjugate reductant relative to the strength of H^+ as an oxidant, and H_2 as a reductant. The E° values of a series of redox pairs also serves as a guide to the relative strengths of these oxidants and reductants. If one redox pair has a larger E° than a second redox pair, the oxidized form of the first will react with the reduced form of the second pair. On the other hand, there will be no observable reaction between the reduced form of the redox pair of higher E° and the oxidized form of a redox pair of lower E°. For example, consider the redox pairs

$$Cl_2 + 2e \rightarrow 2Cl^- \qquad E^\circ = +1.36\ V$$

$$I_2 + 2e \rightarrow 2I^- \qquad E^\circ = +0.45\ V$$

Chlorine, Cl_2, is a stronger oxidant than iodine, I_2. Consequently, if chlorine water is added to a solution containing the I^- anion, reaction occurs and brown I_2 forms. The reaction occurring is

$$Cl_2 + 2e \rightarrow 2Cl^-$$

$$2I^- \rightarrow I_2 + 2e$$

$$\overline{Cl_2 + 2I^- \rightarrow I_2 + 2Cl^-}$$

Since this reaction proceeds virtually to completion, the extent to which I_2 and Cl^- react together (the weaker oxidant and weaker reductant of the redox pairs) is negligible. Iodine is too weak an oxidant to be able to accept electrons from Cl^- to produce Cl_2 and I^-. Note also that if Cl_2 were added to I_2 in solution, or if Cl^- were added to I^- in solution, no redox reactions could occur since there is no species to donate electrons in the first case, and no species to accept the electrons in the second.

Considerable reaction only occurs if the oxidant of one redox pair is added to the reductant of a second redox pair where the E° of the first pair is higher than the E° of the second pair.

It is, however, an over-simplification to imply that the reaction

$$Cl_2 + 2I^- \rightarrow 2Cl^- + I_2$$

proceeds to completion whereas the reaction

$$I_2 + 2Cl^- \rightarrow 2I^- + Cl_2$$

does not occur at all. This is because all redox reactions are equilibrium reactions; more correctly, the extent to which the first reaction proceeds is very considerable (the equilibrium constant is very large) whereas the extent to which the second reaction proceeds is negligible (the equilibrium constant is very small, being the reciprocal of the former equilibrium constant). The further apart the E° values of any two redox pairs considered, the greater is the extent of reaction between stronger oxidant and stronger reductant (the larger is the equilibrium constant), and the smaller is the extent between weaker oxidant and weaker reductant (the smaller the reciprocal equilibrium constant).

EXERCISE

Use the standard redox potentials listed to decide in which of the following could reaction occur to a considerable extent:

$$E^0_{Fe^{3+},Fe^{2+}} = +0.77 \text{ V} \qquad E^0_{Sn^{4+},Sn^{2+}} = +0.15 \text{ V}$$

$$E^0_{Sn^{2+},Sn} = -0.14 \text{ V} \qquad E^0_{Fe^{2+},Fe} = -0.44 \text{ V}$$

(a) Fe^{3+} is added to Sn^{2+};
(b) Fe^{2+} is added to Sn^{2+};
(c) a rod of Fe dips into a solution of Sn^{2+};
(d) a rod of Sn dips into a solution of Fe^{3+};
(e) a rod of Sn dips into a solution of Fe^{2+};
(f) a rod of Sn dips into a solution of Sn^{4+}.

Some substances can undergo self-oxidation and reduction. Thus hydrogen peroxide exhibits two possible reactions:

$$H_2O_2 + 2H^+ + 2e \rightarrow 2H_2O \qquad E^\circ = +1.77 \text{ V}$$

$$O_2 + 2H^+ + 2e \rightarrow H_2O_2 \qquad E^\circ = +0.68 \text{ V}$$

Thus one H_2O_2 molecule can be visualized as an oxidant, itself being reduced to H_2O, and another as a reductant, itself being oxidized to O_2. The equations for this reaction are

$$H_2O_2 + 2H^+ + 2e \rightarrow 2H_2O$$

$$H_2O_2 \rightarrow O_2 + 2H^+ + 2e$$

$$\overline{2H_2O_2 \rightarrow 2H_2O + O_2}$$

However, as some of our readers will know, H_2O_2 can be kept in the bathroom cabinet for some time. It does not appear to decompose to H_2O and O_2. The point is that although this reaction as written can proceed (and does proceed), the *relative* E° *values give no information whatever as to the rate at which the reaction proceeds.* In the absence of catalysts, at room temperature the decomposition of H_2O_2 through the self-oxidation-reduction reaction discussed above proceeds very slowly. The decomposition rate of this inherently unstable molecule may be accelerated by catalysts.

17.7 CONCENTRATION DEPENDENCE OF REDOX POTENTIALS

In obtaining standard redox potentials from cell potentials, the cell potential must be obtained for the case where the activities of oxidizing and reducing species are unity. That is, the partial pressure of oxidizing and reducing gases is 1 atmosphere and the concentration of oxidants and reductants in solution is 1 M (if they behave ideally). Although, as has been pointed out earlier, solutions containing ions do not behave ideally even at quite low concentrations, it has been assumed for the purposes of the following discussion that the activities and molar concentrations of ionic species

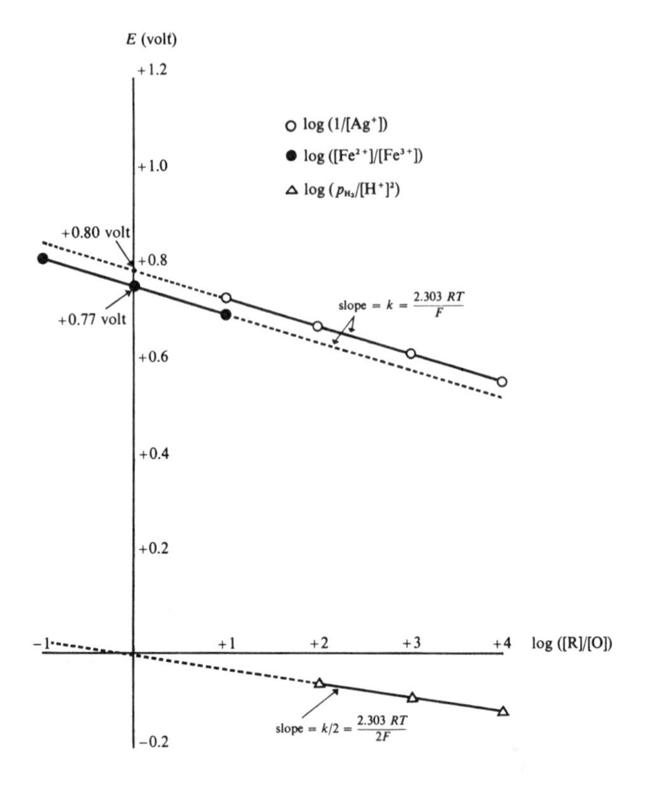

Fig 17.5 Plot of half-cell potentials versus activities of oxidant and reductant for three half-cells.

are identical. The error involved in this assumption does not in any way affect the general validity of the conclusions we shall reach.

Consider the case of a half-cell made up of a rod of silver metal dipping into a solution of silver nitrate. The potential of this half-cell may be measured relative to our arbitrary zero, the standard hydrogen half-cell. It is found that the potential varies with the concentration of silver ions such that there is a linear relation between the cell potential, and the logarithm of the silver ion concentration. This is shown in Figure 17.5; the experimentally-found relation is:

$$E_{Ag^+,Ag} = +0.80 - k \log \frac{1}{[Ag^+]}$$

where k is a constant which is equal to 0.059 and $E_{Ag^+,Ag}$ is the potential of the silver electrode.

We can also study the effects of varying both the partial pressure of hydrogen gas, and the molarity of the hydrogen ions, in the hydrogen half-cell—remembering that if the hydrogen pressure is one atmosphere, and the hydrogen ion molarity is unity, the half-cell is the *standard* hydrogen half-cell. The variation in potential of the hydrogen half-cell relative to a standard hydrogen half-cell is also shown in Figure 17.5, and is given by

$$E_{H^+,H_2} = 0 - \frac{k}{2} \log \frac{p_{H_2}}{[H^+]^2}$$

where k is 0.059, as in the silver ion-silver case.

Finally, consider the case of a half-cell made up by dipping a platinum wire into a solution containing Fe^{3+} and Fe^{2+} ions. The variation in potential relative to a standard hydrogen half-cell is also shown in Figure 17.5 and is given by

$$E_{Fe^{3+},Fe^{2+}} = +0.77 - k \log \frac{[Fe^{2+}]}{[Fe^{3+}]}$$

These three cases exemplify the general rule which has been experimentally observed for half-cell potentials.
viz., given the redox reaction

$$a\mathrm{O} + ze \rightarrow b\mathrm{R}$$

where a mole of the oxidant accept z mole of electrons in producing b mole of reductant, the potential of the half-cell is given by

$$E\,(\text{half-cell}) = E^\circ\,(\text{half-cell}) - \frac{k}{z} \log \frac{a_R^b}{a_O^a}$$

where a_R and a_O are the activities of the reductant and oxidant respectively. Activities will of course be represented by partial pressure in atmosphere for gaseous species, and molar concentration for species in solution. From arguments based on more general considerations, it has been shown that

$$k = \frac{2.303RT}{F}$$

where F is the Faraday constant.
Thus, at 298 K,

$$k = \frac{2.303 \times 8.31 \times 298}{96{,}500} = 0.059$$

and $$E_{\text{half-cell}} = E^\circ_{\text{half-cell}} - \frac{2.303RT}{z\mathrm{F}} \log \frac{a_R^b}{a_O^a}$$

This equation for the potential of a half-cell is known as the Nernst equation.

17.8 RELATION BETWEEN E° AND EQUILIBRIUM CONSTANTS

Partly because of their numerous analytical and other practical applications, it frequently happens that one is concerned with redox reactions which proceed virtually to completion. The equilibrium constants of such reactions will of course be large, and the most convenient way of obtaining these equilibrium constants is from measurements of the *e.m.f.*s ('voltages') of cells. Let us take as a first example a reaction which does not have a large equilibrium constant.

$$Ag^+(aq) + Fe^{2+}(aq) \rightleftharpoons Ag(s) + Fe^{3+}(aq)$$

The equilibrium constant for this reaction is

$$K = \frac{[Fe^{3+}]}{[Ag^+][Fe^{2+}]} = 3.3$$

remembering that the activity of solid silver will be unity. Now the *direction* in which this reaction will actually proceed will be determined by the relative concentrations of the reactants and products. Thus, if, in a particular solution, $[Ag^+] = [Fe^{2+}] = 0.10$ M, while $[Fe^{3+}] = 0.001$ M, then clearly

$$\frac{[Fe^{3+}]}{[Ag^+][Fe^{2+}]} \ll K$$

and the reaction will proceed in the forward direction. This same reaction may be permitted to proceed in a cell, made up as in Figure 17.6. In this case, electrons flow from left to right in the external circuit, since the Ag^+, Ag half-cell has a more positive potential than the Fe^{3+}, Fe^{2+} half-cell. The individual electrode reactions that occur are:

$Ag^+ + e \rightarrow Ag$ (+ve electrode); $E_{Ag^+, Ag} = +0.80 - 0.059 \log \frac{1}{0.10} = +0.74$

$Fe^{2+} \rightarrow Fe^{3+} + e$ (−ve electrode); $E_{Fe^{3+}, Fe^{2+}} = +0.77 - 0.059 \log \frac{0.10}{0.001} = +0.65$

$Ag^+ + Fe^{2+} \rightarrow Ag + Fe^{3+}$ $E_{cell} = E_{Ag^+, Ag} - E_{Fe^{3+}, Fe^{2+}} = +0.09$ volt

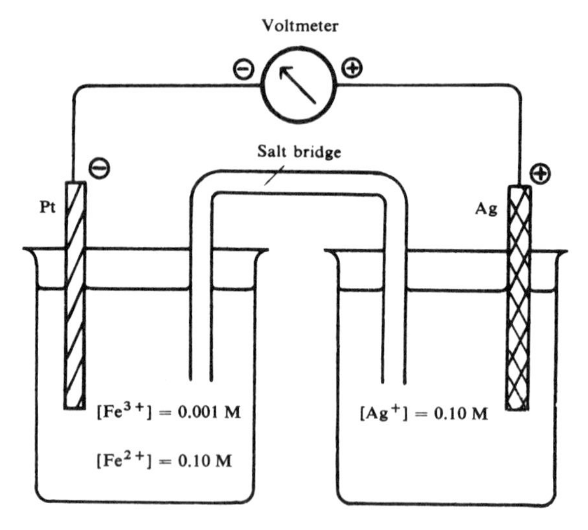

Fig 17.6 Cell for reaction $Ag^+ + Fe^{2+} \rightarrow Fe^{3+} + Ag$.

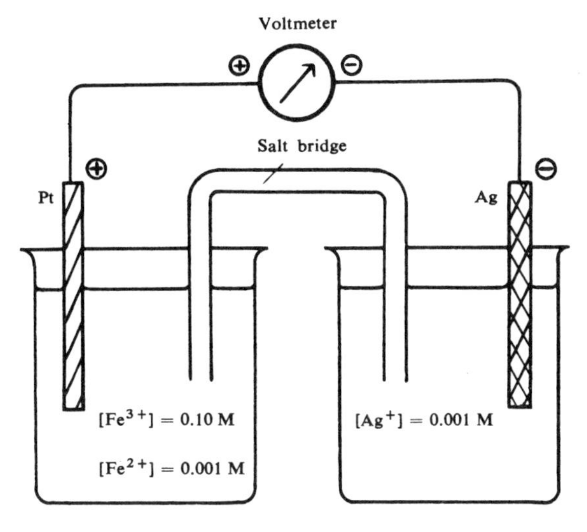

Fig 17.7 Cell for reaction $Fe^{3+} + Ag \rightarrow Fe^{2+} + Ag^{+}$.

where the half-cell e.m.f.s have been calculated using the Nernst equation. On the other hand, we can arrange that the reaction proceeds in the opposite direction by having in a particular solution,

$$[Ag^{+}] = [Fe^{2+}] = 0.001 \text{ M}, \text{ while } [Fe^{3+}] = 0.10 \text{ M}$$

Then

$$\frac{[Fe^{3+}]}{[Ag^{+}][Fe^{2+}]} \gg K$$

and the reaction will proceed in the direction

$$Fe^{3+} + Ag \rightarrow Ag^{+} + Fe^{2+}$$

In the case of the cell, which is shown in Figure 17.7, electrons will flow from right to left in the external circuit, since the Ag^{+}, Ag half-cell now has a more negative potential than the Fe^{3+}, Fe^{2+} half-cell. The individual electrode reactions that occur are:

$Ag \rightarrow Ag^{+} + e$ (−ve electrode); $E_{Ag^{+}, Ag} = +0.80 - 0.059 \log \frac{1}{0.001} = +0.62$

$Fe^{3+} + e \rightarrow Fe^{2+}$ (+ve electrode); $E_{Fe^{3+}, Fe^{2+}} = +0.77 - 0.059 \log \frac{0.001}{0.10} = +0.89$

$Fe^{3+} + Ag \rightarrow Ag^{+} + Fe^{2+}$; $E_{cell} = E_{Ag^{+}, Ag} - E_{Fe^{3+}, Fe^{2+}} = -0.27$ volt

Finally, we can consider the case where in a particular solution

$$[Fe^{2+}] = [Ag^{+}] = 0.01 \text{ M and } [Fe^{3+}] = 0.00033 \text{ M}$$

For these concentrations, $\dfrac{[Fe^{3+}]}{[Ag^{+}][Fe^{2+}]} = 3.3 = K$

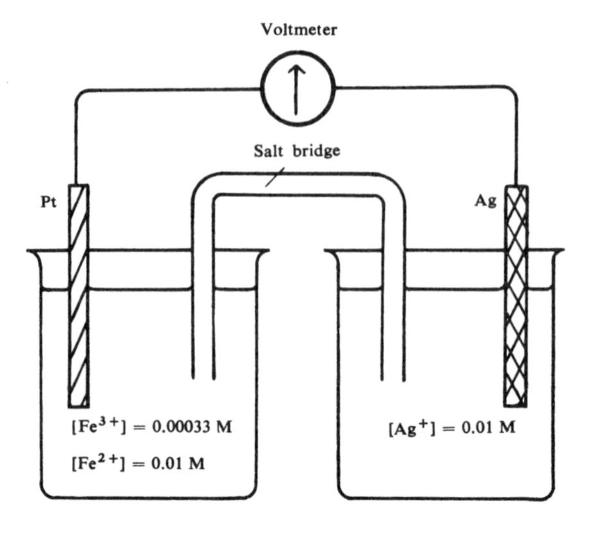

Fig 17.8 Cell in which reaction $Ag^+ + Fe^{2+} \rightleftharpoons Fe^{3+} + Ag$ is at equilibrium.

so that in this solution the system is at equilibrium and no reaction occurs. In the case of the corresponding cell, shown in Figure 17.8, we have

$$E_{Ag^+, Ag} = +0.80 - 0.059 \log \frac{1}{0.01} = 0.918 \text{ volt}$$

$$\text{and } E_{Fe^{3+}, Fe^{2+}} = +0.77 - 0.059 \log \frac{0.01}{0.00033} = 0.918 \text{ volt}$$

i.e. both half-cells have the same potential, so that there is zero potential difference between the electrodes ($E_{cell} = 0$), no nett current flow, and of course no chemical reaction. It becomes clear on considering a number of different examples that we can enunciate a general rule:

When the *e.m.f.* of a cell is zero, the corresponding cell reaction must be at equilibrium.

This fact can be used to relate E°-values to the equilibrium constants of cell reactions. Consider the cell discussed above: if we have *any* set of $[Ag^+]$, $[Fe^{2+}]$ and $[Fe^{3+}]$ concentrations which give a cell *e.m.f.* of zero, then that set of concentrations must necessarily satisfy the relation

$$\frac{[Fe^{3+}]}{[Ag^+][Fe^{2+}]} = K$$

Since we know that $E_{cell} = 0$, it follows that

$$E_{Ag^+, Ag} = E_{Fe^{3+}, Fe^{2+}}$$

$$\therefore \quad E^\circ_{Ag^+, Ag} - 0.059 \log \frac{1}{[Ag^+]} = E^\circ_{Fe^{3+}, Fe^{2+}} - 0.059 \log \frac{[Fe^{2+}]}{[Fe^{3+}]}$$

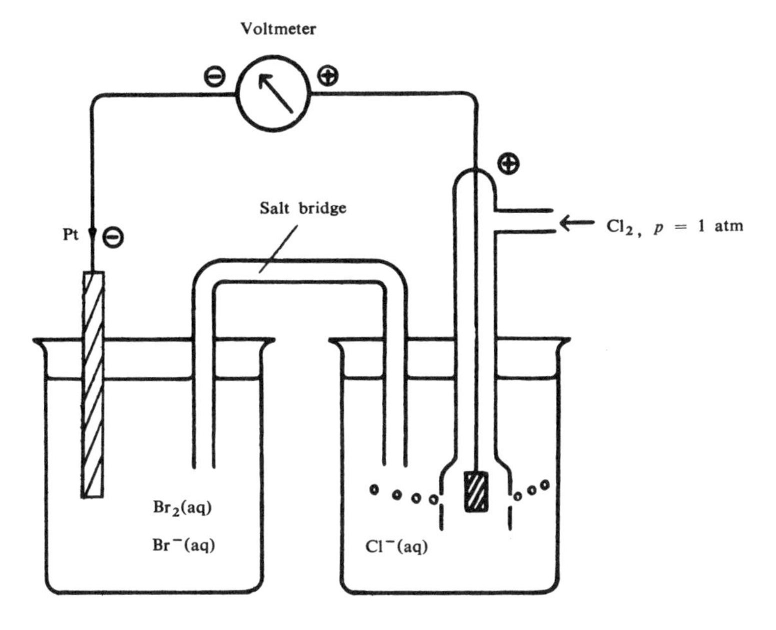

Fig 17.9 Cell for reaction $Cl_2 + 2Br^- \rightarrow Br_2 + 2Cl^-$.

$$\log \frac{[Fe^{3+}]}{[Ag^+][Fe^{2+}]} = \log K = \frac{E^\circ_{Ag^+, Ag} - E^\circ_{Fe^{3+}, Fe^{2+}}}{0.059}$$

$$\therefore \log K = \frac{0.80 - 0.77}{0.059} = \frac{0.03}{0.059} = 0.51$$

$$\therefore K = 3.3$$

This approach is of quite general usefulness in determining the equilibrium constants of cell reactions from tabulated E° values.

EXAMPLE

Use E°-values to calculate the equilibrium constant for the reaction

$$Cl_2(g) + 2Br^-(aq) \rightleftharpoons Br_2(aq) + 2Cl^-(aq)$$

SOLUTION

A cell may be set up as in Figure 17.9,

and
$$E_{Br_2, Br^-} = E^\circ_{Br_2, Br^-} - \frac{0.059}{2} \log \frac{[Br^-]^2}{[Br_2]}$$

$$E_{Cl_2, Cl^-} = E^\circ_{Cl_2, Cl^-} - \frac{0.059}{2} \log \frac{[Cl^-]^2}{p_{Cl_2}}$$

$$E_{cell} = E^\circ_{Cl_2, Cl^-} - E^\circ_{Br_2, Br^-} - \frac{0.059}{2} \log \frac{[Br_2][Cl^-]^2}{[Br^-]^2 p_{Cl_2}}$$

$$\text{At equilibrium } \frac{[Br_2][Cl^-]^2}{p_{Cl_2}[Br^-]^2} = K, \text{ and } E_{cell} = 0$$

$$\frac{0.059}{2} \log K = E^\circ_{Cl_2, Cl^-} - E^\circ_{Br_2, Br^-}$$

$$= 1.36 - 1.09 = 0.27 \text{ volt}$$

$$\therefore \log K = \frac{2 \times 0.27}{0.059} = 9.15$$

$$\therefore K = 1.4 \times 10^9$$

EXERCISE
Use E° values to obtain the equilibrium constant for the reaction

$$Br_2(aq) + Cu(s) \rightleftharpoons Cu^{2+}(aq) + 2Br^-(aq)$$

EXERCISE
From the equilibrium constant obtained in the Example above, calculate the partial pressure of chlorine above a solution that is 0.1 M in both Cl^- and Br^-, and 0.001 M in Br_2.

QUESTIONS

1. State the oxidation numbers of the underlined elements in each of the following compounds:
   $\underline{B}Cl_3$, $Mg\underline{Si}O_3$, $\underline{Mo}F_6$, $\underline{Br}F_3$, $\underline{Br}F_5$, $Na\underline{Cl}O_4$, $NH_4\underline{V}O_3$, $Li_2\underline{U}O_4$, $\underline{U}O_3$, $K_2\underline{Cr}O_4$, $K_2\underline{Cr}_2O_7$, $K_3\underline{Fe}(CN)_6$, $K_4\underline{Fe}(CN)_6$.
2. State the oxidation numbers of the underlined elements in each of the following ions:
   $H\underline{S}O_3^-$, $H\underline{S}O_4^-$, $\underline{S}O_4^{2-}$, $\underline{Mn}O_4^-$, $\underline{Mn}O_4^{2-}$, $\underline{Al}O_2^-$, $\underline{Zn}O_2^{2-}$, $\underline{Sn}O_3^{2-}$.
3. Write partial ionic equations for:
   (a) the reduction of NO_3^- to NO_2
   (b) the reduction of NO_3^- to NO
   (c) the reduction of NO_3^- to N_2
   (d) the oxidation of H_2S to S
   (e) the oxidation of SO_3^{2-} to SO_4^{2-}
   (f) the oxidation of SO_2 to SO_4^{2-}
   (g) the reduction of SO_4^{2-} to SO_2
   (h) the reduction of SO_4^{2-} to S
   (i) the reduction of SO_4^{2-} to H_2S
   (j) the reduction of MnO_4^- to MnO_2
   (k) the reduction of MnO_2 to Mn^{2+}
   (l) the reduction of MnO_4^- to Mn^{2+}
4. Write partial ionic equations for
   (a) the oxidation of manganate ions MnO_4^{2-} to permanganate ions MnO_4^-;
   (b) the reduction of manganate ions to manganese dioxide MnO_2. Hence write an equation for the disproportionation of MnO_4^{2-} to form MnO_2 and MnO_4^- in acid solution.

5. Write balanced overall ionic equations for the following reactions:
   (a) the addition of metallic aluminium to a copper(II) nitrate solution results in the precipitation of copper and the formation of Al^{3+} in solution;
   (b) when metallic copper is added to a dilute solution of nitric acid, gaseous NO is evolved and the solution becomes pale blue;
   (c) potassium permanganate solution is decolorized by acidified iron(II) sulphate solution;
   (d) when SO_2 is bubbled through an acidified solution of $Cr_2O_7^{2-}$, the latter is reduced to Cr^{3+} and the SO_2 is oxidized to SO_4^{2-};
   (e) SO_2 in acid solution reduces iodate IO_3^- to iodide I^-, and is itself oxidized to SO_4^{2-};
   (f) when H_2S is passed into a solution of iron(III) nitrate, a pale yellow precipitate is formed.
6. When NaOCl and $Pb(NO_3)_2$ solutions are boiled together, the reaction which occurs may be represented by the equation:

$$NaOCl + Pb(NO_3)_2 + H_2O \rightarrow PbO_2 + NaCl + 2HNO_3$$

   (a) Write partial and overall *ionic* equations which represent this reaction.
   (b) Identify the oxidant (and its conjugate reductant) and the reductant (and its conjugate oxidant).
   (c) What mass of PbO_2 could precipitate if 2 cm^3 of 1 M $Pb(NO_3)_2$ solution is added to an excess of NaOCl solution (Pb = 207.2).
7. Decide which of the following may be regarded as redox reactions:
   (a) $BaCl_2 + H_2SO_4 \rightarrow BaSO_4 + 2HCl$
   (b) $2Ag + Cl_2 \rightarrow 2AgCl$
   (c) $2FeCl_3 + SnCl_2 \rightarrow 2FeCl_2 + SnCl_4$
   (d) $2H_2 + O_2 \rightarrow 2H_2O$
   (e) $ZnCO_3 \rightarrow ZnO + CO_2$
   (f) $4NH_3 + 5O_2 \rightarrow 4NO + 6H_2O$
   (g) $P_4 + 6H_2 \rightarrow 4PH_3$
   State the criteria for your decisions.
8. Decide which of the following may be regarded as redox reactions:
   (a) $Ag^+ + 2NH_3 \rightarrow Ag(NH_3)_2^+$
   (b) $2CrO_4^{2-} + 2H^+ \rightarrow Cr_2O_7^{2-} + H_2O$
   (c) $2CrO_4^{2-} + 3SO_3^{2-} + H_2O \rightarrow 2CrO_2^- + 3SO_4^{2-} + 2OH^-$
   (d) $2HCO_3^- \rightarrow CO_3^{2-} + H_2O$
   (e) $HPO_3^{2-} + I_2 + OH^- \rightarrow H_2PO_4^- + 2I^-$
   (f) $CaF_2 \rightarrow Ca^{2+} + 2F^-$
   (g) $2Cu^+ \rightarrow Cu^{2+} + Cu$
   State the basis for your decisions.
9. The following equations are incomplete and/or unbalanced. From the information contained in these equations, obtain the balanced overall ionic equation for the reactions. Generally it may be desirable to first write partial ionic equations for oxidation and reduction reactions separately.
   (a) $I_2 + HNO_3 \rightarrow IO^- + NO_2 + H_2O$
   (b) $KI + H_2SO_4 \rightarrow K_2SO_4 + I_2 + H_2S + H_2O$
   (c) $KBr + H_2SO_4 \rightarrow K_2SO_4 + Br_2 + SO_2 + H_2O$
   (d) $I^- + NO_2^- + H^+ \rightarrow I_2 + NO$
   (e) $Fe^{2+} + H_2O_2 + H^+ \rightarrow Fe^{3+} + H_2O$
   (f) $AsO_3^{3-} + BrO_3^- + H^+ \rightarrow AsO_4^{3-} + Br^- + H_2O$
   (g) $OCl^- + I^- + H^+ \rightarrow Cl^- + H_2O + I_2$

(h) $U^{4+} + MnO_4^- + H^+ \rightarrow UO_2^{2+} + Mn^{+2} + H_2O$
(i) $S_2O_8^{2-} + Mn^{2+} + H_2O \rightarrow MnO_2 + SO_4^{2-} + H^+$
(j) $Bi_2O_3 + OCl^- + OH^- \rightarrow BiO_3^- + Cl^- + H_2O$
(k) $H_2O_2 + I^- + H^+ \rightarrow H_2O + I_2$
(l) $V^{2+} + VO^{3+} + H_2O \rightarrow VO^{2+} + H^+$

10. Consider the E° values given for the following cation-metal systems and answer the questions below:

	E° (volt)		E° (volt)
Ag^+, Ag	+0.80	Fe^{2+}, Fe	−0.44
Cu^{2+}, Cu	+0.34	Zn^{2+}, Zn	−0.76
Pb^{2+}, Pb	−0.14	Mg^{2+}, Mg	−2.34

(a) State which species is the strongest oxidant and which is the weakest oxidant.
(b) State which species is the strongest reductant and which is the weakest reductant.
(c) Lead rods are placed in solutions of each of $AgNO_3$, $CuSO_4$, $FeSO_4$ and $MgCl_2$. In which solutions could you expect a coating of another metal on the lead rod? Explain.
(d) Which of the metals—silver, zinc or magnesium—might be coated with lead when immersed in a solution of $Pb(NO_3)_2$?
(e) What would be observed if an iron(II) sulphate solution was stored in a copper vessel? What would be observed if a copper sulphate solution was stored in an iron vessel?

11. It is desired to bring about the oxidation reaction

$$A^+ - e \rightarrow A^{2+} \quad \text{for which } E^\circ = -0.5 \text{ volt}$$

where A represents a particular chemical species. From the table of standard redox potentials (Table 17.1) decide which of the following species may be able to be used to carry out this reaction: Zn metal, 1 M $ZnCl_2$ solution, Cu meta!, 1 M $CuSO_4$ solution, 1 M HCl solution, 1 M $AgNO_3$ solution.

12. A beaker (A) contains 1 M sulphuric acid which is also 0.1 M with respect to iron(III) sulphate, and 0.1 M with respect to iron(II) sulphate. A platinum wire dips into the solution. Another beaker (B) contains 1 M copper(II) sulphate solution with a copper wire dipping into it. A and B are connected by a salt bridge.
(a) If the electrodes in A and B were connected externally by a wire of high resistance, what processes would occur at the two electrodes when the resultant current flows?
(b) Which electrode would be positive?
(c) Briefly explain the function of the salt bridge, supposing it to be a plug of filter paper which has been soaked in concentrated KCl solution.

13. Using the E° values given below, suggest what reactions, if any, one would expect to occur in solution between H_2O_2 in acid solution and the underlined species. (Write your answer in the form A oxidizes B to C and is itself reduced to D, or negligible reaction.)

	E° (volt)
$H_2O_2 + 2H^+ + 2e \rightarrow 2H_2O$	+1.77
$\underline{Fe^{3+}} + e \rightarrow \underline{Fe^{2+}}$	+ 0.77
$O_2 + 2H^+ + 2e \rightarrow H_2O_2$	+ 0.68
$\underline{I_2} + 2e \rightarrow \underline{2I^-}$	+0.54

14. From the standard redox potentials

$$E^{\circ}_{Fe^{3+},Fe^{2+}} = +0.77 \text{ volt}, \qquad E^{\circ}_{Fe^{2+},Fe} = -0.44 \text{ volt}$$

(a) Explain the assertion that Fe^{2+} may act both as an oxidant and as a reductant.
(b) Would you expect metallic iron to dissolve in a 1 M acid solution and produce gaseous hydrogen?
(c) If so, would the Fe be oxidized to Fe^{2+}, or to Fe^{3+}?
(d) Would you expect direct reaction between iron and chlorine to produce $FeCl_2$, or $FeCl_3$? $E^{\circ}_{Cl_2,Cl^-} = +1.36$ volt.

15. (a) Remembering that $E^{\circ}_{H^+,H_2} = 0$, and given $E^{\circ}_{Fe^{3+},Fe^{2+}} = +0.77$ volt, predict what might be expected to happen if gaseous hydrogen is bubbled through a solution containing Fe^{3+} ions.
(b) In fact, when hydrogen is passed into such a solution no observable reaction occurs. Account for this observation.

16. Metallic zinc reacts with $FeCl_3$ solution thus:

$$2Fe^{3+} + Zn \rightarrow Zn^{2+} + 2Fe^{2+}$$

When all the Fe^{3+} has been reduced to Fe^{2+}, any remaining Zn will reduce Fe^{2+} to metallic iron. (Consider the appropriate E° values in Table 17.1.) The second reaction will not proceed to any appreciable extent, however, until the first has been virtually completed. What will be the number of mole of Fe^{3+}, Fe^{2+}, Fe, Zn^{2+}, Zn and Cl^- following reaction between (a) 0.05; (b) 0.10; (c) 0.15; (d) 0.30; and (e) 0.35 mole of metallic zinc and 0.20 mole $FeCl_3$ in solution.

Tabulate answers thus:

	Fe^{3+}	Fe^{2+}	Fe	Zn^{2+}	Zn	Cl^-
(a)						
(b)						
(c)						
(d)						
(e)						

17. Give the formula and approximate number of mole of the solids remaining after evaporation to dryness of solutions containing
(a) 0.1 mole $FeCl_3$ and 0.1 mole $SnCl_2$;
(b) 0.1 mole $FeCl_3$ and 0.04 mole of metallic zinc;
(c) 0.1 mole $FeCl_3$ and 0.4 mole of metallic zinc.
(Consult relevant E° values, and neglect any water of crystallization, and assume that sufficient time is allowed for any reactions occurring to attain equilibrium.)

18. The transuranic element neptunium (Np) may be isolated in the oxidation

states +4 and +5, i.e. Np(IV) and Np(V). These oxidation states react with Fe^{3+} and Fe^{2+} as follows:

(a) Np(IV) + Fe^{3+} → Np(V) + Fe^{2+}

(b) Np(V) + Fe^{2+} → Np(IV) + Fe^{3+}

(i) The rate of oxidation, R, of Np(IV) to Np(V) is given by the rate expression:

$$R = k[\mathrm{Np(IV)}][\mathrm{Fe^{3+}}]$$

where k = 0.058 $M^{-1}\,s^{-1}$ at 25°C.
Calculate the initial rate of oxidation of Np(IV) when 4.5 mg iron(III) chloride is added to 250 cm^3 10^{-5} M Np(IV). What is the rate of formation of Fe^{2+} and Np(V)?

(ii) The rate of reduction, R, of Np(V) to Np(IV) is given by the rate expression:

$$R = k[\mathrm{Np(V)}][\mathrm{Fe^{2+}}]$$

where k = 0.080 $M^{-1}\,s^{-1}$ at 25°C. Calculate the initial rate of reduction of Np(V) when 15 mg crystalline $(NH_4)_2Fe(SO_4)_26H_2O$ is added to 150 cm^3 10^{-5} M Np(V).

(iii) Discuss whether the two reaction mixtures described in (a) and (b) will react completely to their corresponding products.

18

Some Simple Organic Compounds

Carbon occupies a unique place among the elements. It exhibits an extraordinary tendency to form bonds with itself, and thus gives rise to an almost infinite variety of compounds—rings, chains, branched chains. This property of carbon manifests itself most strikingly in living organisms. The bodies of mammals contain quantities of muscle tissue, which is largely a protein, a high molecular weight compound containing carbon. Plants on the other hand contain large amounts of cellulose, which consist of very long chains of carbon-containing rings. Originally it was thought that all complex compounds of carbon could only be made as a result of the metabolic processes of living organisms, hence the term 'organic' chemistry. It is now known that an enormous range of both naturally-occurring and purely 'synthetic' compounds can be readily made in the laboratory and, as a result of the endeavours of the organic chemist, a huge array of economically important compounds has become available. These include drugs, dyestuffs, insecticides, weed-killers and a wide range of synthetic polymers or plastics.

At first sight, organic chemistry seems to consist of a confusing mass of rather formidable looking compounds. However, the properties of a great number of these compounds are determined largely by the chemical properties of small groups of atoms, called *functional groups*, which are present in organic molecules. In this chapter we shall restrict ourselves to a study of the chemistry of some of the simplest possible compounds of carbon, and simultaneously introduce some of the more common functional groups. The simplest organic compound is methane, so we shall start by considering one of its reactions.

18.1 CHLORINATION OF METHANE

Methane is oxidized by chlorine in the gas phase in a stepwise fashion:

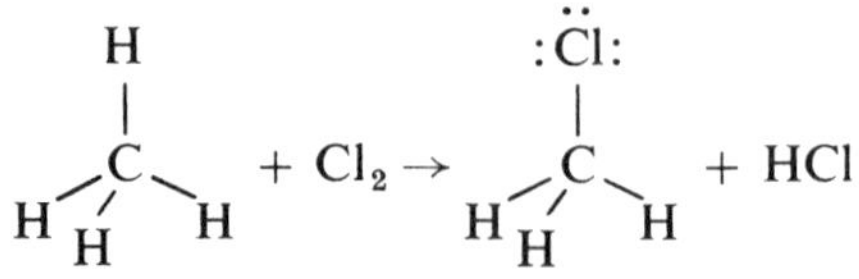

H | H–C(H)–H $+ Cl_2 \rightarrow$:C̈l: | H–C(H)–H $+$ HCl

methane (colourless gas, b.p. $-164°C$) → chloromethane (colourless gas, b.p. $-24°C$)

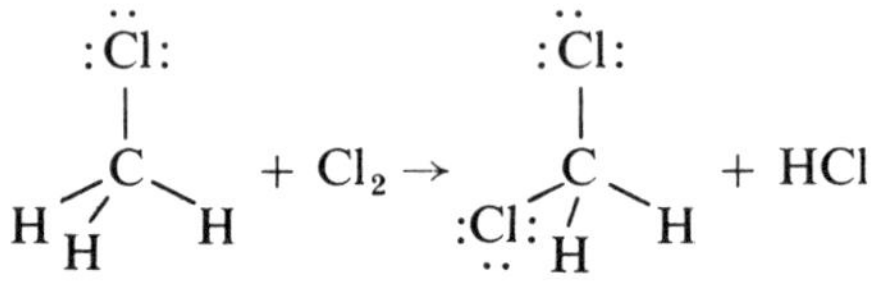

:C̈l: | H–C(H)–H $+ Cl_2 \rightarrow$:C̈l: | :C̈l:–C(H)–H $+$ HCl

chloromethane → dichloromethane (colourless liquid, b.p. $+40°C$)

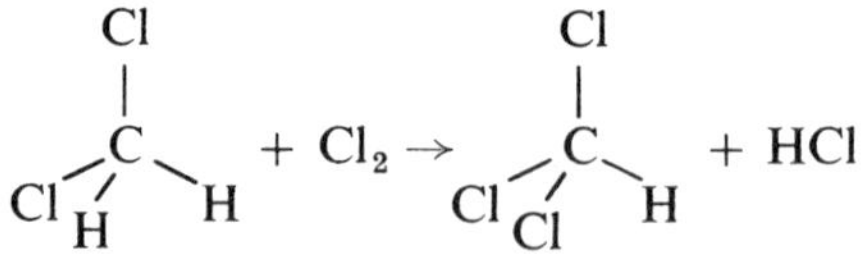

Cl | Cl–C(H)–H $+ Cl_2 \rightarrow$ Cl | Cl–C(Cl)–H $+$ HCl

dichloromethane → chloroform (colourless liquid, b.p. $+61°C$)

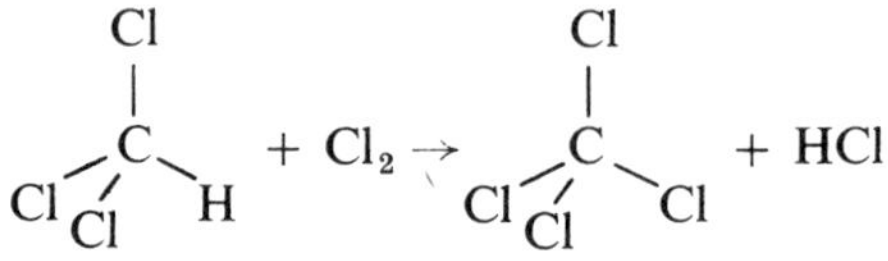

Cl | Cl–C(Cl)–H $+ Cl_2 \rightarrow$ Cl | Cl–C(Cl)–Cl $+$ HCl

chloroform → carbon tetrachloride (colourless liquid, b.p. $+77°C$)

The above reactions proceed smoothly at 300°C, or may be effected at room temperature by the action of ultra-violet light. If a large excess of chlorine is used, only carbon tetrachloride is obtained as a final product. However, if insufficient chlorine is present for complete oxidation, a mixture of CH_3Cl, CH_2Cl_2, $CHCl_3$ and CCl_4 is obtained from which the individual components may be separated by fractional distillation. CH_2Cl_2, $CHCl_3$ and CCl_4 are all widely used industrially as solvents. Carbon tetrachloride is also widely used as a fire extinguisher. All three are only slightly soluble in water.

The electronegativity difference between carbon and chlorine ensures that the carbon-chlorine bonds will take on a dipolar character. Thus in the case of chloromethane the polarity of the molecule is:

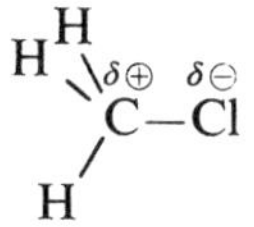

$$\begin{array}{l} H\ \ H \\ \quad \diagdown^{\delta\oplus}\ \ ^{\delta\ominus} \\ \quad\ \ C-Cl \\ \quad \diagup \\ H \end{array}$$

The dipolar nature of this molecule makes it susceptible to attack by certain negative ions in solution. Chloromethane is slightly soluble in water, and in aqueous solutions containing hydroxide ions, we obtain:

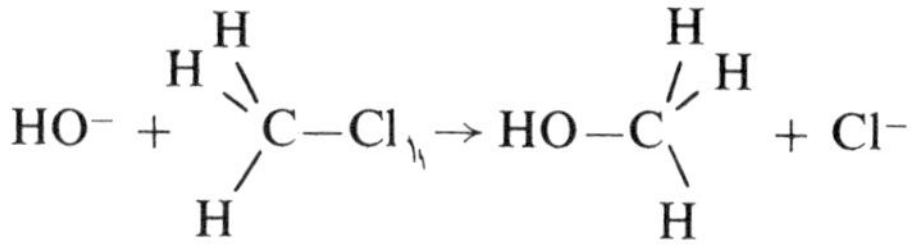

$$HO^- + H_3C-Cl \rightarrow HO-CH_3 + Cl^-$$

methanol
(colourless liquid,
b.p. +64°C)

18.2 METHANOL

We may write the equation for a preparation of methanol as:

$$CH_3Cl \xrightarrow{OH^-(aq)} CH_3OH$$

Methanol is widely used as a solvent, and may be prepared industrially by the direct catalytic oxidation of methane at about 200°C.

$$2CH_4 + O_2 \xrightarrow[200^\circ]{\text{catalyst}} 2CH_3OH$$

Methanol is miscible in all proportions with water and it behaves as a very weak acid

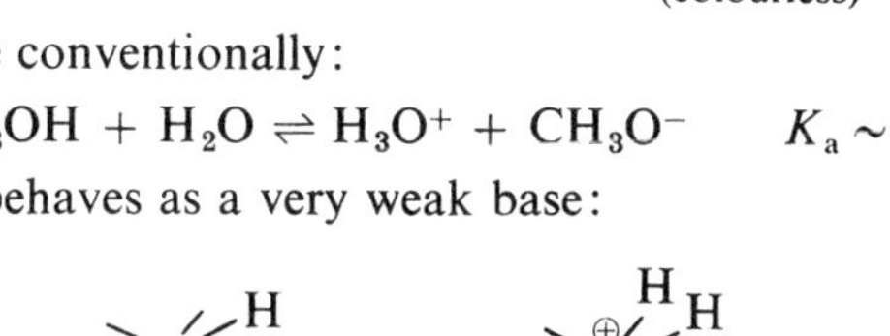

$$H_3C-\ddot{O}H + H_2O \rightleftharpoons H_3O^+ + H_3C-\ddot{O}:^{\ominus}$$

methoxide ion
(colourless)

or, written more conventionally:

$$CH_3OH + H_2O \rightleftharpoons H_3O^+ + CH_3O^- \qquad K_a \sim 10^{-16}$$

Methanol also behaves as a very weak base:

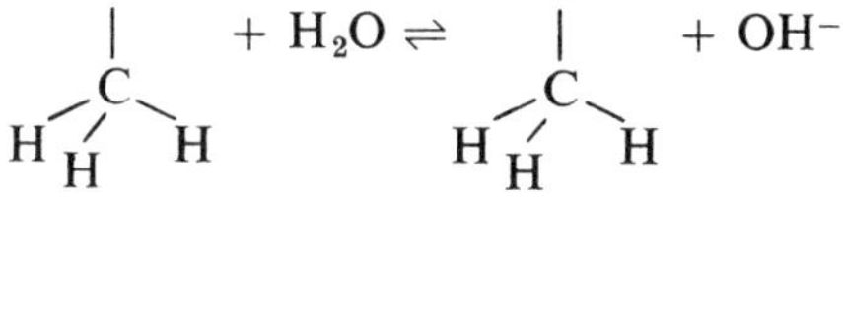

$$H_3C-\ddot{O}-H + H_2O \rightleftharpoons H_3C-\overset{\oplus}{O}H_2 + OH^-$$

or, again, more conventionally:

$$CH_3OH + H_2O \rightleftharpoons CH_3\overset{+}{O}H_2 + \overset{-}{O}H \qquad K_b \sim 10^{-16}$$

Thus methanol is both a weaker acid, and a weaker base, than water itself. (Water, $K_a \sim 10^{-15}$.)

Methanol may be oxidized by heating the vapour in the presence of copper metal at 300°. The heated copper acts as a dehydrogenating agent, and oxidizes the methanol by removal of hydrogen.

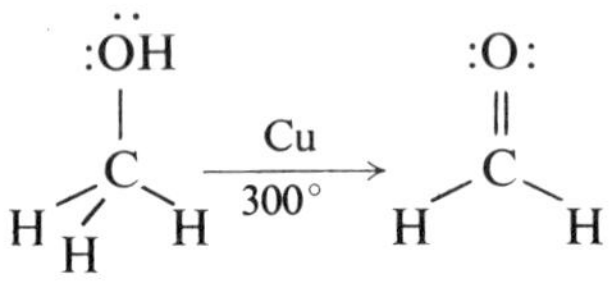

$$H_3C{-}\ddot{O}H \xrightarrow[300^\circ]{Cu} H_2C{=}\ddot{O}:$$

formaldehyde (colourless gas, b.p. −21 °C)

Notice that in this section we have twice used a chemical equation which is not stoichiometrically balanced, both in describing the reaction of methyl chloride with hydroxide ions, and the oxidation of methanol over hot copper. This practice is commonly used when writing down equations for organic reactions in cases where the precise stoichiometry is unimportant, and we are concerned solely with the chemical nature of the *organic* reactant and the *organic* product. The conditions under which the reaction is carried out, together with any reagents used in effecting the reaction, are shown above or below the arrow which indicates the direction of the chemical change. Further examples of this practice will continue to appear in the text.

18.3 METHYLAMINE

This substance may be prepared by reacting chloromethane with ammonia, using ordinary alcohol as the solvent:

$$CH_3Cl + NH_3 \rightarrow \underset{\text{methylamine}}{CH_3NH_2} + H^+ + Cl^-$$

The reaction is analogous to the preparation of methanol from chloromethane, with the ammonia molecule rather than the hydroxide ion replacing the chlorine atom. The reaction is not as simple as it looks, since the product, CH_3NH_2, can react further with chloromethane, producing dimethylamine, and eventually trimethylamine, thus:

$$CH_3Cl + CH_3NH_2 \rightarrow NH(CH_3)_2 + H^+ + Cl^-$$

$$CH_3Cl + NH(CH_3)_2 \rightarrow N(CH_3)_3 + H^+ + Cl^-$$

Methylamine is a weak base

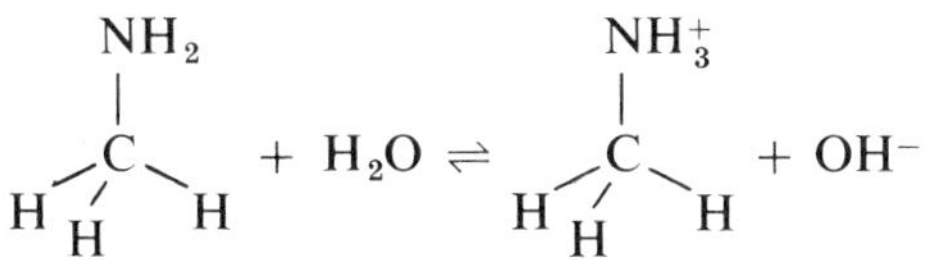

$$H_3C{-}NH_2 + H_2O \rightleftharpoons H_3C{-}NH_3^+ + OH^-$$

methylammonium ion

or, more simply

$$CH_3NH_2 + H_2O \rightarrow CH_3NH_3^+ + OH^- \qquad K_b = 4 \times 10^{-4}$$

Compare the basicities of the compounds:

Compound	CH_3NH_2	NH_3	CH_3OH	H_2O
K_b	4×10^{-4}	1.8×10^{-5}	$\sim 10^{-16}$	$\sim 10^{-15}$

Thus methylamine is a slightly stronger base than ammonia, and forms ionic compounds analogous to the ammonium compounds, e.g., methylammonium chloride is an ionic lattice containing Cl^- ions and $CH_3NH_3^+$ ions.

18.4 FORMALDEHYDE

As we have seen in section 18.2, formaldehyde is a planar molecule with a double bond between carbon and oxygen. Its preparation may be written:

$$CH_3.OH \xrightarrow[300^\circ]{Cu} H.CHO$$

Formaldehyde may also be obtained by the hydrolysis of dichloromethane:

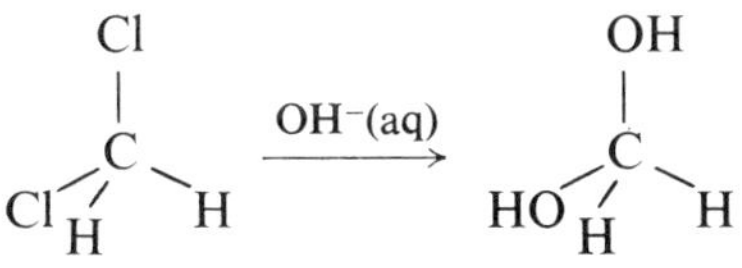

$$CH_2Cl_2 \xrightarrow{OH^-(aq)} CH_2(OH)_2$$

methane diol (colourless)

where methane diol can be obtained as discrete molecules only in aqueous solution. When this solution is evaporated to dryness, a white solid of empirical formula CH_2O remains. This is a polymer of formaldehyde which decomposes to formaldehyde on heating to about 200°C.

Formaldehyde is an efficient antiseptic and disinfectant and is widely used as a preservative for biological specimens in the form of a 40% aqueous solution in water, known as *formalin*, in which the formaldehyde has been largely converted to methane diol.

$$H_2C{=}O + H_2O \rightarrow CH_2(OH)_2$$

or:

$$\text{H.CHO} \xrightarrow{\text{aq}} \text{CH}_2\text{(OH)}_2$$

Indeed, because of this reaction formaldehyde is extremely soluble in water.

Formaldehyde, when mixed with oxygen in the gas phase and passed over a platinum catalyst, is oxidized to formic acid:

$$H_2C{=}O \xrightarrow[150^\circ]{\text{Pt, } O_2} HO{-}CH{=}O$$

formic acid (colourless liquid b.p. +100°C)

When treated with aqueous acidified potassium dichromate solution, formaldehyde is rapidly oxidized to carbon dioxide:

$$H_2C{=}O \xrightarrow[Cr_2O_7^{2-}(aq)]{H^+(aq)} O{=}C{=}O$$

18.5 FORMIC ACID

We have seen that formic acid may be prepared by the oxidation of formaldehyde:

$$\text{H.CHO} \xrightarrow[150^\circ]{\text{Pt, } O_2} \text{H.CO.OH}$$

It is also formed when chloroform is heated with concentrated aqueous sodium or potassium hydroxide and the resultant solution is acidified.

$$CHCl_3 \xrightarrow{OH^-(aq)} HO{-}CH{=}O$$

Formic acid is miscible in all proportions with water, and is a weak acid:

$$HO{-}CH{=}O + H_2O \rightleftharpoons {}^{\ominus}O{-}CH{=}O + H_3O^+$$

formate ion

or more conventionally:

$$H.CO.OH + H_2O \rightleftharpoons H_3O^+ + H.CO.O^- \qquad K_a = 2.4 \times 10^{-4}$$

Thus formic acid is a much stronger acid than methanol (methanol $K_a \sim 10^{-16}$).

The formate ion, $H.CO.O^-$, occurs in such compounds as sodium and potassium formate, which are ionic lattices containing formate anions, and metal cations.

The industrial preparation of formic acid proceeds via sodium formate, which is made when sodium hydroxide is heated under pressure with carbon monoxide:

$$OH^- + CO \rightarrow H\text{—}C(=O)\text{—}O^{\ominus}$$

Formic acid is oxidized by aqueous acidified potassium dichromate to give carbon dioxide:

$$HO\text{—}C(=O)\text{—}H \xrightarrow[Cr_2O_7^{2-}(aq)]{H^+(aq)} O{=}C{=}O$$

18.6 CONCLUSIONS

In this chapter we have discussed the preparation and properties of a series

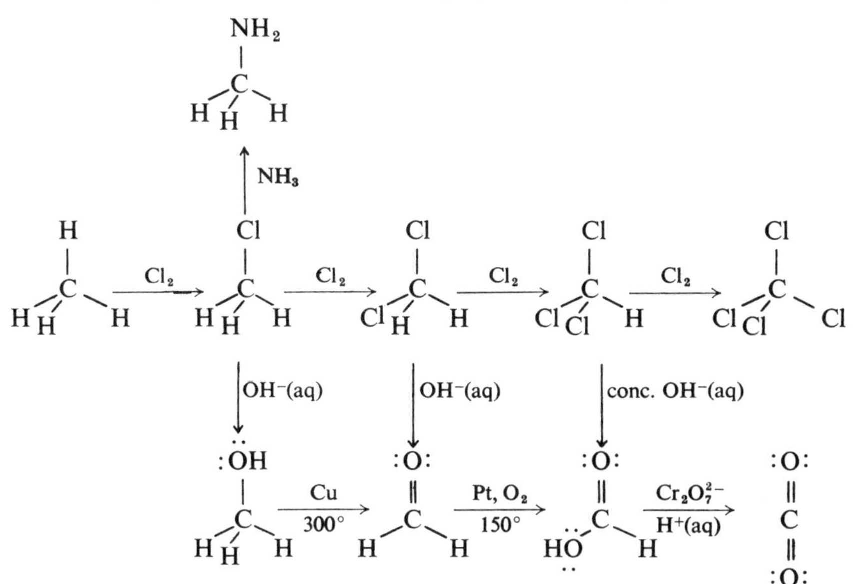

Fig 18.1 Steps in the oxidation of methane.

of simple organic compounds, each containing one carbon atom, by a number of successive oxidation reactions. The sequence we have followed is shown in Figure 18.1.

The horizontal arrows all indicate oxidation reactions, and the vertical arrows indicate non-redox reactions. We could characterize the oxidation states of carbon in these compounds by formal oxidation numbers thus:

Oxidation no.	−4	−2	0	+2	+4
Compounds	CH_4	CH_3Cl	CH_2Cl_2	$CHCl_3$	CCl_4
		CH_3OH	H.CHO	H.CO.OH	CO_2

However, oxidation number formalism is not generally useful in organic chemistry because we will normally be dealing with large molecules which contain many carbon atoms in a variety of different oxidation states and the use of oxidation numbers becomes quite unreal and hopelessly complicated.

It has been found by experience that the chemistry of large organic molecules is best described in terms of the various atoms or groups of atoms which are bonded to the carbon atoms in the large molecules. These atoms or groups of atoms are given the generic name of *functional groups*. The discussion of simple organic molecules in this chapter has served to introduce by name some of the more important functional groups. The properties of large organic molecules are determined by the properties of functional groups within the molecule.

The important functional groups introduced to date, together with their names, are:

Group	*Name*	*Group*	*Name*
$-\ddot{\underset{\cdot\cdot}{Cl}}:$	chloro	$>C=\ddot{\underset{\cdot\cdot}{O}}$	carbonyl
$-\ddot{\underset{\cdot\cdot}{O}}H$	hydroxy	$-C(=\ddot{O}:)-\ddot{\underset{\cdot\cdot}{O}}H$	carboxyl
$-\ddot{N}H_2$	amino		

QUESTIONS

1. Write structural formulae, showing all outer-shell electrons, for chloromethane, chloroform, formic acid, methanol, formaldehyde, dichloromethane, carbon tetrachloride, methane.

2. By writing chemical equations, show how you would prepare:
   (a) formic acid from formaldehyde;
   (b) formaldehyde from methanol;
   (c) dichloromethane from chloromethane;
   (d) carbon dioxide from formic acid;
   (e) carbon tetrachloride from chloroform;
   (f) chloromethane from methane;
   (g) methanol from chloromethane;
   (h) chloroform from dichloromethane;
   (i) sodium formate from formic acid;
   (j) methylamine from methanol.
3. Name and write structural formulae for:
   CH_3OH, H.CHO, CH_2Cl_2, $CHCl_3$, $CH_2(OH)_2$, H.CO.OH, CO_2, H.CO.ONa, CH_3O^-.
4. Arrange the following molecules in order of
   (a) increasing acid strength
   (b) increasing basic strength
   formic acid, methanol, formaldehyde. Write equations where appropriate showing these molecules acting as acids and bases.
5. By writing chemical equations, write down as many ways as you can for carrying out the following preparations:
   (a) formic acid from dichloromethane;
   (b) formaldehyde from methane;
   (c) formic acid from chloromethane;
   (d) methanol from methane;
   (e) potassium formate from chloroform;
   (f) formic acid from methane;
   (g) methylamine from methane.
6. Write down structural formulae for the following functional groups:
   hydroxy, carboxyl, carbonyl, chloro, amino.
7. Write *balanced* stoichiometric equations for the following reactions:
   (a) oxidation of formaldehyde to carbon dioxide by bromine in aqueous solution;
   (b) oxidation of formic acid to carbon dioxide by potassium permanganate in an acidified aqueous medium;
   (c) hydrolysis of dichloromethane to methane diol in aqueous potassium hydroxide;
   (d) oxidation of formaldehyde to carbon dioxide by potassium dichromate in an acidified aqueous medium.
8. Calculate:
   (a) the concentration of formate ions in an aqueous solution which is 0.20 M in formic acid and 0.10 M in H_3O^+;
   (b) the concentrations of both hydrogen ions and hydroxide ions in an aqueous solution which is 0.20 M in formic acid and 0.20 M in sodium formate.
   K_a (formic acid) $= 2.4 \times 10^{-4}$.

19

Hydrocarbons

In this chapter we shall consider the origin and properties of some of the vast range of compounds composed of carbon and hydrogen alone, known as hydrocarbons. Naturally-occurring hydrocarbons in petroleum and coal deposits provide the starting materials for the preparation of the bulk of the world's synthetic organic compounds. Broadly speaking, hydrocarbons can be grouped into three main classes.

(a) Alkanes, or saturated hydrocarbons. These form the bulk of petroleum deposits, and consist of straight chains, branched chains or rings of carbon atoms, with each carbon atom forming four single covalent bonds. The term 'saturated' refers to the absence of double or triple bonds in these hydrocarbons—every carbon atom in an alkane forms four tetrahedrally-arranged single bonds. Cycloalkanes are the special class of alkanes in which the carbon atoms are formed into a ring.

(b) Alkenes and alkynes, or unsaturated hydrocarbons. Alkenes contain a carbon to carbon double bond, as in ethylene (see section 9.8). Alkynes contain a carbon to carbon triple bond.

(c) Aromatic hydrocarbons. These are cyclic hydrocarbons which have a characteristic bonding arrangement whereby some of the bonding electrons are not localized in individual bonds, but are shared by a whole group of atoms. We shall consider only the most important of the aromatic hydrocarbons, benzene.

In the following two chapters, the necessity arises of naming the many different compounds referred to. In recent years, organic chemists have tended increasingly to use a naming system recommended by the International Union of Pure and Applied Chemistry (IUPAC). However, particularly for the smaller common organic molecules, the older trivial names are well entrenched and must be known by all chemists. We have adopted the principle of following what seems to be common practice by using 'systematic' names in all cases where there is not a particularly strongly entrenched trivial name. In many cases, both trivial and systematic names are given side by side. A brief summary of the IUPAC recommendations is given in an Appendix.

19.1 ALKANES

(1) Structure and occurrence. Carbon has a remarkable tendency to bond to itself, and forms an apparently infinite series of colourless hydrides of general formula C_nH_{2n+2} where n is any positive integer, e.g.

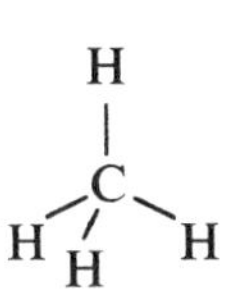

methane, CH_4;
n = 1

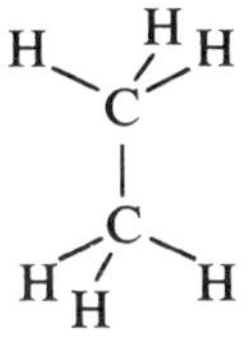

ethane, $CH_3.CH_3$;
n = 2 (gas, b.p. −89°C)

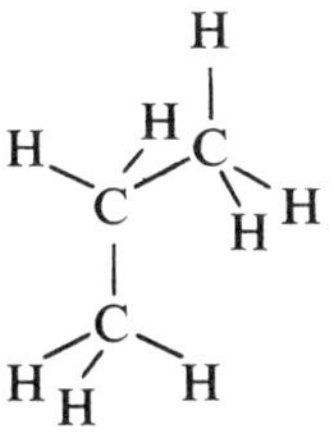

propane, $CH_3.CH_2.CH_3$;
n = 3 (gas, b.p. −44.5°C)

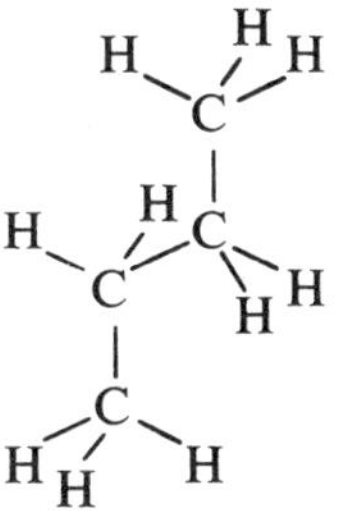

normal butane, $CH_3.CH_2.CH_2.CH_3$;
n = 4 (gas, b.p. −0.5°C)

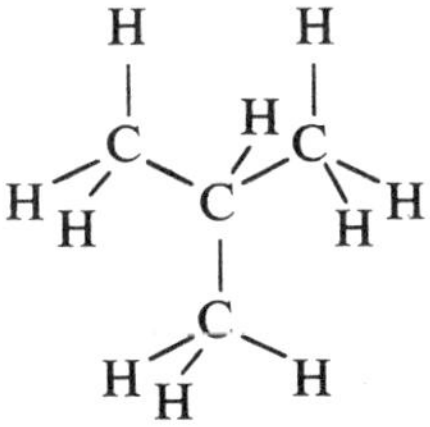

isobutane, $CH(CH_3)_3$;
n = 4 (gas, b.p. −10°C)

The infinite series of hydrides is an example of a *homologous series.*

Any series of organic compounds in which each successive member of the series differs by CH_2 from the previous one is called a homologous series.

The members of this particular series of hydrocarbons are called *alkanes*, or *saturated hydrocarbons*.

Notice that butane, of molecular formula C_4H_{10}, exists in two distinct forms with different boiling-points.

Two compounds are said to be *isomers* if they have the same molecular formulae but have different chemical or physical properties.

In this case, the difference in properties arises from the different ways in which the carbon atoms are bonded together, and this sort of isomerism is therefore called *structural isomerism*. The number of possible isomers of an alkane rises rapidly as the number of carbon atoms in the alkane rises.

The names of the first ten alkanes in the homologous series are given in the following table.

Table 19.1

n =	*Formula*	*Name*
1	CH_4	methane
2	C_2H_6	ethane
3	C_3H_8	propane
4	C_4H_{10}	butane
5	C_5H_{12}	pentane
6	C_6H_{14}	hexane
7	C_7H_{16}	heptane
8	C_8H_{18}	octane
9	C_9H_{20}	nonane
10	$C_{10}H_{22}$	decane

The alkanes may be formally regarded as derivatives of methane, the first member of the homologous series, whereby the hydrogen atoms of methane have been replaced by hydrocarbon groupings.

Thus ethane arises when *one* hydrogen in methane is replaced by a CH_3 group, propane when *two* hydrogens in methane are replaced by CH_3 groups. Such hydrocarbon 'groups' are called *alkyl groups*, and are given specific names. They have a general formula C_nH_{2n+1}, and the general formula is often given the symbol R. Under the heading 'Structural formula' the various alkyl groups are shown as derivatives of the CH_3 or methyl group.

Notice the abbreviations, which we shall use hereafter for normal (n–), secondary (s–) and tertiary (t–). These alkyl groups are often regarded as functional groups in the same sense as the functional groups listed in section 18.7. Thus *any* hydrocarbon may be represented as R—H or RH; alternatively R—R′ represents a hydrocarbon where R and R′ may be different alkyl groups.

The alkanes occur naturally in petroleum deposits, which consist of mixtures of alkanes from CH_4 to about $C_{40}H_{82}$, together with smaller amounts of non-saturated organic compounds. The alkanes from CH_4 to C_4H_{10} are gaseous at room temperature and pressures, and form the major part of natural gas. The remaining constituents of petroleum deposits may

Table 19.2

n	*R*	*Common name*	*Structural formula*
1	$CH_3\cdot$	methyl	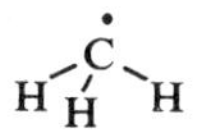 $\dot{C}$ with H, H, H
2	$CH_3.CH_2\cdot$	ethyl	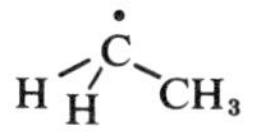 $\dot{C}$ with H, H, CH_3
3	$CH_3.CH_2.CH_2\cdot$	normal propyl (n-propyl)	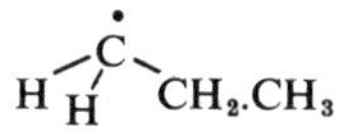 $\dot{C}$ with H, H, $CH_2.CH_3$
3	$(CH_3)_2CH\cdot$	isopropyl	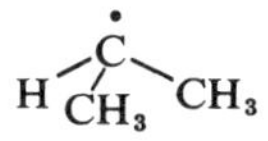 $\dot{C}$ with H, CH_3, CH_3
4	$CH_3.CH_2.CH_2CH_2\cdot$	normal butyl (n-butyl)	 $\dot{C}$ with H, H, $CH_2.CH_2.CH_3$
4	$(CH_3)_2.CH.CH_2\cdot$	isobutyl	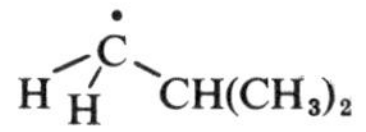 $\dot{C}$ with H, H, $CH(CH_3)_2$
4	$CH_3.CH_2.\underset{\bullet}{C}H.CH_3$	secondary butyl (s-butyl)	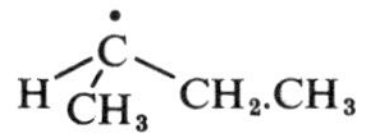 $\dot{C}$ with H, CH_3, $CH_2.CH_3$
4	$(CH_3)_3C\cdot$	tertiary butyl (t-butyl)	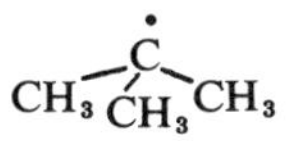 $\dot{C}$ with CH_3, CH_3, CH_3

be separated into fractions by distillation, and have the names and uses indicated in the following table.

Table 19.3

Name	*B.P. (°C)*	*Hydrocarbon fraction*	*Use*
Natural gas	< 20	C_1–C_4	Fuel
Petroleum ether	20–100	C_5–C_7	Solvent
Petrol	70–200	C_6–C_{11}	Fuel
Kerosene	200–300	C_{12}–C_{16}	Lighting, fuel
Lubricating oil	> 300	C_{16}–C_{20}	Lubricant
Greases, vaseline	> 300	C_{18}–C_{22}	Lubricant
Paraffin wax	> 300	C_{20}–C_{30}	Candles, waxed paper

As well as the chain alkanes mentioned to date, cycloalkanes are present in petroleum deposits, e.g.

cyclohexane (C_6H_{12}) methyl cyclohexane

Various other cyclic hydrocarbons, although they do not occur in nature, have been synthesized in the laboratory, e.g.

cyclopropane cyclobutane cyclopentane

One important reaction of the cyclohexanes is in the process known as *catalytic reforming*, in which the aromatic compounds benzene and toluene are formed by the catalytic reduction of the cyclohexane and methyl cyclohexane from petroleum deposits.

$$\text{cyclohexane} \xrightarrow{\text{catalyst}} \text{benzene} + 3H_2$$

$$\text{methyl cyclohexane} \xrightarrow{\text{catalyst}} \text{toluene} + 3H_2$$

(2) Some chemical properties. One of the most important and, at the same time, complex reactions of alkanes is their thermal decomposition to alkenes and either lower molecular weight alkanes or hydrogen. This 'cracking' process, as it is called, is of great importance in providing fuel suitable for operating the high-compression engines of the modern motor car. The so-called 'straight run' petrol, obtained from a preliminary fractional distillation of raw petroleum, contains hydrocarbons in the C_4 to C_{12} range. In order to improve the combustion properties for use in cars, a catalytic cracking process is employed which leads to decomposition of the higher molecular weight alkanes to give alkenes and lower molecular weight alkanes. The actual type of product obtained by cracking is partly determined by the sort of catalyst used, and the temperature of the cracking. The cracking reaction is quite sensitive to temperature, and this can be illustrated in the two cases of the cracking of ethane and propane, viz.,

$$\underset{\text{ethane}}{CH_3CH_3} \rightleftharpoons \underset{\text{ethylene}}{CH_2{=}CH_2} + H_2$$

$$K_p = \frac{p_{CH_2=CH_2}\,p_{H_2}}{p_{CH_3CH_3}}$$

T (°C)	25	425	625	825	925
K_p	10^{-17}	5×10^{-4}	8×10^{-2}	1.4	7.5

$$\underset{\text{propane}}{CH_3.CH_2.CH_3} \rightleftharpoons \underset{\text{ethylene}}{CH_2{=}CH_2} + \underset{\text{methane}}{CH_4}$$

$$K_p = \frac{p_{CH_2=CH_2}\,p_{CH_4}}{p_{CH_3CH_2CH_3}}$$

T (°C)	25	125	225	325	425
K_p	10^{-7}	4×10^{-4}	5×10^{-2}	1.3	13

Notice that the extent of cracking increases markedly with rise in temperature.

EXERCISE

Calculate the equilibrium percentage conversion of propane to ethylene and methane at 125°C and 325°C respectively, when 1 mole of propane is introduced into a 10^{-5} m^3 vessel, in the presence of a suitable catalyst. Explain why the catalyst is necessary.

Quite apart from the usefulness of cracking in the production of fuels, it is also the only significant commercial source of ethylene. Ethylene is used as the starting point in the commercial production of an enormous range of synthetic organic compounds.

At low temperatures, the alkanes do not exhibit many interesting chemical properties. Because of their essentially non-polar nature, they do not readily undergo reactions with ionic reagents in solution. Their best known chemical property is their ready oxidation. Thus a mixture of methane and oxygen explodes when heated:

$$CH_4 + 3O_2 \rightarrow CO_2 + 4H_2O$$

The heat released in the reactions of the C_6 to C_{11} alkanes with oxygen provides the energy which drives our motor cars. The oxidation of hydrocarbons with oxygen is noteworthy in that, in the absence of catalysts, it proceeds extremely slowly at room temperatures, although when it is 'triggered' by heating or applying a spark the reaction proceeds explosively.

The oxidation of alkanes with chlorine (or bromine) proceeds much more smoothly than with oxygen. We have already considered the chlorination of methane, and it is found that all alkanes are chlorinated in an analogous fashion. Thus, in the presence of comparatively small amounts of chlorine:

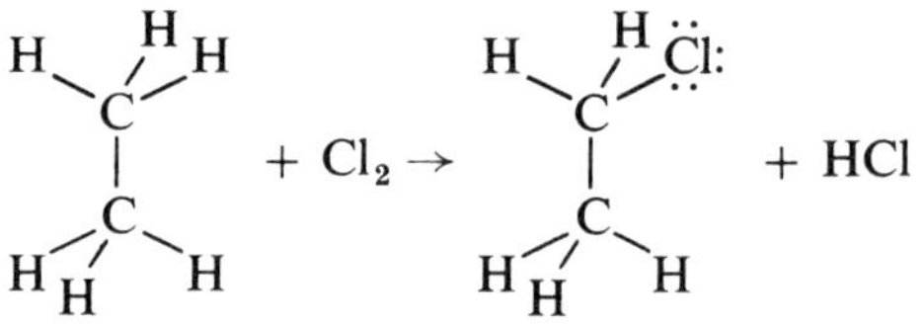

$$CH_3{-}CH_3 + Cl_2 \rightarrow CH_3{-}CH_2Cl + HCl$$

ethane

chloroethane, ethyl chloride (colourless gas, b.p. +12.5°C

or with propane:

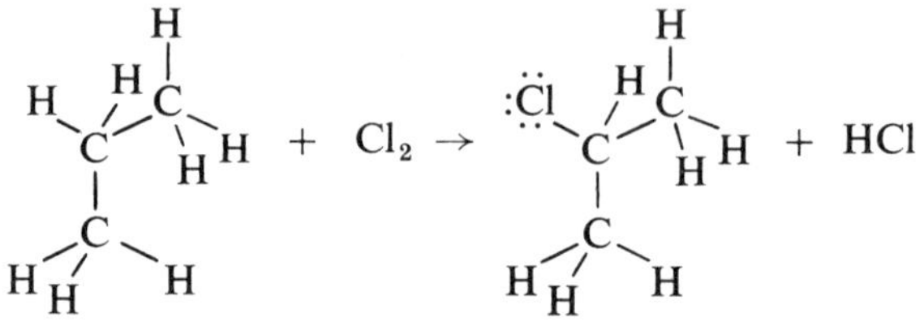

$$CH_3{-}CH_2{-}CH_3 + Cl_2 \rightarrow CH_3{-}CHCl{-}CH_3 + HCl$$

propane

2-chloropropane, isopropyl chloride (colourless liquid)

This equation may be more briefly written:

$$CH_3.CH_2.CH_3 + Cl_2 \rightarrow CH_3.CHCl.CH_3 + HCl$$

Occurring simultaneously with the production of isopropyl chloride, we have

$$CH_3.CH_2.CH_3 + Cl_2 \rightarrow CH_3.CH_2.CH_2Cl + HCl$$

1-chloropropane, n-propyl chloride (colourless liquid)

In general we can write, for the chlorination of an alkane:

$$\underset{\text{alkane}}{RH} + Cl_2 \rightarrow \underset{\substack{\text{alkyl}\\ \text{chloride}}}{RCl} + HCl$$

where R represents any of the alkyl groups, of which those in Table 19.3 are examples.

As well as the products containing one chlorine atom, molecules containing two or more chlorine atoms are also formed (compare the chlorinâtion of methane). The various products may be separated by fractional distillation. It must be remembered that the chlorination of any alkane will generally give a mixture of chlorinated products, which must be separated by fractionation.

Bromination will occur under similar conditions to chlorination, i.e.

$$RH + Br_2 \rightarrow RBr + HBr$$

Here again a mixture of products is obtained, which must be separated by fractional distillation.

19.2 ALKENES AND ALKYNES

(1) Structure and preparation The simplest member of each of these classes of unsaturated hydrocarbon is ethylene (ethene) and acetylene (ethyne) respectively. The valence structures are

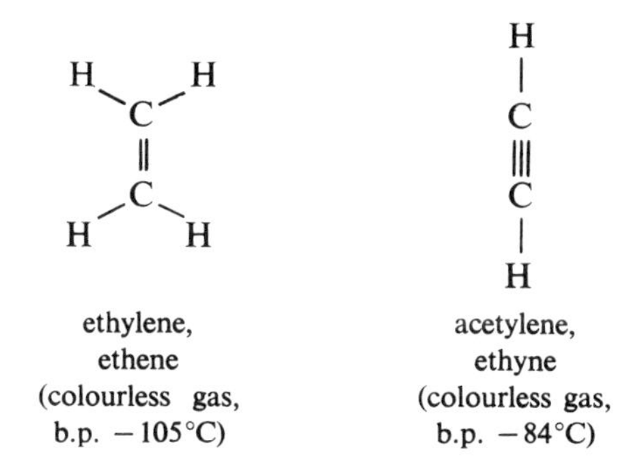

ethylene, ethene (colourless gas, b.p. −105°C)

acetylene, ethyne (colourless gas, b.p. −84°C)

Ethylene is a planar molecule, with a carbon to carbon double bond (see section 9.8) whilst acetylene is a linear molecule with a carbon to carbon triple bond. Ethylene is produced in large quantities from the catalytic cracking of petroleum, but may be produced in the laboratory by the dehydration of ethanol:

$$CH_3.CH_2.OH \xrightarrow{H_2SO_4} CH_2{=}CH_2$$

The production of acetylene from calcium carbide and water is a well-established process. The carbide is made directly from carbon, in the form of coke, and calcium oxide

$$CaO + 3C \rightarrow CaC_2 + CO$$

Calcium carbide contains the negative ion $\overset{\ominus}{C}\equiv\overset{\ominus}{C}$ which reacts with water:

$$\overset{\ominus}{C}\equiv\overset{\ominus}{C} + 2H_2O \rightarrow CH\equiv CH + 2OH^-$$

i.e.

$$CaC_2 + 2H_2O \rightarrow CH\equiv CH + Ca(OH)_2$$

Acetylene is also formed directly from the partial oxidation of methane:

$$6CH_4 + O_2 \rightarrow 2CH\equiv CH + 2CO + 4H_2$$

This reaction is useful because the CO and H_2 may be combined catalytically to give methanol

$$CO + 2H_2 \rightarrow CH_3OH$$

There exists a homologous series of alkenes and alkynes, derived from the parents ethylene and acetylene by the substitution of alkyl groups for the hydrogens on the parent compound. Thus we have:

$CH_3.CH{=}CH_2$	propene
$CH_3.CH_2.CH{=}CH_2$	1-butene
$CH_3.CH{=}CH.CH_3$	2-butene
$CH_3.C\equiv CH$	propyne
$CH_3.CH_2.C\equiv CH$	1-butyne
$CH_3.C\equiv C.CH_3$	2-butyne

(2) Some chemical reactions. The most important single chemical characteristic of the close-to-room-temperature behaviour of unsaturated hydrocarbons is their ability to undergo addition reactions to form, eventually, saturated compounds. From the equilibrium data given in section 19.1(2), it is apparent that at room temperature, ethylene will tend to react spontaneously with hydrogen to give the corresponding alkane. In practice, these reduction reactions are immeasurably slow except in the presence of a catalyst, so

$$CH_2{=}CH_2 + H_2 \xrightarrow{\text{catalyst}} CH_3.CH_3$$

and also,

$$CH\equiv CH + 2H_2 \xrightarrow{\text{catalyst}} CH_3.CH_3$$

Also of great importance is the addition of polar molecules, e.g.,

$$CH_2{=}CH_2 + HCl \rightarrow \underset{\text{ethyl chloride}}{CH_3.CH_2.Cl}$$

$$CH_2{=}CH_2 + \underset{\text{sulphuric acid}}{SO_2(OH)_2} \rightarrow \underset{\text{ethyl hydrogen sulphate}}{CH_3.CH_2.O.SO_2(OH)}$$

and also the addition of halogens

$$CH_2{=}CH_2 + Br_2 \rightarrow CH_2Br.CH_2Br$$

1, 2-dibromoethane

Acetylene undergoes similar reactions

$$CH{\equiv}CH + HBr \rightarrow CH_2{=}CHBr$$

bromoethene

$$CH_2{=}CHBr + HBr \rightarrow CH_3.CHBr_2$$

1, 1-dibromoethane

Notice that the product with two bromines on one carbon atom is obtained. Compare the reaction of ethylene with bromine given above. These two dibromoethanes provide a further example of structural isomerism.

A most important property of unsaturated molecules lies in their ability to bond to each other to form polymers. The simplest example of this is given by ethylene itself which can be polymerized catalytically under conditions of high pressure,

$$CH_2{=}CH_2 \rightarrow {-}CH_2.CH_2.CH_2.CH_2.CH_2.CH_2{-}$$

The polymer produced may be represented $(-CH_2.CH_2-)_n$, where n may be as high as 20,000; this is essentially a very long chain alkane. The basic unit $CH_2{=}CH_2$ from which the polymer is formed is called the monomer. This particular polymer is called polyethylene, or polythene, and is widely used in the manufacture of packaging films and containers. Analogous polymerizations occur with derivatives of ethylene in which various functional groups are substituted for the hydrogens. Some of these, together with the popular names of the resulting polymers, are listed below:

$$\underset{\text{vinyl chloride}}{H_2C{=}CHCl} \xrightarrow{\text{catalyst}} {-}CH_2.\underset{\displaystyle Cl}{\underset{|}{CH}}.CH_2.\underset{\displaystyle Cl}{\underset{|}{CH}}.CH_2.\underset{\displaystyle Cl}{\underset{|}{CH}}{-}$$

vinyl chloride

polyvinyl chloride, PVC
(plastic pipes, gramophone records, fabrics)

$$\underset{\text{vinyl acetate}}{H_2C{=}CH(O.CO.CH_3)} \rightarrow {-}CH_2.\underset{\displaystyle O.CO.CH_3}{\underset{|}{CH}}.CH_2.\overset{\displaystyle O.CO.CH_3}{\overset{|}{CH}}.CH_2.\underset{\displaystyle O.CO.CH_3}{\underset{|}{CH}}.CH_2{-}$$

vinyl acetate

polyvinyl acetate, PVA
(fabrics)

$$CF_2{=}CF_2 \xrightarrow{\text{catalyst}} -CF_2.CF_2.CF_2.CF_2.CF_2-$$

tetrafluoroethylene — polytetrafluoroethylene, Teflon (non-stick cookware, greaseless bearings)

19.3 AN AROMATIC HYDROCARBON—BENZENE

(1) Structure and preparation. Benzene has the empirical formula C_6H_6, and consists of a planar ring of carbon atoms with one hydrogen bonded to each carbon. The evolution of a satisfactory description of the bonding in benzene has been one of the triumphs of modern bonding theory. A valence structure that is consistent with the basic rules laid down to date is

```
          H
          |
    H     C     H
     \  //  \  /
       C      C
       |      ||
       C      C
     /  \\  /  \
    H     C     H
          |
          H
```

Such a structure suggests the presence of three double bonds in the benzene molecule, and therefore, by analogy with the unsaturated molecule ethylene, one might expect benzene to undergo addition reactions with, say, hydrogen bromide and bromine. However, benzene can be boiled with concentrated hydrobromic acid for hours, and no reaction occurs. Certainly benzene reacts with bromine, in the presence of iron(III) chloride, but no addition occurs, merely a substitution reaction (cf. bromination of alkanes):

```
          H                                        Br
          |                                        |
    H     C     H                            H     C     H
     \  //  \  /                              \  //  \  /
       C      C                                 C      C
       |      ||     + Br2    --FeCl3-->        |      ||
       C      C                                 C      C
     /  \\  /  \                              /  \\  /  \
    H     C     H                            H     C     H
          |                                        |
          H                                        H
                                              bromobenzene
```

This property of benzene of being so different in chemical properties from alkenes is characteristic of the properties associated with aromatic hydrocarbons. This lack of reactivity in benzene is related to the nature of the bonding between the carbon atoms. The valence structure can be described as a 'resonance hydrid' (see section 9.9) of two equivalent canonical forms:

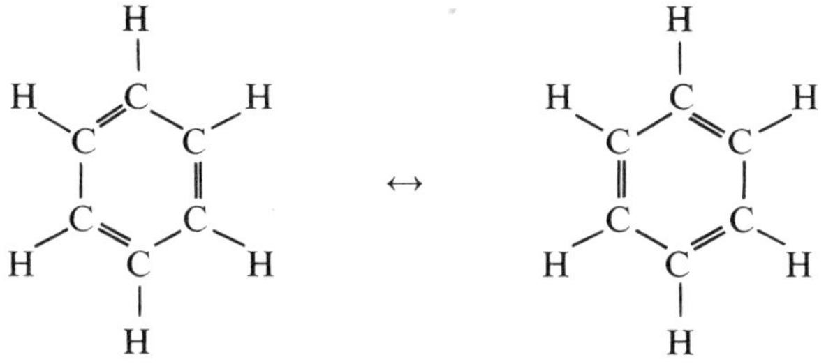

and the carbon-carbon bonds are neither single nor double, but are intermediate between the two.

Because of the great importance of the benzene ring, one is frequently called upon to represent it on paper, and a number of shorthand ways of representing it has been evolved. It is now customary to omit the hydrogens and write simply:

As we have noted in section 9.9 it is always 'correct' to write a conventional valence structure for a molecule, since the experienced chemist will recognize the presence of resonance between equivalent canonical forms. Another representation which lays emphasis on the existence of this resonance is simply:

Benzene is produced as one of a large number of products when coal is destructively distilled. It is also produced from petroleum by the catalytic dehydrogenation of cyclohexane, by the process known as 'catalytic reforming':

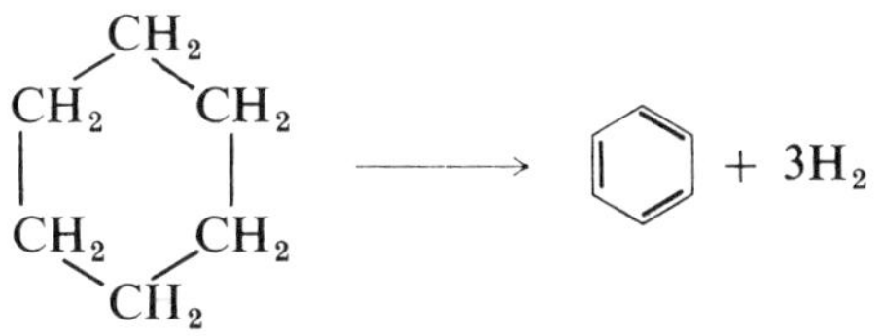

$$C_6H_{12} \longrightarrow C_6H_6 + 3H_2$$

Some benzene is formed when acetylene is polymerized:

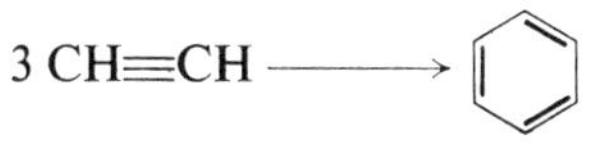

$$3\,CH{\equiv}CH \longrightarrow C_6H_6$$

(2) Some chemical reactions. As has already been noted, the most common reactions observed for benzene are substitution reactions. For example, in a mixture of nitric and sulphuric acids, nitric acid dissociates to form the nitronium ion:

$$2HNO_3 \text{ (in } H_2SO_4) \rightleftharpoons NO_2^+ + NO_3^- + H_2O$$

and this ion reacts with benzene

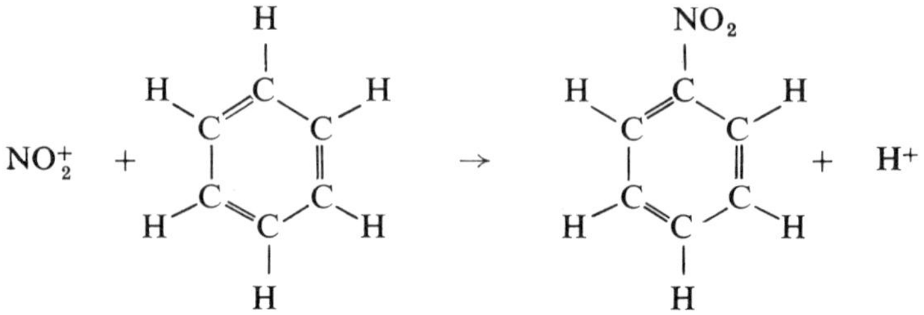

$$NO_2^+ + C_6H_6 \rightarrow C_6H_5NO_2 + H^+$$

nitrobenzene

where the NO_2^+ ion has displaced H^+ from benzene. This reaction would normally be written more briefly as:

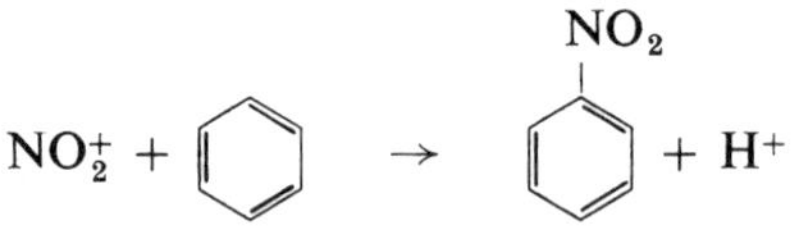

$$NO_2^+ + C_6H_6 \rightarrow C_6H_5NO_2 + H^+$$

Chlorine and bromine react in the presence of iron(III) chloride, displacing H^+, e.g.,

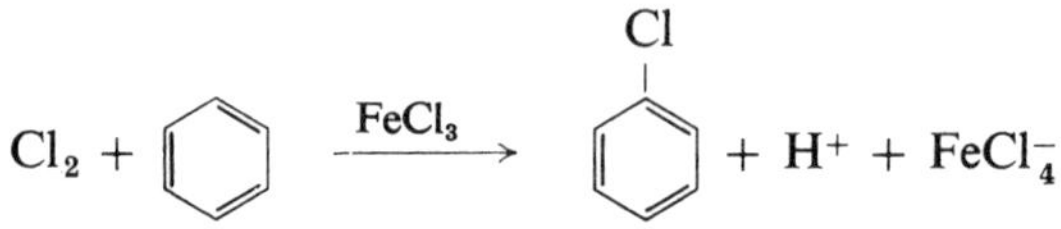

$$Cl_2 + C_6H_6 \xrightarrow{FeCl_3} C_6H_5Cl + H^+ + FeCl_4^-$$

chlorobenzene

Sulphur trioxide reacts in fuming sulphuric acid:

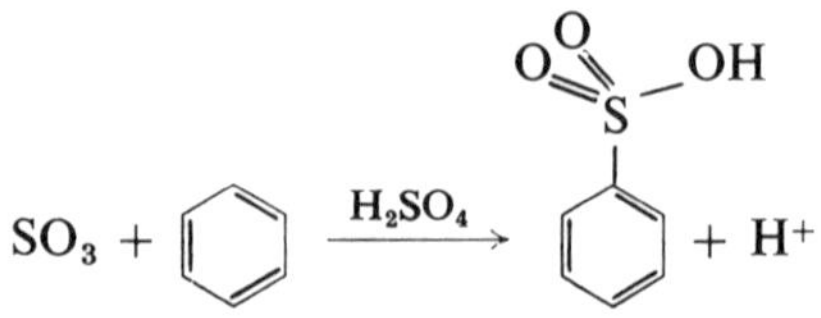

$$SO_3 + C_6H_6 \xrightarrow{H_2SO_4} C_6H_5SO_2OH + H^+$$

benzene sulphonic acid

One very important reaction of benzene is the reaction in which a mixture of an alkyl chloride and aluminium chloride brings about the substitution of an alkyl group on the benzene ring:

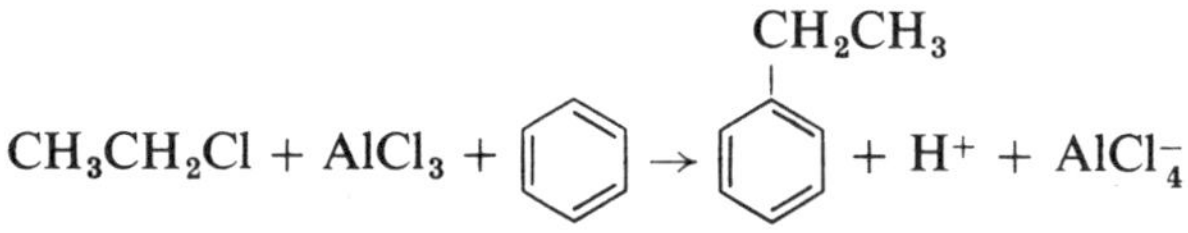

$$CH_3CH_2Cl + AlCl_3 + C_6H_6 \rightarrow C_6H_5CH_2CH_3 + H^+ + AlCl_4^-$$

ethyl benzene

Ethyl benzene can be dehydrogenated to give styrene:

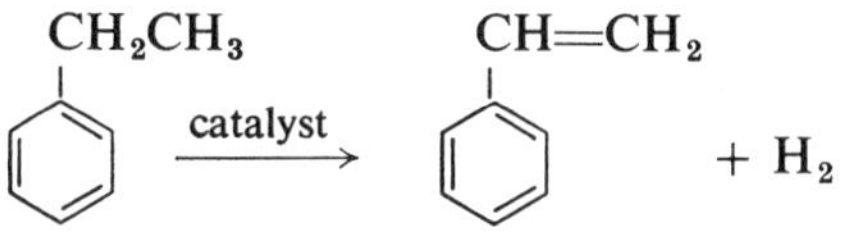

$$C_6H_5CH_2CH_3 \xrightarrow{\text{catalyst}} C_6H_5CH{=}CH_2 + H_2$$

styrene

Styrene is the basic unit in the polymer polystyrene, which is frequently used in making plastic toys. It is also important as a unit in some synthetic rubbers. The polystyrene links together in the same way as polyethylene:

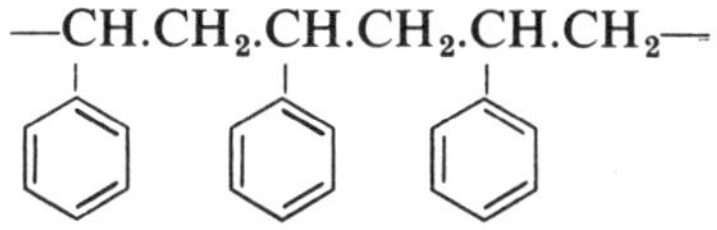

$$-CH(C_6H_5).CH_2.CH(C_6H_5).CH_2.CH(C_6H_5).CH_2-$$

polystyrene

QUESTIONS

1. Name the following hydrocarbons:
   (a) $CH_3CH_2CH_3$
   (b) $(CH_3)_2CHCH_3$
   (c) CH_3CH_3
   (d) C_2H_2
   (e) C_6H_6
   (f) $CH_3CH{=}CH_2$
   (g) $CH_3C{\equiv}CCH_3$
2. Write the structural formulae for the following hydrocarbons:
   (a) ethane;
   (b) ethylene;
   (c) acetylene;
   (d) n-pentane;
   (e) cyclopentane;
   (f) 2-butane;
   (g) 1-butyne;
   (h) benzene;
   (i) styrene.
3. A compound has the formula C_4H_8. Write *three* possible structures the compound could have, and name each.
4. A hydrocarbon has eight carbon atoms. What would be the molecular formula of the compound if it is
   (a) an alkane;

(b) an alkene;
(c) an alkyne;
(d) a cycloalkane;
(e) a cycloalkene;
(f) a saturated benzene derivative.

5. Give the structural formulae of possible dibromo-compounds which could be formed from
(a) ethane
(b) ethylene
(c) 1-butane
6. When 3.333 mg of a hydrocarbon is burned, 10.182 mg of carbon dioxide and 5.010 mg of water are produced. Measurements of the vapour density of the compound indicate a molecular weight of approximately 70. Determine the molecular formula of the hydrocarbon, and suggest *three* possible structural formulae the compound could have.
7. Using any non-organic reagents necessary, how would you prepare the following chlorinated derivatives of hydrocarbons:
(a) ethyl chloride from ethane
(b) ethyl chloride from ethylene
(c) 1,1-dichloroethane from acetylene
(d) chlorobenzene from benzene
8. Two compounds, A and B, each have the molecular formula C_6H_8. Compound A has no methyl groups in the molecule. When hydrogenated with excess hydrogen in the presence of a suitable catalyst only 1 mole of hydrogen was absorbed. Compound B when hydrogenated under similar conditions was converted to a compound of formula C_6H_{12}.

   Give the structural formulae for compound A and possible structural formulae for compound B and explain the hydrogenation reactions in each case.

20

Functional Group Chemistry

As implied in chapter 18, a generally useful means of ordering the complexity of organic chemistry is to recognize regularities in the chemical behaviour of various functional groups. Thus the presence of a hydroxy group (.OH) in a molecule is readily recognizable from the properties that group bestows on the molecule. The present chapter is concerned with presenting an introductory account of the properties of some of the more common functional groups. We shall briefly consider the chemistry of these groups both when attached to alkyl chains (aliphatic compounds) and to the benzene ring (aromatic compounds).

20.1 THE HYDROXY GROUP

(1) Preparation and uses of aliphatic alcohols. Just as we could regard the homologous series of alkanes as being derived from methane, in which the hydrogens had been replaced by alkyl groups, so we have a homologous series of hydroxy compounds, called alcohols, which are derived from methanol by the successive addition of CH_2 units. In the case of alcohols, it is useful to consider the three possible homologous series arising from the successive substitution of alkyl groups for the three hydrogens attached to carbon in methanol. Thus we have:

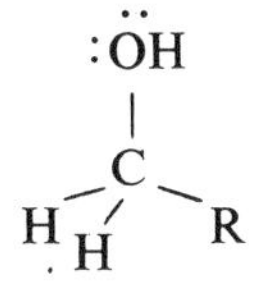

one substitution,
a primary alcohol: $R.CH_2.OH$

$$\begin{array}{c} :\ddot{O}H \\ | \\ H-C(R')-R \end{array}$$

two substitutions,
a secondary alcohol: R.CH(OH).R′

$$\begin{array}{c} :\ddot{O}H \\ | \\ R''-C(R')-R \end{array}$$

three substitutions,
a tertiary alcohol: R.C.(OH).R″ (with R′ on C)

where R, R′ and R″ may represent the same or different alkyl groups.

Alcohols may be prepared by the reaction of alkyl chlorides with hydroxide ions in water, thus:

$$HO^- + H_2C(CH_3)-\ddot{Cl}: \rightarrow H\ddot{O}-CH_2(CH_3) + Cl^-$$

chloroethane, ethyl chloride → ethanol, ethyl alcohol (liquid, b.p. +78.1 °C)

or expressed more simply:

$$CH_3.CH_2.Cl \xrightarrow{OH^-(aq)} CH_3.CH_2OH$$

Ethanol is a primary alcohol which is present in common alcoholic beverages. Last century, it was prepared almost exclusively by the fermentation of sugars by yeast. Today, in addition to this source of ethanol, it is also made by the direct hydration of ethylene, obtained by cracking petroleum, thus:

$$CH_2{=}CH_2 + H_2O \xrightarrow{H_2SO_4} CH_3.CH_2.OH$$

It is widely used industrially as a solvent and as a starting material in the preparation of many important synthetic compounds.

Another important primary alcohol is normal propanol:

$$CH_3.CH_2.CH_2.Cl \xrightarrow{OH^-(aq)} CH_2.CH_2.CH_2.OH$$

1-chloropropane, n-propyl chloride (liquid) → 1-propanol, n-propanol (liquid)

A typical *secondary* alcohol is prepared by the reaction of isopropyl chloride with hydroxide ions:

$$HO^- + \underset{CH_3}{\overset{CH_3}{H\!-\!C}}\!-\!Cl \rightarrow HO\!-\!\underset{CH_3}{\overset{CH_3}{C}}\!-\!H + Cl^-$$

2-chloropropane, isopropyl chloride (liquid) → 2-propanol, isopropanol (liquid)

n-propanol and isopropanol both have the empirical formula C_3H_8O, but different structures. They are thus structural isomers. It is apparent that alcohols may be prepared quite generally by the reaction of alkyl chlorides with hydroxide ions.

$$R.Cl \xrightarrow{OH^-(aq)} R.OH$$

We must realize that R represents *any* alkyl group; in reactions we have written down already, R has been

CH_3-	$CH_3.CH_2-$	$(CH_3)_2CH-$	$CH_3.CH_2CH_2-$
methyl	ethyl	isopropyl	n-propyl

(2) Properties of aliphatic alcohols. We have seen that ethyl chloride will react with hydroxide ions in an aqueous medium to yield ethanol and chloride ions. If we wish to convert ethanol to ethyl chloride the reaction may be effectively reversed by adding dry HCl to ethanol:

$$CH_3.CH_2.OH \xrightarrow[HCl]{dry} CH_3.CH_2.Cl$$

When HCl is added to ethanol, the reaction occurring is:

$$CH_3CH_2\!-\!O\!-\!H + HCl \rightarrow CH_3CH_2\!-\!\overset{\oplus}{O}H_2 + Cl^-$$

The ion $CH_3.CH_2.\overset{+}{O}H_2$ then reacts with chloride ions yielding $CH_3.CH_2Cl$ and water. This then is a general method for converting an alcohol to an alkyl chloride:

$$ROH \xrightarrow[HCl]{dry} R.Cl$$

This reaction is best represented as an equilibrium

$$ROH + H^+ + Cl^- \rightleftharpoons RCl + H_2O$$

$$K = \frac{[RCl][H_2O]}{[ROH][H^+][Cl^-]}$$

Where $[H_2O]$ is low, and $[H^+]$ and $[Cl^-]$ are high, the position of equilibrium lies to the right and RCl forms readily. However, if $[H_2O]$ is high, as when water is the solvent, and $[H^+]$ is low as in alkaline solution, the equilibrium lies to the left and RCl readily hydrolyses (see sections 18.1, 18.2 and 19.2).

Alkyl bromides and iodides cannot be conveniently prepared from the corresponding hydrohalic acids, and are instead prepared by the action of either PBr_3 or PI_3. Thus:

$$ROH \xrightarrow{PBr_3} RBr$$

Alcohols can be readily oxidized. Thus:

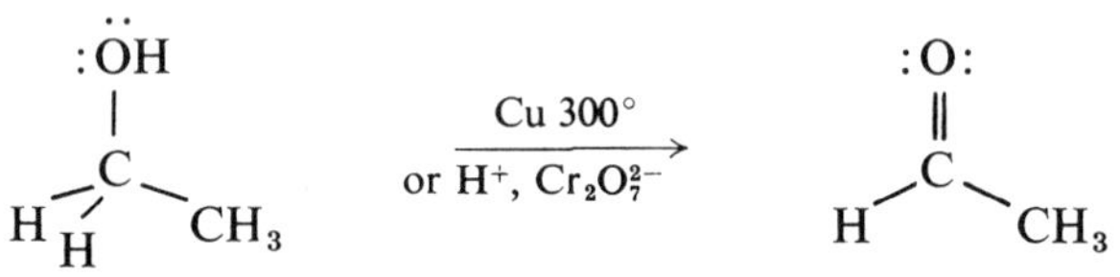

$$H(H)C(\ddot{O}H)CH_3 \xrightarrow[\text{or } H^+,\ Cr_2O_7^{2-}]{\text{Cu } 300°} H\text{–}C(=\ddot{O})\text{–}CH_3$$

ethanol

acetaldehyde,
ethanal
(colourless liquid,
b.p. +21 °C)

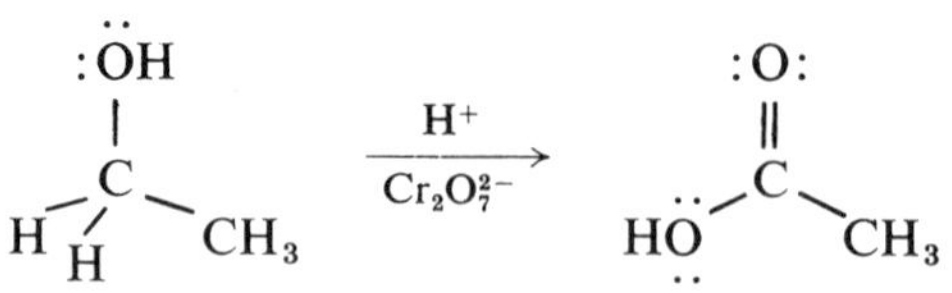

$$H(H)C(\ddot{O}H)CH_3 \xrightarrow[Cr_2O_7^{2-}]{H^+} H\ddot{O}\text{–}C(=\ddot{O})\text{–}CH_3$$

ethanol

acetic acid,
ethanoic acid
(colourless liquid,
b.p. +118 °C)

or 1-propanol:

$$CH_3.CH_2.CH_2.OH \xrightarrow[\text{or } H^+,\ Cr_2O_7^{2-}]{\text{Cu, } 300°} CH_3.CH_2.CHO$$

propionaldehyde,
propanal
(liquid)

$$CH_3.CH_2.CH_2.OH \xrightarrow[Cr_2O_7^{2-}]{H^+} CH_3.CH_2.CO.OH$$

propionic acid,
propanoic acid
(liquid)

Acidified dichromate may be used to oxidize alcohols either to aldehydes or acids, depending on the reaction conditions used. Thus in general, for a *primary alcohol*

$$R.CH_2.OH \xrightarrow[Cr_2O_7^{2-}]{H^+} R.CHO \text{ (aldehyde)}$$

$$R.CH_2.OH \xrightarrow[Cr_2O_7^{2-}]{H^+} R.CO.OH \text{ (carboxylic acid)}$$

In the case of isopropanol:

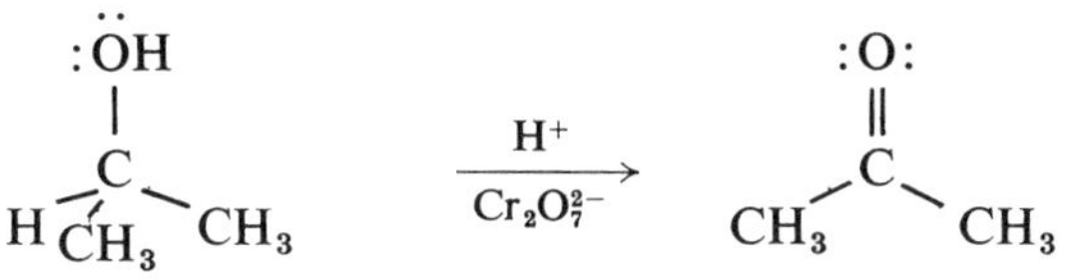

$$\underset{\text{2-propanol}}{(CH_3)_2CH\ddot{O}H} \xrightarrow[Cr_2O_7^{2-}]{H^+} \underset{\text{acetone, (liquid) propanone}}{CH_3C(=\ddot{O})CH_3}$$

or:

$$CH_3.CH(OH).CH_3 \xrightarrow[Cr_2O_7^{2-}]{H^+} \underset{\text{acetone}}{CH_3.CO.CH_3}$$

In general we have for a *secondary* alcohol:

$$R.CH(OH).R \xrightarrow[Cr_2O_7^{2-}]{H^+} \underset{\text{(ketone)}}{R.CO.R'}$$

Tertiary alcohols are difficult to oxidize directly, and when they are, carbon-carbon bonds are broken, and a mixture of oxidation products is obtained.

All alcohols are quite weak bases and weak acids, with K_as and K_bs similar to methanol (see section 18.3). Alcohols react with alkali metals in an analogous fashion to water. Thus, with water we have:

$$Na(s) + H_2O(l) \rightarrow Na^+(aq) + OH^-(aq) + \tfrac{1}{2}H_2$$

while with an alcohol,

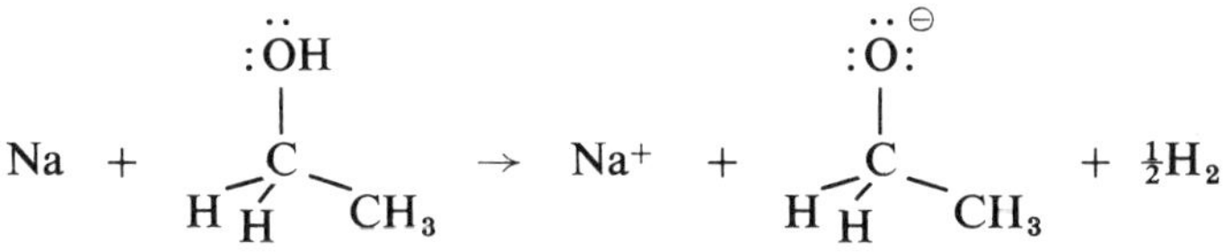

$$Na + CH_3CH_2\ddot{O}H \rightarrow Na^+ + \underset{\text{ethoxide ion}}{CH_3CH_2\ddot{O}^{\ominus}} + \tfrac{1}{2}H_2$$

or, in general,

$$Na + ROH \rightarrow Na^+ + \underset{\text{alkoxide ion}}{^-OR} + \tfrac{1}{2}H_2$$

Thus the addition of any alkali metal to an alcohol yields an alcoholic solution of sodium ions and alkoxide ions.

(3) Preparation and uses of phenol. The preparation of the benzene analogue of the alcohols, phenol, is not particularly easy in the laboratory. Industrially, it is produced by the reaction of chlorobenzene with sodium hydroxide at high pressure and temperature:

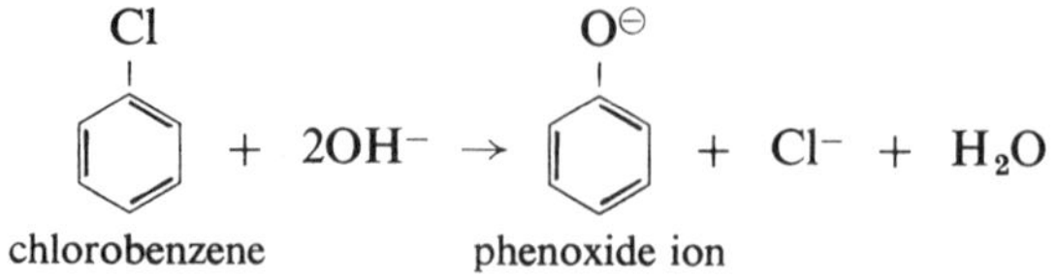

$$C_6H_5Cl + 2OH^- \rightarrow C_6H_5O^\ominus + Cl^- + H_2O$$

chlorobenzene — phenoxide ion

The high temperature and pressure necessary are in sharp contrast to the ease with which alkyl halides are hydrolysed in water. The phenoxide ion may then be converted to phenol by reaction with acid:

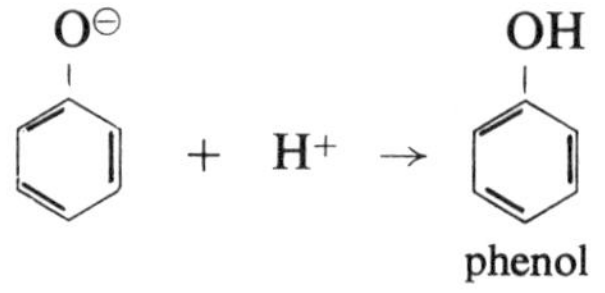

$$C_6H_5O^\ominus + H^+ \rightarrow C_6H_5OH$$

phenol

Some phenol is also obtained directly from the distillation of coal. A usable laboratory preparation, however, is via benzene sulphonic acid:

$$C_6H_5SO_2(OH) \xrightarrow[350°C]{NaOH} C_6H_5O^\ominus$$

benzene sulphonic acid

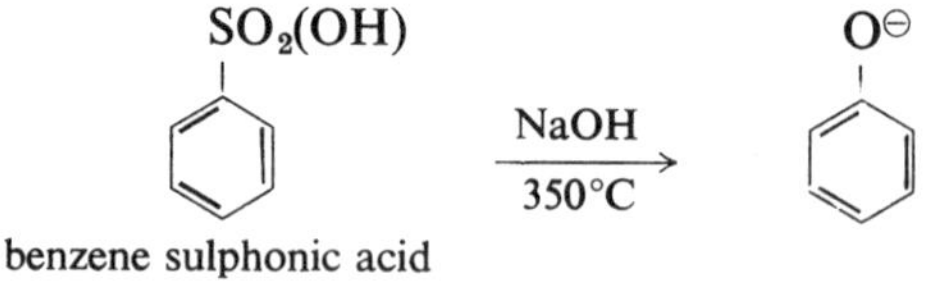

Phenol is still sometimes used as a disinfectant, but a major industrial use is in the production of phenol-formaldehyde resins, which are discussed below.

(4) Properties of phenol. A striking difference between phenol and aliphatic alcohols lies in the higher acid strength of phenol. Phenol dissociates in water according to

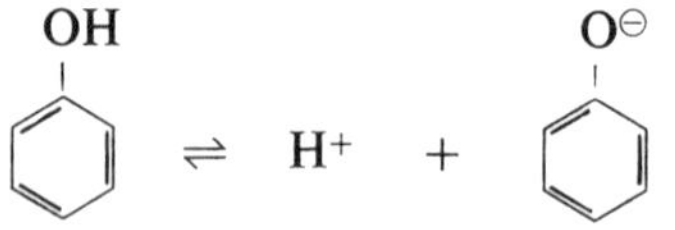

$$C_6H_5OH \rightleftharpoons H^+ + C_6H_5O^\ominus \qquad K_a = 1.3 \times 10^{-10}$$

Compare the K_a of ethanol of *ca.* 10^{-16}.

Phenol is extremely reactive chemically and is the starting point for the synthesis of a large range of compounds. One of the early plastics was made by reacting phenol with formaldehyde to form a long chain polymer.

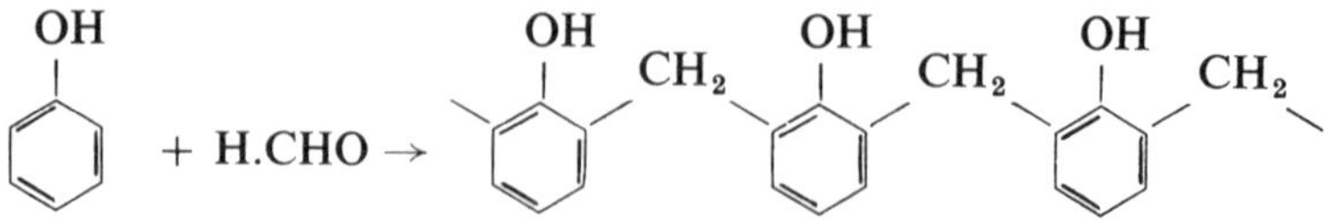

$$C_6H_5OH + H.CHO \rightarrow \text{—}C_6H_3(OH)\text{—}CH_2\text{—}C_6H_3(OH)\text{—}CH_2\text{—}C_6H_3(OH)\text{—}CH_2\text{—}$$

In a cross-linked form, this material becomes Bakelite, used in the manufacture of many moulded articles.

Phenol is also the source of a number of compounds used medicinally. One of these is aspirin:

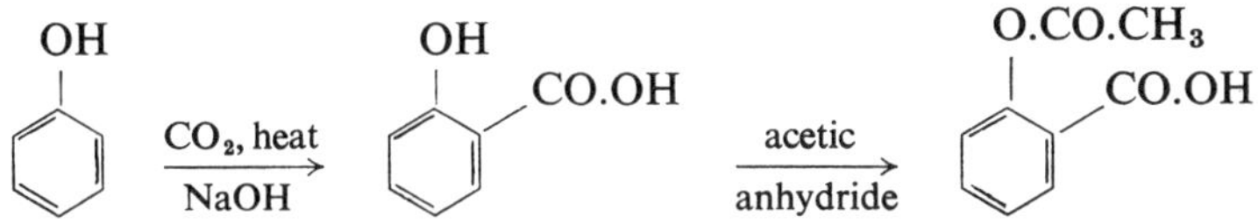

$$C_6H_5OH \xrightarrow[NaOH]{CO_2,\ heat} C_6H_4(OH)CO.OH \xrightarrow[anhydride]{acetic} C_6H_4(O.CO.CH_3)CO.OH$$

salicylic acid — acetyl salicylic acid (aspirin)

(5) Glycol and glycerol. There is a number of important compounds containing more than one hydroxy group. Glycol is prepared from ethylene by the action of hypochlorous acid, followed by hydrolysis:

$$CH_2{=}CH_2 + HOCl \rightarrow CH_2(OH){-}CH_2Cl$$

ethylene chlorhydrin

$$CH_2(OH){-}CH_2Cl + OH^- \rightarrow CH_2(OH){-}CH_2(OH)$$

glycol,
ethane 1, 2-diol

Glycol is used as an antifreeze agent in motor-car cooling systems, and as a component in certain polymers (see section 20.6).

Glycerol is a constituent of naturally-occurring fats, and has the structure

$$\begin{array}{l} CH_2.OH \\ | \\ CH.OH \\ | \\ CH_2.OH \end{array}$$

Both glycol and glycerol are miscible with water in all proportions.

20.2 THE AMINO GROUP

(1) Preparation of aliphatic amines. One of the most generally useful methods of amine preparation is by the direct action of ammonia on an alkyl halide:

$$RCl + NH_3 \rightarrow \underset{\text{primary amine}}{RNH_2} + H^+ + Cl^-$$

A drawback with the process is that it is difficult to control conditions so as to stop with one substitution as shown above. The primary amine formed readily reacts further

$$RCl + RNH_2 \rightarrow \underset{\text{secondary amine}}{R.NH.R} + H^+ + Cl^-$$

and then to completion

$$RCl + R_2.NH \rightarrow \underset{\text{tertiary amine}}{R_3N} + H^+ + Cl$$

Notwithstanding this disadvantage, the method shown here is useful, provided the products can be separated by distillation or some other suitable technique. Thus, to take a particular example

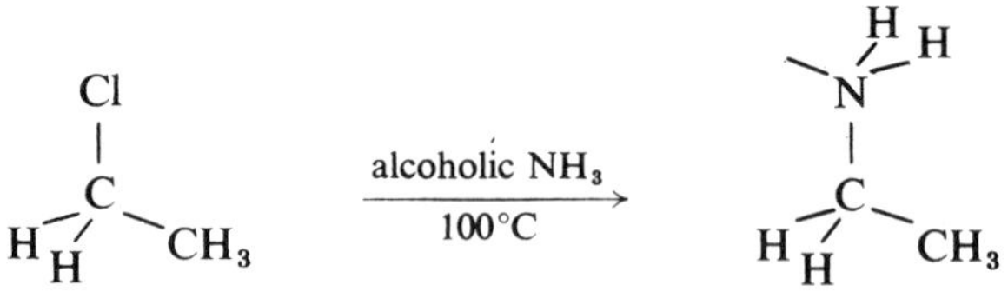

$$CH_3CH_2Cl \xrightarrow[100°C]{\text{alcoholic } NH_3} CH_3CH_2NH_2$$

ethylamine, aminoethane (liquid, b.p. +17°C)

(2) Properties of aliphatic amines. A characteristic property of amines is their basicity, which is similar to ammonia. We have already compared the basicity of methylamine with both ammonia and methanol. Below are shown the K_b values of a primary, secondary and tertiary amine:

Amine	CH_3NH_2	$(CH_3)_2NH$	$(CH_3)_3N$
K_b	4.4×10^{-4}	5.2×10^{-4}	5.5×10^{-5}

One very important property of amines is their ability to react with carboxylic acids to form amides. This property is dealt with in section 20.7.

(3) Preparation and properties of aniline. The aromatic analogues of alkyl amines cannot be conveniently prepared directly from ammonia and, say, chlorobenzene, although this is sometimes done industrially under extreme conditions. In the laboratory, and also industrially, a standard procedure is to reduce nitrobenzene:

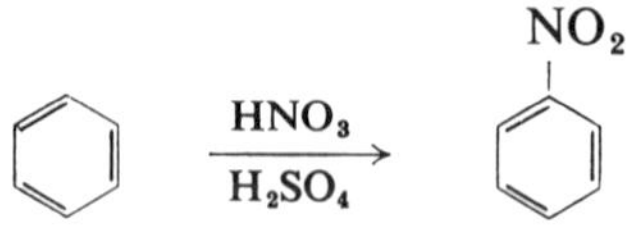

$$C_6H_6 \xrightarrow[H_2SO_4]{HNO_3} C_6H_5NO_2$$

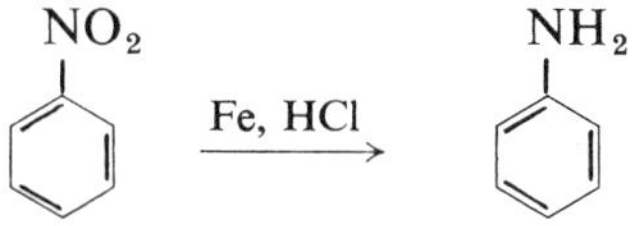

$$C_6H_5NO_2 \xrightarrow{\text{Fe, HCl}} C_6H_5NH_2$$

The basic strength of aniline is a good deal lower than that of the alkylamines. The relationships are shown below:

Amine	CH_3NH_2	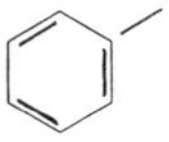 $C_6H_5NH_2$		
K_b	4×10^{-4}	4×10^{-10}		
Acid	$CH_3NH_3^+$	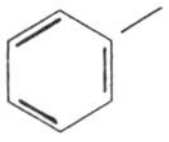 $C_6H_5NH_3^+$	CH_3OH	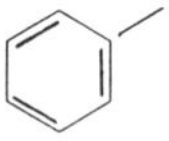 C_6H_5OH
K_a	2.5×10^{-11}	2.5×10^{-5}	$\sim 10^{-16}$	$\sim 10^{-10}$

20.3 ETHERS

Preparation and properties. We have seen that alcohols react with alkali metals to form solutions containing alkoxide ions, e.g.

$$Na(s) + CH_3.OH \rightarrow Na^+ + {}^-OCH_3 + \tfrac{1}{2}H_2$$

methoxide ion

Alkoxide ions can react with alkyl halides in the same way as can hydroxide ions, e.g.

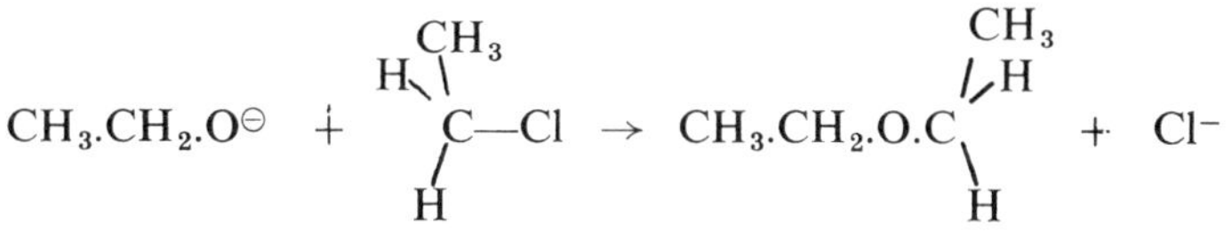

$$CH_3.CH_2.O^{\ominus} + CH_3CH_2Cl \rightarrow CH_3.CH_2.O.CH_2CH_3 + Cl^-$$

chloroethane; diethyl ether (liquid, b.p. +34.5°C)

or, in general,

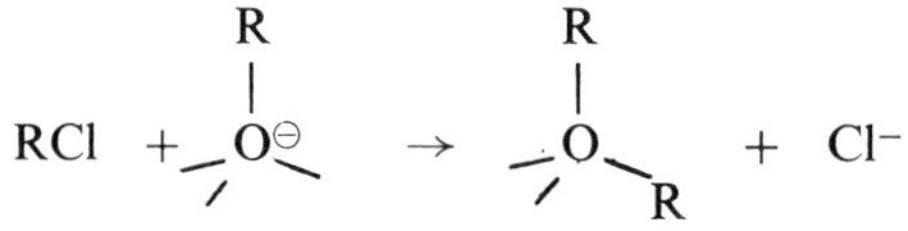

$$RCl + R\text{—}O^{\ominus} \rightarrow R\text{—}O\text{—}R + Cl^-$$

ether

Diethyl ether is the 'ether' commonly employed as a solvent in laboratories, and as an anaesthetic for minor surgical operations. Although 'ether' can be prepared as indicated above, in practice it is prepared by the action of sulphuric acid on ethanol:

$$2CH_3.CH_2.OH \xrightarrow[200°C]{H_2SO_4} CH_3.CH_2.O.CH_2.CH_3$$

Ethers are very weak bases with K_bs slightly greater than alcohols.

20.4 THE CARBONYL GROUP

(1) Preparation of aliphatic aldehydes and ketones. We must now consider the two homologous series derivable from formaldehyde by successive replacement of the hydrogens of formaldehyde by alkyl groups. When one hydrogen is replaced, we obtain a homologous series of compounds called *aldehydes*:

$$\overset{:\ddot{O}:}{\underset{H\quad R}{\overset{\|}{C}}}$$

one substitution, an aldehyde: R.CHO

When both hydrogens are replaced with alkyl groups, we obtain a *ketone*

$$\overset{:\ddot{O}:}{\underset{R\quad R'}{\overset{\|}{C}}}$$

two substitutions, a ketone: R.CO.R′

An aldehyde may be prepared by the oxidation of a *primary* alcohol. Thus, as we saw in section 20.1(2):

$$CH_3.CH_2.OH \xrightarrow[Cr_2O_7^{2-}]{H^+} CH_3.CHO$$

ethanol → acetaldehyde, ethanal

$$CH_3.CH_2.CH_2.OH \xrightarrow[Cr_2O_7^{2-}]{H^+} CH_3.CH_2.CHO$$

n-propanol, 1-propanol → propionaldehyde, propanal (colourless liquid)

$$CH_3.CH_2.CH_2.CH_2.OH \xrightarrow[Cr_2O_7^{2-}]{H^+} CH_3.CH_2.CH_2.CHO$$

n-butanol, 1-butanol → n-butyraldehyde, butanal (colourless liquid)

or, in general:

$$R.CH_2.OH \xrightarrow[Cr_2O_7^{2-}]{H^+} R.CHO$$

Ketones may be prepared by the oxidation of *secondary* alcohols:

$$CH_3.CH(OH).CH_3 \xrightarrow[Cr_2O_7^{2-}]{H^+} CH_3.CO.CH_3$$

isopropanol, 2-propanol → acetone, propanone (colourless liquid, b.p. +56°C)

$$CH_3.CH(OH).CH_2.CH_3 \xrightarrow[Cr_2O_7^{2-}]{H^+} CH_3.CO.CH_2.CH_3$$

s-butanol, 2-butanol → methyl ethyl ketone, 2-butanone (liquid)

In general:

$$RCH_2OH \xrightarrow{\text{oxidation}} RCHO$$

primary alcohol → aldehyde

$$R.CH(OH)R' \xrightarrow{\text{oxidation}} R.CO.R'$$

secondary alcohol → ketone

(2) Properties of aliphatic aldehydes and ketones. Aldehydes are easily oxidized to carboxylic acids, e.g.

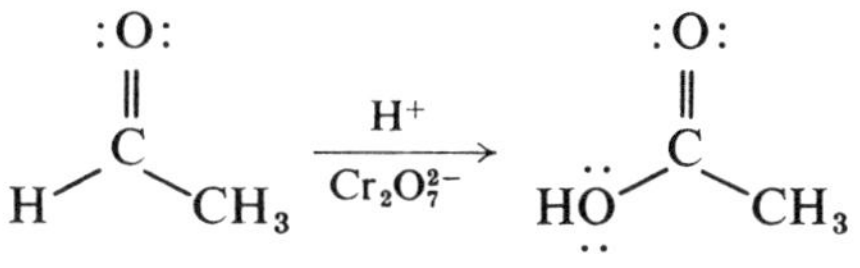

$$H{-}\overset{\overset{:O:}{\|}}{C}{-}CH_3 \xrightarrow[Cr_2O_7^{2-}]{H^+} H\ddot{\underset{..}{O}}{-}\overset{\overset{:O:}{\|}}{C}{-}CH_3$$

acetic acid (colourless liquid, b.p. +118°C)

or, more generally:

$$R.CHO \xrightarrow{\text{oxidation}} R.CO.OH$$

aldehyde → carboxylic acid

It is because of this reaction that one must take care in preparing aldehydes by oxidizing alcohols with acidified dichromate. Although dichromate will oxidize an alcohol to an aldehyde, the aldehyde so formed can be itself oxidized by dichromate, and the final reaction product may be a carboxylic acid.

Ketones are rather difficult to oxidize and, when they are, carbon-carbon bonds are broken, and a mixture of oxidation products results.

Both aldehydes and ketones can be reduced to alcohols either by catalytic hydrogenation, or by the action of lithium aluminium hydride ($LiAlH_4$) in ethereal solution. Aldehydes yield primary alcohols on reduction, ketones yield secondary alcohols, e.g.

$$\underset{\text{acetaldehyde}}{CH_3CHO} \xrightarrow{LiAlH_4} \underset{\text{ethanol}}{CH_3.CH_2.OH}$$

$$\underset{\text{acetone}}{CH_3.CO.CH_3} \xrightarrow{LiAlH_4} \underset{\text{isopropanol, 2-propanol}}{CH_3.CH(OH).CH_3}$$

More generally we could write:

$$\underset{\text{aldehyde}}{R.CHO} \xrightarrow{LiAlH_4} \underset{\text{primary alcohol}}{R.CH_2.OH}$$

$$\underset{\text{ketone}}{R.CO.R'} \xrightarrow{LiAlH_4} \underset{\text{secondary alcohol}}{R.CH(OH).R'}$$

20.5 THE CARBOXYL GROUP

(1) Preparation of aliphatic carboxylic acids. There is but one homologous series of compounds derived from formic acid—when its single hydrogen attached to carbon is replaced by an alkyl group. A carboxylic acid can be prepared by the direct oxidation of either a primary alcohol, or an aldehyde, e.g.

$$\underset{\text{ethanol}}{H_2C(\ddot{O}H)CH_3} \xrightarrow[Cr_2O_7^{2-}]{H^+} \underset{\text{acetic acid, ethanoic acid}}{H\ddot{O}-C(=\ddot{O})-CH_3}$$

or, more simply:

$$CH_3.CH_2.OH \xrightarrow[Cr_2O_7^{2-}]{H^+} CH_3.CO.OH$$

Also:

$$\underset{\text{acetaldehyde}}{H-C(=\ddot{O})-CH_3} \xrightarrow[H^+]{Cr_2O_7^{2-}} \underset{\text{acetic acid}}{H\ddot{O}-C(=\ddot{O})-CH_3}$$

and

$$\underset{\text{propionaldehyde}}{CH_3.CH_2.CHO} \xrightarrow[H^+]{Cr_2O_7^{2-}} \underset{\text{propionic acid, propanoic acid}}{CH_3.CH_2.CO.OH}$$

Thus we may write quite generally:

$$\underset{\text{aldehyde (or primary alcohol)}}{R.CHO\ (\text{or}\ R.CH_2.OH)} \xrightarrow{\text{oxidation}} \underset{\text{carboxylic acid}}{R.CO.OH}$$

(2) Some chemical properties. The carboxylic acids are all weak acids, and produce small concentrations of hydrogen ions when dissolved in water, e.g.

$$\underset{\text{acetic acid}}{HO{-}C({=}O){-}CH_3} + H_2O \rightleftharpoons \underset{\text{acetate ion}}{{}^{\ominus}O{-}C({=}O){-}CH_3} + H_3O^+$$

The K_as for the first few members of the homologous series are:

Formula	*Acid*	K_a	*Corresponding anion*
H.CO.OH	formic	24×10^{-5}	formate
$CH_3CO.OH$	acetic	1.75×10^{-5}	acetate
$CH_3.CH_2.CO.OH$	propionic	1.22×10^{-5}	propionate
$CH_3.CH_2.CH_2.CO.OH$	n-butyric	1.5×10^{-5}	n-butyrate
$(CH_3)_2CH.CO.OH$	isobutyric	1.4×10^{-5}	isobutyrate

The general trend is to slowly decreasing acid strength in the series. The carboxylate ions are found in ionic solids in compounds formed with the alkali and alkaline earth metals, e.g. sodium formate, sodium propionate.

The acidity of the OH group in carboxylic acids is obviously very much greater than the acidity of the OH group in simple alcohols (where $K_a \sim 10^{-16}$). This is due to the close proximity of the carbonyl grouping in the carboxylic acids. It appears that the carbonyl group assists in the removal of electrons from the O-H bond in the hydroxy group, and the bond is thereby weakened making it easier to remove hydrogen as H^+. It is instructive to consider the effects of chloro-substitution of the hydrogens of acetic acid on the K_a.

Formula	*Acid*	K_a	*Corresponding anion*
$CH_3CO.OH$	acetic	1.8×10^{-5}	acetate
$CH_2Cl.CO.OH$	chloracetic	1.5×10^{-3}	chloracetate
$CHCl_2.CO.OH$	dichloracetic	5.1×10^{-2}	dichloracetate
$CCl_3.CO.OH$	trichloracetic	2.0×10^{-1}	trichloracetate

The high electronegativity of chlorine has a dramatic effect on the ease with which the carboxyl group releases a proton. This may again be interpreted as being at least partially due to a weakening of the O-H bond in the carboxyl group by withdrawal of electrons from the region between the oxygen and the hydrogen, thus making it easier to remove the hydrogen as a proton.

Carboxylic acids are difficult to oxidize, and oxidation results in the breaking of carbon-carbon bonds, and a mixture of reaction products is produced. The acids are also fairly difficult to reduce, but they are converted to primary alcohols by the action of lithium aluminium hydride:

$$CH_3.CO.OH \xrightarrow{LiAlH_4} CH_3.CH_2.OH$$

or, in general:

$$\underset{\text{carboxylic acid}}{R.CO.OH} \xrightarrow{LiAlH_4} \underset{\text{primary alcohol}}{R.CH_2.OH}$$

One very important reaction of carboxylic acids is that with alcohols:

$$\underset{\text{acetic acid}}{HO{-}C({=}O){-}CH_3} + \underset{\text{methanol}}{CH_3.OH} \xrightarrow[]{\text{trace } H_2SO_4} \underset{\substack{\text{methyl acetate}\\ \text{(colourless liquid)}}}{CH_3O{-}C({=}O){-}CH_3} + H_2O$$

Methyl acetate is an *ester*, and esters are formed by the reaction of an alcohol with a carboxylic acid, usually in the presence of a trace of a strong acid, which acts as a catalyst. In general then:

$$\underset{\text{carboxylic acid}}{R{-}C({=}O){-}OH} + \underset{\text{alcohol}}{R'OH} \xrightarrow{\text{trace of strong acid}} \underset{\text{ester}}{R{-}C({=}O){-}OR'} + H_2O$$

(3) Preparation and properties of benzoic acid. Benzoic acid may be prepared by the vigorous oxidation of toluene

$$\underset{\text{toluene}}{C_6H_5{-}CH_3} \xrightarrow[Cr_2O_7^{2-}]{H^+} \underset{\substack{\text{benzoic acid}\\ \text{(white solid,}\\ \text{m.p. 120°C)}}}{C_6H_5{-}CO.OH}$$

Benzoic acid dissociates in water to hydrogen ion and the benzoate anion; the K_a is 6.5×10^{-5}, very similar to acetic acid. Benzoic acid forms esters with alcohols in the same way as aliphatic carboxylic acids. Thus:

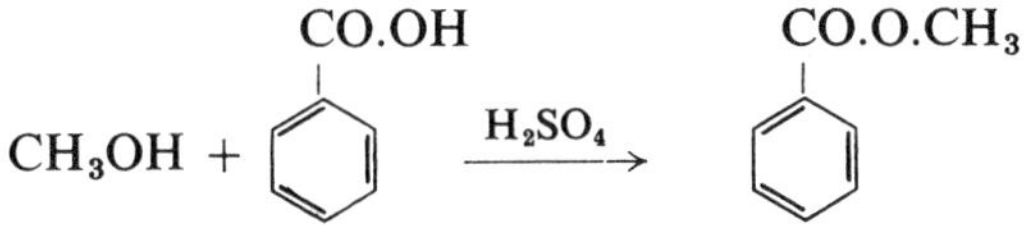

$$CH_3OH + C_6H_5.CO.OH \xrightarrow{H_2SO_4} C_6H_5.CO.O.CH_3$$

methyl benzoate

20.6 ESTERS

(1) Preparation of esters. Esters are formed when carboxylic acids react with alcohols in the presence of a strong acid catalyst. Many esters have pleasant fruity odours, and are often used in the preparation of perfumes and artificial flavourings, e.g. ethyl butanoate gives an apple flavouring and n-octyl acetate, orange flavouring. The general reaction for formation of an ester is:

$$R.CO.OH + R'OH \xrightarrow{H_2SO_4} R.CO.OR' + H_2O$$

Some examples of simple esters are:

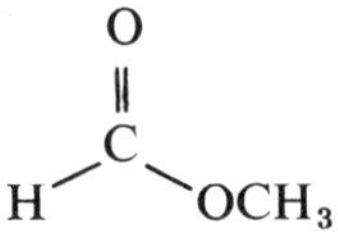

$H-C(=O)-OCH_3$
methyl formate
($H.CO.OCH_3$)

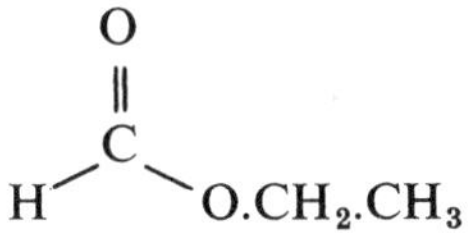

$H-C(=O)-O.CH_2.CH_3$
ethyl formate
($H.CO.O.CH_2.CH_3$)

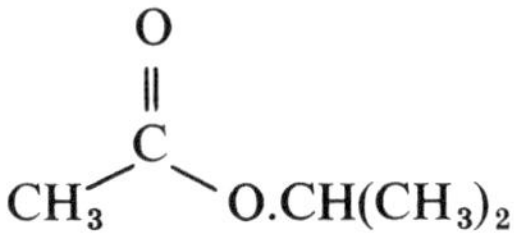

$CH_3-C(=O)-O.CH(CH_3)_2$
isopropyl acetate
($CH_3.CO.O.CH(CH_3)_2$)

$CH_3-C(=O)-O.CH_2CH_2CH_2.CH_3$
n-butyl acetate
($CH_3.CO.O.CH_2.CH_2.CH_2.CH_3$)

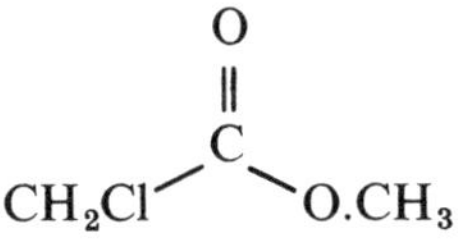

$CH_2Cl-C(=O)-O.CH_3$
methyl chloroacetate
($CH_2Cl.CO.O.CH_3$)

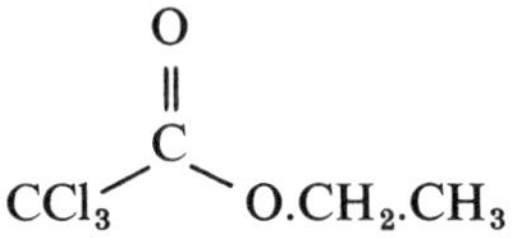

$CCl_3-C(=O)-O.CH_2.CH_3$
ethyl trichloroacetate
($CCl_3.CO.O.CH_2.CH_3$)

EXERCISE
Write down the carboxylic acid and alcohol from which each of the above esters could have been prepared.

(2) Properties of esters. Let us consider the reaction of an alcohol with a carboxylic acid in a little more detail. Strictly, the reaction should be written as an equilibrium, e.g.

$$CH_3.CO.OH + CH_3.CH_2.OH \rightleftharpoons CH_3.CO.O.CH_2.CH_3 + H_2O$$

ethyl acetate

so we have

$$K = \frac{[CH_3.CO.O.CH_2.CH_3][H_2O]}{[CH_3.CO.OH][CH_3.CH_2.OH]}$$

hence
$$\frac{[CH_3.CO.O.CH_2.CH_3]}{[CH_3.CO.OH][CH_3.CH_2.OH]} = \frac{K}{[H_2O]}$$

Any strong acid present merely catalyses the reactions and speeds up the attainment of equilibrium.

When pure ethanol is mixed with pure acetic acid in the presence of the catalyst, $K \gg [H_2O]$, hence the ester forms at the expense of acetic acid and ethanol. The ester may then be obtained pure by fractional distillation of the resulting mixture.

However, when an ester is placed in an acidified aqueous solution, $[H_2O]$ is quite large, $K \ll [H_2O]$, and acetic acid and ethanol form at the expense of the ester, i.e. we say the ester hydrolyses.

$$CH_3.CO.O.CH_2.CH_3 \xrightarrow{H^+(aq)} CH_3.CO.OH + CH_3.CH_2.OH$$

In the presence of hydroxide ions, the hydrolysis of an ester proceeds virtually to completion according to:

$$\underset{\text{ethyl acetate}}{CH_3.CO.O.CH_2.CH_3} + OH^- \rightarrow \underset{\text{acetate ion}}{CH_3.CO.O^-} + \underset{\text{ethanol}}{CH_3.CH_2.OH}$$

Thus esters are *hydrolysed* in aqueous solutions either by an acid-catalysed reaction, or by direct reaction with hydroxide ions:

$$R.CO.OR' \xrightarrow{H^+(aq)} R.CO.OH + R'.OH$$

$$R.CO.OR' \xrightarrow{OH^-(aq)} R.CO.O^- + R'.OH$$

(3) Fats, soaps and polymeric esters. Fats are high molecular weight naturally-occurring esters formed from glycerol acting as the alcohol, and a wide range of long chain carboxylic acids. Some of the commoner carboxylic acids involved are:

$CH_3(CH_2)_{10}CO.OH$	lauric acid
$CH_3(CH_2)_{12}CO.OH$	myristic acid
$CH_3(CH_2)_{14}CO.OH$	palmitic acid
$CH_3(CH_2)_{16}CO.OH$	stearic acid

Most of the naturally-occurring carboxylic acids found in fats have an even number of carbon atoms. A simple fat, found in palm oil, is glyceryl tripalmitate, which on hydrolysis yields glycerol and palmitic acid:

$$\begin{array}{l} CH_2.O.CO.(CH_2)_{14}.CH_3 \\ | \\ CH.O.CO.(CH_2)_{14}.CH_3 \\ | \\ CH_2.O.CO.(CH_2)_{14}.CH_3 \end{array} \rightarrow \begin{array}{l} CH_2.OH \\ | \\ CH.OH \\ | \\ CH_2OH \end{array} + 3CH_3(CH_2)_{14}.CO.OH$$

Salts of the high molecular weight carboxylic acids have been used for centuries as 'soaps'. Thus when palm oil is hydrolysed with hydroxide ions in aqueous solution ('saponified'), sodium palmitate is produced, which is a soap.

$$CH_3(CH_2)_{14}.CO.\overset{\ominus}{O}\overset{\oplus}{Na}$$

sodium palmitate

The sodium salt of any long chain fatty acid is, similarly, a soap.

Some polymeric esters are much used industrially, e.g. when a molecule with two carboxyl groups forms an ester with one with two hydroxy groups, long chain molecules are formed. Thus, by combining the following two molecules:

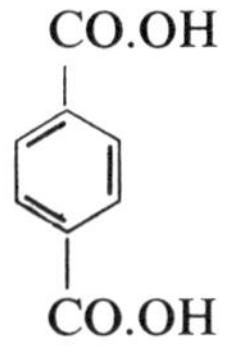

$HOOC-C_6H_4-COOH$ (CO.OH groups para on a benzene ring)

terephthalic acid

$CH_2.OH-CH_2.OH$

glycol

we obtain the long chain

$$—O.CH_2.CH_2.O.CO—C_6H_4—CO.O.CH_2.CH_2.O.CO—C_6H_4—CO—$$

Fibres made from this polymer are called Dacron or Terylene and are used in the production of garments.

20.7 THE AMIDE GROUP

(1) Preparation and properties of aliphatic amides. Amides may be prepared by the direct action of excess ammonia on the pure carboxylic acids at high temperatures:

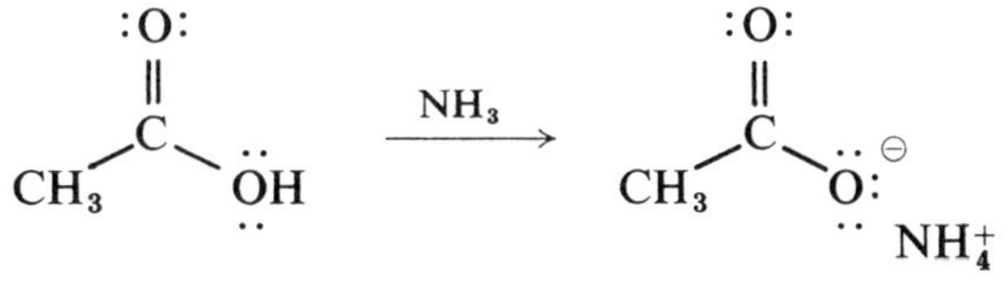

$$CH_3C(=\ddot{O}:)\ddot{O}H \xrightarrow{NH_3} CH_3C(=\ddot{O}:)\ddot{O}:^{\ominus}\ NH_4^+$$

acetic acid → ammonium acetate

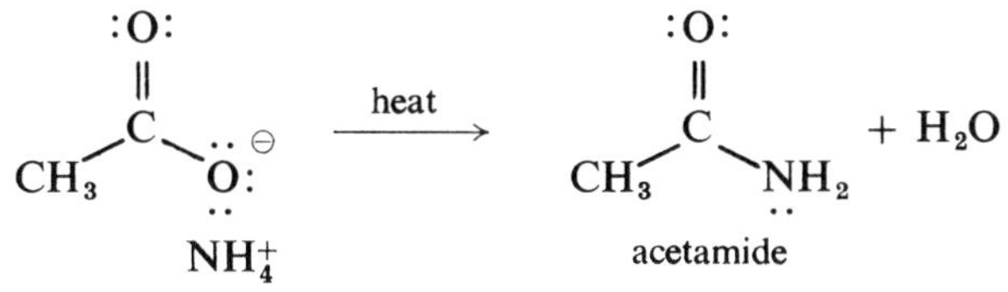

$$CH_3C(=\ddot{O}:)\ddot{O}:^{\ominus}\ NH_4^+ \xrightarrow{heat} CH_3C(=\ddot{O}:)\ddot{N}H_2 + H_2O$$

acetamide

or, simply

$$CH_3.CO.OH \xrightarrow{NH_3,\ heat} CH_3.CO.NH_2$$

Also:

$$\underset{\text{formic acid}}{H.CO.OH} \xrightarrow{NH_3,\ heat} \underset{\text{formamide}}{H.CO.NH_2}$$

or in general:

$$R.CO.OH \xrightarrow{NH_3,\ heat} \underset{\text{acid amide}}{R.CO.NH_2}$$

However, in aqueous solutions, when excess water is present, these simple amides hydrolyse slowly in the presence of an acid catalyst:

$$CH_3.CO.NH_2 \xrightarrow{H^+(aq)} CH_3.CO.OH + NH_4^+$$

Amines, as well as ammonia, can react with carboxylic acids to give amides, e.g.

$$CH_3-C(=\ddot{O}:)-\ddot{O}H \xrightarrow{CH_3.NH_2} \underset{\text{methyl acetamide}}{CH_3-C(=\ddot{O}:)-\ddot{N}H.CH_3}$$

or in general:

$$R.CO.OH + R'.NH_2 \rightarrow R.CO.NH.R' + H_2O$$

Amides have no acidic properties, but are very weak bases, e.g.

$$CH_3-C(=\ddot{O}:)-\ddot{N}H_2 + H_2O \rightleftharpoons CH_3-C(=\overset{\oplus}{:OH})-\ddot{N}H_2 + OH^-$$

or:

$$CH_3.CO.NH_2 + H_2O \rightleftharpoons CH_3.C(\overset{+}{O}H)NH_2 + OH^- \quad K_b = 3 \times 10^{-15}$$

$$\text{compare } CH_3.NH_2 \quad K_b = 4 \times 10^{-4}$$

The basicity of the $-NH_2$ group in amides is obviously very much less than that of the $-NH_2$ group in amines, and this is due to the proximity of the carbonyl group in the amides. (Compare the greatly increased *acidity* of the –OH group in carboxylic acids with the acidity of the –OH group in simple alcohols.)

(2) Polymeric amides and proteins. The reaction between amines and carboxylic acids is of great importance in protein synthesis. Proteins are made

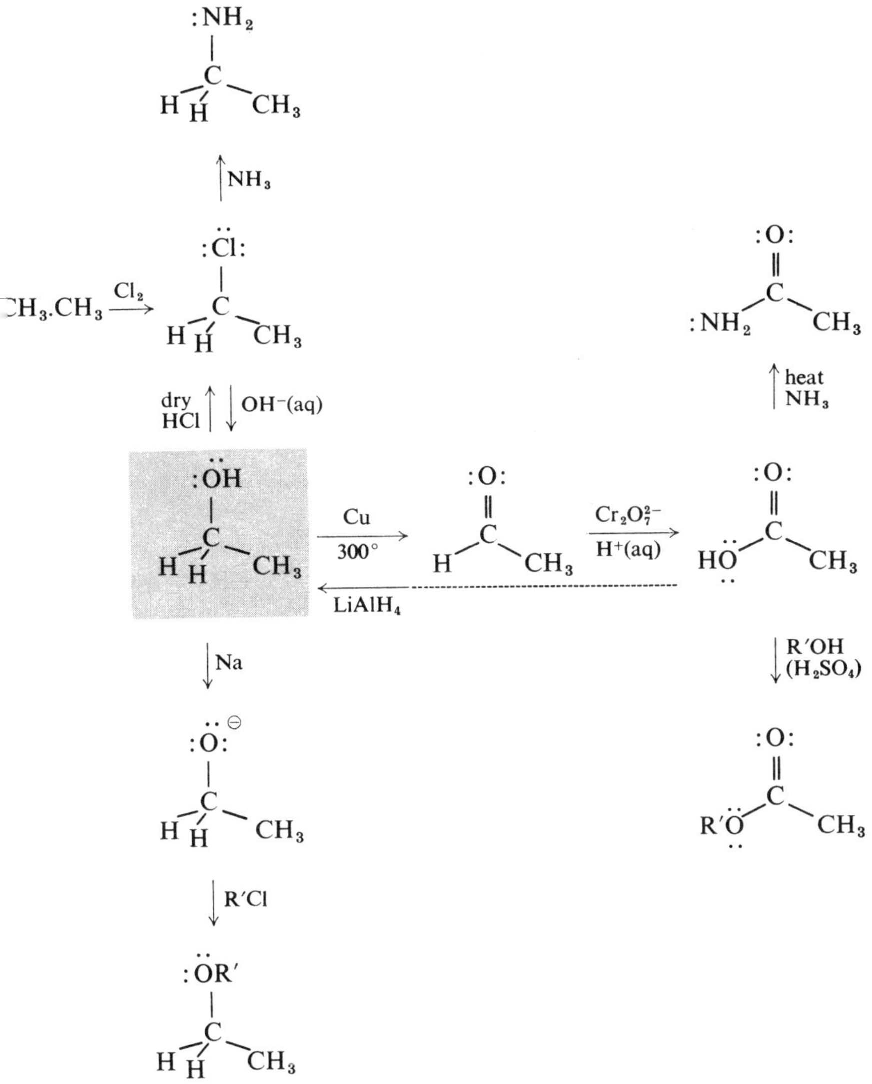

$:NH_2$ — C — H, H, CH_3

↑ NH_3

$:\ddot{Cl}:$ — C — H, H, CH_3

$CH_3.CH_3 \xrightarrow{Cl_2}$

dry HCl ↑ ↓ $OH^-(aq)$

$:\ddot{O}H$ — C — H, H, CH_3 $\xrightarrow[300°]{Cu}$ $:O:$ ‖ C — H, CH_3 $\xrightarrow[H^+(aq)]{Cr_2O_7^{2-}}$ $:O:$ ‖ C — $H\ddot{O}$, CH_3

← $LiAlH_4$ (dashed)

↓ Na

$:\ddot{O}:^{\ominus}$ — C — H, H, CH_3

↓ R′Cl

$:\ddot{O}R'$ — C — H, H, CH_3

↑ heat NH_3

$:O:$ ‖ C — $:NH_2$, CH_3

↓ R′OH (H_2SO_4)

$:O:$ ‖ C — $R'\ddot{O}$, CH_3

Fig 20.1

Summary: starting with the primary alcohol ethanol. Horizontal arrows show redox reactions, vertical arrows non-redox reactions. The scheme may be generalized by replacing the methyl group in ethanol with an R group.

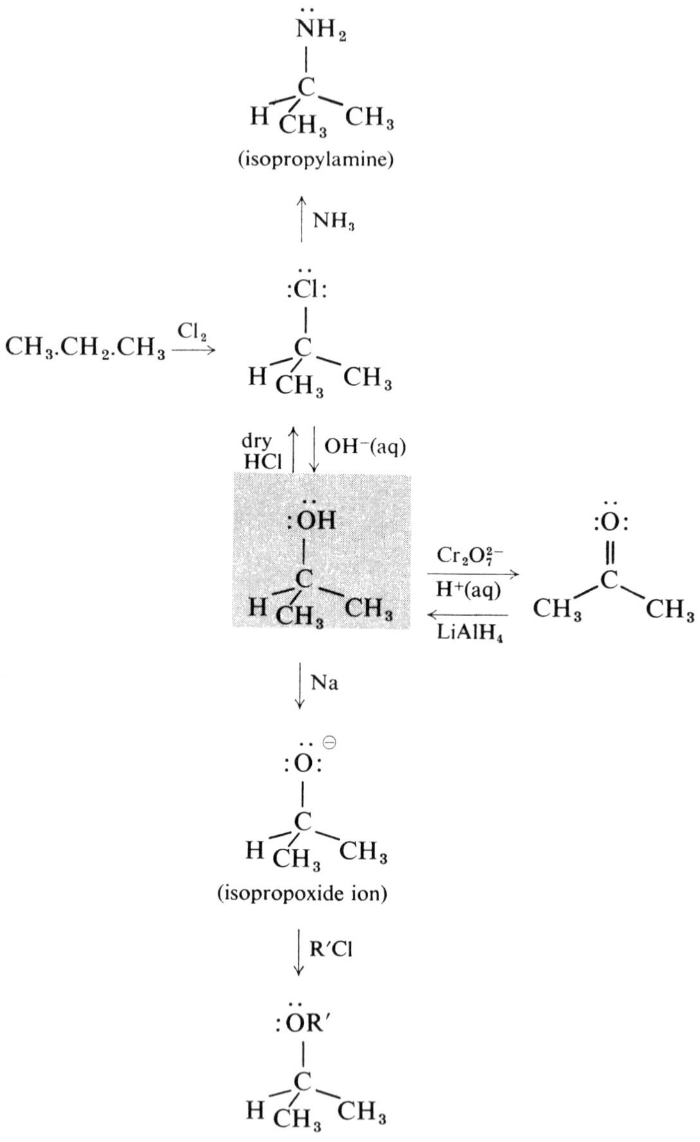

$\ddot{N}H_2$ — C(H)(CH_3)(CH_3)

(isopropylamine)

↑ NH_3

$:\ddot{Cl}:$ — C(H)(CH_3)(CH_3)

$CH_3.CH_2.CH_3 \xrightarrow{Cl_2}$

dry HCl ↑ ↓ $OH^-(aq)$

$:\ddot{O}H$ — C(H)(CH_3)(CH_3)

$\xrightarrow[H^+(aq)]{Cr_2O_7^{2-}}$ $CH_3C(=\ddot{O}:)CH_3$

$\xleftarrow{LiAlH_4}$

↓ Na

$:\ddot{O}:^{\ominus}$ — C(H)(CH_3)(CH_3)

(isopropoxide ion)

↓ R′Cl

$:\ddot{O}R'$ — C(H)(CH_3)(CH_3)

Fig 20.2

Summary: starting with the secondary alcohol isopropanol. Horizontal arrows show redox reactions, vertical arrows non-redox reactions. The scheme may be generalized by replacing the methyl groups in isopropanol with R groups.

from amino acids, which are molecules containing both an amino group and a carboxyl group. Two simple amino acids are:

$$CH_3.CH(NH_2).CO.OH \qquad NH_2.CH_2.CO.OH$$

alanine   glycine

Thus alanine can form a protein-like structure when the amino grouping of one molecule reacts with the carboxyl grouping of a neighbour, and a very long chain, or *polymer*, is built up. Thus with alanine we would obtain:

$$\begin{array}{c} \overset{\displaystyle CH_3}{\overset{|}{}}\overset{\displaystyle CH_3}{\overset{|}{}}\overset{\displaystyle CH_3}{\overset{|}{}} \\ —NH.CO.CH.NH.CO.CH.NH.CO.CH— \end{array}$$

Such a chain is called a *polypeptide* chain, and proteins consist of such chains, formed from one definite arrangement of the twenty-six naturally-occurring amino acids. Nylon is a general term for certain synthetic polypeptides which have found numerous commercial uses. A common form of Nylon is prepared by linking the following two molecules:

$$HO.CO(CH_2)_4.CO.OH \qquad \text{adipic acid}$$

$$NH_2.(CH_2)_6.NH_2 \qquad \text{hexamethylene diamine}$$

which gives:

$$—(CH_2)_4.CO.NH(CH_2)_6.NH.CO.(CH_2)_4.CO.NH(CH_2)_6NH.CO(CH_2)_4—$$

QUESTIONS

1. A compound has the formula C_3H_8O. Give the structural formulae and names of the *three* possible structures it could have.
2. A compound has the formula C_3H_6O. There are no carbon-carbon double bonds in the molecule. Give the possible structural formulae and names for the compound.
3. A compound has the formula $C_3H_6O_2$. There are no carbon-carbon double bonds in the molecule. Give the possible structural formulae and names for the compound.
4. Write structural formulae for:
   (a) ethyl chloride;
   (b) methanol;
   (c) isopropanol;
   (d) n-propanol;
   (e) n-butyl chloride;
   (f) s-butyl chloride.
5. Name the following molecules:
   $CH_3.CH_2.CH_2.OH$; CH_3CHO; $(CH_3)_2.CH.CHO$; $CH_3.CO.CH_3$; $H.CHO$; CH_3Cl; $CH_3.CH_2.CH_2Cl$; $CH_3.CH_2CHO$; $CH_3.CO.CH_2.CH_3$; $(CH_3)_2.CH.CH_3$; $CH_3.CH_2.CH_2.CHO$; $CH_3.CH_2.Cl$.

6. Using any non-organic reagents necessary, show with equations how you would prepare:
   (a) acetaldehyde from ethanol;
   (b) acetone from isopropanol;
   (c) formaldehyde from methanol;
   (d) methyl ethyl ketone from s-butanol.
7. Name the following compounds:
   $CH_3.CO.NH_2$; $CH_3.O.CH_3$; $CH_3.NH_2$; $CH_3.CH_2.O.CH_3$; $H.CO.NH_2$; $CH_3.CO.OH$; $CH_3.CO.O.CH_3$; $CH_2Cl.CO.O.CH_2CH_3$; $(CH_3)_2.CH.CO.O.CH(CH_3)_2$; $(CH_3)_2.CH.CO.OK$; $(CH_3.CO.O)_2Ca$; $CH_3.NH_3.Cl$; $CH_3.CH_2O.Cs$; $CH_3.CH_2.CH_2CO.OH$; $(CH_3)_2CH.OH$; CH_2Cl_2.
8. Using any non-organic reagents necessary show, with equations, how you would prepare:
   (a) propionamide from propionic acid;
   (b) methyl formate from methanol and formic acid;
   (c) ethylamine from ethyl chloride;
   (d) dimethyl ether from methanol;
   (e) methyl ethyl ether from methanol and ethanol;
   (f) propionic acid from propionaldehyde;
   (g) isobutyric acid from isobutanol;
   (h) sodium acetate from acetic acid;
   (i) potassium propionate from methyl propionate;
   (j) acetamide from acetic acid;
   (k) isobutyraldehyde from isobutanol.
9. Using any non-organic reagents necessary, show with equations how you would prepare:
   (a) ethanol from ethane;
   (b) propionaldehyde from propane;
   (c) acetone from propane;
   (d) propionic acid from propane;
   (e) ethyl chloride from ethyl acetate;
   (f) methyl formate from methanol;
   (g) acetamide from methyl acetate;
   (h) acetamide from ethyl formate;
   (i) methylamine from methane;
   (j) methylamine from methyl propionate;
   (k) diethyl ether from ethyl chloride;
   (l) ethyl acetate from ethane;
   (m) sodium formate from formaldehyde;
   (n) methylammonium chloride from methane;
   (o) sodium ethoxide from ethyl acetate;
   (p) methylammonium formate from methane.
10. (a) Define homology.
    (b) Write down structural formula for four successive aldehydes in a homologous series.
11. (a) Define structural isomerism.
    (b) Find which of the following pairs of molecules are isomers: propionaldehyde, methyl ethyl ether, diethyl ether, acetone, methyl formate, s-butanol, n-propanol, acetic acid.

12. Using any non-organic reagents necessary, show with equations how you would prepare:
    (a) propionaldehyde from propionic acid;
    (b) ethyl acetate from acetic acid;
    (c) di-n-propyl ether from propionaldehyde;
    (d) ethylammonium acetate from acetaldehyde;
    (e) propionamide from propyl bromide.

21

Mechanisms of Chemical Reactions

21.1 THE IDEA OF A REACTION MECHANISM

Usually, the first piece of quantitative information obtained about a chemical reaction is its stoichiometry, e.g. consider the gas phase reactions:

$$H_2 + I_2 \rightarrow 2HI \tag{1}$$

$$H_2 + Br_2 \rightarrow 2HBr \tag{2}$$

Next, it is generally a straightforward matter to measure the *rate* of a chemical reaction, and to determine its *rate law*. Reactions (1) and (2) above have rate laws:

$$R_1 = k_1[H_2][I_2]$$

$$R_2 = k_2[H_2][Br_2]^{\frac{1}{2}}$$

where R_1 and R_2 are the rates of production of the relevant hydrogen halide in M s^{-1}. The rate law shown for reaction (2) is obeyed during the initial stages of reaction between hydrogen and bromine. As the product HBr accumulates, it interferes in the reaction; a more complicated rate law is thus necessary to describe the reaction rate in systems containing appreciable amounts of HBr. For simplicity then, we shall assume that in the systems we are studying the reaction products are being continuously removed from the system.

One of the chemist's tasks is to account for the difference between the rate laws for such similar reactions as (1) and (2). To explain the difference, it is

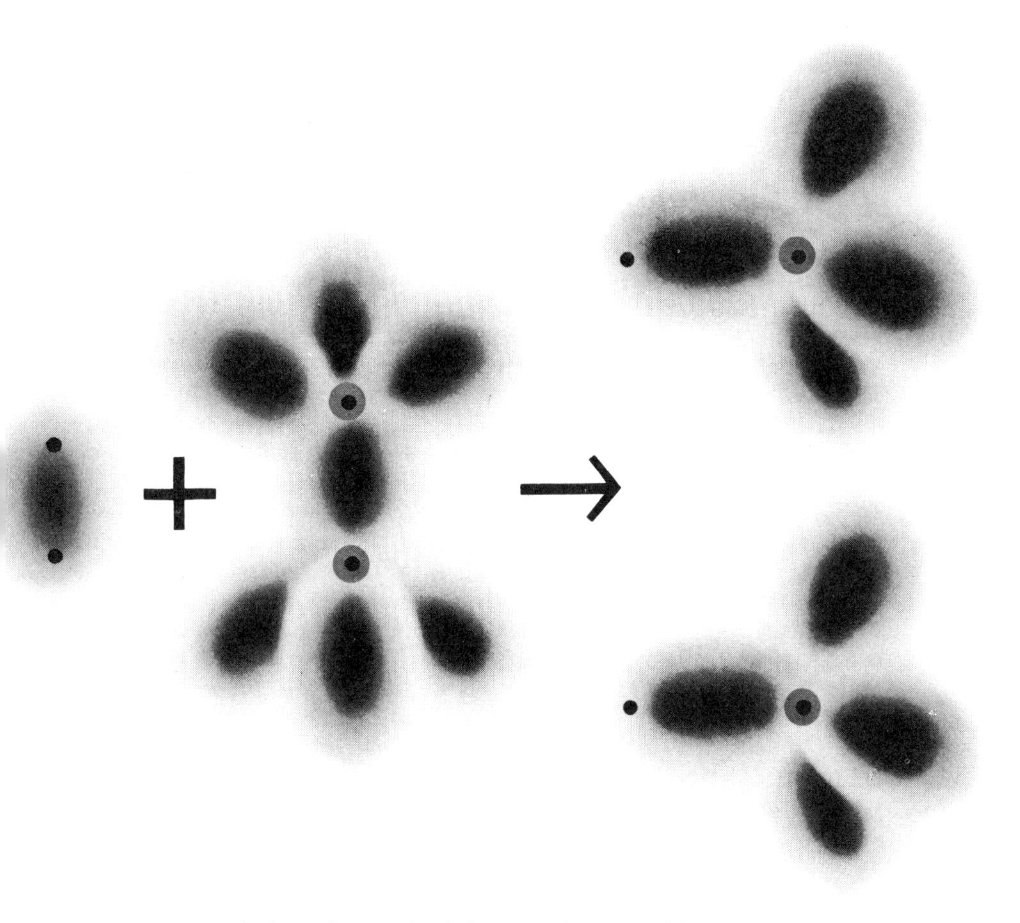

Fig 21.1 Charge cloud diagrams of H_2, I_2 and HI.

necessary to consider the mechanisms of the two reactions. Every chemical reaction involves a rearrangement of atoms; charge cloud diagrams of the reactants and products in reaction (1) are shown in Figure 21.1, and it can be clearly seen that the molecules of hydrogen and iodine must in some way split apart so that two molecules of hydrogen iodide can form. The precise way in which the atoms of the reactant molecules rearrange themselves to give the product molecules is called the 'mechanism' of the reaction. In the next section we will discuss in some detail just how the required rearrangement of atoms might occur in the case of one particular simple reaction.

21.2 COLLISIONS BETWEEN MOLECULES

The first and most important item of information about a reaction, which must be known before a completely satisfactory reaction mechanism can be postulated, is its rate law. One particularly simple reaction is that between nitrogen dioxide, NO_2, and carbon monoxide, CO, to give nitric oxide, NO, and carbon dioxide, CO_2, i.e.

$$CO(g) + NO_2(g) \rightarrow CO_2(g) + NO(g)$$

The rate law for this reaction, at temperatures around 400°C, is

$$R = k[CO][NO_2]$$

and it is this simple fact which suggests to us a mechanism for the reaction.

In the gaseous state, carbon monoxide and nitrogen dioxide both exist as large numbers of independent molecules in continuous random motion. We have already seen in chapter 3 that the general properties of molecules in the gas phase are adequately described by the kinetic theory of gases. One consequence of the kinetic theory is that the pressure exerted by a gas is due to the collisions of individual gas molecules with the walls of the containing vessel. Furthermore, at a given temperature, the total number of wall-collisions per unit time is proportional to the number of gas molecules per unit volume of gas, i.e. the number of wall-collisions per unit time is proportional to the partial pressure, and hence to the molar concentration of the gas. As well as colliding with walls, gas molecules also collide with each other. Each molecule of a gas has a definite volume, and hence a definite cross-sectional area, and it can be shown that the number of collisions undergone by a given gas molecule per unit time is also proportional to the partial pressure and hence the molar concentration of the gas. For a mixture of two different gases, say carbon monoxide and nitrogen dioxide, it can be shown that the number of collisions per unit time between, in this case, carbon monoxide and nitrogen dioxide molecules, $Z_{CO\text{-}NO_2}$, is given by

$$Z_{CO\text{-}NO_2} \propto [CO][NO_2]$$

Using the known masses and sizes of carbon monoxide and nitrogen dioxide molecules, the kinetic theory of gases tells us that, at 400°C,

$$Z_{CO\text{-}NO_2} = 5 \times 10^{35}[CO][NO_2] \text{ collisions dm}^{-3}\text{ s}^{-1}$$

This result is based solely on the assumption that gaseous carbon monoxide and nitrogen dioxide behave as ideal gases.

Let us now postulate that the chemical reaction which occurs between CO and NO_2 to yield CO_2 and NO requires that a CO and an NO_2 molecule should come into collision so that an oxygen atom can be transferred from being bonded to nitrogen (in NO_2) to being bonded to carbon (in CO_2). If this be so, we would expect the reaction rate to be proportional to the number of CO-NO_2 collisions per unit time, i.e. we would expect

$$R \propto Z_{CO\text{-}NO_2}$$

Hence

$$R \propto [CO][NO_2]$$

which is of the same form as the experimentally-observed rate law. We shall therefore tentatively conclude that the CO-NO_2 reaction proceeds by collisions between CO and NO_2 molecules in which an oxygen atom is transferred from NO_2 to CO, given NO and CO_2 as products. This is an example of a reaction proceeding by a single simple collisional step. A step such as this is described as a *bimolecular* step, since it involves the collision of two molecules.

21.3 EFFECTS OF TEMPERATURE ON A SIMPLE REACTION

If we accept the one-step mechanism

$$CO + NO_2 \rightarrow CO_2 + NO$$

for the CO-NO_2 reaction, we must explain why it is that not every collision between CO and NO_2 leads to reaction. Thus if $[CO] = [NO_2] = 0.01$ M the reaction rate is 1.74×10^{-5} M s^{-1} at 360°C. Yet if every collision between the molecules were to lead to reaction, the rate would be 1.2×10^8 M s^{-1} at this temperature. This means that only 1 in 7×10^{12} collisions between CO and NO_2 is effective in bringing about reaction at 360°C. If we raise the temperature to 465°C, the reaction rate becomes 5.9×10^{-4} M s^{-1} with the same reactant concentrations, whereas if every collision were effective, the rate would be 1.3×10^8 M s^{-1}. At this higher temperature, however, the fraction of collisions which is 'effective' has risen appreciably to 1 in 2×10^{11}. Some more data of the 'collision efficiency' of the reaction are given in Table 21.1.

Table 21.1

$[CO] = [NO_2] = 0.01$ M

T *(K)*	*Rate constant*, k_T *(M^{-1} s^{-1})*	*Rate assuming all collisions lead to reaction (M s^{-1})*	*Actual rate (M s^{-1})*	*Proportion of collisions leading to reaction*
633	0.174	1.19×10^8	0.174×10^{-4}	1 in 70×10^{11}
667	0.76	1.22×10^8	0.76×10^{-4}	1 in 20×10^{11}
710	2.51	1.26×10^8	2.51×10^{-4}	1 in 5×10^{11}
738	5.88	1.29×10^8	5.88×10^{-4}	1 in 2×10^{11}

These data can be understood if we assume that the colliding CO and NO_2 molecules must possess a good deal of excess energy before they can react. This is reasonable, since in order that an oxygen atom may be transferred between the two colliding molecules, a good deal of distortion of the original molecule is necessary in order to break an N-O bond and form a C-O bond. The energy necessary to effect this change can only come from the kinetic energy of the molecules taking part in the collision.

We saw in chapter 3 that the fraction of molecules in a gas possessing a high energy rises rapidly as the temperature of the gas increases. If only high-energy molecules are responsible for chemical reaction between CO and

NO_2, the marked effect of temperature on their reaction rate can be immediately understood; indeed, the greatly increased rates at higher temperatures simply reflect the greater proportion of high-energy molecules present in a gas at high temperatures.

For a simple bimolecular reaction such as the CO-NO_2 reaction, the amount of energy required for an 'effective' collision can be estimated from experimentally-determined rate constants. For any homogeneous reaction with rate constant k_T at T K, a plot of $\log k_T$ against T^{-1} is usually a straight line. When this is so, the slope of the straight line so obtained is said to be $(-E_a/2.303R)$ where R is the gas constant, and E_a is called the *activation energy* for the reaction in question. For a single-step reaction, such as our CO-NO_2 example, it is believed that E_a is very nearly equal to the amount of energy that is necessary in order that a collision should be 'effective'. For the CO-NO_2 reaction, $E_a = 132$ kJ mol^{-1}.

EXERCISE
From the data in Table 21.1 determine E_a for the reaction $CO + NO_2 \rightarrow CO_2 + NO$. Plot the logarithm of the rate constants (k_T) in the second column, against the reciprocal of the absolute temperature. The slope of the straight line is $(-E_a/2.303)$.

21.4 SOME REACTION MECHANISMS

(i) The reaction $CH_3Cl + OH^- \rightarrow CH_3OH + Cl^-$ occurs in aqueous solution, and has the rate law

$$R = k[CH_3Cl][OH^-]$$

It is therefore reasonable to suggest that the reaction proceeds in a simple collisional step. Since the reaction is in solution, it will not be as simple as the collisional step in the gas phase we have just discussed. In solution, all ionic species are solvated, and the solvent plays a most important role in determining the actual rates of reactions in solution.

(ii) The reaction $H_2 + Br_2 \rightarrow 2HBr$ occurs in the gas phase, and has the rate law

$$R = k[H_2][Br_2]^{\frac{1}{2}}$$

It is apparent that this reaction cannot have the same mechanism as the CO-NO_2 reaction, since such a mechanism would lead to the simple rate law

$$R = k[H_2][Br_2]$$

We must therefore ask ourselves what mechanism could give rise to the rate law

$$R = k[H_2][Br_2]^{\frac{1}{2}}$$

We can find out only by trial and error, i.e. by suggesting reaction mechanisms, and testing them to see if they are consistent with the observed rate law. It turns out that the only reasonable suggestion is the three-step mechanism

$$Br_2 \rightleftharpoons Br + Br \qquad (1)$$

$$Br + H_2 \rightarrow HBr + H \qquad (2)$$

followed almost immediately by

$$H + Br_2 \rightarrow HBr + Br \qquad (3)$$

i.e. the mechanism we are suggesting is that the bromine atoms in equilibrium with molecular bromine, which are present in very low concentrations, react at a definite rate with hydrogen molecules to yield HBr molecules and hydrogen atoms. The H atoms thus formed then react virtually instantaneously with other bromine molecules, giving HBr molecules and regenerating bromine atoms. The nett effect therefore is that the small equilibrium concentration of bromine atoms in the system 'catalyses' the formation of HBr as shown in steps (2) and (3). The sum of steps (2) and (3) is simply

$$Br + H_2 + Br_2 \rightarrow 2HBr + Br$$

i.e. the bromine atoms simply 'catalyse' the reaction. Charge cloud pictures of steps (1), (2) and (3) are shown in Figure 21.2.

Let us now verify that this mechanism is consistent with the observed rate law. The rate of production of HBr, i.e. the reaction rate, will be proportional to the rate of step (2)

$$R \propto [H_2][Br]$$

since this step is directly responsible for the production of HBr. However, we know from the equilibrium law

$$K = \frac{[Br]^2}{[Br_2]}$$

$$\therefore\ [Br] = [Br_2]^{\frac{1}{2}} \times K^{\frac{1}{2}}$$

$$\text{i.e. } [Br] \propto [Br_2]^{\frac{1}{2}}$$

Therefore

$$R \propto [H_2][Br_2]^{\frac{1}{2}}$$

which is in agreement with the observed rate law.

(iii) The reaction $H_2 + I_2 \rightarrow 2HI$ occurs in the gas phase, and has the rate law

$$R = k[H_2][I_2]$$

It was long thought that this reaction proceeded via a direct collision of hydrogen and iodine, similar to the CO-NO_2 reaction discussed earlier. More recently however, a careful investigation of the reaction has shown that the mechanism is more probably

$$I_2 \rightleftharpoons 2I \qquad (1)$$

$$I + H_2 \rightleftharpoons H_2I \qquad (2)$$

$$I + H_2I \rightarrow 2HI \qquad (3)$$

That this mechanism is consistent with the observed rate law may be shown

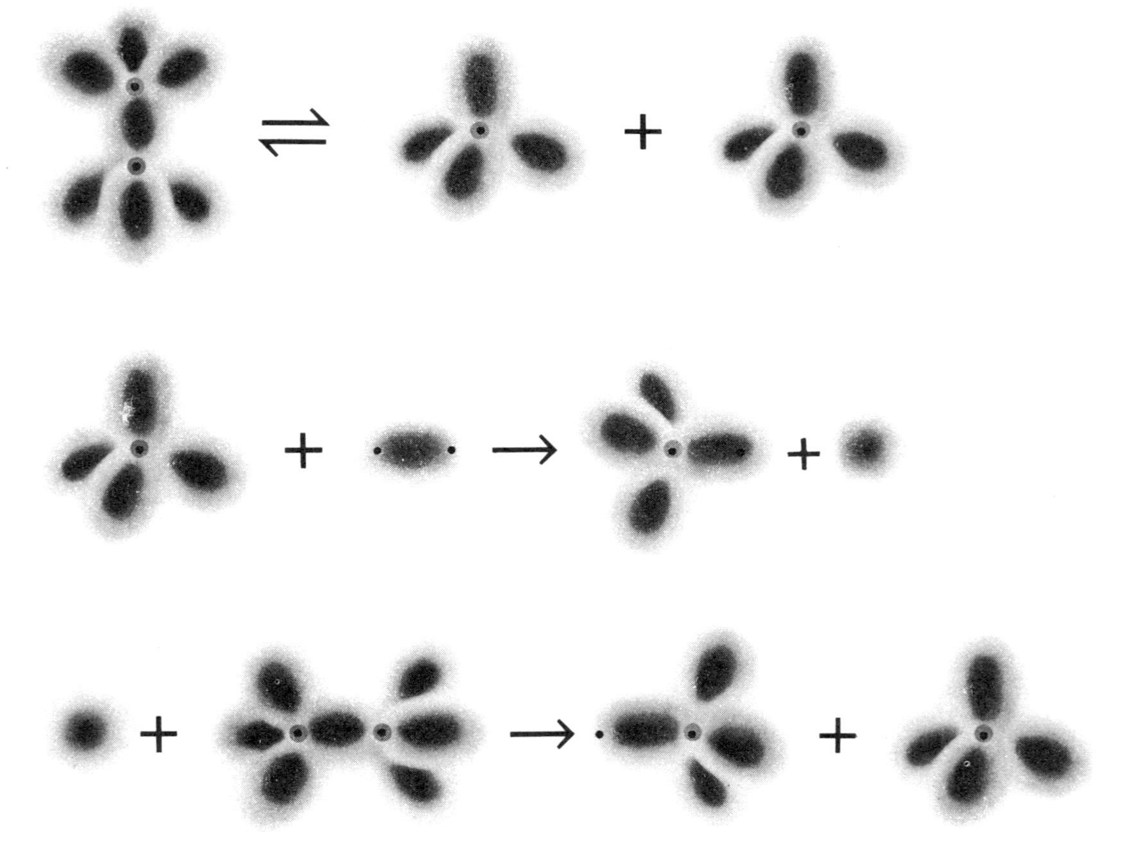

Fig 21.2 Charge cloud pictures of the steps in the mechanism of the reaction $H_2 + Br_2 \rightarrow 2HBr$.

by noting that the overall reaction rate is proportional to the rate of step (3), i.e.

$$R \propto [H_2I][I]$$

Steps (1) and (2) are equilibria, so

$$K_1 = \frac{[I]^2}{[I_2]} \text{ and } K_2 = \frac{[H_2I]}{[I][H_2]}$$

It follows that $[I] = \sqrt{K_1[I_2]}$

and

$$[H_2I] = K_2[H_2][I]$$
$$= K_2[H_2]\sqrt{K_1[I_2]}$$
$$\therefore R \propto [H_2I][I] \propto K_2[H_2]\sqrt{K_1[I_2]}.\sqrt{K_1[I_2]}$$
$$\propto K_1K_2[H_2][I_2]$$

which is in agreement with the observed rate law.

One obvious question which arises at this point is why do the H_2–Br_2 and the H_2–I_2 reactions go by different mechanisms? This is a question we cannot answer satisfactorily in this book, but it should be realised that *both* mechanisms are available to each reaction. Apparently one halogen atom is sufficient to bring about HBr formation in the H_2-Br_2 reaction, whereas two halogen atoms are necessary to bring about the HI formation in the H_2-I_2 case. This does *not* mean that a few HI molecules will not be formed from *one* single iodine atom in the H_2-I_2 reaction, nor that some HBr molecules will not be formed from *two* bromine atoms in the H_2-Br_2 case. All we can say is that one mechanism dominates in H_2-I_2, and another dominates in H_2-Br_2.

21.5 CONCLUSIONS

It would be beyond the scope of this text to embark on an exhaustive examination of reaction mechanisms. In this chapter, the only evidence we have used in discussing reaction mechanisms has been the reaction rate law. This is a most important fact to have, but many other pieces of evidence are necessary in order to arrive at the mechanisms of the majority of complex reactions. The reactions we have discussed have all had fairly simple mechanisms. Two have had single-step mechanisms involving a collision between two entities (*bimolecular collisions*). Such reaction steps are called bimolecular steps. These reactions were:

$$CO + NO_2 \rightarrow CO_2 + NO \quad \text{(gas phase)}$$

$$OH^- + CH_3Cl \rightarrow CH_3OH + Cl^- \quad \text{(solution)}$$

The other examples chosen had three-step mechanisms. In each case, reaction was initiated by the initial decomposition of a single molecule (Br_2 and I_2) in what is called a *unimolecular step*, followed by two steps leading to production of the hydrogen halide.

A large number of the reactions described in this text have quite complicated mechanisms—indeed in many cases, reaction mechanisms are unknown. It is possible, however, to make one quite useful generalization about reaction mechanisms, which is:

Almost all reactions proceed by a series of one or more uni- or bimolecular steps.

Steps involving a collision between three entities (*termolecular steps*) are extremely rare, and no reaction is known in which collisions between four or more entities are important in propagating the reaction.

One final point worth making is that the mode of operation of catalysts for chemical reactions can usually be understood in terms of reaction mechanism. A catalyst operates by making available an 'easy' reaction mechanism whereby the reactants can be converted to products.

QUESTIONS

1. It is sometimes said that the rates of most simple chemical reactions 'increase by a factor of 2 or 3 for every 10° rise in temperature'. Suggest reasons why reaction rate rises so rapidly with increasing temperature.
2. The reaction $H_2 + Cl_2 \rightarrow 2HCl$ has the rate law $R = k[H_2][Cl_2]^{\frac{1}{2}}$. A student has suggested that the mechanism is

$$H_2 \rightleftharpoons 2H$$

$$Cl_2 \rightleftharpoons 2Cl$$

$$H + Cl \rightarrow HCl$$

   Show that this mechanism is impossible by verifying that it would lead to the rate law

$$R \propto [H_2]^{\frac{1}{2}}[Cl_2]^{\frac{1}{2}}$$

   Suggest an alternative mechanism for the reaction.
3. A student has suggested the following mechanism for the reaction $H_2 + Br_2 \rightarrow 2HBr$:

$$H_2 \rightleftharpoons 2H$$

$$H + Br_2 \rightarrow HBr + Br$$

   followed very rapidly by

$$Br + H_2 \rightarrow HBr + H$$

   Show that this suggested mechanism is inconsistent with the observed rate law for this reaction by verifying that it leads to the rate law

$$R \propto [H_2]^{\frac{1}{2}}[Br_2]$$

4. The reaction $CH_3I + OH^- \rightarrow CH_3OH + I^-$ in aqueous solution has the rate law

$$R = k[CH_3I][OH^-]$$

   Suggest a mechanism for this reaction.

22

The Periodic Table

As the result of the investigations of many chemists it was possible, in 1869, for the Russian chemist, D. Mendeleef, and the German chemist, L. Meyer, to announce a very far-reaching generalization about the physical and chemical properties of the elements, viz. that if the elements are arranged in order of atomic weights, elements with similar properties recur at regular intervals, or periodically. In the first decade of this century the basis for ordering the elements was changed from atomic weight to atomic number and it is this fundamental criterion which we still use. Mendeleef was able to express this periodicity in the form of a table which grouped elements with similar properties together. Such a table is called a *periodic table*.

This discovery of the periodic variation in properties was empirical but investigation of atomic structure and bonding theory has revealed its theoretical basis. We now realize that the atomic number dictates the number of electrons in an atom and it is the periodic recurrence of similar electron configurations which leads to a periodic recurrence of chemical properties. We cannot, in our present state of knowledge, deduce all of the vast number of observed properties of the elements from their known electron configurations. We can, however, correlate a great many experimentally observed properties with electron configuration, and it is in this way that the periodic table provides a valuable framework for the study of the elements.

22.1 ATOMIC STRUCTURE AND THE PERIODIC TABLE

It was seen in chapter 6 that if the atomic number of an element is known, the electron configuration of the atom can be derived by imagining a process in which electrons are successively added about the nucleus of an atom of the element, so that each successive electron goes into the subshell of lowest

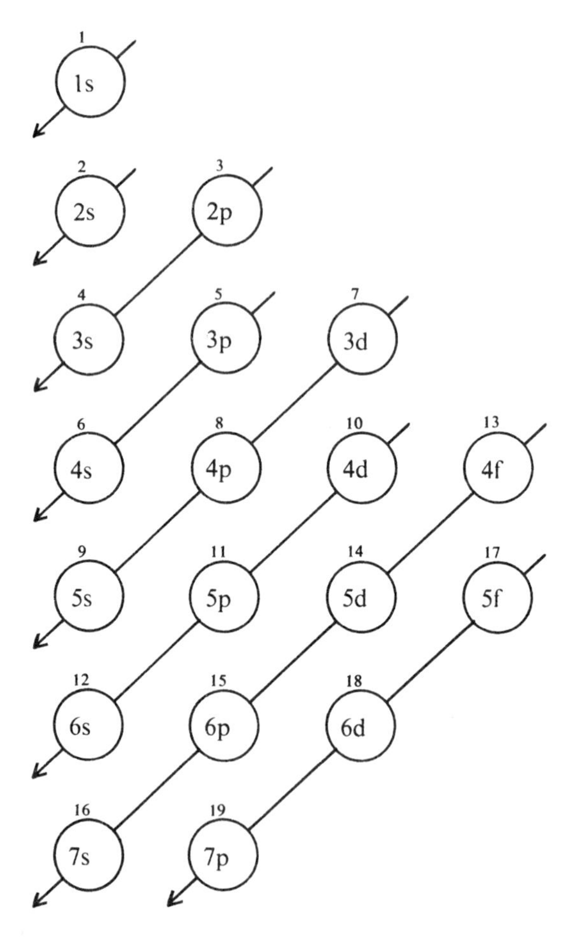

Fig 22.1 The order of filling of orbitals.

energy available. The order of subshells, in terms of increasing energy, is an experimentally determined sequence, and for the majority of elements is given by the order shown in Figure 22.1.

If the electron configurations of the known elements are written down we can recognize sets of elements, each with some common feature. Thus lithium ($1s^22s^1$), sodium ($1s^22s^22p^63s^1$) and potassium ($1s^22s^22p^63s^23p^64s^1$) have the common feature that their subshell of highest energy is a half-filled s-subshell. All of these elements have atoms with one outer-shell electron. In a similar way we find that fluorine ($1s^22s^22p^5$), chlorine ($1s^22s^2 2p^63s^23p^5$) and bromine ($1s^22s^22p^63s^23p^63d^{10}4s^24p^5$) all have seven outer-shell electrons (s^2p^5).

It is possible to arrange the elements in horizontal rows (called *periods*) in order of increasing atomic number, in such a way that elements with

I	II											III	IV	V	VI	VII	VIII
1 H																	2 He
3 Li	4 Be											5 B	6 C	7 N	8 O	9 F	10 Ne
11 Na	12 Mg											13 Al	14 Si	15 P	16 S	17 Cl	18 Ar
19 K	20 Ca	21 Sc	22 Ti	23 V	24 Cr	25 Mn	26 Fe	27 Co	28 Ni	29 Cu	30 Zn	31 Ga	32 Ge	33 As	34 Se	35 Br	36 Kr
37 Rb	38 Sr	39 Y	40 Zr	41 Nb	42 Mo	43 Tc	44 Ru	45 Rh	46 Pd	47 Ag	48 Cd	49 In	50 Sn	51 Sb	52 Te	53 I	54 Xe
55 Cs	56 Ba	57 La	72 Hf	73 Ta	74 W	75 Re	76 Os	77 Ir	78 Pt	79 Au	80 Hg	81 Tl	82 Pb	83 Bi	84 Po	85 At	86 Rn
87 Fr	88 Ra	89 Ac															

58 Ce	59 Pr	60 Nd	61 Pm	62 Sm	63 Eu	64 Gd	65 Tb	66 Dy	67 Ho	68 Er	69 Tm	70 Yb	71 Lu
90 Th	91 Pa	92 U	93 Np	94 Pu	95 Am	96 Cm	97 Bk	98 Cf	99 Es	100 Fm	101 Md	102 No	103 Lw

Fig 22.2 A form of the periodic table.

similar electron configurations fall into the same vertical columns (called *groups*). Such an arrangement is a periodic table. There are a number of such tables but one useful form is that shown in Figure 22.2.

22.2 THE STRUCTURE OF THE PERIODIC TABLE

We will now see how the classification shown in Figure 22.2 groups elements with similar electron configurations together.

(a) In the first period (H to He) the 1s subshell is filled. An s subshell contains one orbital and this period therefore contains two elements only.

(b) In the second period (Li to Ne) the 2s subshell is filled first and then the 2p subshell. A p subshell contains three orbitals and hence the total number of elements in this period is eight.

(c) In the third period (Na to Ar) the 3s subshell is filled first and then the 3p subshell fills. The total number of elements in this period is eight. In the first two periods the completion of a period coincided with the filling of the K- and L-shells, respectively. In the third period the M-shell is not filled completely. The M-shell contains the 3s, 3p and 3d subshells. A d subshell contains five orbitals, hence the M-shell contains nine orbitals (eighteen electrons). In all of the remaining periods we also find that the completion of a period does not mean that all the shells involved have been filled completely.

(d) In the fourth period (K to Kr) we find that the 4s subshell has lower energy than the 3d subshell. Hence for the first two elements (K, Ca) the 4s subshell is filled. Then the 3d subshell fills. This involves ten elements (Sc to Zn) and these are known as *transition elements*. Notice that with completion of the 3d subshell the M-shell is now filled. The 4p subshell is then filled to complete the period. The total number of elements in this period is eighteen.

(e) In the fifth period (Rb to Xe) a similar process occurs to that found in the fourth period. The 5s, 4d and 5p orbitals are filled in succession. We find a second series of transition elements, corresponding to the filling of the 4d subshell. The total number of elements in this period is eighteen.

(f) In the sixth period (Cs to Rn) the 6s subshell fills first. In the next element (lanthanum) an electron enters the 5d subshell, and then in the next element (cerium) an electron enters the 4f subshell. An f subshell contains seven orbitals. The group of fourteen elements involved in the filling of the 4f subshell (Ce to Lu) is known as the *rare earth elements* or *lanthanides*. They are very similar chemically and for convenience are classified as a series outside the body of the table. With completion of the 4f subshell the 5d subshell now fills to yield a third series of transition elements, and then the 6p subshell fills to complete the period. The total number of elements in this period is thirty-two.

(g) In the seventh period (Fr to Lw) the 7s subshell fills first. Thereafter a somewhat similar situation to that found in the sixth period is believed to exist. The determination of the sequence of subshell filling is difficult with these elements, but a series of fourteen elements (Th to Lw) is grouped as a second rare earth series (also known as the *actinides*). This series, like the first, is classified outside the body of the table. The period is incomplete and contains only seventeen elements.

The periodic table we have discussed has been arranged so that elements with similar electron configurations fall into the same groups. There are eight groups in which elements are involved in the filling of s or p subshells. These groups are usually represented by roman numerals from I to VIII. The transition elements, in which d subshells are filling, are placed as a block between groups II and III. The lanthanides and actinides, in which f subshells are filling are classified outside the body of the table.

Group I elements (Li to Fr) have one outer-shell electron (s^1). These elements are often called the *alkali metals*. Group II elements (Be to Ra) have two outer-shell electrons (s^2). These elements are often called the *alkaline earth metals*. Group VII elements (F to At) have seven outer-shell electrons (s^2p^5). These elements are often called the *halogens*. Group VIII elements (He to Rn) have eight outer-shell electrons (s^2p^6), except helium which has two (s^2). These elements are often called the *rare* or *noble gases*. Hydrogen with one valence electron shows so many unique properties that it is best not to classify it with any particular group. The outer-shell structures of the elements in the eight groups are shown in Table 22.1.

As it has been pointed out that electron configurations recur in a periodic

Table 22.1
The outer-shell structures of group elements

Group	*Elements*	*Number of outer-shell electrons*	*Electronic structure of outer-shell electrons*
	H	1	s^1
I	Li, Na, K, Rb, Cs, Fr	1	s^1
II	Be, Mg, Ca, Sr, Ba, Ra	2	s^2
III	B, Al, Ga, In, Tl	3	s^2p^1
IV	C, Si, Ge, Sn, Pb	4	s^2p^2
V	N, P, As, Sb, Bi	5	s^2p^3
VI	O, S, Se, Te, Po	6	s^2p^4
VII	F, Cl, Br, I, At	7	s^2p^5
VIII	Ne, Ar, Kr, Xe, Rn	8	s^2p^6
	He	2	s^2

manner we might expect properties to show a similar periodic recurrence. This is in general found to be so, but the periodicity is not always obvious. In subsequent chapters we will look for similarities in the elements within a group (remember that they have similar electron configurations) and also gradations across periods (remember that there are regular changes in electron configurations).

22.3 SOME ATOMIC PROPERTIES AND THE PERIODIC TABLE

Many characteristic atomic properties show periodic variation with respect to atomic number. We will now investigate some of these properties and see how this periodic variation is shown within the periodic table.

(1) Ionization energy. This is discussed in detail in chapter 6. In Figure 22.3 the first ionization energy of each element has been plotted as a function of atomic number.

The periodic variation in this property is quite clearly shown by the form of the graph. Notice that within a period the value of the first ionization energy tends to rise as we move from a group I element to a group VIII element. This trend may be at least partially understood in terms of increasing 'core charge' of the group elements within a given period. The 'core charge' increases from +1 to +8 across all periods and hence we would expect a greater amount of energy to be required to remove an electron from a group VIII element than from a group I element. This general trend is found but we can see that the explanation is not complete, because the first ionization energy of boron is less than that of beryllium, and the first ionization energy of oxygen is less than that of nitrogen.

EXERCISE

Write down the electron configuration (using the subshell notation) of beryllium, boron, nitrogen and oxygen. Are there any features in these electron configurations which might correlate with the reversal of trends in the values of the ionization energies?

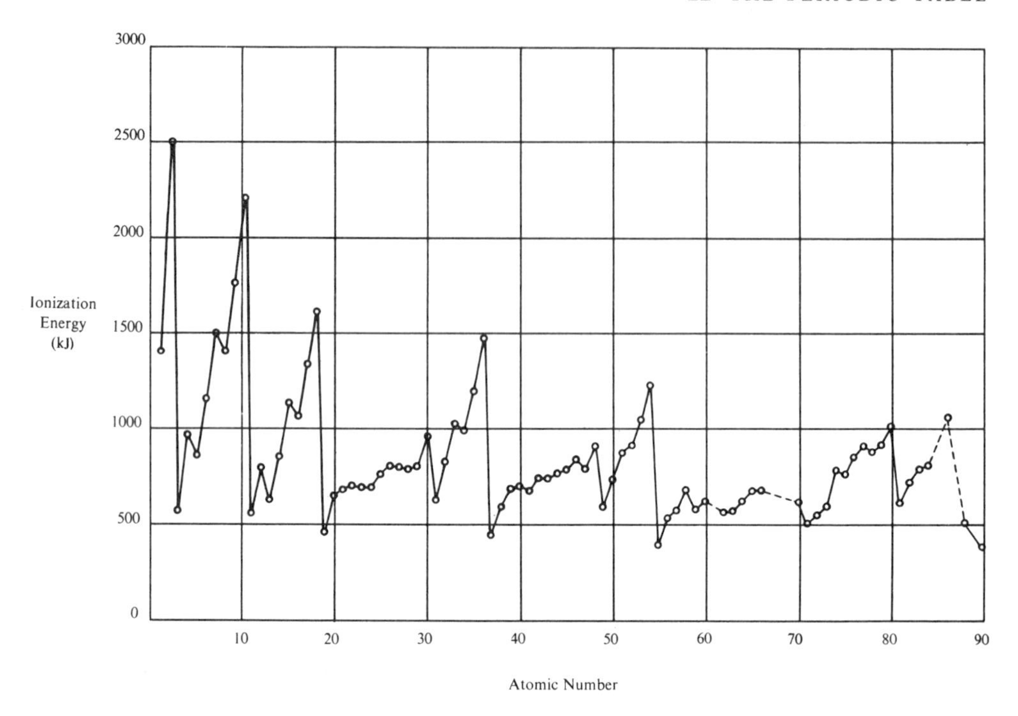

Fig 22.3 Variation in first ionization energy with atomic number.

Within a group the first ionization energy decreases down the group. This trend is probably due to the increasing size of the atoms. As the outer-shell electrons on the average get further away from the nucleus we would expect that less energy would be required to remove an electron from the outer shell. Combination of the trends within periods and trends within groups would lead one to predict caesium to have the lowest ionization energy and helium to have the highest ionization energy. This is experimentally observed to be true.

(2) Electron affinity. The electron affinity of an atom is the energy released when an electron is added to a gaseous atom. The experimental determination of electron affinities is difficult. Values have been determined by mass spectrometry for a few elements and in other cases values have been assigned indirectly. Notice that the electron affinity of an atom will be equal to the ionization energy of the appropriate negative ion. Some values for electron affinities are listed in Table 22.2.

It is difficult to draw generalizations from the limited data available, but it appears that electron affinities tend to increase across a period as the atomic number increases. Trends within groups are not clear. Notice that high values tend to be shown by the non-metallic elements.

(3) Electronegativity. The electronegativity of an atom is a numerical measure

Table 22.2

Electron affinities of some elements (kJ mol^{-1})

H 72.4					
	B 31.8	C 120	N 4.8	O 142	F 349
			P 74.3	S 207	Cl 364
					Br 343
					I 314

of the electron-attracting power of an atom. This is discussed in detail in chapter 8. A set of electronegativities based on Pauling's scale is given in Figure 22.4.

This same set of electronegativity values is presented graphically in Figure 22.5, where the electronegativities have been plotted as a function of atomic number. Notice the general trends in electronegativity values. The values decrease down a group and increase across a period. The most electronegative elements are those in the top right-hand corner of the periodic table, and the least electronegative are those in the bottom left-hand corner.

In chapter 11 electronegativity was correlated with the structure of elements and compounds. It is useful to summarize these correlations:

(a) elements with low electronegativity (and low ionization energy and electron affinity) tend to be metals, e.g. K, Mg;

(b) elements with high electronegativity (and high ionization energy and electron affinity) tend to be non-metals, e.g. F, O;

H 2.1							He –
Li 1.0	Be 1.5	B 2.0	C 2.5	N 3.0	O 3.5	F 4.0	Ne –
Na 0.9	Mg 1.2	Al 1.5	Si 1.8	P 2.1	S 2.5	Cl 3.0	Ar –
K 0.8	Ca 1.0	Ga 1.6	Ge 1.8	As 2.0	Se 2.4	Br 2.8	Kr –
Rb 0.8	Sr 1.0	In 1.7	Sn 1.8	Sb 1.9	Te 2.1	I 2.5	Xe –
Cs 0.7	Ba 0.9	Tl 1.8	Pb 1.8	Bi 1.9	Po 2.0	At 2.2	Rn –
Fr 0.7	Ra 0.9						

Fig 22.4 Electronegativity values for main-group elements.

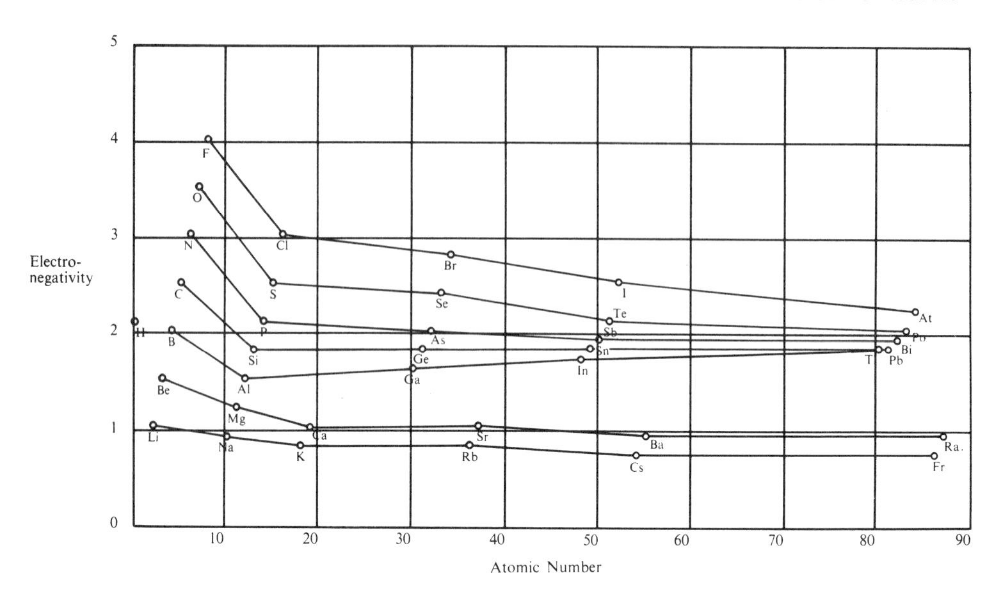

Fig 22.5 Variation in electronegativity with atomic number.

(c) in a compound a high electronegativity difference between constituent elements tends to result in bonding predominantly ionic in character, e.g. NaF;

(d) in a compound a low electronegativity difference between elements of high electronegativity tends to result in bonding predominantly covalent in character, e.g. SiC;

(e) a low electronegativity difference between elements of low electronegativity tends to indicate bonding predominantly metallic in character, e.g. copper-zinc (brass).

(4) Sizes of atoms and ions. The assigning of a meaningful, quantitative measure for the sizes of atoms and ions bristles with difficulties. However, radii have been assigned to atoms and ions, and provided these are used carefully they can be useful in interpreting chemical properties. Some atomic and ionic radii are set out in Figure 22.6.

Inspection of Figure 22.6 shows some general trends:

(a) Atomic radii increase in value down a group, and the radii of ions of like charge also increase in value down a group. This trend is to be expected as the outer-shell electrons get further away from the nucleus as we pass down a group.

(b) Atomic radii tend to decrease across a period. This trend may be correlated with the fact that the atomic number is rising whilst a given outer shell is filling (ignoring the intrusion of the transition and rare earth elements).

(c) The ionic radius of a positive ion is less than the corresponding atomic

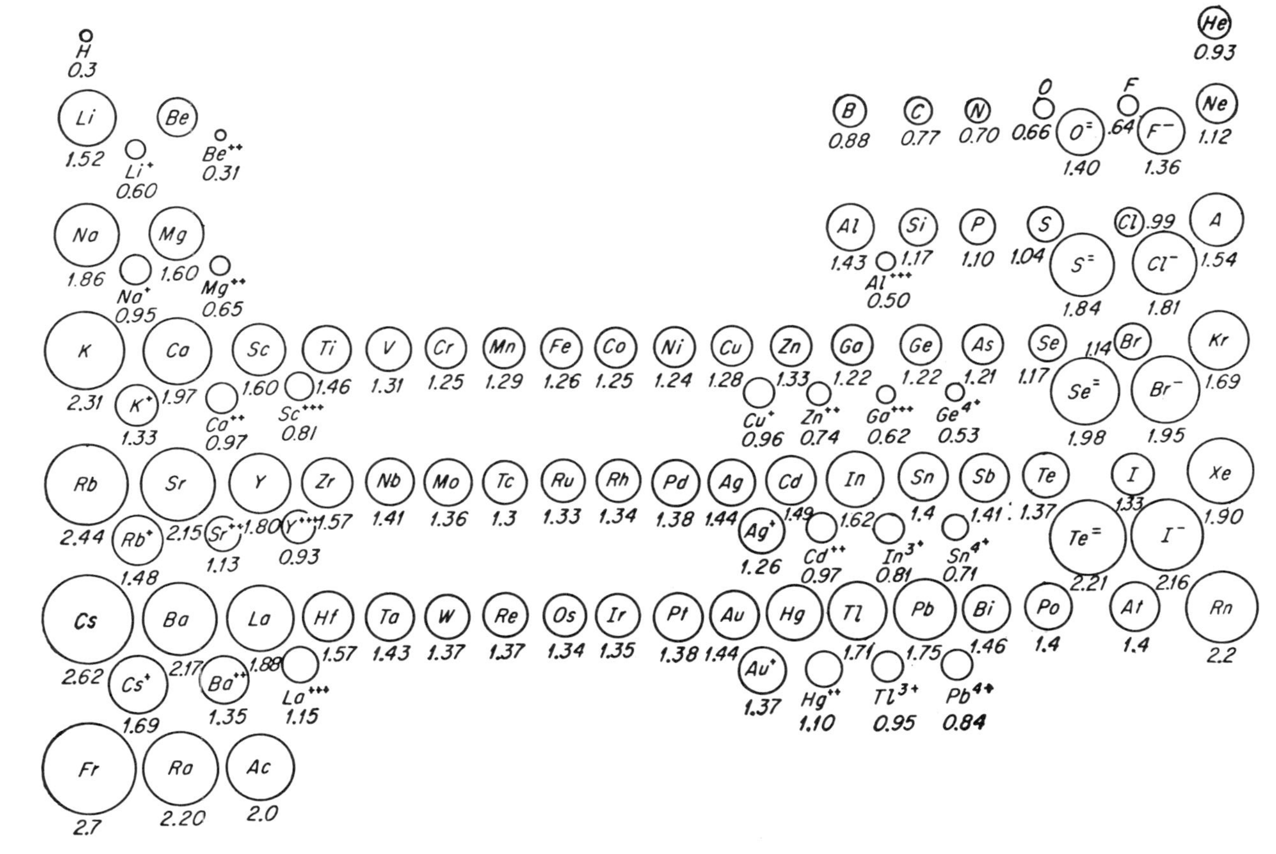

Fig 22.6 Variation in atomic and ionic radii with atomic number. Radii measured in Angstrom units (Å), where 1 Å = 10^{-10} m.

radius of the atom. When one or more electrons are removed there is a decrease in the repulsion between charge clouds and the result is a contraction of the overall electron charge cloud.

(d) The ionic radius of a negative ion is greater than the corresponding atomic radius of the atom. When one or more electrons are added there is an increase in the repulsion between charge clouds and the result is an expansion of the overall electron charge cloud.

(5) Atomic weights. The atomic weights of the elements increase regularly across a period as the atomic number increases, and also increase regularly down a group as the atomic number increases. This regular change is not a periodic variation but this is not unexpected, since we are here dealing with a nuclear property rather than with a consequence of electron configuration. This regular change is shown graphically in Figure 22.7.

In some cases this regularity is interrupted and inversions in the sequence of atomic weights occur. Thus in the case of argon and potassium, the atomic numbers are respectively 18 and 19, whilst the atomic weights are respectively 39.948 and 39.102. Argon has a high proportion of the isotope $^{40}_{18}Ar$, whilst potassium has a high proportion of the isotope $^{39}_{19}K$. Since atomic weight is the weighted mean of the appropriate relative isotopic masses, the atomic weight of argon exceeds that of potassium. Similar inversions occur for the elements cobalt, nickel, and for the elements tellurium, iodine.

22.4 SOME CHEMICAL PROPERTIES AND THE PERIODIC TABLE

The periodic table can be used as a classification of the chemical properties of the elements and their compounds. In subsequent chapters we will study this use of the periodic table in detail, as it is applied to inorganic compounds. In this section we will consider two quite general chemical properties which will be studied for specific groups of compounds in later chapters.

(1) Formulae The formulae of similar types of compounds formed by elements within a given group are usually similar. Thus in group I we find that all of these elements react with fluorine. The atoms of these elements can all lose their single outer-shell electron and form the appropriate positive ion (M^+). The fluorine atoms form fluoride ions (F^-). Thus all of these fluorine compounds consist of ionic lattices in which the positive and negative ions occur in the mole ratio 1 : 1. All of these compounds have the common empirical formula, MF.

In group V we find that a series of hydrides, of general formula XH_3, exists. All of the atoms of the elements in this group have a common five-electron outer shell (s^2p^3). The three hydrogen atoms each share an electron with one of the electrons in the outer shell of the group V atom and an octet of electrons is built up in the outer shell of each group V atom. The four charge clouds present will repel to approximately a tetrahedral disposition and these molecules would be expected to be pyramidal. This structure is, in fact, confirmed by experiment.

The common stoichiometry shown in the compounds formed by the elements within a group is an important feature of the periodic classification,

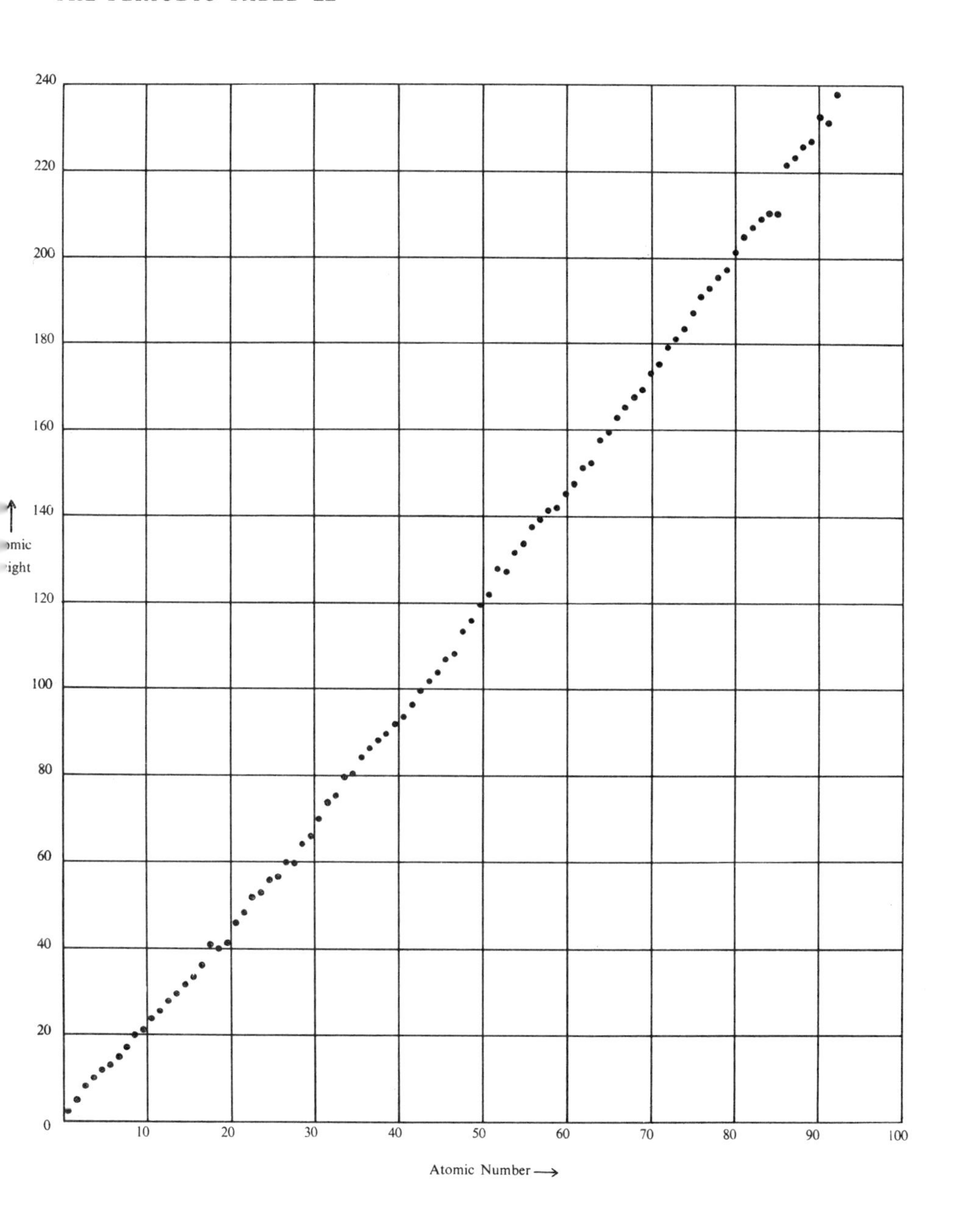

Fig 22.7 Variation in atomic weight with atomic number.

but we must exercise caution. Thus in group VI we find that sulphur and selenium form oxyacids of general formula H_2XO_4. However, tellurium, in the same group, forms no compound of this formula, but there is a chemically related compound of formula $Te(OH)_6$. Although the stoichiometry of these compounds differs, notice that in all of them the oxidation number of the group VI element is +6, emphasizing their common nature.

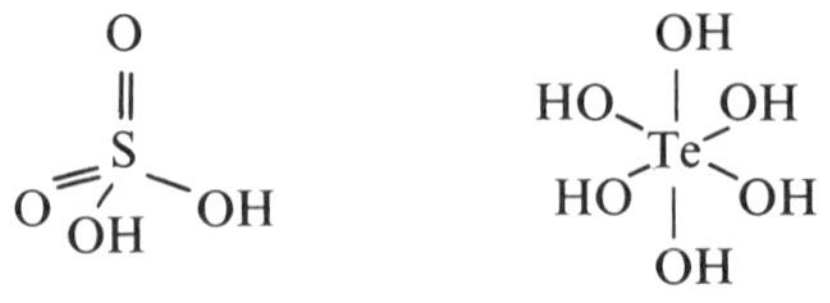

O
||
S
O OH OH

OH
HO OH
Te
HO OH
OH

Fig 22.8 Valence structures of sulphuric acid and telluric acid.

(2) Stability. In comparing a set of compounds formed within a group or across a period we often compare their stability. When we use the adjective 'stable' to describe a chemical compound, we mean that, under stated conditions, the compound is not decomposed into appreciable amounts of decomposition products. The concept of stability is widely used in chemistry and it is important to appreciate its significance and on what basis a compound is judged to be stable. We may recognize two criteria which must be applied when a compound's stability is assessed:

(a) *the equilibrium criterion*,
(b) *the rate criterion*.

The equilibrium criterion is a direct consequence of the equilibrium law which specifies the *maximum* concentration of decomposition products which can be formed from a compound at a stated temperature. A knowledge of the equilibrium constant for decomposition of a compound enables us to specify the concentration of these decomposition products. Consider the example of the decomposition of hydrogen chloride:

$$2HCl(g) \rightleftharpoons H_2(g) + Cl_2(g)$$

At 1000°C, the equilibrium constant is 5×10^{-9}. It may be calculated that, at equilibrium, for every mole of HCl we have 7×10^{-5} mole of H_2.

A compound may be described as stable when the equilibrium concentration of decomposition products is not appreciable relative to the equilibrium concentration of the compound. It is difficult to assign a definite value to the equilibrium concentration ratio just as it is difficult to sharply distinguish 'strong' from 'weak' electrolytes. However, we can say that hydrogen chloride is a stable compound at 1000°C.

When a compound is judged to be stable on the basis of the equilibrium law it may be described as inherently stable.

Hydrogen chloride may therefore be described as inherently stable at 1000°C.

It should be noted quite clearly that one *cannot* say that hydrogen chloride *is* an inherently stable compound *without stating the temperature at which the judgment is made*. A compound may be inherently stable at one temperature but inherently unstable at another temperature. For instance at 2000°C the equilibrium constant of the decomposition of HCl is approximately 10^{-2} and we may calculate that at equilibrium for every mole of HCl we have 10^{-1} mole of H_2. *Hydrogen chloride may therefore be described as inherently unstable at* 2000°C.

One other feature of the decomposition of HCl is important. At 1000°C the rate at which HCl decomposes is very high yet the compound is still inherently stable. The rate of the back reaction is, of course, also very high in the equilibrium system. The crucial factor in deciding on the inherent stability of HCl is the maximum concentration of H_2 (and Cl_2) produced by the decomposition, not the fact that H_2 and Cl_2 are formed rapidly. At 25°C, HCl decomposes at an immeasurably slow rate *and* the equilibrium criterion shows that it is inherently stable.

Let us now consider the analogous decomposition of hydrogen iodide at 1000°C:

$$2HI(g) \rightleftharpoons H_2(g) + I_2(g)$$

At 1000°C the equilibrium constant is 6×10^{-2}. It may be calculated that, at equilibrium, for every mole of HI we have 0.25 mole of H_2. Thus hydrogen iodide may be described as an inherently unstable compound at 1000°C. Moreover, this equilibrium concentration of H_2 (and I_2) is rapidly formed at 1000°C and in conformity with usual practice, we can simply describe HI as unstable at 1000°C.

We may summarize the criteria which we have developed so far to decide on the inherent stability of a compound. If the equilibrium constant for the decomposition of a compound is so small that only a very small proportion of decomposition products may be formed at a given temperature, then that compound may be referred to as inherently stable whether the decomposition rate is rapid (e.g. HCl at 1000°C) or whether this rate is very slow (e.g. HCl at 25°C). If the equilibrium constant for decomposition is large and the compound decomposes rapidly, the compound may be referred to as unstable. These criteria are summarized in the following table.

Table 22.3
Stability criteria

Equilibrium constant for decomposition	*Rapid rate of decomposition*	*Slow rate of decomposition*
K small	Inherently stable	Inherently stable
K large	Unstable	Metastable

One difficulty arises when a compound has a large equilibrium constant for decomposition—and as such might be described as 'inherently unstable'—yet its rate of decomposition at the stated temperature is so slow that no significant concentration of products are formed in a long time (say weeks). In these circumstances, such a compound is described as *metastable*. In practical terms a metastable compound may be stored unchanged for weeks without apparent signs of decomposition. There are many examples of metastable compounds. At normal temperatures diamond is inherently unstable with respect to the carbon allotrope graphite, yet the rate of transformation of diamond to graphite is immeasurably slow at room tempera-true. Hence the slightly inaccurate saying, 'Girls, diamonds are for ever.' At 25°C HI is inherently unstable with respect to H_2 and I_2, yet it remains unchanged for long periods of time. Similarly the equilibrium constant for decomposition of nitric oxide to nitrogen and oxygen is 10^{15} at 25°C, yet its decomposition at this temperature is negligibly small.

Inherently unstable compounds *and* metastable compounds are similar in that both types of compounds do not form any appreciable concentrations of decomposition products in a period of, say, weeks at a stated temperature. Not infrequently, the equilibrium constant for decomposition of a compound is unknown and if it does not form appreciable concentrations of decomposition products, the chemist cannot classify it as either inherently stable or as metastable. In these circumstances, it is usual practice to describe the compound simply as *stable*. This situation is represented in Table 22.3 by the coloured area covering both types of compounds.

QUESTIONS

1. Why are the periods in the periodic table sometimes short and sometimes long?
2. Why are physical and chemical properties of elements not perfect periodic functions of atomic number?
3. Discuss the distribution of metals and non-metals in the periodic table in terms of electronegativity, ionization energy, and electron affinity.
4. In the first periodic table due to Mendeleef (1869) there was no rare gas group present. Why was the existence of this group not predicted by Mendeleef?
5. The atomic radii of Fe, Co, Ni show a much smaller variation in size than do those of Na, Mg, Al. Suggest a reason why this may be so.
6. Predict which ion in each pair is the larger: (a) Mg^{2+} or Ca^{2+}; (b) K^+ or Ca^{2+}; (c) Na^+ or F^-; (d) Cu^{2+} or Cu^+.
7. Which of the alkaline earth metals has:
   (a) the highest ionization energy for the gaseous atom;
   (b) smallest number of electrons per atom;
   (c) smallest positive ion (X^{2+});
   (d) a positive ion isoelectronic with the bromide ion?
8. Is it possible for an element to be discovered which would be placed between sulphur and chlorine in the periodic table? Discuss.
9. Why do the rare earth elements show such similar chemical properties?
10. In older forms of the periodic table manganese was classified as a sub-group of the halogens. What justification is there for this? What objections can be raised?
11. Which element has the higher electronegativity in each pair: (a) P or Si; (b) Li or Be; (c) Ca or Cs; (d) Sc or Cr; (e) Ag or I; (f) Rn or Ra?
12. Cadmium has two electrons in its outermost shell, as does magnesium. Why are these elements not classified in the same group?
13. Suggest properties, other than those discussed in this chapter, which might vary periodically. Plot the data for some of these properties, for the first 20 elements.

14. Before the discovery of isotopy the order of potassium and argon by atomic weights was inverted and it was expected that the atomic weights were experimentally inaccurate. Why was this inversion made?
15. Hydrogen is sometimes placed in periodic tables at the head of *both* groups I and VII. How can this be justified?

23

Elements of the Second and Third Periods

In this chapter we shall be concerned with the chemical properties of the elements in the second and third periods of the periodic table. To illustrate the trends in chemical behaviour in these periods, we will consider the elements themselves and restrict the compounds studied to the appropriate hydrides, fluorides, oxides, and hydroxy compounds. In considering the

I	II											III	IV	V	VI	VII	VIII
1 H																	2 He
3 Li	4 Be											5 B	6 C	7 N	8 O	9 F	10 Ne
11 Na	12 Mg											13 Al	14 Si	15 P	16 S	17 Cl	18 Ar
19 K	20 Ca	21 Sc	22 Ti	23 V	24 Cr	25 Mn	26 Fe	27 Co	28 Ni	29 Cu	30 Zn	31 Ga	32 Ge	33 As	34 Se	35 Br	36 Kr
37 Rb	38 Sr	39 Y	40 Zr	41 Nb	42 Mo	43 Tc	44 Ru	45 Rh	46 Pd	47 Ag	48 Cd	49 In	50 Sn	51 Sb	52 Te	53 I	54 Xe
55 Cs	56 Ba	57 La	72 Hf	73 Ta	74 W	75 Re	76 Os	77 Ir	78 Pt	79 Au	80 Hg	81 Tl	82 Pb	83 Bi	84 Po	85 At	86 Rn
87 Fr	88 Ra	89 Ac															

Table 23.1

Physical properties of elements in second and third periods

	(Second period)							
	Li	*Be*	*B*	*C*	*N*	*O*	*F*	*Ne*
Electron configuration	2.1	2.2	2.3	2.4	2.5	2.6	2.7	2.8
Outer-shell configuration	$2s^1$	$2s^2$	$2s^22p^1$	$2s^22p^2$	$2s^22p^3$	$2s^22p^4$	$2s^22p^5$	$2s^22p^6$
Electronegativity	1.0	1.5	2.0	2.5	3.0	3.5	4.0	—
First ionization energy ($kJ\ mol^{-1}$)	519	900	799	1090	1400	1310	1680	2080
	(Third period)							
	Na	*Mg*	*Al*	*Si*	*P*	*S*	*Cl*	*Ar*
Electron configuration	2.8.1	2.8.2	2.8.3	2.8.4	2.8.5	2.8.6	2.8.7	2.8.8
Outer-shell configuration	$3s^1$	$3s^2$	$3s^23p^1$	$3s^23p^2$	$3s^23p^3$	$3s^23p^4$	$3s^23p^5$	$3s^23p^6$
Electronegativity	0.9	1.2	1.5	1.8	2.1	2.5	3.0	—
First ionization energy ($kJ\ mol^{-1}$)	494	737	577	786	1060	990	1250	1520

oxides and hydroxy compounds we will arbitrarily restrict ourselves to the compound of highest available oxidation number. The outer-shell configuration of the elements in these periods varies from s^1 on the extreme left-hand side of the periodic table to s^2p^6 on the extreme right-hand side. The electronegativities of the elements increase steadily from left to right, and the first ionization energies show a general tendency to increase from left to right.

23.1 THE ELEMENTS

The structures of the most stable modifications of the elements show a broad gradation from metallic structures for atoms of low electronegativity and low ionization energy, through essentially covalent lattices, to discrete molecules for the atoms of high electronegativity and high ionization energy.

In the second period the first two elements, lithium and beryllium, are metals, while in the third period the first three elements, sodium, magnesium and aluminium, are metallic. By comparison with the transition metals, these metals are characterized by comparatively low melting points. When vaporized around the boiling point, lithium and sodium form some diatomic molecules in the gas phase. At higher temperatures, the vapours consist of individual atoms.

Covalent lattice structures are observed for boron, carbon and silicon. It was pointed out in the previous chapter that carbon exists as both a diamond and graphite modification, the latter being stable at room temperature. Silicon has a structure resembling diamond, and crystalline boron, which has a complex structure, is nearly as hard as diamond. These elements have extremely high melting and boiling points.

Phosphorus and sulphur, whilst also being solids at room temperature, exist as discrete molecules, P_4 and S_8, weakly bonded in lattices by dispersion forces. When heated, these weak forces are overcome, but initially the units P_4 and S_8 persist. Thus P_4 is known in the gas phase, and S_8 molecules are present in liquid sulphur near the melting point. At higher temperatures, dissociation into smaller molecular units, and ultimately to atoms, occurs.

Nitrogen, oxygen and fluorine in the second period, and chlorine in the third, exist as gaseous, diatomic molecules at room temperature. The noble gases, neon and argon, are known only as individual atoms.

Table 23.2

Further physical properties of elements in second and third periods

	(Second period)							
	Li	*Be*	*B*	*C*	*N*	*O*	*F*	*Ne*
Physical state (room temperature)	Solid	Solid	Solid	Solid	Gas	Gas	Gas	Gas
M.P.(°C)	180.5	1283	2027	—	−210.1	−218.8	−219.6	−248.6
B.P.(°C)	1331	2477	3927	(3800)	−195.8	−183.0	−187.9	−246.0
	(Third period)							
	Na	*Mg*	*Al*	*Si*	*P*	*S*	*Cl*	*Ar*
Physical state (room temperature)	Solid	Solid	Solid	Solid	Solid	Solid	Gas	Gas
M.P.(°C)	98	650	660	1423	44.2	119	−101	−189.4
B.P.(°C)	890	1120	2447	2680	(280) white	444.60	−34.05	−185.9

23.2 THE HYDRIDES

The structures of these compounds range from crystalline ionic lattices to gaseous covalent molecules.

The hydrides of the metals generally form essentially ionic lattices comprising metal cations and hydride, H^- anions. Thus LiH and NaH comprise Li^+ cations and H^- anions, and Na^+ cations and H^- anions, respectively. MgH_2 is largely ionic, and the rather unstable BeH_2 shows some ionic character. The nature of bonding in solid AlH_3 is unknown, but is probably a network lattice with a fair amount of negative charge remaining on the hydrogen.

Boron hydride is unusual in that the expected formula would be BH_3; these molecules appear to dimerize readily to B_2H_6 with the structure indicated in Figure 23.1.

As discussed in chapter 9, the related ion BH_4^- is tetrahedral, the boron

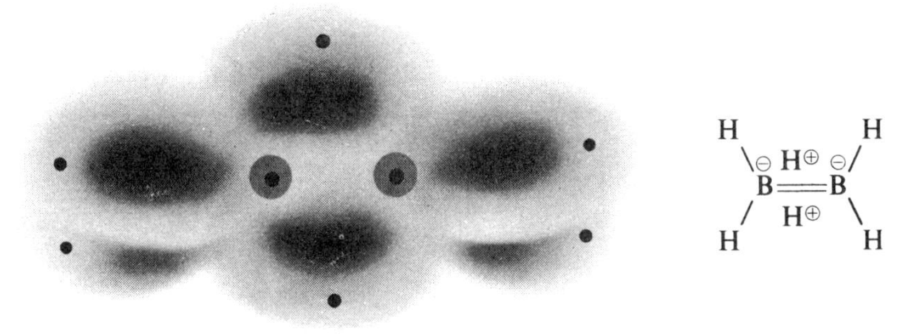

Fig 23.1 Charge cloud representation and valence structure of diborane, B_2H_6.

atom having four electron pairs in its outer shell in this anion. Likewise, aluminium forms the tetrahedral anion AlH_4^-.

The remaining hydrides are molecular. The stereochemistry of the CH_4, NH_3, OH_2, and HF molecules has been considered in chapter 9, and those of SiH_4, PH_3, SH_2 and ClH are similar. Some of these molecular hydrides exhibit acidic properties; the strength of these molecular hydrides as acids is a crude measure of the ability of the molecules to release protons. It is with the highly electronegative elements that acidic properties are evident, i.e. the acid strength increases across the period.

The basic strength of the molecular hydrides depend upon the existence of non-bonding pairs which can accept protons. In the second period, NH_3, OH_2 and FH have such non-bonding pairs but CH_4 has not. The ease of addition of a proton is measured by the K_b of the hydride, which decreases as the electronegativity of the central atom, from N to F, increases.

Table 23.3

Acid-base properties of molecular hydrides

	K_a	K_b
NH_3	*ca.* 10^{-30}	10^{-5}
OH_2	2×10^{-16}	2×10^{-16}
FH	7×10^{-4}	—

By contrast, both LiH and NaH, the ionic crystalline hydrides, are sources of the very strong base H^-. This base hydrolyses completely in aqueous solution:

$$H^- + H_2O \rightarrow H_2 + OH^-$$

23.3 THE FLUORIDES

As for the hydrides, we find that the fluorides range in structure from crystalline ionic lattices to gaseous covalent molecules.

Table 23.4
Physical properties of hydrides in second and third periods

	(Second period)						
	LiH	*BeH_2*	*B_2H_6*	*CH_4*	*NH_3*	*OH_2*	*FH*
Physical state (room temperature)	Solid	Solid	Gas	Gas	Gas	Liquid	Liquid
M.P.(°C)	680	Decomp.	−165.5	−182.5	−77.74	0.00	−83.07
B.P.(°C)	—	—	−92.5	−161.5	−33.40	100.0	+19.9
	(Third period)						
	NaH	*MgH_2*	*AlH_3*	*SiH_4*	*PH_3*	*SH_2*	*ClH*
Physical state (room temperature)	Solid	Solid	Solid	Gas	Gas	Gas	Gas
M.P.(°C)	Decomp.	Decomp.	Decomp.	−184.7	−133.75	−85.53	−114.19
B.P.(°C)	—	—	—	−111.4	−87.72	−60.31	−85.03

The fluorides of the metals form essentially ionic lattices comprising metal cations and fluoride, F^-, anions. Thus LiF and NaF are lattices comprised of Li^+ and Na^+ cations with F^- anions. BeF_2 and MgF_2 are essentially ionic lattices, as is AlF_3, but BF_3 consists of discrete molecules involving essentially covalent bonds. The remaining fluorides are composed of discrete molecules.

In the sequence of fluorides of the second period, viz. BF_3, CF_4, NF_3, OF_2, and F_2, the number of outer-shell electrons, for any atom, never exceeds eight. In the case of boron the number of outer-shell electrons, in the boron atom, is six. The geometry of these molecules may be predicted from the charge cloud repulsion theory and is shown in Figure 23.2.

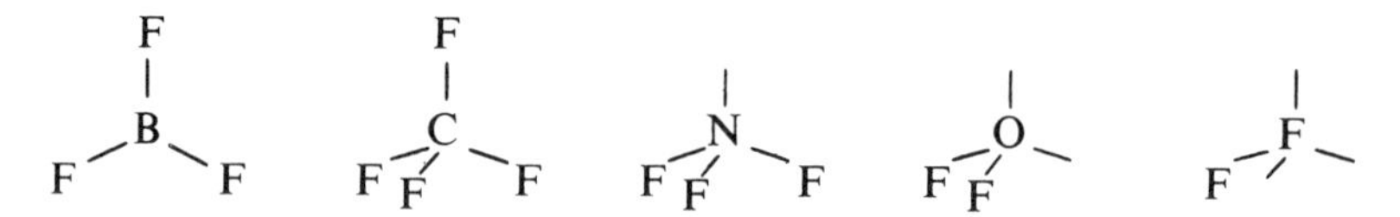

Fig 23.2 Valence structures of BF_3, CF_4, NF_3, OF_2, F_2.

It should be noted that the arrangement of electron pairs around a boron atom becomes tetrahedral when a fourth pair is added, as in the fluoroborate ion, BF_4^-. Similarly, although the BeF_2 molecule in the gas phase is linear, the BeF_4^{2-} ion, which is isoelectronic with the BF_4^- ion, is also tetrahedral.

In the third period the number of electron pairs in the outer shell of these atoms may exceed four, so that in the sequence of fluorides SiF_4, PF_3, SF_2, and ClF we find four pairs of electrons in the outer shell, but in addition the fluorides PF_5, SF_4, SF_6 and ClF_3 exist, involving five or six electron pairs in the outer shell of the central atom. The geometry of all the fluorides mentioned may be predicted from the charge cloud repulsion theory and some examples are shown in Figure 23.3.

Fig 23.3 Valence structures of PF_3, SF_4, SF_6.

In addition, the isoelectronic ions AlF_6^{3-}, SiF_6^{2-}, and PF_6^-, are formed showing an octahedral arrangement of bonding electron pairs about the central atom.

Table 23.5

Physical properties of some fluorides in second and third periods

	(Second period)						
	LiF	*BeF$_2$*	*BF$_3$*	*CF$_4$*	*NF$_3$*	*OF$_2$*	*F$_2$*
Physical state (room temperature)	Solid	Solid	Gas	Gas	Gas	Gas	Gas
M.P.(°C)	845	803	−128.7	−183.7	−208.5	−223.9	−219.6
B.P.(°C)	1681	—	−99	−182.0	−129.1	−144.9	−187.9
	(Third period)						
	NaF	*MgF$_2$*	*AlF$_3$*	*SiF$_4$*	*PF$_3$*	*SF$_2$*	*ClF*
Physical state (room temperature)	Solid	Solid	Solid	Gas	Gas	Gas	Gas
M.P.(°C)	995	1263	1290	−90.3	−151.5	—	−155.6
B.P.(°C)	1704	2227	(1257 subl.)	(−95.5 subl.)	−101.2	−35	−100.3
				PF$_5$	*SF$_4$*	*SF$_6$*	*ClF$_3$*
Physical state (room temperature)				Gas	Gas	Gas	Liquid
M.P.(°C)				−83	—121	−50.7	−82.6
B.P.(°C)				−75	−40.4	(−63.7 subl.)	12.1

23.4 THE OXIDES

The oxides show a wider range of structures than we have seen in the hydrides and fluorides; in fact we find practically every division of our structural classification exemplified. For convenience we will consider each period separately.

(1) Second period. Lithium oxide, Li_2O, and beryllium oxide, BeO, may be regarded as essentially ionic lattices. Boron oxide, B_2O_3, is a layer lattice with covalent bonds between the boron and oxygen atoms in the sheets. Carbon dioxide, CO_2, and fluorine monoxide, F_2O, consist of discrete molecules, whilst dinitrogen pentoxide, N_2O_5, in the solid state is ionic ($NO_2^+.NO_3^-$). In the gaseous state molecular N_2O_5 has the valence structure shown in Figure 23.4.

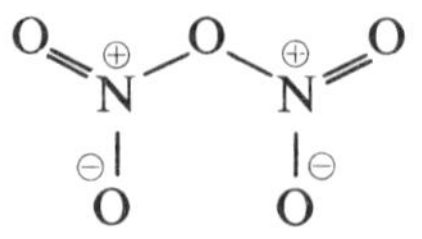

Flg 23.4 Valence structure of N_2O_5 molecule.

There is a steady decrease in the charge on the oxygen in the oxides as we pass from left to right in the period. Thus in Li_2O we have O^{2-} ions in the lattice, whereas in F_2O, the oxygen probably has a small positive charge, since fluorine is more electronegative than oxygen. This difference in negative charge apparently leads to a profound difference in the way in which the oxides react with water, acids and bases.

Thus Li_2O reacts vigorously with water since it makes available the very strong base O^{2-}:

$$O^{2-} + H_2O \rightarrow 2OH^-$$

or

$$LiO(s) + H_2O \rightarrow 2Li^+(aq) + 2OH^-(aq)$$

BeO does not react in this way, but acids stronger than water, e.g. the hydrogen ion, can react:

$$BeO(s) + 2H^+(aq) \rightarrow Be^{2+}(aq) + H_2O$$

BeO has 'made available' O^{2-} ions for reaction with hydrogen ions. BeO also reacts with bases:

$$BeO(s) + 2OH^-(aq) + H_2O \rightarrow Be(OH)_4^{2-}(aq)$$

where $Be(OH)_4^{2-}$ is called the beryllate ion.

B_2O_3 does not react with water in this way, rather it attaches water to itself:

$$B_2O_3 + 3H_2O \rightarrow 2B(OH)_3$$

forming the weak acid $B(OH)_3$, boric acid. B_2O_3 reacts with bases to form the borate ion $B(OH)_4^-$

$$B_2O_3(s) + 2OH^-(aq) + 3H_2O \rightarrow 2B(OH)_4^-(aq)$$

CO_2 dissolves in water and reacts slightly to form carbonic acid, H_2CO_3, which exists only in solution:

$$CO_2(aq) + H_2O \rightleftharpoons CO(OH)_2$$

At equilibrium a solution of carbon dioxide in water contains only about 1 per cent of the CO_2 in the form of carbonic acid. CO_2 will react with hydroxide ions to form the bicarbonate, HCO_3^-, and carbonate, CO_3^{2-}, ions.

N_2O_5 consists of an ionic lattice of nitronium ions, NO_2^+, and nitrate ions, NO_3^-. It reacts vigorously with water to produce a solution of nitric acid. The NO_2^+ ion reacts with water:

$$NO_2^+ + H_2O \rightarrow 2H^+ + NO_3^-$$

or
$$N_2O_5(s) + H_2O \rightarrow 2H^+(aq) + 2NO_3^-(aq)$$

N_2O_5 will react with hydroxide ions to form nitrate, NO_3^-, ions.

F_2O does not react readily with water and shows no acidic properties.

(2) Third period. Sodium oxide, Na_2O, magnesium oxide, MgO, and aluminium oxide, Al_2O_3, may be regarded as essentially ionic lattices. Silicon dioxide, SiO_2, is a covalent network lattice. Phosphorus pentoxide, P_4O_{10}, and chlorine heptoxide, Cl_2O_7, consist of discrete molecules. Sulphur trioxide, SO_3, in the solid state, is composed of a chain lattice of SO_3 molecules, or an aggregate of rings of SO_3 molecules.

We have already noted that the elements in this period may have more than four electron pairs in the outer shell. In P_4O_{10} the phosphorus atoms have five electron pairs, in SO_3 the sulphur atoms have six electron pairs, and in Cl_2O_7 the chlorine atoms have seven electron pairs in their respective outer shells. The valence structures for these molecules are shown in Figure 23.5.

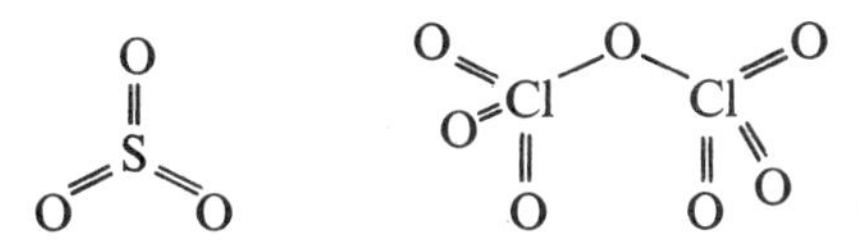

Fig 23.5 Valence structure of SO_3, Cl_2O_7.

Again we note a transition from the highly charged O^{2-} ions in the Na_2O lattice to the anticipated zero charge in Cl_2O_7

Na_2O reacts vigorously in water:

$$Na_2O(s) + H_2O \rightarrow 2Na^+(aq) + 2OH^-(aq)$$

MgO reacts with water to form the sparingly soluble $Mg(OH)_2$

$$MgO(s) + H_2O \rightarrow Mg(OH)_2(s)$$

$$Mg(OH)_2(s) \rightleftharpoons Mg^{2+}(aq) + 2OH^-(aq)$$

Al_2O_3 will not react with water, nor, when very pure, will it normally react with acids or bases. However, when freshly prepared by dehydrating $Al(OH)_3$ it reacts both with hydrogen ions and hydroxide ions:

$$Al_2O_3(s) + 6H^+(aq) \rightarrow 2Al^{3+}(aq) + 3H_2O,$$

$$Al_2O_3(s) + 2OH^-(aq) + 3H_2O \rightarrow 2Al(OH)_4^-(aq)$$

where $Al(OH)_4^-$ is called the aluminate ion.

SiO_2 will not react with water or acids, but reacts with fused sodium hydroxide:

$$SiO_2(s) + 2OH^- \rightarrow SiO_3^{2-} + H_2O$$

where SiO_3^{2-} is called the silicate ion.

The remaining oxides react readily with water to form the appropriate acids, for example

$$P_4O_{10} + 6H_2O \rightarrow 4PO(OH)_3$$

where H_3PO_4 (phosphoric acid) is a triprotic acid, which yields in solution the ions $H_2PO_4^-$, HPO_4^{2-}, PO_4^{3-};

$$SO_3 + H_2O \rightarrow SO_2(OH)_2$$

where H_2SO_4 (sulphuric acid) is a diprotic acid, which yields in solution the ions HSO_4^-, SO_4^{2-};

$$Cl_2O_7 + H_2O \rightarrow 2ClO_3(OH)$$

where $HClO_4$ (perchloric acid) is a monoprotic acid, which yields in solution the ion ClO_4^-.

These three oxides will react with hydroxide ions to form the various oxyanions which have been listed.

Table 23.6

Physical properties of some oxides in second and third periods.

(Second period)	Li_2O	BeO	B_2O_3	CO_2	N_2O_5	O_2	F_2O
Physical state (room temperature)	Solid	Solid	Solid	Gas	Solid	Gas	Gas
M.P.(°C)	1727	2547	450	−56.6	30	−218.8	−223.9
B.P.(°C)	—	—	—	(−78.5 subl.)	45–50	−183.0	−144.8

(Third period)	Na_2O	MgO	Al_2O_3	SiO_2	P_4O_{10}	SO_3	Cl_2O_7
Physical state (room temperature)	Solid	Solid	Solid	Solid	Solid	Solid	Liquid
M.P.(°C)	920	3802	2027	1710	422	16.85	−91.5
B.P.(°C)	—	—	—	2590	—	44.8	83

23.5 A CLASSIFICATION OF OXIDES

On the basis of the way in which oxides react with water, and also with acids and bases, they are sometimes loosely classified as acidic, basic amphoteric, or neutral.

(1) Basic oxides. These may react in one or both of two ways:

(a) with water to yield a solution containing hydroxide ions, together with the relevant cation of the element concerned. The oxide releases the strong base O^{2-} which reacts with water:

$$O^{2-} + H_2O \rightarrow 2OH^-$$

Examples are Li_2O, K_2O, CaO, thus

$$K_2O(s) + H_2O \rightarrow 2K^+(aq) + 2OH^-(aq)$$

(b) with aqueous solutions containing hydrogen ions to yield a solution containing cations of the element concerned. The oxide makes O^{2-} ions available for reaction with hydrogen ions. The oxides considered in (a) above will also react in this way.
Examples are CuO, Ag_2O, thus

$$CuO(s) + 2H^+(aq) \rightarrow Cu^{2+}(aq) + H_2O$$

(2) Acidic oxides. These may react in one or both of two ways:
(a) with water to yield a hydroxy compound which is an acid.
This acid may be weak or strong, for example,

$$B_2O_3(s) + 3H_2O \rightarrow 2B(OH)_3 \qquad K_{a1} = 6 \times 10^{-10}$$

$$Cl_2O_7(l) + H_2O \rightarrow 2ClO_3(OH) \qquad K_a = ca.\ 10^{+10}$$

Other examples are SO_3, P_4O_{10}, N_2O_5, CO_2, CrO_3.
(b) with aqueous solutions containing hydroxide ions, or with fused hydroxides to yield oxygen-containing anions. The oxides considered in (a) above will also react in this way. An example of this second type is SiO_2:

$$SiO_2(s) + 2OH^-(aq) \rightarrow SiO_3^{2-}(aq) + H_2O$$

(3) Amphoteric oxides. These are oxides which show both acidic and basic properties as defined above, e.g. Al_2O_3, BeO

Acidic properties $BeO(s) + H_2O + 2OH^-(aq) \rightarrow Be(OH)_4^{2-}(aq)$

Basic properties $BeO(s) + 2H^+(aq) \rightarrow Be^{2+}(aq) + H_2O$

(4) Neutral oxides. These show no acidic or basic properties, e.g. F_2O, CO.

If we now refer back to the discussion of oxides in section 23.4, we see that on the left of both periods we find basic oxides (Li_2O, Na_2O, MgO), then moving to the right we find amphoteric oxides (BeO, Al_2O_3), and then further to the right a succession of acidic oxides (B_2O_3, SiO_2, CO_2, N_2O_5, P_4O_{10}, SO_3, Cl_2O_7), with finally F_2O, a neutral oxide, ending the sequence. This change in character may be correlated with the decreasing charge on the oxygen as we go from left to right in a period.

23.6 THE HYDROXY COMPOUNDS

Sodium hydroxide, NaOH, and beryllium hydroxide, $Be(OH)_2$, may be regarded as ionic networks. Lithium hydroxide, LiOH, magnesium hydroxide, $Mg(OH)_2$, and aluminium hydroxide, $Al(OH)_3$, are layer lattices which may be regarded as involving ions in the layers. Boric acid, $B(OH)_3$, consists of layers of $B(OH)_3$ molecules hydrogen-bonded together. The remaining hydroxides may be regarded as being composed of discrete molecules. It may be noted that the hydroxides of elements to the right of each period exist in partially dehydrated form. Thus instead of silicic acid, $Si(OH)_4$, being followed by $P(OH)_5$ we find the partially dehydrated form $PO(OH)_3$—phosphoric acid, and similarly in the second period, instead of finding

$C(OH)_4$, we find $CO(OH)_2$—carbonic acid. It may be noted that no oxyacids of fluorine exist.

Table 23.7

Formulae of some hydroxy compounds in second and third periods

Second period	$LiOH$	$Be(OH)_2$	$B(OH)_3$	$CO(OH)_2$ (H_2CO_3)	$NO_2(OH)$ (HNO_3)		
Third period	$NaOH$	$Mg(OH)_2$	$Al(OH)_3$	$Si(OH)_4$	$PO(OH)_3$ (H_3PO_4)	$SO_2(OH)_2$ (H_2SO_4)	$ClO_3(OH)$ ($HClO_4$)

$LiOH$ and $NaOH$ are both soluble and release OH^- ions into solution. $Mg(OH)_2$ is sparingly soluble, but does release some OH^- ions. $Be(OH)_2$ and $Al(OH)_3$ are both amphoteric.

$$Al(OH)_3(s) + 3H^+(aq) \rightarrow Al^{3+}(aq) + 3H_2O$$

$$Al(OH)_3(s) + OH^-(aq) \rightarrow Al(OH)_4^-(aq)$$

Further to the right the hydroxy compounds now release protons to any available base, rather than releasing OH^- ions. They are now functioning as acids, e.g. nitric acid:

$$NO_2(OH)(l) + H_2O \rightarrow H_3O^+(aq) + NO_3^-(aq)$$

The acid strength of the hydroxy compounds increases across the period as shown in Table 23.8.

Table 23.8

Acidity constants (K_{a1}) of some hydroxy compounds

Compound	*Formula*	K_{a1}
Boric acid	H_3BO_3	6×10^{-10}
Carbonic acid	H_2CO_3	4.5×10^{-7}
Nitric acid	HNO_3	10^{+1}
Phosphoric acid	H_3PO_4	8×10^{-3}
Sulphuric acid	H_2SO_4	*ca.* 10^{+9}
Perchloric acid	$HClO_4$	*ca.* 10^{+10}

23.7 SUMMARY

In this chapter we have observed gradations in valence structures, bond type, structural classification, and acid strength as we have passed from compounds of atoms of low electronegativity to compounds of high electronegativity, i.e. as we have passed from left to right across a period. We shall now examine the generalizations we can make as a result of these observations.

(a) *Structure and bond type.* The gradation in these may be correlated with change in electronegativity across the period, from low values on the left of the periods to high values on the right. Thus we find that atoms of low electronegativity tend to form compounds in which they have given up electrons to some other atom. Such compounds are often ionic lattices in which the atom of higher electronegativity has acquired a negative charge, e.g. NaF, LiH. Reactions of such compounds are thus those of the appropriate positive and negative ions. In contrast to this, compounds between atoms of high electronegativity usually exist as discrete molecules, e.g. ClF, or as covalently bonded network lattices, e.g. SiO_2. The bonding, therefore, in any series of compounds of a set of atoms, e.g. the hydrides or fluorides, becomes less ionic and more covalent as we go from atoms of low to atoms of high electronegativity.

(b) *Valence structures.* In the second period, in all the compounds considered, the central atom never has more than four electron pairs in the outer shell and there must therefore exist a gradation in valence structures for any given set of compounds, e.g. the fluorides. Even in the case of dinitrogen pentoxide, N_2O_5, where nitrogen is showing an oxidation state of +5, there are no more than eight electrons in the outer shell of each atom, as is shown by examining the valence structure for N_2O_5.

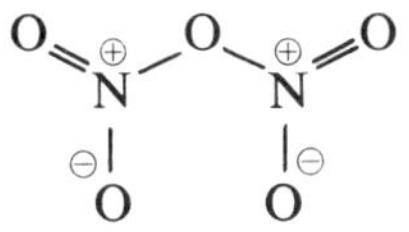

Fig 23.6 Valence structure of N_2O_5 molecule.

Remember that the oxidation number classification is an arbitrary one to which it is often difficult to assign a clear physical meaning. In the third and later periods we may have more than four electron pairs in an outer shell, for example, we find five, six or seven electron pairs in PF_5, SF_6, and $HClO_4$, respectively.

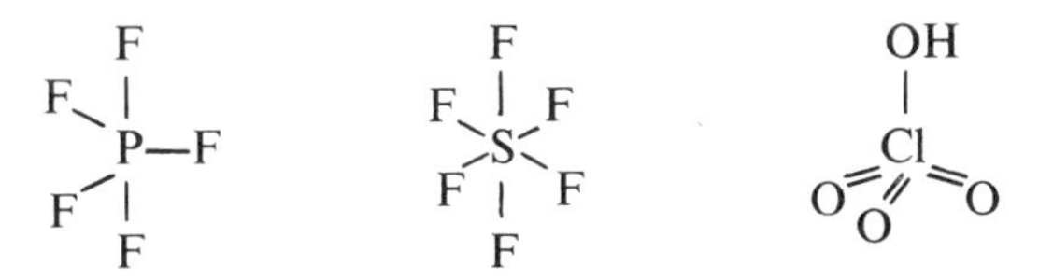

Fig 23.7 Valence structures of PF_5, SF_6, $HClO_4$.

(c) *Acid strengths.* The acid strengths of the molecular hydrides and molecular hydroxy compounds increase with increasing electronegativity of the atoms bonded to the —H or —OH. We may consider that as the electronegativity of the atom attached to the —H or —OH increases, negative charge will tend to reside more on the attached atom and less on the —H or —OH. Hence we might imagine it becomes increasingly easy to remove —H as H^+. This is not a very satisfactory way of explaining the observed acidities of the hydrides and hydroxy compounds, although it emphasizes *one* important factor which decides acid strengths. When considering reasons for the variation in the acid strengths of the hydrides and hydroxy compounds of the atoms within a group of the periodic table, we shall see that another factor—bond strength—may outweigh electronegativity considerations.

QUESTIONS

1. Silicon melts at 1423°C whilst sulphur melts at 119°C. Account for this difference in melting points in terms of the structures of the two elements.
2. Sodium hydroxide is an ionic solid which crystallizes with a sodium chloride type structure:
   (a) Describe with the aid of a diagram the structure of solid hydroxide.
   (b) How much energy would be released when 6.000 g of sodium hydroxide was formed from the elements? The standard heat of formation of sodium hydroxide is −426.8 kJ.
3. By discussing three properties of sodium, magnesium, aluminium, and silicon, show that the properties of elements vary with the atomic numbers of the elements.
4. Using Table 23.4, plot the boiling points of the hydrides from CH_4 to FH, and from SiH_4 to ClH, against the molecular weights of the hydrides. Account for any differences in the slope between these two curves.
5. (a) What are the oxidation numbers of sulphur in sulphurous acid, H_2SO_3, and in sulphuric acid, H_2SO_4?
   (b) For the system

$$SO_4^{2-} + 4H^+ + 2e \rightarrow SO_2 + 2H_2O \qquad E^\circ = +0.20 \text{ volt}$$

   The E° values for the following systems are given:

$$Fe^{3+} + e \rightarrow Fe^{2+} \qquad E^\circ = +0.77 \text{ volt}$$

$$Sn^{4+} + 2e \rightarrow Sn^{2+} \qquad E^\circ = +0.15 \text{ volt}$$

$$I_2 + 2e \rightarrow 2I^- \qquad E^\circ = +0.54 \text{ volt}$$

$$Cr^{3+} + e \rightarrow Cr^{2+} \qquad E^\circ = -0.41 \text{ volt}$$

   Decide which of the species Fe^{3+}, Sn^{4+}, I_2 and Cr^{3+} could be reduced by SO_2.
   (c) Write a balanced equation for the reaction between SO_2 and chlorine, to give chloride and sulphate.
6. (a) Which elements in the second period and third period are (i) strong oxidants; (ii) strong reductants.
   (b) Can the trend in oxidizing/reducing properties of the elements in the second and third periods be correlated with trends in electronegativity? Discuss.
7. If hydrogen chloride is dissolved in water the products are hydrogen ion and chloride ion, whilst if sodium hydride is dissolved in water the products are sodium hydroxide and hydrogen gas. Account for this different in behaviour in terms of the structure of these two compounds.
8. An enthusiastic but not always accurate research worker announces that he has prepared gaseous nitrogen pentafluoride. Why should we adopt a cautious attitude to such a claim?
9. Discuss the gradation in the geometry of the hydrides: SiH_4, PH_3, SH_2, ClH.

10. Suggest valence structures for the oxides, in the gaseous state: (a) N_2O_5; (b) Cl_2O_7; (c) SO_3; (d) F_2O.
11. A solution containing 0.1 M H_3PO_3 is tested with indicators and the pH is found to be 1.48. Calculate the K_a for this acid assuming it to be monobasic. Would one expect this to be a stronger or weaker acid than H_3BO_3? Explain.
12. Calculate the heat energy required to vaporize 2.50 dm^3 of water at 100°C. The density of water at 100°C is 0.9584 g cm^{-3} and the ΔH of vaporization is 40.80 kJ mol^{-1} at 100°C.
13. The solubility product of magnesium hydroxide is 6×10^{-12}. Calculate the pH of a saturated solution of magnesium hydroxide.
14. A fluoride of sulphur is found to have the percentage composition by mass: 29.63% sulphur and 70.37% fluorine. 0.100 g of the gaseous fluoride occupies a volume of 22.1 cm^3 at 20°C, and 766 mmHg. Find the molecular formula of the fluoride. Suggest a valence structure for this fluoride.
15. At a certain temperature the equilibrium constant (K) for the reaction

$$H_2(g) + Br_2(g) \rightleftharpoons 2HBr(g)$$

is 10^9. A 2000 cm^3 vessel contains an equilibrium mixture of the three gases. It is found that the vessel contains 10^{-4} mole of H_2 and 10^{-2} mole of Br_2. What is the equilibrium partial pressure of HBr?

24

Group I and II Elements

The elements of group I (the alkali metals) possess an outer shell containing one electron (s^1). The elements and their compounds are chemically very similar. The elements of group II (the alkaline earth metals) possess an outer shell containing two electrons (s^2). The elements and their compounds are chemically very similar, but the first element, beryllium, shows some significant differences from the others. The elements of both groups are found combined with other elements in mineral ores. They occur in combination in the form of positive ions. The alkali metals occur as halide salts in sea

I	II											III	IV	V	VI	VII	VIII
1 H																	2 He
3 Li	4 Be											5 B	6 C	7 N	8 O	9 F	10 Ne
11 Na	12 Mg											13 Al	14 Si	15 P	16 S	17 Cl	18 Ar
19 K	20 Ca	21 Sc	22 Ti	23 V	24 Cr	25 Mn	26 Fe	27 Co	28 Ni	29 Cu	30 Zn	31 Ga	32 Ge	33 As	34 Se	35 Br	36 Kr
37 Rb	38 Sr	39 Y	40 Zr	41 Nb	42 Mo	43 Tc	44 Ru	45 Rh	46 Pd	47 Ag	48 Cd	49 In	50 Sn	51 Sb	52 Te	53 I	54 Xe
55 Cs	56 Ba	57 La	72 Hf	73 Ta	74 W	75 Re	76 Os	77 Ir	78 Pt	79 Au	80 Hg	81 Tl	82 Pb	83 Bi	84 Po	85 At	86 Rn
87 Fr	88 Ra	89 Ac															

Table 24.1

Physical properties of group I and II elements

Element	*Atomic weight*	*Electron configuration*	*Outer-shell structure*	*Electro-negativity*	*Percentage abundance in earth's crust*
Lithium	6.939	2.1	$2s^1$	1.0	6.5×10^{-3}
Sodium	22.9898	2.8.1	$3s^1$	0.9	2.83
Potassium	39.102	2.8.8.1	$4s^1$	0.8	2.59
Rubidium	85.47	2.8.18.8.1	$5s^1$	0.8	0.031
Caesium	132.905	2.8.18.18.8.1	$6s^1$	0.7	7×10^{-4}
Beryllium	9.0122	2.2	$2s^2$	1.5	6×10^{-4}
Magnesium	24.132	2.8.2	$3s^2$	1.2	2.09
Calcium	40.08	2.8.8.2	$4s^2$	1.0	3.63
Strontium	87.62	2.8.18.8.2	$5s^2$	1.0	0.030
Barium	137.34	2.8.18.18.8.2	$6s^2$	0.9	0.025

water, and as halides, carbonates, sulphates, nitrates, borates in salt deposits formed by the drying up of inland seas. They are also found as complex insoluble silicates. The alkaline earth metals are found in combination as carbonates, sulphates, and silicates. Francium, the last element in group I, is radioactive and is formed in the radioactive decay series derived from actinium, and also in appropriate nuclear reactions. Radium, the last element in group II, is formed in the radioactive decay series derived from uranium ($^{238}_{92}U$). Each is chemically similar to the other members of its group and will not be considered in the following discussion.

24.1 STRUCTURE OF THE ELEMENTS

The elements are all metals and are good conductors of heat and of electricity. Thus the electrical resistivities of the elements range from 3.4×10^{-6} Ω cm to 6.2×10^{-5} Ω cm, whilst the resistivity of a typical non-metal, sulphur, is 4×10^{15} Ω cm. The densities and melting points of the alkali metals are all lower than those of the alkaline earth metals in the same period, but both are lower than the values for a typical transition element. Thus the melting points for sodium, magnesium, and iron are 98°, 651°, and 1535°C, respectively; whilst the densities are 0.97, 1.74, and 7.77 g cm^{-3} respectively. All of the metals in these groups are soft except beryllium and magnesium.

The bonding between atoms in sodium metal is weaker than that between magnesium atoms in magnesium metal, which in turn is weaker than that between iron atoms in iron metal. This change in the strength of the bonding is reflected in the values for the heat of reaction of the process

$$M(s) \rightarrow M(g)$$

For sodium the value is +109 kJ, for magnesium it is +150 kJ, and for iron it is +414 kJ.

The redox potential for the system

$$2H_2O + 2e \rightarrow 2OH^- + H_2 \qquad [OH^-] = 10^{-7}\,M$$

is -0.41 volt. The standard redox potentials for the group I and II elements range from -1.70 volt for the system

$$Be^{2+} + 2e \rightarrow Be$$

to -3.02 volt for the system

$$Cs^+ + e \rightarrow Cs$$

We would therefore predict that water would oxidize all of the group I and II metals. Thus sodium would be oxidized as follows:

$$Na \rightarrow Na^+ + e \qquad \times 2$$
$$2H_2O + 2e \rightarrow 2OH^- + H_2$$
$$\overline{2Na + 2H_2O \rightarrow 2Na^+ + 2OH^- + H_2}$$

The products of reaction would be hydrogen gas and sodium hydroxide solution. If we treat the above reaction as an equilibrium system

$$2Na + 2H_2O \rightleftharpoons 2Na^+ + 2OH^- + H_2$$

we can calculate the equilibrium constant. The value is found to be about 10^{78}. This physically means that the reaction will effectively give complete conversion of reactants to products. The measured rate of reaction, in this case, is considerable and so the sodium reacts rapidly and completely with water. All of the alkali metals react rapidly and completely with water in a similar way. We would similarly expect magnesium to be oxidized by water as follows:

$$Mg \rightarrow Mg^{2+} + 2e$$
$$2H_2O + 2e \rightarrow 2OH^- + H_2$$
$$\overline{Mg + 2H_2O \rightarrow Mg^{2+} + 2OH^- + H_2}$$

The products of reaction would be hydrogen and insoluble magnesium hydroxide ($K_s\,Mg(OH)_2 = 6 \times 10^{-12}$). The calculated equilibrium constant for this system is about 10^{66}. Again we expect complete conversion of reactants to products. However, in this case the first product of the reaction is the oxide which forms a coherent film over the metal, preventing the metal coming in further contact with water and thus substantially reducing the rate of reaction. In the terminology of chapter 22 one might say that the magnesium-water system is 'metastable' with respect to its conversion to magnesium hydroxide and hydrogen. A similar situation occurs when beryllium is reacted with water.

At higher temperatures magnesium does react more vigorously with water, presumably because the magnesium oxide film is disrupted in some way. Calcium, strontium, barium all react with water completely, and with increasing rapidity, at normal temperatures.

The standard redox potential for the system

$$2H^+ + 2e \rightarrow H_2 \qquad [H^+] = 1.000\ M, p_{H_2} = 1\ atm$$

is exactly zero by definition. Since the standard redox potentials of all group I and II metals are negative we would predict that the hydrogen ions from an acid would oxidize all of these elements. In all cases the rate of reaction is high and so acids completely and rapidly react with all group I and II metals. For example, if magnesium is added to a hydrochloric acid solution then the reaction is

$$Mg \rightarrow Mg^{2+} + 2e$$
$$2H^+ + 2e \rightarrow H_2$$
$$\overline{Mg + 2H^+ \rightarrow Mg^{2+} + H_2}$$

EXERCISE

The standard redox potential for the system

$$Hg^{2+} + 2e \rightarrow Hg$$

is +0.85 volt, and for the system

$$Mg^{2+} + 2e \rightarrow Mg$$

is −2.34 volt. If magnesium is added to a 1.0 M $Hg(NO_3)_2$ solution, may a reaction occur? Write down an ionic equation for any such reaction.

When reacting in the solid state the elements also act as strong reductants. Thus potassium is often used to reduce metal chlorides to metal, e.g. the element titanium may be formed by the reaction

$$TiCl_4 + 4K \rightarrow Ti(s) + 4KCl$$

The course of such reactions depends to an appreciable extent on the nature of the products. For example, sodium monoxide is reduced to metallic sodium by magnesium metal at a moderately high temperature (above the boiling point of sodium, but below that of magnesium). The more volatile sodium escapes as sodium vapour and is condensed:

$$Na_2O(s) + Mg(s) \rightarrow MgO(s) + 2Na(g)$$

This method for preparing sodium depends, in this case, upon the relative volatility of the two metals. One cannot predict the course of this reaction from a knowledge of standard redox potentials, which refer to solution equilibria.

The preparation of the elements by chemical reaction presents difficulties. The appropriate positive ion must be reduced by addition of electrons and few chemical substances can achieve this. The usual method is to reduce the ions by electrolysis of the molten salt (usually the chloride). At the negative electrode the reduction action is

$$M^+ + e \rightarrow M$$

or

$$M^{2+} + 2e \rightarrow M$$

24.2 FEATURES OF THE COMPOUNDS

All of the elements of group I have an outer shell containing one electron (s^1) whilst all of the elements of group II have an outer shell containing two electrons (s^2). Because of the common outer-shell configurations of the elements within each group we expect the elements within each group to show similar chemical properties.

All of the elements in these groups are metals and one expects that they will show low values for electronegativity. This is found to be so. When they react with elements of high electronegativity (e.g. oxygen, sulphur, or the halogens) one expects compounds to form which show predominantly ionic bonding. In fact the vast majority of the compounds of group I and II elements do show ionic bonding. With the elements of low electronegativity (other metals) they may form alloys in which the bonding is predominantly metallic, an example being the alloy formed between sodium and mercury, viz. sodium amalgam. A few compounds are formed in which the bonding may be regarded as covalent. Thus lithium forms a compound lithium ethyl, $Li(C_2H_5)$, for which this is true, and in the vapour of alkali metals small proportions of diatomic molecules may be detected.

When sodium reacts with fluorine the product is a lattice of sodium ions and fluoride ions. A detailed analysis of the energetics of the reaction is complex, but we can say that the formation of the sodium ion is consistent with the low value of the first ionization energy (494 kJ), whilst the strong bonding between the ions in the lattice is shown by the high value for the lattice energy (891 kJ). The first ionization energy for barium is 502 kJ, which is about the same as that of sodium, yet the barium fluoride lattice contains doubly charged ions (Ba^{2+}). Apparently the lattice energy released on the aggregation of barium ions and fluoride ions, in the mole ratio 1 : 2, is more than enough to compensate for the higher amount of energy required to form doubly charged ions. The lattice energy for barium fluoride is 2368kJ.

The doubly charged ions of group II elements are smaller than the iso-electronic ions of group I. Thus the ionic radius of Na^+ is 0.98 Å whilst that of Mg^{2+} is 0.78 Å. This may be explained as being due to the higher nuclear charge of the group II ion (Mg has $Z = 12$) than that of the group I ion (Na has $Z = 11$). The lithium ion (Li^+) has the smallest ionic radius (0.78 Å) of the ions formed by group I elements. Because of this, lithium may show some distinct differences in properties from other members of group I. Thus it is found that the lattice energy of lithium fluoride (992 kJ) is the highest for the fluorides formed in this group. The hydration energy of the fluoride ion is much higher than that of other ions in the group. Thus we find for the process

$$Li^+(g) + aq \rightarrow Li^+(aq) \qquad \Delta H = 514 \text{ kJ}$$

and for

$$Na^+(g) + aq \rightarrow Na^+(aq) \qquad \Delta H = 420 \text{ kJ}$$

The question of hydration energies is discussed in detail in chapter 10.

Similarly the beryllium ion (Be^{2+}) has the lowest ionic radius (0.34 Å) of the ions formed by group II elements. The beryllium ion has a much higher hydration energy than the other ions. Again some differences in properties are found between this first member of a group and the others. Thus beryllium chloride consists of a lattice of infinite chains formed by tetrahedral $BeCl_4$ groups sharing opposite edges, whereas magnesium chloride forms a layer lattice and the other chlorides form three-dimensional network lattices.

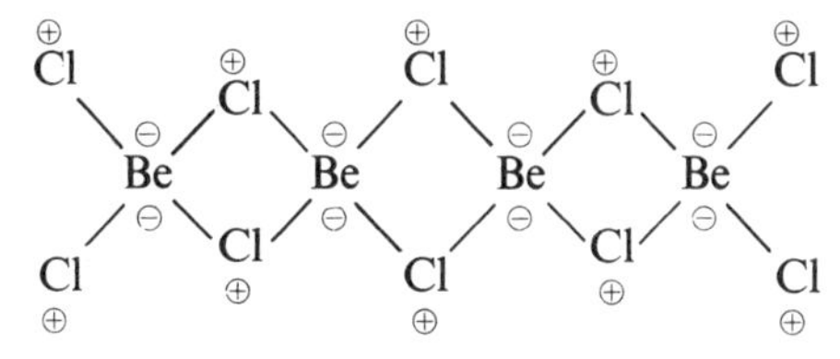

Fig 24.1 Valence structure of solid $BeCl_2$.

Beryllium chloride has a lower boiling point (488°C) than magnesium chloride (1412°C). The beryllium chloride lattice is more readily disrupted yielding dimeric molecules (Be_2Cl_4) in the vapour, whereas magnesium chloride yields clusters of $MgCl_2$ units. Further, beryllium hydroxide is amphoteric whilst the other hydroxides are basic only. The beryllium ion has a greater tendency than the other ions to form complex ions such as the beryllium fluoride ion, BeF_4^{2-}.

EXERCISE

Write a valence structure for the BeF_4^{2-} ion and predict its stereochemistry.

24.3 THE HYDRIDES

These compounds, with the exception of BeH_2, may be formed by heating the metal and hydrogen. In this type of reaction hydrogen is acting as an oxidizing agent. Thus if hydrogen is passed over gently heated lithium the following reaction takes place:

$$2Li + H_2 \rightarrow 2LiH$$

The compounds are largely ionic with the negative hydride ion (H^-) present in the lattice. If molten lithium hydride is electrolysed hydrogen appears at the positive electrode and lithium at the negative electrode, as shown:

$$Li^+ + e \rightarrow Li$$

$$2H^- \rightarrow H_2 + 2e$$

When added to water the hydrides react vigorously:

$$H^- + H_2O \rightarrow H_2 + OH^-$$

Inspection shows that this is an acid-base reaction, the hydride ion being a base with its conjugate acid being the hydrogen molecule (H_2). However, this reaction is also an oxidation-reduction reaction in which the hydride ion acts as a reductant and water acts as an oxidant:

$$H^- \rightarrow \tfrac{1}{2}H_2 + e$$
$$H_2O + e \rightarrow \tfrac{1}{2}H_2 + OH^-$$
$$\overline{H^- + H_2O \rightarrow H_2 + OH^-}$$

This reaction has been discussed in more detail in section 17.2.

24.4 THE OXIDES

Three types of oxides are formed by the elements of groups I and II.

(1) Oxides containing the oxide ion. All of the group I metals can form oxides of the type M_2O. These are ionic lattices containing the oxide ion (O^{2-}). When added to water the strong base, the oxide ion, reacts to form the hydroxide ion. The reaction is vigorous and the product in each case is the appropriate metal hydroxide

$$O^{2-} + H_2O \rightarrow 2OH^-$$

All of the group II metals can form oxides of the type MO. These oxides all have high melting points and involve strong bonding between ions. These group II oxides are ionic and show the 6 : 6 co-ordination of the sodium chloride lattice, except for beryllium oxide which shows a tetrahedral disposition of beryllium and oxygen atoms. Neither beryllium oxide nor magnesium oxide reacts with water to any significant extent, but the others react vigorously to form the appropriate hydroxide, for example

$$CaO + H_2O \rightarrow Ca(OH)_2$$

(2) Oxides containing the peroxide ion. All of the group I and II metals, except beryllium, can form peroxides under suitable conditions. These oxides consist of ionic lattices containing the metal ions and peroxide ions (O_2^{2-}). The peroxide ion involves four charge clouds about each oxygen.

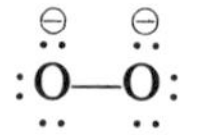

$$:\overset{\ominus}{\ddot{\underset{\cdot\cdot}{O}}}—\overset{\ominus}{\ddot{\underset{\cdot\cdot}{O}}}:$$

Fig 24.2 Valence structure of the peroxide ion O_2^{2-}.

The peroxide ion is a strong base (basicity constant K_{b1}); it is a stronger base than the HO_2^- ion (basicity constant K_{b2}).

$$O_2^{2-} + H_2O \rightleftharpoons OH^- + HO_2^- \qquad K_{b1}(> K_{b2})$$
$$HO_2^- + H_2O \rightleftharpoons OH^- + H_2O_2 \qquad K_{b2} = 6 \times 10^{-3}$$

Thus these peroxides in aqueous solution give appreciably alkaline solutions. In fact, all the alkali metal peroxides are soluble in water but the alkaline earth peroxides are insoluble in water. They will, however, react with acids to form hydrogen peroxide:

$$BaO_2(s) \rightleftharpoons Ba^{2+}(aq) + O_2^{2-}(aq)$$
$$2H^+(aq) + O_2^{2-}(aq) \rightleftharpoons H_2O_2(aq)$$

(3) Oxides containing the superoxide ion. The group I metals, except lithium, can form oxides containing the superoxide ion (O_2^-), e.g. potassium forms potassium superoxide, KO_2. There is some evidence for the existence of superoxides of calcium and barium. The superoxide ion contains an odd number of electrons (seventeen) and must involve at least one singly-occupied charge cloud. The superoxides are coloured and are all powerful oxidants. They react vigorously with water to form hydrogen peroxide and oxygen:

$$2O_2^- + 2H_2O \rightarrow O_2 + H_2O_2 + 2OH^-$$

All of the group II metals burn in oxygen to form oxides of the general formula XO. In air they may form in addition some nitride of general formula X_3N_2. All of the group I metals can form oxides of the type M_2O, but not all form them on burning in oxygen, or air. Sodium forms a peroxide, whilst potassium, rubidium and caesium form superoxides:

$$2Li + \tfrac{1}{2}O_2 \rightarrow Li_2O \qquad Rb + O_2 \rightarrow RbO_2$$

$$2Na + O_2 \rightarrow Na_2O_2 \qquad Cs + O_2 \rightarrow CsO_2$$

$$K + O_2 \rightarrow KO_2$$

24.5 SOLUBILITY OF COMPOUNDS

The factors influencing the solubility of compounds have been discussed in chapter 16. The analysis of the factors affecting trends in the solubilities of group I and II compounds are complex and will not be attempted. In fact, experiment shows that all group I compounds are soluble in water. Sodium and potassium compounds are, for this reason, widely used as sources of negative ions.

The trends in solubility in group II compounds are difficult to analyse. For example we find that the hydroxides become more soluble down the group, whilst the reverse is true for the sulphates. We may summarize the trends in solubility of group II compounds as follows:

chlorides, nitrates—all soluble
carbonates—all insoluble
sulphates—Be, Mg soluble; Ca, Sr, Ba insoluble
hydroxides—Be, Mg insoluble; Ca, Sr, Ba soluble.

24.6 GENERAL SUMMARY

The overall chemistry of these two groups is marked by its simplicity due to the close resemblance of all elements within each group, only the first member in each group tending to show some differences. In compounds, the oxidation state of the metal is constant, and in fact the chemistry of each group is largely that of the appropriate M^+ or M^{2+} ions.

QUESTIONS

1. Write the formulae of the following:
   (a) the hydride of radium;
   (b) three oxides of caesium;
   (c) two oxides of strontium.

2. For the elements of groups I and II state which atom:
   (a) has the largest ionic radius;
   (b) the lowest value for the first ionization energy;
   (c) the least soluble hydroxide;
   (d) the least soluble sulphate;
   (e) the lowest electronegativity.
3. Write valence structures and show the stereochemistry of the following:
   (a) BeF_4^{2-} (b) $Mg(H_2O)_6^{2+}$ (c) BeH_2
   (d) Be_2Cl_4 (e) $Mg(C_2H_5)_2$ (f) MgF_4^{2-}
4. In sodium vapour diatomic molecules can be detected. In iodine vapour diatomic molecules are present. When the molecules come together to form the solid in each case, we obtain a metal in one case, and a volatile, non-conducting solid in the other case. Explain, as far as you can, why the solids formed have such different structures and properties.
5. Describe the arrangement of ions in solid sodium chloride and in solid caesium chloride. Explain why the arrangements differ.
6. When hydrated magnesium chloride ($MgCl_2.6H_2O$) is heated it yields a basic chloride (Mg(OH)Cl) and hydrogen chloride. When hydrated barium chloride ($BaCl_2.2H_2O$) is heated it yields the anhydrous salt. Suggest a reason for this difference in behaviour.
7. Write a general account of the chemistry of radium and its chief compounds, predicting the properties from the trends shown in group II.
8. The magnesium ion has the same electron configuration as neon. Compare the properties of these two species and account, as far as you can, for any differences.
9. When sodium carbonate is added to a magnesium salt solution, a basic carbonate ($Mg(OH)_2.MgCO_3.xH_2O$) precipitates. Explain why this is so.
10. Predict the reaction of the following compounds, in aqueous solution, towards litmus:
    (a) sodium bisulphate;
    (b) potassium carbonate;
    (c) magnesium sulphate;
    (d) calcium chloride;
    (e) barium sulphate.
11. Write a general account of the chemistry of francium and its chief compounds, predicting the properties from the trends shown in group I.
12. Strontium occurs in the form of four isotopes: 0.6% of relative mass 83.94, 9.9% of relative mass 85.94, 7% of relative mass 86.94, 82.5% of relative mass 87.93. From these data calculate the atomic weight of strontium.
13. 0.257 g of impure sodium peroxide was dissolved in water and the volume was made up to 250 cm^3. 50 cm^3 of this solution exactly reduced 8.07 cm^3 of 0.02 M $KMnO_4$. Calculate the percentage of the impure material which was in fact sodium peroxide.
14. (a) Why is the solubility product of calcium carbonate not equal to the square of its solubility in water?
    (b) Why does calcium carbonate precipitate if carbon dioxide is passed into calcium hydroxide solution, but not if it is passed into calcium chloride solution?
15. Calculate the pH of a saturated strontium hydroxide solution ($K_s\,Sr(OH)_2 = 3.2 \times 10^{-4}$).

16. How many mole of strontium hydroxide can be dissolved in 1000 cm^3 of each of the following:
    (a) pure water;
    (b) 1.0 M NaOH;
    (c) 1.0 M $SrCl_2$.
17. Calculate the concentrations of each ion in each of the following mixtures:
    (a) 20 cm^3 of 0.2 M $CaCl_2$ and 40 cm^3 of 0.2 M $SrCl_2$;
    (b) 10 cm^3 of 0.005 M HCl and 40 cm^3 of 0.01 M $Ba(OH)_2$;
    (c) 20 cm^3 of 0.1 M H_2SO_4 and 40 cm^3 of 0.05 M $SrCl_2$.
    (K_s $SrSO_4$ = 7.6×10^{-7})
18. (a) To 10 cm^3 of 0.1 M $MgCl_2$ is added 10 cm^3 of 2 M NH_3 solution. To 10 cm^3 of 0.1 M $SrCl_2$ is added 10 cm^3 of 2 M NH_3 solution. Will a precipitate form in either solution? (K_b NH_3 = 1.8×10^{-5}; K_s $Mg(OH)_2$ = 5.5×10^{-12}; K_s $Sr(OH)_2$ = 3.2×10^{-4})
    (b) Explain why solid magnesium hydroxide and calcium hydroxide dissolve in 2 M NH_4Cl.

25

Group VII Elements

The elements of group VII possess a common outer-shell configuration of seven electrons (s^2p^5). They are collectively known as the *halogens* (Greek—salt-formers) as they are all very reactive non-metals which can combine directly with metals to form 'salts'. Fluorine indeed is exceptionally reactive, combining directly with nearly all known elements. The halogens occur largely in the form of halide ions in minerals. Fluorine is found, for example, in the widely distributed mineral fluorite (CaF_2); chlorine is found in sodium chloride; bromine in soluble bromides in sea-water, and iodine as soluble iodide in sea-water, also as sodium iodate ($NaIO_3$) in

I	II											III	IV	V	VI	VII	VIII
1 H																	2 He
3 Li	4 Be											5 B	6 C	7 N	8 O	9 F	10 Ne
11 Na	12 Mg											13 Al	14 Si	15 P	16 S	17 Cl	18 Ar
19 K	20 Ca	21 Sc	22 Ti	23 V	24 Cr	25 Mn	26 Fe	27 Co	28 Ni	29 Cu	30 Zn	31 Ga	32 Ge	33 As	34 Se	35 Br	36 Kr
37 Rb	38 Sr	39 Y	40 Zr	41 Nb	42 Mo	43 Tc	44 Ru	45 Rh	46 Pd	47 Ag	48 Cd	49 In	50 Sn	51 Sb	52 Te	53 I	54 Xe
55 Cs	56 Ba	57 La	72 Hf	73 Ta	74 W	75 Re	76 Os	77 Ir	78 Pt	79 Au	80 Hg	81 Tl	82 Pb	83 Bi	84 Po	85 At	86 Rn
87 Fr	88 Ra	89 Ac															

Table 25.1
Physical properties of group VII elements

Element	*Atomic weight*	*Electron configuration*	*Outer-shell configuration*	*Electro-negativity*	*Percentage abundance in earth's crust*
Fluorine	18.998	2.7	$2s^22p^5$	4.0	0.06–0.09
Chlorine	35.453	2.8.7	$3s^23p^5$	3.0	0.0314
Bromine	79.909	2.8.18.7	$4s^24p^5$	2.8	1.6×10^{-4}
Iodine	126.904	2.8.18.18.7	$5s^25p^5$	2.5	3×10^{-5}

sodium nitrate deposits. Chemically the elements show considerable similarity, with fluorine sometimes showing marked differences from the general trends shown by the group. The last element, astatine (At), is radioactive and available only in minute amounts. It will be omitted from the following discussion.

25.1 STRUCTURE OF THE ELEMENTS

The elements all exist as diatomic molecules involving four electron pairs in the outer shell of each atom. The molecules persist in the gas, liquid and solid phases. In the gas phase the elements are all coloured, pungent, and very poisonous. Fluorine gas is yellow, chlorine gas is greenish yellow, bromine vapour is dark red, and iodine vapour is purple. Fluorine and chlorine exist as gases at normal temperatures, whilst bromine is a dark red liquid, and iodine is a black, lustrous, crystalline solid. In iodine crystals the diatomic molecules are arranged in layers, with weak dispersion forces acting between molecules. This structure accounts for the flakiness of the iodine crystal. The other halogens in the solid state display a similar structure. The magnitudes of the dispersion forces increase with increase in molecular size, hence we might expect a rise in the melting point, boiling point and heat of vaporization as we pass down the group. This is experimentally observed to be true, the data being shown in Table 25.2.

If the elements are heated their total energy rises, and the molecules may acquire sufficient energy to break the bonds and form free halogen atoms. At a given temperature the equilibrium

$$X_2(g) \rightleftharpoons 2X(g)$$

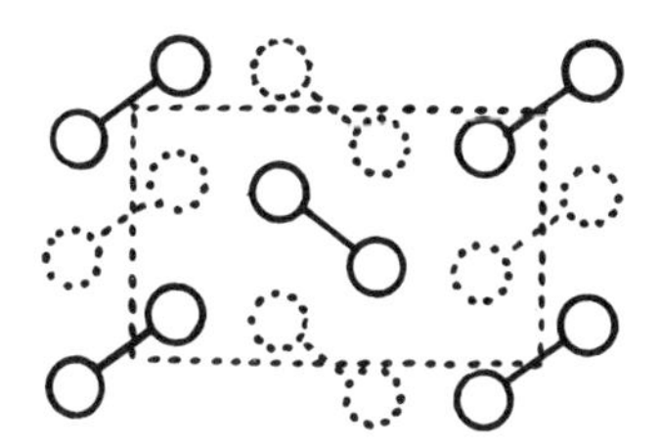

Fig 25.1 Structure of the iodine crystal.

Table 25.2

Further physical properties of group VII elements

Element	*Melting point (°C)*	*Boiling point (°C)*	*Heat of vaporization (kJ)*	*Bond dissociation energy (kJ) ($X_2 \rightarrow 2X$)*	*Equilibrium constant (K) at 1000°C ($X_2 \rightleftharpoons 2X$)*
Fluorine	−218	−188	3.347	158	4.9×10^{-3}
Chlorine	−101	−34.1	9.204	242	1×10^{-10}
Bromine	−7.3	58.8	15.99	192	8.1×10^{-5}
Iodine	−114	184	31.18	149	8.1×10^{-3}

is set up. The known equilibrium constants for the reaction (at 1000°C) are given in Table 25.2. It can be seen that the equilibrium constant rises from chlorine to iodine and this may be correlated with the decrease in the value of the bond dissociation energy. The equilibrium constant and bond dissociation energy for fluorine are less than might be expected from the trend shown by other members of the group. No simple explanation is available for this irregularity in trend.

The standard redox potentials for the halogens

$$X_2 + 2e \rightarrow 2X^-$$

are listed in Table 25.3. The halogens are all quite strong oxidants. This may be seen by comparing their E°s with other well-known oxidizing and reducing agents.

Table 25.3

Standard redox potentials

Oxidant		*Reductant*		*Half-reaction*	*Standard redox potential (volt)*
↑	F_2	F^-	│	$F_2 + 2e \rightarrow 2F^-$	+2.87
│	MnO_4^-, H^+	Mn^{2+}	│	$MnO_4^- + 8H^+ + 5e \rightarrow Mn^{2+} + 4H_2O$	+1.52
Increasing	Cl_2	Cl^-	Increasing	$Cl_2 + 2e \rightarrow 2Cl^-$	+1.36
oxidizing	Br_2	Br^-	reducing	$Br_2 + 2e \rightarrow 2Br^-$	+1.09
strength	Fe^{3+}	Fe^{2+}	strength	$Fe^{3+} + e \rightarrow Fe^{2+}$	+0.77
│	I_2	I^-	│	$I_2 + 2e \rightarrow 2I^-$	+0.54
│	S, H^+	H_2S	↓	$S + 2H^+ + 2e \rightarrow H_2S$	+0.14

It can be seen that in this sequence of systems fluorine is the strongest oxidant and hydrogen sulphide is the strongest reductant. Thus all of the halogens will oxidize hydrogen sulphide.

$$X_2 + 2e \rightarrow 2X^-$$

$$H_2S \rightarrow S + 2H^+ + 2e$$

$$\overline{H_2S + X_2 \rightarrow 2H^+ + 2X^- + S}$$

However, chlorine and bromine will oxidize iron(II), but iodine reacts only to a very slight extent. Thus for the reaction

$$2Fe^{2+} + Cl_2 \rightleftharpoons 2Fe^{3+} + 2Cl^- \qquad K = 5 \times 10^{20}$$

but for the reaction

$$2Fe^{2+} + I_2 \rightleftharpoons 2Fe^{3+} + 2I^- \qquad K = 8 \times 10^{-8}$$

Clearly the iron(III) ion will oxidize the iodide ion, as may be predicted from Table 25.3.

$$2Fe^{3+} + 2I^- \rightleftharpoons 2Fe^{2+} + I_2 \qquad K = 1/(8 \times 10^{-8}) \simeq 1 \times 10^7$$

Within the halogen group itself we can see that fluorine can oxidize chloride ion to chlorine. However, in aqueous solution the fluorine also oxidizes water

$$2H_2O + 2F_2 \rightarrow 4HF + O_2$$

so that the preparation of chlorine by oxidizing chloride ion with fluorine is not practicable. However, chlorine will oxidize iodide ion. These reactions may be used to prepare the appropriate halogens.

$$\begin{array}{l} Cl_2 + 2e \rightarrow 2Cl^- \\ 2Br^- \rightarrow Br_2 + 2e \\ \hline Cl_2 + 2Br^- \rightarrow Br_2 + Cl^- \end{array} \qquad \begin{array}{l} Cl_2 + 2e \rightarrow 2Cl^- \\ 2I^- \rightarrow I + 2e \\ \hline 2Cl_2 + 2I^- \rightarrow I_2 + 2Cl^- \end{array}$$

Table 25.3 also shows that, in acidic solution, chloride, bromide and iodide ions are oxidized by potassium permanganate. These reactions provide a further method for preparing the appropriate halogen from halide ion. The halide ion is often obtained by using a hydrohalic acid solution. Manganese dioxide may also be used as the oxidant.

$$\begin{array}{r} MnO_4^- + 8H^+ + 5e \rightarrow Mn^{2+} + 4H_2O \\ 2Cl^- \rightarrow Cl_2 + 2e \\ \hline 2MnO_4^- + 10Cl^- + 16H^+ \rightarrow 2Mn^{2+} + 8H_2O + 5Cl_2 \end{array}$$

In summary we may note that the strongest reductant amongst the halide ions is iodide, and the weakest reductant is fluoride. Thus we find that whilst iodide ion will reduce Fe^{3+} none of the other halide ions can accomplish this. Indeed the fluoride ion is so difficult to oxidize that the preparation of elemental fluorine is very difficult. To prepare fluorine the fluoride ion is oxidized solution of potassium fluoride in liquid hydrogen fluoride. The oxidation is performed by electrolysis, the fluoride ion being oxidized at the positive electrode

$$2F^- \rightarrow F_2 + 2e$$

The oxidation reduction reactions discussed earlier have referred to aqueous solution. The halogens will oxidize many metals in direct reaction. In the preparation of fluorine this presents a real problem since fluorine will oxidize any metal (or glass) container. However, fluorine will oxidize copper or nickel to the appropriate fluoride which then forms a thin unreactive film over the metal. This reduces the rate of reaction and copper or nickel vessels can be used for storing, preparing and reacting fluorine.

25.2 FEATURES OF THE COMPOUNDS

Most of the elements will form compounds with the halogens and many of these compounds will be dealt with in subsequent chapters on the various groups. In forming compounds where the electronegativity difference is small, the bonding shown will be predominantly covalent. In such compounds fluorine will have a maximum of four electron pairs in the outer shell, as in hydrogen fluoride, HF. In the case of the other halogens, the maximum number of electron pairs in the outer shell may exceed four. Thus in chlorine trifluoride (ClF_3) there are five electron pairs in the outer shell of the chlorine atom, in bromine pentafluoride (BrF_5) there are six electron pairs in the outer shell of the bromine atom, and in iodine heptafluoride (IF_7) there are seven electron pairs in the outer shell of the iodine atom.

In forming compounds where the electronegativity difference is large, network solids or layer lattices are formed, in which the bonding is predominantly ionic. Thus when potassium reacts with chlorine, electrons are lost from potassium atoms yielding potassium ions, whilst chlorine atoms gain electrons to form chloride ions. The ions form an ionic network (KCl). Magnesium chloride, by contrast, is a layer lattice.

25.3 THE HALOGEN HYDRIDES

All of the halogens can combine with hydrogen to form hydrogen halides:

$$H_2 + X_2 \rightleftharpoons 2HX$$

The extent of reaction decreases down the group. Fluorine reacts explosively and almost completely, whilst iodine reacts slowly and partially under the same conditions. The mechanisms of the reaction of hydrogen with iodine and with bromine have been discussed in chapter 21. The mechanism of reaction involving chlorine resembles that involving bromine. The hydrogen halides may also be prepared by reacting a high boiling point acid with a halide salt. Concentrated sulphuric acid may be used with fluorides and chlorides.

$$CaF_2 + H_2SO_4 \rightarrow CaSO_4 + 2HF$$

$$NaCl + H_2SO_4 \rightarrow NaHSO_4 + HCl$$

With bromides and iodides, concentrated sulphuric acid is unsuitable because the ions are oxidized to the halogen, e.g.:

$$H_2SO_4 + 2H^+ + 2Br^- \rightarrow 2H_2O + SO_2 + Br_2$$

However, phosphoric acid may be used:

$$NaBr + H_3PO_4 \rightarrow NaH_2PO_4 + HBr$$

The bonding in the hydrogen halides is covalent but the electronegativity difference indicates that the molecules are dipolar. In hydrogen fluoride the bond is highly polar and, in fact, strong hydrogen bonding exists between the molecules. This accounts for the very high boiling point and comparatively high melting point compared with other members of the group. The bonding is sufficiently strong to allow aggregates of HF to persist in the vapour. The highly polar hydrogen fluoride liquid acts as an important non-aqueous solvent. It will dissolve salts such as sodium sulphate as well as many organic compounds.

The equilibrium constants (at 1000°C) for the dissociation

$$2HX \rightleftharpoons H_2 + X_2$$

are listed in Table 25.4. Clearly the hydrides become less stable down the group. It is also clear that the bond dissociation energy decreases down the group and this trend, at least partially, explains the trend in stability.

Table 25.4

Physical properties of the halogen hydrides

Hydride	*Boiling point (°C)*	*Melting point (°C)*	*Acidity constant* (K_a)	*Bond dissociation energy (kJ)* $HX(g) \rightarrow H(g) + X(g)$	*Equilibrium constant* (K) *at 1000°C* ($2HX \rightleftharpoons H_2 + X_2$)
Hydrogen fluoride	+19.5	−83	10^{-4}	562	1×10^{-24}
Hydrogen chloride	−85	−115	10^{+7}	431	5×10^{-9}
Hydrogen bromide	−67	−87	10^{+9}	366	6×10^{-6}
Hydrogen iodide	−35	−51	10^{+11}	298	6×10^{-2}

When the hydrides are in the pure state or dissolved in some inert solvent (such as toluene) they do not conduct electricity as there are no ions present. In aqueous solution, however, ions are formed as the proton is donated to the base water

$$HX + H_2O \rightleftharpoons H_3O^+ + X^-$$

and such solutions will conduct electricity. The solutions are referred to as hydrohalic acids, e.g. hydrochloric acid is the solution of hydrogen chloride in water. Inspection of acidity constants shows that hydrogen fluoride is a comparatively weak acid, about as strong as formic acid. Hydrogen chloride, hydrogen bromide and hydrogen iodide are very strong, being almost completely present in solution in the form of ions. In fact for these three acids acidity constants have been difficult to measure exactly. It is not easy to give a simple explanation of the trends in the strengths of these acids; however, we can say that the bond dissociation energy is an important factor. Other factors involved are the energy changes associated with the reaction of the acids, and their ions, with water. A further discussion of trends in acid strengths of non-metal hydrides is given in chapter 23.

Aqueous solutions of all of these acids contain hydrogen ions, hence aqueous solutions will oxidize metals with negative values for $E°$. For example, for the systems

$$Mg^{2+} + 2e \rightarrow Mg \qquad E° = -2.34 \text{ volt}$$

and

$$2H^{+} + 2e \rightarrow H_2 \qquad E° = 0.00 \text{ volt}$$

Hence when magnesium is added to hydrochloric acid the following reaction occurs:

$$\begin{array}{r} Mg \rightarrow Mg^{2+} + 2e \\ 2H^{+} + 2e \rightarrow H_2 \\ \hline Mg + 2H^{+} \rightarrow Mg^{2+} + H_2 \end{array}$$

Although hydrogen fluoride is a weak acid, hydrofluoric acid is a very reactive material. It will readily etch glass:

$$SiO_2 + 4HF \rightarrow SiF_4 + 2H_2O$$

hence it is stored in rubber or wax bottles. It also attacks flesh vigorously and painfully.

25.4 INTERHALOGEN COMPOUNDS

The halogens can react directly with one another to form a series of interhalogen compounds, some of the higher members being formed by the reaction of halogens with lower interhalogens:

$$I_2 + 3Cl_2 \rightarrow 2ICl_3$$

$$ICl + Cl_2 \rightarrow ICl_3$$

All of the halogens have high values for the electronegativity and in the interhalogens the electronegativity differences range from zero in halogen molecules (hence no polarity) to quite high in a compound such as iodine

Table 25.5

Interhalogen compounds

F	*Cl*	*Br*	*I*	
F_2(g)	ClF(g) ClF_3(g)	BrF(g) BrF_3(l)	IF_5(l) IF_7(s)	*F*
	Cl_2(g)	ClBr(l)	ClI(s, l) Cl_3I(l)	*Cl*
		Br_2(l)	BrI(s)	*Br*
			I_2(s)	*I*

monochloride (ICl), which is appreciably polar. In between these extremes various degrees of polarity are shown. The interhalogens are often coloured, for example iodine monochloride is red, and they may exist as gases, liquids or solids at room temperature. They are chemically very reactive; for example, chlorine trifluoride (ClF_3) can oxidize antimony to antimony pentafluoride (SbF_5) more readily than can fluorine itself. The known interhalogens are listed in Table 25.5, with the physical state at 25°C indicated.

The number of electron pairs in the outer shell of the 'central' halogen atom can be:

(a) four, as in chlorine monofluoride, ClF.
(b) five, as in chlorine trifluoride, ClF_3.
(c) six, as in bromine pentafluoride, BrF_5.
(d) seven, as in iodine heptafluoride, IF_7.

These examples are shown in Table 25.6.

Table 25.6

Stereochemistry of the interhalogen compounds

Total number of electron pairs	*Arrangement of electron pairs*	*Number of bonding pairs*	*Number of non-bonding pairs*	*Shape of molecule*	*Examples*
5	Trigonal bipyramid	3	2	 F F Cl— F	ClF_3, ICl_3
6	Octahedral	5	1	 F F F Br F F	BrF_5, IF_5
7	Pentagonal bipyramid	7	0	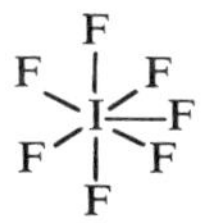 F F F I—F F F F	IF_7

Most of the interhalogens exist as discrete molecules, but some are partially ionized. These latter will conduct slightly in the molten state, or when dissolved in organic solvents. Examples of partially ionized interhalogens are

$$2IF_5 \rightleftharpoons IF_4^+ + IF_6^-$$

$$2ICl \rightleftharpoons I^+ + ICl_2^-$$

$$2ICl_3 \rightleftharpoons ICl_2^+ + ICl_4^-$$

The shapes of these ions may also be deduced from the charge cloud repulsion theory. For example, ICl_4^- is planar:

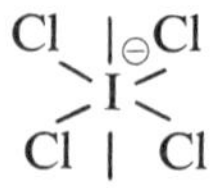

Cl Cl
I
Cl Cl

Fig 25.2 Valence structure of the ICl_4^- ion.

The halide ions can add on additional halogen atoms to form polyhalide ions. Thus iodine reacts with iodide ion to form the tri-iodide ion (I_3^-).

$$I_2(aq) + I^-(aq) \rightleftharpoons I_3^-(aq)$$

The valence structure of the tri-iodide ion is similar to that of the ICl_2^- ion, viz. a linear arrangement.

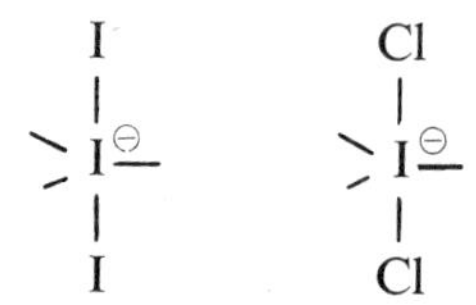

Fig 25.3 Valence structures of the I_3 and ICl_2^- ions.

Bromine has a slight tendency to form the tribromide ion (Br_3^-) and chlorine an even slighter tendency to form the trichloride ion (Cl_3^-).

25.5 HALOGEN OXYACIDS

The halogens do not directly combine with oxygen, but a range of halogen oxides can be prepared by indirect methods. Examples are fluorine monoxide (F_2O), chlorine heptoxide (Cl_2O_7), iodine pentoxide (I_2O_5). These oxides are generally unstable and their reactions are usually quite complex. They will react with water to form oxyacids:

$$Cl_2O_7 + H_2O \rightarrow 2HClO_4$$

$$I_2O_5 + H_2O \rightarrow 2HIO_3$$

The halogens, with the exception of fluorine, form a series of oxyacids. The known oxyacids and their acidity constants are listed in Table 25.7. Within these series of acids we may note the following trends:

(a) For a given acid type the acidity constant increases as the electronegativity of the halogen atom increases, e.g. in the series HOI, HOBr, HOCl. This trend may be interpreted as being due to the weakening of the OH bond as the increasingly electronegative halogen atoms cause electron withdrawal, thus making it easier for the hydrogen to be removed as proton by water.

(b) For a given halogen the acidity constant increases as the number of oxygen atoms bonded directly to the halogen atom increases, e.g. in the series HOCl, $HClO_2$, $HClO_3$, $HClO_4$. This dramatic change in acidity constant is not easily interpreted, but one factor is again the progressive

Table 25.7
The halogen oxyacids

Hypohalous		*Halous*		*Halic*		*Perhalic*	
HOCl Cl(OH)	10^{-7}	$HClO_2$ ClO(OH)	10^{-2}	$HClO_3$ $ClO_2(OH)$	10^{+1}	$HClO_4$ $ClO_3(OH)$	10^{+10}
HOBr Br(OH)	10^{-9}	—		$HBrO_3$ $BrO_2(OH)$	strong	—	
HOI I(OH)	10^{-10}	—		HIO_3 $IO_2(OH)$	2×10^{-1}	H_5IO_6 $IO(OH)_5$	5×10^{-4}

weakening of the OH bond due to electron withdrawal as the number of electronegative oxygen atoms increase, thus making it easier for the hydrogen to be removed as a proton by water.

The valence structures and stereochemistry of the chlorine oxyacids are indicated in Figure 25.4.

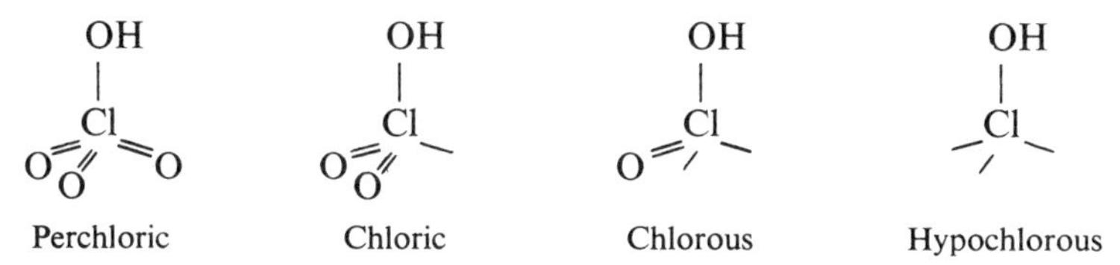

Fig 25.4 Valence structures of the chlorine oxyacids.

Hypochlorous acid (HClO) is formed when chlorine reacts with water. This may be represented, at 25°C:

$$Cl_2(g) \rightleftharpoons Cl_2(aq) \qquad K_1 = 0.06$$

$$Cl_2(aq) + H_2O \rightleftharpoons H^+ + Cl^- + HOCl \qquad K_2 = 4 \times 10^{-4}$$

and for the overall reaction

$$Cl_2(g) + H_2O \rightleftharpoons H^+ + Cl^- + HOCl \qquad K_3 = K_1 \times K_2 = 2.4 \times 10^{-5}$$

In alkaline solution the HOCl and H^+ react with hydroxide ions:

$$H^+ + OH^- \rightleftharpoons H_2O$$

$$HOCl + OH^- \rightleftharpoons OCl^- + H_2O$$

so that the overall equation becomes

$$Cl_2(g) + 2OH^- \rightleftharpoons Cl^- + OCl^- + H_2O \qquad K_4 = 7.5 \times 10^{15}$$

The hypochlorite ion in basic solution is unstable and reacts as follows:

$$3OCl^- \rightleftharpoons 2Cl^- + ClO_3^- \qquad K_5 = 10^{27}$$

However, the rate of this reaction is low and equilibrium is established slowly. We may regard the hypochlorite ion as being 'metastable'. Similarly for the chlorate ion in solution the following equilibrium is set up:

$$4ClO_3^- \rightleftharpoons Cl^- + 3ClO_4^- \qquad K_6 = 10^{29}$$

Again the rate of reaction is slow and the chlorate ion is 'metastable'. If solid potassium chlorate is carefully heated it does form a perchlorate:

$$4KClO_3 \rightarrow 3KClO_4 + KCl$$

On stronger heating the perchlorate decomposes:

$$KClO_4 \rightarrow KCl + 2O_2$$

The overall equation is

$$2KClO_3 \rightarrow 2KCl + 3O_2$$

The oxyacids and derived oxyions can act as oxidants. For example, hypochlorous acid can oxidize chloride ion

$$HOCl + H^+ + e \rightarrow \tfrac{1}{2}Cl_2 + H_2O$$

$$Cl^- \rightarrow \tfrac{1}{2}Cl_2 + e$$

$$\overline{HOCl + H^+ + Cl^- \rightarrow Cl_2 + H_2O}$$

25.6 THE METAL HALIDES

There is a very large number of metal halides which can be prepared by direct oxidation of the metal by the appropriate halogen, for example

$$2Fe + 3Cl_2 \rightarrow 2FeCl_3$$

$$Sn + 2I_2 \rightarrow SnI_4$$

All of the halides of the alkali metals are ionic network solids and have fairly high melting and boiling points, e.g. for sodium fluoride the melting point is 995°C and the boiling point 1704°C. These networks show 6 : 6 co-ordination except for CsCl, CsBr, CsI where an 8 : 8 co-ordination is shown. For other metal halides, it is found that the fluorides often differ in structure from the other halides. Fluorine has a high value for the electronegativity and in many cases the fluorides, of general formula MF_2 or MF_3, formed with metals (which have low electronegativities) are ionic networks, whereas the corresponding chlorides, bromides, and iodides of the same general formula may consist of layer or chain lattices. Thus iron(III) fluoride has a network structure, whereas the chloride, bromide, and iodide form layer lattices.

The metal halides are generally soluble in water. In this respect the fluorides may behave rather differently from the other halides. Thus, whereas the chloride, bromide and iodide of silver are insoluble, silver fluoride is extremely soluble. By comparison, the chlorides, bromides and iodides of the alkaline earth metals are soluble, whereas the fluorides are insoluble. In summary the sparingly soluble halides are: (a) those of mercury(I) (Hg_2^{2+}), silver (Ag^+), and copper(I) (Cu^+), (b) the group II fluorides and lead(II) fluoride (PbF_2), and in addition $PbCl_2$, $PbBr_2$ and PbI_2 are sparingly soluble in cold water, but much more soluble in hot solution.

The halogen chemistry of solutions of these metal halides is largely the chemistry of the appropriate halide ions. Thus a solution of any chloride will react with silver nitrate solution to yield a white precipitate of silver chloride, soluble in ammonia solution:

$$Ag^+(aq) + Cl^-(aq) \rightleftharpoons AgCl(s)$$

$$AgCl(s) + 2NH_3(aq) \rightleftharpoons Ag(NH_3)^+(aq) + Cl^-(aq)$$

A solution of any bromide will yield a cream precipitate of silver bromide partially soluble in ammonia solution

$$Ag^+(aq) + Br^-(aq) \rightleftharpoons AgBr(s)$$

and a solution of any iodide will yield a yellow precipitate of silver iodide, 'insoluble' in ammonia solution:

$$Ag^+(aq) + I^-(aq) \rightleftharpoons AgI(s)$$

Silver fluoride is soluble in water. The solution of the chloride, bromide, and iodide of silver in ammonia solution is discussed in detail in chapter 16. These reactions may be used to distinguish the halide ions.

25.7 SUMMARY

The halogens are a group of very electronegative non-metals which can act as powerful oxidants. The oxidizing strength diminishes down the group. They oxidize metals to form a wide range of metal halides of varied structures. In these halides the halogens show an oxidation state of -1. They show positive oxidation states of $+1$, $+3$, $+5$, and $+7$ in their various oxyacids and oxyions (with the exception of fluorine which always shows an oxidation state of -1). The oxyacids and oxyions are all oxidants ultimately yielding halide ion or free halogen. The halogens can also react with one another to form a range of interhalogens, and with hydrogen to form the halogen hydrides, which are acids.

QUESTIONS

1. Chlorine, bromine, and iodine will all oxidize hydrogen sulphide to sulphur. Chlorine and bromine will oxidize nitrous acid to nitric acid, iodine will not. Chlorine will oxidize bromide ion to bromine. Write equations for all chemical reactions and show how the three halogens may be arranged in order of increasing oxidizing strength using these data.
2. What experimental evidence can be produced for believing that chlorine is a diatomic molecule?
3. Chlorine can form chlorine trifluoride, bromine can form bromine pentafluoride, and iodine can form iodine heptafluoride. Write valence structures for these molecules and show their stereochemistry.
4. Fluorine may be prepared by electrolysis of potassium hydrogen fluoride (KHF_2) dissolved in liquid hydrogen fluoride at a temperature of 65°C in a steel cell.

(a) Why is liquid hydrogen fluoride used, rather than water, as the solvent?
(b) Since fluorine is very reactive how can steel apparatus be used?

5. If molten iodine monochloride is electrolysed, iodine is formed at the negative electrode and chlorine at the positive electrode. Explain these observations.
6. Discuss the differences in properties of hydrogen fluoride from the other hydrogen halides. Account for these differences as far as you can.
7. In preparing the hydrogen halides one may make use of the reaction of the sodium halide salt with concentrated sulphuric acid

$$NaX + H_2SO_4 \rightarrow NaHSO_4 + HX$$

This method works for HF and HCl, but not for HBr or HI. Why? If phosphoric acid (H_3PO_4) is used instead of sulphuric acid all of the hydrogen halides may be prepared. Account for this.

8. Iodine is practically insoluble in water but quite soluble in carbon tetrachloride, but potassium iodide is insoluble in this solvent. Account for these differences.
9. In each of the following unbalanced equations identify the oxidant and the reductant. Write partial ionic equations for the oxidizing action and for the reducing action, and then write the oxidation-reduction equation:
(a) $Cr_2O_7^{2-} + I^- + H^+ \rightarrow Cr^{3+} + I_2 + H_2O$
(b) $O_3 + Cl^- + H^+ \rightarrow O_2 + Cl_2 + H_2O$
(c) $SO_4^{2-} + Br^- + H^+ \rightarrow Br_2 + SO_2 + H_2O$
(d) $ClO_2^- + OH^- \rightarrow ClO^- + ClO_3^- + H_2O$
(e) $Cl_2 + OH^- \rightarrow Cl^- + ClO^- + H_2O$
(f) $HSO_3^- + IO_3^- + H^+ \rightarrow SO_4^{2-} + I_2 + H_2O$
(g) $ClO^- \rightarrow Cl^- + ClO_3^-$
(h) $HNO_3 + I_2 \rightarrow HIO_3 + NO + H_2O$
10. Account for the reaction to litmus (acidic, alkaline, or neutral) of the following salts in aqueous solution:
(a) potassium chloride; (b) sodium hypochlorite;
(c) iron(III) chloride; (d) sodium fluoride.
11. Explain how one may deduce the shapes of the following molecules and ions:
(a) BrF_3; (b) ClO_2^-; (c) ClO_4^-; (d) IF_5; (e) ICl_2^-; (f) BrF_5.
12. If gaseous silicon tetrafluoride is added to water, silicon dioxide precipitates and the strong acid, fluorosilicic acid, forms. The solution of this may be filtered off. If a solution containing barium ion is added to the acid solution, barium fluorosilicate precipitates,

$$3SiF_4(g) + 2H_2 \rightarrow SiO_2(s) + 2H_2SiF_6(aq)$$

$$Ba^{2+}(aq) + SiF_6^{2-}(aq) \rightarrow BaSiF_6(s)$$

(a) If 500 cm³ of silicon tetrafluoride, at 10°C, 765 mmHg, is added to water and the acid solution made up to 500 cm³, what is the molarity of the H_2SiF_6 solution?
(b) If to the solution in (a) 20 cm³ of 0.1 M $BaCl_2$ solution is added, what is the mass of precipitate obtained and what is the final molarity of the solution with respect to H_2SiF_6?

13. To 200 cm³ of 0.10 M $Ba(OH)_2$ is added 300 cm³ of 0.05 M $HClO_4$. To this solution is added 50 cm³ of M KCl. Will potassium perchlorate precipitate from this solution at 15°C if the solubility product of $KClO_4$ at 15°C is 1.2×10^{-2}?
14. Silicon tetrachloride is a liquid of specific gravity 1.49. When added to water it reacts

$$SiCl_4(l) + 2H_2O \rightarrow SiO_2(s) + 4HCl(aq)$$

5.0 cm^3 of liquid $SiCl_4$ is added to water. The solution of HCl is filtered off and made up to 400 cm^3. What is the molarity of the HCl solution obtained? What volume of 0.02 M $AgNO_3$ is needed to precipitate all the chloride ion as silver chloride?

15. A chlorine and fluorine compound has the percentage composition 37.4% chlorine, and 61.6% fluorine. 0.181 g of the gaseous compound occupies a volume of 46.5 cm^3 at 20°C, 770 mmHg. Calculate the molecular formula of the compound.
16. 10.0 cm^3 of a solution of 0.001 M NaBr, 0.001 M Na_2SO_4 is added to 10.0 cm^3 of 0.01 M $AgNO_3$. Will any precipitate form? Explain.
(K_s Ag_2SO_4 = 1.2×10^{-5}, K_s AgBr = 5.0×10^{-13}.)
17. An aqueous solution of 0.1 M HF is 7.8% dissociated. Calculate the acidity constant (K_a) for the acid.
18. A white anhydrous compound, thought to be silver diammine sulphate, was subjected to complete analysis as follows: 6.460 g of the compound was dissolved in distilled water and made up to 500 cm^3 with distilled water. 25 cm^3 of this solution was carefully neutralized and then 25 cm^3 0.150 M NaCl solution was added. The excess chloride which had not reacted with the silver was titrated with 0.100 M $AgNO_3$ using potassium chromate as an indicator. For three such determinations the average volume of silver nitrate required was 20.47 cm^3.

    In another set of determinations, 25 cm^3 of the original solution was treated with excess sodium hydroxide and the evolved ammonia allowed to react with 50 cm^3 0.1120 M HCl. The hydrochloric acid which had not reacted with the ammonia was titrated with 0.105 M NaOH. For three such determinations, the average volume of sodium hydroxide required to neutralize the excess hydrochloric acid was 20.86 cm^3.

    In a third set of determinations, 25 cm^3 of the original solution was acidified with dilute nitric acid and excess barium nitrate added. The resultant barium sulphate was filtered, washed and ignited to constant weight. The average mass of the barium sulphate was 0.1995 g for three such determinations.

    Calculate (a) the percentages of silver, ammonia and sulphate in the compound; (b) the empirical formula of the compound.

26

Group VI Elements

The elements in this group possess a common outer-shell configuration of six electrons (s^2p^4). Within the group there is a gradation in the nature of the elements from those which are non-metallic, e.g. oxygen, sulphur, to the predominantly metallic element polonium. Oxygen differs quite markedly in many properties from the other members of the group. This difference of the first member of the group has already been noticed with lithium in group I, beryllium in group II, and fluorine in group VII and is true for other groups as well. Sulphur and selenium by comparison are very

I	II											III	IV	V	VI	VII	VIII
1 H																	2 He
3 Li	4 Be											5 B	6 C	7 N	8 O	9 F	10 Ne
11 Na	12 Mg											13 Al	14 Si	15 P	16 S	17 Cl	18 Ar
19 K	20 Ca	21 Sc	22 Ti	23 V	24 Cr	25 Mn	26 Fe	27 Co	28 Ni	29 Cu	30 Zn	31 Ga	32 Ge	33 As	34 Se	35 Br	36 Kr
37 Rb	38 Sr	39 Y	40 Zr	41 Nb	42 Mo	43 Tc	44 Ru	45 Rh	46 Pd	47 Ag	48 Cd	49 In	50 Sn	51 Sb	52 Te	53 I	54 Xe
55 Cs	56 Ba	57 La	72 Hf	73 Ta	74 W	75 Re	76 Os	77 Ir	78 Pt	79 Au	80 Hg	81 Tl	82 Pb	83 Bi	84 Po	85 At	86 Rn
87 Fr	88 Ra	89 Ac															

Table 26.1
Physical properties of group VI elements

Element	*Atomic weight*	*Electron configuration*	*Outer-shell structure*	*Electro-negativity*	*Percentage abundance in earth's crust*
Oxygen	15.9994	2.6	$2s^22p^4$	3.5	46.6
Sulphur	32.064	2.8.6	$3s^23p^4$	2.5	0.052
Selenium	78.96	2.8.18.6	$4s^24p^4$	2.4	9×10^{-6}
Tellurium	127.60	2.8.18.18.6	$5s^25p^4$	2.1	1.8×10^{-7}
Polonium	—	2.8.18.32.18.6	$6s^26p^4$	2.0	3×10^{-14}

similar indeed in their properties. Oxygen is the most abundant element in the earth's crust. It occurs free in the atmosphere and combined in many minerals in the form of oxides or oxyions. Sulphur occurs free, sometimes in large deposits, and in many sulphide (e.g. iron pyrites, FeS_2) and sulphate (e.g. gypsum, $CaSO_4.2H_2O$) minerals. Selenium and tellurium occur in combined form in many native sulphides (e.g. iron pyrites) in trace amounts. Polonium occurs in the radioactive mineral pitchblende. It is a disintegration product of radium and is radioactive. It resembles bismuth as well as tellurium in much of its chemistry. We will not discuss the detailed chemistry of polonium in this chapter.

26.1 STRUCTURE OF THE ELEMENTS

In comparison with groups I, II and VII the elements of group VI show a wide variation in structure, and the gradation from the non-metallic element oxygen to the metallic element polonium is clearly shown.

(1) Oxygen. Oxygen exists in the gaseous, liquid, and solid states as discrete molecules (O_2). At high temperatures sulphur, selenium and tellurium vapours also contain diatomic molecules. By subjecting oxygen to an electrical discharge it may be converted to another modification called ozone (O_3):

$$3O_2 \rightarrow 2O_3$$

Ozone may be represented by the following valence structure:

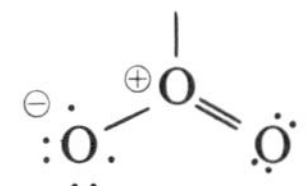

Fig 26.1 Valence structure of ozone molecule, O_3.

The molecule is experimentally found to be angular and the two bonds are found to be equivalent. The true state of the molecule is a resonance hybrid of two equivalent structures. Ozone is more reactive than oxygen itself. For example, it can oxidize silver to silver oxide, whereas this cannot be directly accomplished by oxygen.

(2) Sulphur and selenium. Sulphur can exist in a number of modifications, two important ones being rhombic sulphur (S_α) and monoclinic sulphur

(S_β). S_α is stable at room temperature and S_β is stable above 96°C. In both modifications (allotropes) the sulphur atoms are organized into a puckered ring of eight atoms, the packing of these rings being slightly different in each modification.

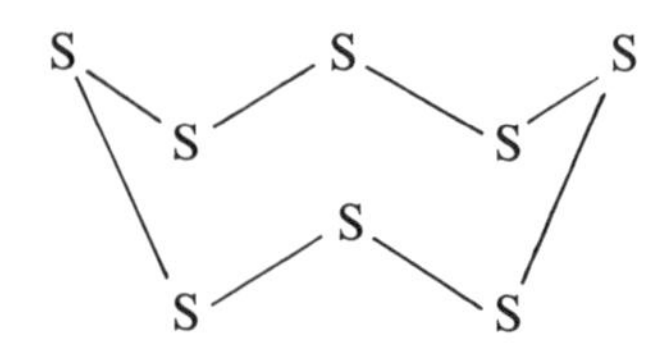

Fig 26.2 Valence structure of the S_8 molecule.

Selenium can exist in two modifications, Se_α and Se_β. These both contain Se_8 rings and the packing of these rings is similar to the packing of the S_8 rings in S_α and S_β. Se_α and Se_β are metastable with respect to a third grey, or 'metallic', form. This consists of spiral chains of selenium atoms with dispersion forces acting between adjacent chains. Grey selenium is a flaky material, like graphite, and it also conducts electricity, but has a higher resistivity than a typical metal, such as copper. The resistivity is lowered in the presence of light, hence the use of selenium in photo-cells and similar devices.

(3) Tellurium. There is only one form, called grey tellurium. This closely resembles grey selenium in structure.

(4) Polonium. This is a radioactive metal which emits α-particles. It is very rare but is now prepared in milligram amounts by the neutron bombardment of ^{290}Bi. Its electrical resistivity is comparable with that of typical metals.

26.2 FEATURES OF THE COMPOUNDS

In compounds formed between group VI elements and other elements having high electronegativities the bonding is predominantly covalent. In such compounds, oxygen can have a maximum of four electron pairs in the outer shell, as in fluorine monoxide, F_2O.

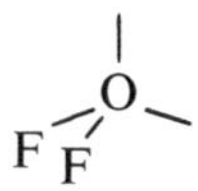

O
F F

Fig 26.3 Valence structure of the F_2O molecule.

The other elements can have more than four electron pairs in the outer shell. For example, in hydrogen telluride (H_2Te) there are four electron pairs, but in selenium tetrafluoride (SeF_4) there are five electron pairs, and in sulphur hexafluoride (SF_6) there are six electron pairs.

Fig 26.4 Valence structures of H_2Te, SeF_4, SF_6 molecules.

These compounds illustrate the range of oxidation states attained. An oxidation state of -2 is shown by each element (e.g. H_2S), oxidation states of $+4$ (e.g. SO_2) and $+6$ (e.g. SO_3) are shown by the elements from sulphur to polonium, whilst an oxidation state of $+2$ may be shown if the elements are bonded to other very electronegative elements (e.g. TeF_2, OF_2).

In forming compounds with elements of low electronegativity (metals) the bonding is expected to be predominantly ionic. Metals for which this is true are all the group I metals and most of the group II metals. The resulting network lattices contain the appropriate group VI ion (X^{2-}) and the appropriate group I or II ion (M^+ or M^{2+}). The formation of the metal cation and the negative group VI ion requires the overall absorption of energy, so that these negative ions are only present in ionic lattices, where the requisite stabilizing energy is provided by the formation of the lattice. Many metals have electronegativities comparable with at least some of the group VI elements and the electronegativity difference is not very great. In such cases the bonding may be intermediate in character between ionic and covalent. Thus zinc sulphide forms networks which can be regarded as made up of ions but the bonding is not purely ionic. Many of the selenium and tellurium compounds are of this type.

Oxygen can form a peroxide ion, O_2^{2-}, as in sodium peroxide, Na_2O_2, and a superoxide ion, O_2^-, as in potassium superoxide, KO_2. The sulphide ion can react with sulphur atoms to form polysulphide ions containing two to six sulphur atoms linked in a chain, e.g. Na_2S_5. These negative ions are examples of inorganic polymers. The simplest polysulphide ion is the persulphide ion, S_2^{2-}, as in FeS_2, which may be compared with the peroxide ion, O_2^{2-}. Polyselenides and polytellurides can also be formed, but the chains in the latter tend to be shorter (usually two or three atoms).

26.3 THE HYDRIDES

The general formula of these compounds is H_2X, with the group VI atom showing an oxidation state of -2. They all have four electron pairs in the outer shell of the group VI atom. By the charge cloud repulsion theory we would expect the four charge clouds to assume a tetrahedral configuration resulting in an angular molecule. This is confirmed by experiment.

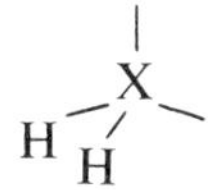

Fig 26.5 Valence structure of H_2X.

As the electronegativity of the group VI atom decreases so does the polarity of the molecules. In the case of water the oxygen atom is very electronegative and strong hydrogen bonding exists between molecules. As a consequence the melting point, boiling point and heat of vaporization are very much greater than for hydrogen sulphide. This has been discussed in detail in chapter 10. For the hydrides other than water, dispersion forces increase down the group as molecular size increases, and at the same time the bond polarity decreases. The effect of the dispersion forces is sufficiently strong for the melting point, boiling point and heat of vaporization to rise from H_2S to H_2Te.

Table 26.2

Physical properties of the group VI hydrides

Hydride	*Melting point (°C)*	*Boiling point (°C)*	*Heat of vaporization (kJ)*	*Heat of formation (ΔH_f° in kJ)*	*Acidity constant* (K_a)
Water	0.0	100	40.8	−241.8	2×10^{-16}
Hydrogen sulphide	−85.6	−60.75	18.7	−20.2	10^{-7}
Hydrogen selenide	−65.4	−41.5	19.9	+85.8	10^{-4}
Hydrogen telluride	−51	−1.8	*ca.* 24	+154.4	10^{-3}

The acidity constants increase down the group. The explanation of this trend is complex, but some of the factors involved have been discussed in chapter 25. It is worth noting that the heat of formation correlates with the trend in acidity; however, it must be stressed that this is but one factor involved in an analysis of the energetics which will lead to a satisfactory explanation of the observed trends. Inspection of the values of the standard redox potentials, listed in Table 26.3, shows that the hydrides become increasingly stronger reductants down the group.

Water can act as a reductant with sufficiently strong oxidants, such as chlorine, bromine, and fluorine. For example, for the system

$$F_2 + 2e \rightarrow 2F^- \qquad E^\circ = 2.87 \text{ volt}$$

hence we expect fluorine to oxidize water and this is found to occur:

$$2F_2 + 2H_2O \rightarrow 4HF + O_2$$

Hydrogen sulphide is a common strong reductant. In agreement with Table 26.3 it will reduce Fe^{3+} and also hydrogen peroxide

$$H_2O_2 + 2H^+ + 2e \rightarrow 2H_2O$$
$$H_2S \rightarrow S + 2H^+ + 2e$$
$$H_2O_2 + H_2S \rightarrow 2H_2O + S$$

$$Fe^{3+} + e \rightarrow Fe^{2+}$$
$$H_2S \rightarrow S + 2H^+ + 2e$$
$$2Fe^{3+} + H_2S \rightarrow 2Fe^{2+} + 2H^+ + S$$

As has been discussed in chapter 25, it will also reduce all of the halogens.

Hydrogen selenide and hydrogen telluride are even stronger reductants than hydrogen sulphide.

Water can also act as an oxidant, since for the system

$$2H_2O + 2e \rightarrow 2OH^- + H_2$$

Table 26.3
Standard redox potentials

Oxidant		Reductant		Half-reaction	Standard redox potential (volt)
↑	H_2O_2, H^+	H_2O		$H_2O_2 + 2H^+ + 2e \rightarrow 2H_2O$	+1.77
	O_2, H^+	H_2O		$\frac{1}{2}O_2 + 2H^+ + 2e \rightarrow H_2O$	+1.23
Increasing	Fe^{3+}	Fe^{2+}	Increasing	$Fe^{3+} + e \rightarrow Fe^{2+}$	+0.77
oxidizing	O_2, H^+	H_2O_2	reducing	$O_2 + 2H^+ + 2e \rightarrow H_2O_2$	+0.68
strength	S, H^+	H_2S	strength	$S + 2H^+ + 2e \rightarrow H_2S$	+0.14
	Se, H^+	H_2Se		$Se + 2H^+ + 2e \rightarrow H_2Se$	−0.40
	Te, H^+	H_2Te	↓	$Te + 2H^+ + 2e \rightarrow H_2Te$	−0.72

where $[OH^-] = 10^{-7}$ M (i.e. pure water) the redox potential is —0.41 volt. For sodium the E° of the system

$$Na^+ + e \rightarrow Na$$

is −2.71 volt, hence water will oxidize sodium

$$2H_2O + 2e \rightarrow 2OH^- + H_2$$
$$Na \rightarrow Na^+ + e \qquad \times 2$$
$$\overline{2Na + 2H_2O \rightarrow 2Na^+ + 2OH^- + H_2}$$

From Table 26.3 we can predict that oxygen should oxidize hydrogen sulphide to sulphur:

$$\tfrac{1}{2}O_2 + 2H^+ + 2e \rightarrow H_2O$$
$$H_2S \rightarrow S + 2H^+ + 2e$$
$$\overline{H_2S + \tfrac{1}{2}O_2 \rightarrow S + H_2O}$$

The equilibrium constant for this reaction is about 10^{37}. However, the rate of reaction is only moderate and hydrogen sulphide solutions may be used for some time before they 'go off'.

The other hydride of oxygen, hydrogen peroxide, has the following valence structure and stereochemistry:

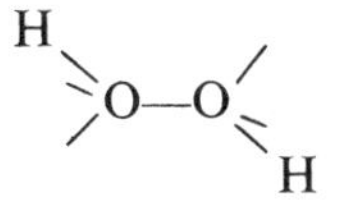

H–O–O–H (structure diagram)

Fig 26.6 Valence structure of H_2O_2.

It is a strong oxidant, for example oxidizing the iodide ion to iodine, in acid solution

$$H_2O_2 + 2H^+ + 2e \rightarrow 2H_2O$$
$$2I^- \rightarrow I_2 + 2e$$
$$\overline{H_2O_2 + 2H^+ + 2I^- \rightarrow I_2 + 2H_2O}$$

It can also act as a reductant with sufficiently strong oxidants, for example acidified potassium permanganate solution:

$$MnO_4^- + 8H^+ + 5e \rightarrow Mn^{2+} \rightarrow 4H_2O$$
$$H_2O_2 \rightarrow O_2 + 2H^+ + 2e$$
$$\overline{2MnO_4^- + 5H_2O_2 + 6H^+ \rightarrow 2Mn^{2+} + 5O_2 + 8H_2O}$$

Reference to Table 26.3 shows that it can itself undergo an oxidation reduction reaction:

$$H_2O_2 + 2H^+ + 2e \rightarrow 2H_2O$$
$$H_2O_2 \rightarrow O_2 + 2H^+ + 2e$$
$$\overline{2H_2O_2 \rightarrow 2H_2O + O_2}$$

The equilibrium constant for this reaction is about 10^{37}. The rate of reaction in pure solutions is quite low and hydrogen peroxide solutions can be stored for long periods without appreciable decomposition. The reaction can be catalysed, for example by finely divided platinum, and then the rate rises dramatically and equilibrium is attained very quickly. When weak hydrogen peroxide solution is placed on a cut, an enzyme in the blood catalyses the decomposition and this is why bubbles of gas are seen. Concentrated hydrogen peroxide solutions can be catalytically decomposed:

$$H_2O_2(l) \rightarrow H_2O(g) + \tfrac{1}{2}O_2(g) \qquad \Delta H = -997.2\,\text{kJ}$$

the products being steam and oxygen. This reaction is employed as a propellant system in rockets, the gases being ejected very rapidly and providing the requisite thrust.

26.4 THE OXIDES

There is a range of oxides formed by the group VI elements. Two important series are the dioxides, XO_2, in which an oxidation state of +4 is shown, and trioxides, XO_3, in which an oxidation state of +6 is shown.

(1) Dioxides. All of the elements (except oxygen) burn in air to form a dioxide:

$$X + O_2 \rightarrow XO_2$$

Sulphur dioxide is a colourless, pungent gas. It exists as discrete SO_2 molecules in the gas, liquid and solid phases. The valence structure of the molecule may be represented

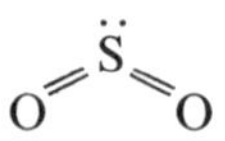

S
O O

Fig 26.7 Valence structure of SO_2.

The molecule is angular as expected from the charge cloud repulsion theory. The valence structure may be compared with ozone, which is also an angular molecule. Selenium dioxide is a white, volatile solid which forms infinite chains weakly bonded together. In the vapour, discrete SeO_2 molecules exist which structurally resemble SO_2.

Tellurium dioxide is a white, non-volatile solid which crystallizes in two forms, both apparently ionic lattices.

(2) Trioxides. Sulphur trioxide is obtained by the direct oxidation of sulphur dioxide by oxygen, usually in the presence of a platinum or vanadium pentoxide catalyst:

$$2SO_2 + O_2 \rightleftharpoons 2SO_3.$$

The valence structure for the molecule is

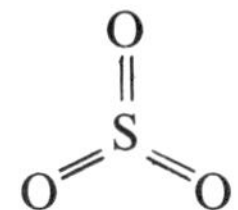

Fig 26.8 Valence structure of SO_3.

The charge cloud repulsion theory predicts a planar, triangular molecule and this has been experimentally found to be correct, in the vapour state. In the solid state the molecules polymerize and may form linear chains which have dispersion forces between adjacent layers. This form is asbestos-like in appearance.

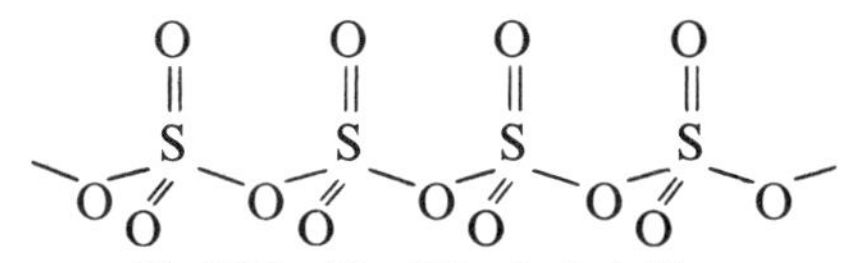

Fig 26.9 The SO_3 chain lattice.

It is also possible to obtain a form containing cyclic molecules, consisting of three SO_3 molecules linked as shown in Figure 26.10.

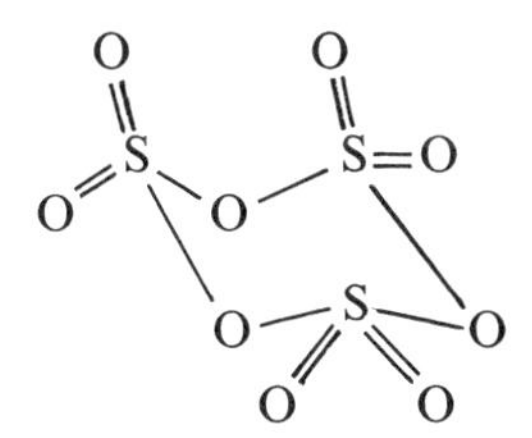

Fig 26.10 The cyclic form of SO_3.

In both structures the oxygen atoms are tetrahedrally disposed about the sulphur atoms. Selenium trioxide exists in two solid forms structurally resembling SO_3, whilst the structure of TeO_3 is uncertain.

26.5 THE OXYACIDS

A wide range of oxyacids is formed by the group VI elements. Two important series of acids are the —ous acids H_2XO_3, in which the oxidation state of +4 is shown, and the —ic acids H_2XO_4, in which an oxidation state of +6 is shown.

(1) Acids of the +4 oxidation state. These acids are formed by reaction of the dioxide with water. TeO_2, however, is only slightly soluble in water:

$$H_2O + XO_2 \rightleftharpoons H_2XO_3$$

These acids may be regarded as being diprotic. It should be noted, however, that there is very little experimental evidence to show that molecular H_2SO_3 exists in solution. The first acidity constant K_a for H_2SO_3 in fact refers to the reaction

$$H_2O + SO_2 \rightleftharpoons H^+ + HSO_3^-$$

Strictly, the K_{a1} for H_2SO_3 is not comparable with the K_a for an acid like acetic acid, which does exist in molecular form in solution. The K_{a1} for H_2SO_3 is nevertheless quoted and used, and the acid is sometimes formulated as H_2SO_3 for convenience.

EXERCISE

The K_{a1} for carbonic acid H_2CO_3 is frequently quoted as 4.5×10^{-7}. This, in fact, refers to the equilibrium

$$CO_2 + H_2O \rightleftharpoons H^+ + HCO_3^-$$

Given that the equilibrium constant for the reaction

$$CO_2 + H_2O \rightleftharpoons H_2CO_3$$

has been measured, and found to be 0.0037, show that the acidity constant $K_a(H_2CO_3)$ of H_2CO_3, defined by the equation

$$H_2CO_3 \rightleftharpoons H^+ + HCO_3^-$$

is *ca.* 1.3×10^{-4}.

Selenous acid can be crystallized as a white solid, whilst very little is known about tellurous acid.

Table 26.4

Oxyacids of formula H_2XO_3

Acid	*Formula*	K_{a1}	K_{a2}
Sulphurous	H_2SO_3 $SO(OH)_2$	1.3×10^{-2}	1×10^{-7}
Selenous	H_2SeO_3 $SeO(OH)_2$	2.5×10^{-3}	1×10^{-8}
Tellurous	H_2TeO_3 $TeO(OH)_2$	2×10^{-3}	—

None of these acids is strong and gradation in acid strength is not clearly marked.

The valence structures for the sulphite and bisulphite ions are shown:

Fig 26.11 Valence structures of SO_3^{2-} and HSO_3^- ions.

The sulphites of the alkali metals, and the ammonium ion, are soluble in water. Most other sulphites are insoluble in water, e.g. calcium sulphites. However, insoluble sulphites will dissolve in sulphurous acid solution:

$$CaSO_3(s) \rightleftharpoons Ca^{2+}(aq) + SO_3^{2-}(aq)$$

$$SO_3^{2-}(aq) + H_2SO_3(aq) \rightleftharpoons 2HSO_3^-(aq)$$

The soluble sulphites yield alkaline solutions due to the hydrolysis of the base SO_3^{2-}

$$SO_3^{2-} + H_2O \rightleftharpoons HSO_3^- + OH^- \qquad K_b = 1.6 \times 10^{-7}$$

Bisulphite solutions are acidic, as the amphoteric HSO_3^- is a stronger acid than it is a base:

$$HSO_3^- + H_2O \rightleftharpoons H_3O^+ + SO_3^{2-} \qquad K_a = 10^{-7}$$

$$HSO_3^- + H_2O \rightleftharpoons OH^- + H_2SO_3 \qquad K_b = 10^{-12}$$

EXERCISE

(a) By considering the relevant equilibrium constants decide whether solutions of the ampholyte HCO_3^- are acidic or alkaline;

(b) account for the difference in shape of the sulphite ion SO_3^{2-} and the carbonate ion CO_3^{2-}.

Sulphurous acid, and the sulphite ion, are reductants. For the system

$$SO_4^{2-} + 4H^+ + 2e \rightarrow H_2SO_3 + H_2O \qquad E^\circ = +0.20 \text{ volt}$$

Thus sulphurous acid will reduce potassium dichromate:

$$Cr_2O_7^{2-} + 14H^+ + 6e \rightarrow 2Cr^{3+} + 7H_2O \qquad E^\circ = +1.36 \text{ volt}$$

$$H_2SO_3 + H_2O \rightarrow SO_4^{2-} + 4H^+ + 2e$$

$$Cr_2O_7^{2-} + 3H_2SO_3 + 2H^+ \rightarrow 2Cr^{3+} + 3SO_4^{2-} + 4H_2O$$

With very strong reductants sulphurous acid can act as an oxidant. For example for the system

$$S + 2H^+ + 2e \rightarrow H_2S \qquad E^\circ = +0.14 \text{ volt}$$

whilst for the system

$$H_2SO_3 + 4H^+ + 4e \rightarrow S + 3H_2O \qquad E^\circ = +0.45 \text{ volt}$$

hence if H_2S is passed into H_2SO_3 solution the following reaction occurs:

$$H_2SO_3 + 4H^+ + 4e \rightarrow S + 3H_2O$$
$$H_2S \rightarrow S + 2H^+ + 2e$$
$$\overline{H_2SO_3 + 2H_2S \rightarrow 3H_2O + 3S}$$

If sodium sulphite is heated with sulphur, sodium thiosulphate is formed:

$$S + SO_3^{2-} \rightarrow S_2O_3^{2-}$$

This reaction may be compared with the slow oxidation of sulphite ion to sulphate ion by atmospheric oxygen:

$$\tfrac{1}{2}O_2 + SO_3^{2-} \rightarrow SO_4^{2-}$$

The thiosulphate ion is structurally analogous to the sulphate ion, a sulphur atom replacing an oxygen atom in the sulphate ion.

(2) Acids of the +6 oxidation state. These acids may be prepared by the reaction of the trioxide and water:

$$H_2O + XO_3 \rightarrow H_2SO_4$$

Table 26.5

Oxyacids of formula H_2XO_4

Acid	*Formula*	K_{a1}	K_{a2}
Sulphuric	H_2SO_4 $SO_2(OH)_2$	10^{+9}	10^{-2}
Selenic	H_2SeO_4 $SeO_2(OH)_2$	10^{+3}	—
Telluric	H_6TeO_6 $Te(OH)_6$	10^{-7}	—

Sulphuric and selenic acids are strong, but telluric acid is a weak dibasic acid. Structurally it differs from the others having six OH groups arranged octahedrally about the central tellurium atom. It is interesting to notice that periodic acid ($IO(OH)_5$) has a similar structure and it also differs from the other perhalic acids (HXO_4) in structure.

We cannot draw any conclusions about the trend in acidity of the H_2XO_4 acids, but if we compare H_2SO_4 with H_2SO_3, or H_2SeO_4 with H_2SeO_3 there is a marked decrease in acidity. This is due to the decrease in the number of bonded oxygen atoms:

$XO_2(OH)_2$	$XO(OH)_2$
—ic acid	—ous acid

As the number of bonded oxygen atoms increases the OH bonds are weakened due to electron withdrawal by the electronegative oxygen atoms, thus making it easier for the hydrogen atoms to be removed as protons by water.

Sulphuric acid is a dibasic acid and forms two ionic species in solution, the sulphate ion and the bisulphate ion:

$$H_2SO_4 + H_2O \rightleftharpoons H_3O^+ + HSO_4^-$$

$$HSO_4^- + H_2O \rightleftharpoons H_3O^+ + SO_4^{2-}$$

The bisulphate ion is itself a moderately strong acid and solutions of bisulphate compounds are acidic, whereas solutions of sulphate compounds are nearly neutral. All bisulphates are soluble in water, and most sulphates are soluble, except those of Ca^{2+}, Sr^{2+}, Ba^{2+}, and Pb^{2+}.

Valence structures of the sulphate and bisulphate ions are shown in Figure 26.12.

Fig 26.12 Valence structures of SO_4^{2-} and HSO_4^- ions.

Hot concentrated sulphuric acid is quite a strong oxidant. It will not oxidize chloride or fluoride ion

$$NaCl(s) + H_2SO_4(l) \rightarrow NaHSO_4(s) + HCl(g)$$

but it will oxidize bromide and iodide ions, yielding halogens and being itself reduced to SO_2 (and perhaps S or H_2S),

$$\begin{array}{r} H_2SO_4 + 2H^+ + 2e \rightarrow 2H_2O + SO_2 \\ 2Br^- \rightarrow Br_2 + 2e \\ \hline H_2SO_4 + 2H^+ + 2Br^- \rightarrow Br_2 + 2H_2O + SO_2 \end{array}$$

It will oxidize many metals to the appropriate positive ion, being itself reduced to SO_2:

$$Zn + H_2SO_4 + 2H^+ \rightarrow Zn^{2+} + 2H_2O + SO_2$$

The concentrated acid also acts as a powerful dehydrating agent removing water molecules from sugar ($C_{12}H_{22}O_{11}$) and formic acid:

$$C_{12}H_{22}O_{11} \xrightarrow{\text{conc. } H_2SO_4} 12C$$

$$H.CO.OH \xrightarrow{\text{conc. } H_2SO_4} CO$$

The dilute acid will oxidize metals with a more negative redox potential than hydrogen:

$$\begin{array}{r} 2H^+ + 2e \rightarrow H_2 \\ Mg \rightarrow Mg^{2+} + 2e \\ \hline Mg + 2H^+ \rightarrow Mg^{2+} + H_2 \end{array}$$

26.6 SUMMARY

In this group there is a marked trend towards metallic character in the elements as one moves down the group. Oxygen and sulphur, chemically and physically, behave as non-metals. Selenium and tellurium show a transition towards the metallic state in the structure of the elements, although chemically they are definitely non-metallic. Polonium is definitely a metal. Oxygen can have no more than four electron pairs in its outer shell, but sulphur, selenium, and tellurium can have more than four pairs. These elements can show a maximum oxidation state of +6. Chemically there is a striking resemblance between sulphur and selenium. The resemblance between selenium and tellurium is not quite so marked, whilst oxygen and polonium show some distinct differences from other members of the group.

QUESTIONS

1. Compare and contrast hydrogen sulphide with hydrogen chloride with respect to physical properties, acidity, stability, oxidation-reduction properties. Where possible account for these differences.
2. The elements and compounds of group VI show marked similarities in their physical and chemical properties, and also some marked differences. Discuss this statement.
3. A solution of 0.1 M Na_2S is more basic than a solution of 0.1 M Na_2Se. Explain why this is so.
4. (i) Give valence structures and suggest the stereochemistry of the following:
   (a) SCl_2
   (b) S_2Cl_2
   (c) SO_2Cl_2
   (d) $SeCl_4$
   (e) SF_6
   (f) $SOCl_2$
   (g) SOF_4
   (ii) The molecule S_2F_2 has been found to exist in two different forms. Suggest the valence structures and stereochemistry of these forms.
5. Discuss the electronic structure and geometry of the ions SO_4^{2-}, $S_2O_3^{2-}$, HSO_4^-, SO_3^{2-}, HSO_3^-
6. In what physical and chemical properties would polonium be expected to resemble bismuth?
7. Barium sulphite is soluble in dilute hydrochloric acid but barium sulphate is insoluble. Explain.
8. In each of these unbalanced chemical equations pick the oxidant and the reductant. Write a partial ionic equation for the oxidizing action and for the reducing action, and hence write the final balanced equation.
   (a) $I_2 + S_2O_3^{2-} \rightarrow I^- + S_4O_6^{2-}$
   (b) $Mn^{2+} + S_2O_8^{2-} + H_2O \rightarrow MnO_4^- + SO_4^{2-} + H^+$
   (c) $Cr_2O_7^{2-} + H_2S + H^+ \rightarrow 2Cr^{3+} + H_2O + S$
   (d) $SeO_3^{2-} + H_2O + Cl_2 \rightarrow SeO_4^{2-} + Cl^- + H^+$
   (e) $H_2SeO_3 + H_2S \rightarrow H_2O + Se + S$
   (f) $Zn + SO_4^{2-} + H^+ \rightarrow Zn^{2+} + SO_2 + H_2O$
   (g) $H_2Te + Fe^{3+} \rightarrow Fe^{2+} + H^+ + Te$
   (h) $SeO_4^{2-} + I^- + H^+ \rightarrow SeO_2 + H_2O + I_2$

9. These two chemical reactions illustrate different aspects of the properties of sulphuric acid:

$$\underset{\text{sugar}}{C_{12}H_{22}O_{11}} \xrightarrow{\text{conc. } H_2SO_4} 12C + 11H_2O$$

$$C + 2H_2SO_4 \rightarrow CO_2 + SO_2 + 2H_2O$$

What aspects are being emphasized?

10. In the manufacture of sulphuric acid, sulphur dioxide is reacted with oxygen to form gaseous sulphur trioxide

$$2SO_2 + O_2 \rightarrow 2SO_3 \qquad \Delta H = -187.2 \text{ kJ}$$

If we desire to obtain as high a yield of SO_3 as possible, what conditions of temperature and pressure are theoretically required?
At temperatures below 400°C the rate of reaction is very slow. How does this affect the choice of conditions?
What factors other than temperature and pressure may be modified to result in a high equilibrium yield of SO_3?

11. When hydrogen sulphide is passed into a solution of iron(III) chloride a white precipitate forms. When hydrogen sulphide is passed into a slightly alkaline solution of iron(II) chloride a black precipitate forms. Identify these two precipitates and comment on the difference between the two reactions.

12. Explain how the occurrence of hydrogen bonding in water accounts for the sharp difference in physical properties between water and hydrogen sulphide.

13. Discuss sulphur trioxide under the headings:
    (a) preparation;
    (b) valence structure of the molecule;
    (c) stereochemistry of the molecule in the gas phase;
    (d) stereochemistry of the substance in the solid phase.

14. How could you prepare a solution 0.015 M with respect to H^+ from concentrated sulphuric acid which contains 98% H_2SO_4 and has a specific gravity of 1.84?

15. The equilibrium constant for the reaction (at 800°C)

$$SO_2 + \tfrac{1}{2}O_2 \rightleftharpoons SO_3$$

is 31. Calculate the mass of sulphur trioxide in equilibrium with 1 mole of sulphur dioxide and 0.5 mole of oxygen, at this temperature, if the volume of the vessel is 1000 cm^3.

16. (a) The first dissociation constant for H_2S is 9×10^{-8}. Calculate the approximate pH of a 0.05 M solution.
    (b) In a 0.1 M H_2S solution what must be the pH if we wish the concentration of sulphide ion to be 10^{-21} M? The second dissociation constant for H_2S is 1×10^{-14}.
    (c) What is the least value for the molarity of cadmium ion in the above solution for a cadmium sulphide precipitate to form? (K_s of CdS = 4×10^{-28}.)

17. 2000 cm^3 of hydrogen sulphide at 1 atm, 100°C, is mixed with 5000 cm^3 of oxygen at 1 atm, 100°C in a 7000 cm^3 vessel. The mixture is reacted

$$2H_2S + 3O_2 \rightarrow 2H_2O + 2SO_2$$

What is the final pressure in the vessel at 100°C?

18. A sample of the residue from the action of a limited amount of sulphuric acid on crude common salt contained sodium sulphate, sodium bisulphate, sodium chloride and a small amount of insoluble matter. Three 0.500 g samples were taken and treated as follows:
    (a) Titrated with exactly decimolar sodium hydroxide solution to the methyl orange end point; 29.17 cm^3 of alkali required.
    (b) Titrated with 0.05 M silver nitrate solution, of which 10.00 cm^3 was required for complete precipitation of the chloride.
    (c) Treated with excess barium chloride; the precipitate, when washed and dried, weighed 0.8428 g.

    Calculate the percentages of the anhydrous salts present in the sample.
19. A compound containing the elements oxygen, sulphur, and chlorine is found to react readily with water to give a solution of sulphuric and hydrochloric acids. 6.75 g of the compound was reacted with water and the resulting solution made up to 1000 cm^3.

    50.0 cm^3 of this solution when treated with excess barium chloride solution gave a precipitate which weighed 0.5835 g. 50.0 cm^3 of the solution required 20.0 cm^3 of 0.5 M sodium hydroxide for neutralizing. Calculate the percentages of sulphur, chlorine and oxygen in the compound.

    When 0.5 g of the compound was vaporized it occupied 112 cm^3 at 100°C, 770 mmHg.

    What is the molecular formula and *accurate* molecular weight of the compound?

27

Group V Elements

The elements and compounds in this group display a quite wide variation in physical and chemical properties. There is a clear trend from distinct non-metals, nitrogen and phosphorus, through arsenic and antimony which exhibit intermediate characteristics, to bismuth which is essentially a true metal. A second general feature of the group is that nitrogen, whilst showing some compounds formally similar to those of the other elements, has rather different properties when compared with the rest of the group as a whole. There is an abrupt drop in electronegativity on passing from nitrogen to

I	II											III	IV	V	VI	VII	VIII
1 H																	2 He
3 Li	4 Be											5 B	6 C	7 N	8 O	9 F	10 Ne
11 Na	12 Mg											13 Al	14 Si	15 P	16 S	17 Cl	18 Ar
19 K	20 Ca	21 Sc	22 Ti	23 V	24 Cr	25 Mn	26 Fe	27 Co	28 Ni	29 Cu	30 Zn	31 Ga	32 Ge	33 As	34 Se	35 Br	36 Kr
37 Rb	38 Sr	39 Y	40 Zr	41 Nb	42 Mo	43 Tc	44 Ru	45 Rh	46 Pd	47 Ag	48 Cd	49 In	50 Sn	51 Sb	52 Te	53 I	54 Xe
55 Cs	56 Ba	57 La	72 Hf	73 Ta	74 W	75 Re	76 Os	77 Ir	78 Pt	79 Au	80 Hg	81 Tl	82 Pb	83 Bi	84 Po	85 At	86 Rn
87 Fr	88 Ra	89 Ac															

Table 27.1
Physical properties of group V elements

Element	*Atomic weight*	*Electron configuration*	*Outer-shell configuration*	*Electro-negativity*	*Percentage abundance in earth's crust*
Nitrogen	14.008	2.5	$2s^2 2p^3$	3.0	4.6×10^{-3}
Phosphorus	30.975	2.8.5	$3s^2 3p^3$	2.1	0.118
Arsenic	74.91	2.8.18.5	$4s^2 4p^3$	2.0	5×10^{-4}
Antimony	121.76	2.8.18.18.5	$5s^2 5p^3$	1.9	1×10^{-4}
Bismuth	209.00	2.8.18.32.18.5	$6s^2 6p^3$	1.8	2×10^{-5}

the other elements. Further, the electronegativity values of some of these later group V elements vary depending on their mode of measurement, and correlation of chemical behaviour with such a factor cannot readily be achieved.

Nitrogen occurs abundantly in the atmosphere, but it is also present in certain solid nitrate deposits, notably of $NaNO_3$. The other elements occur in nature exclusively in the form of solids. Phosphorus occurs in a number of complex minerals, one of which, rock phosphate, is generally represented as $Ca_3(PO_4)_2$, but in fact is far more complex. Arsenic and antimony are found mainly in sulphide ores, frequently together. Their existence as sulphide ore minerals reflects a tendency towards metallic character, and their frequent occurrence together highlights a similar chemical behaviour. Bismuth exists as the native metal, and in sulphide and oxide ores.

27.1 STRUCTURE OF THE ELEMENTS

As in group VI we find here a wide variation in structure, and a gradation from the non-metallic element nitrogen to the metallic element bismuth is clearly shown.

(1) Nitrogen. Nitrogen exists as discrete diatomic molecules in the gaseous liquid, and solid states. Despite its high electronegativity, nitrogen is a relatively unreactive element. For example, it does not readily react with oxygen, nor with the alkali metals. In nitrogen this may be attributed to the strong triple bond between the atoms in the molecule, and is consistent with a high bond dissociation energy. For

$$N_2(g) \rightleftharpoons 2N(g) \qquad \Delta H = +946 \text{ kJ}$$

and the K, at 25°C, is 10^{-120}. Diatomic molecules of phosphorus, P_2, and arsenic, As_2, may be detected only in the gas phase at high temperatures.

(2) Phosphorus. Phosphorus in the vapour state consists of discrete tetratomic molecules. The dissociation

$$P_4(g) \rightleftharpoons 2P_2(g)$$

is not appreciable until temperatures approaching 1000°C, and for further dissociation into atoms much higher temperatures are required. The structure is one in which P atoms are arranged at the vertices of a tetrahedron, derived from four tetrahedrally arranged charge clouds for each P atom, three of which are bonded to other P atoms, the fourth containing a non-bonding electron pair. As shown in Figure 27.1 the tetrahedral distribution is somewhat distorted, the P—P—P bond angle being as low as 60°, in comparison with the theoretical tetrahedral angle of 109.5°.

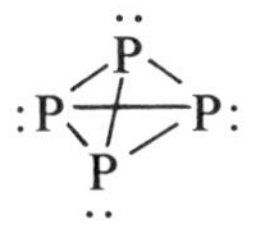

Fig 27.1 The P_4 molecule.

In the solid state, three main forms of phosphorus exist—white, red, and black. White phosphorus is very reactive and comprises P_4 molecules in both liquid and solid. The structure of the red form is unknown. Black phosphorus is a flaky layer lattice, not unlike graphite. Both black and red phosphorus are much less reactive than white phosphorus.

(3) Arsenic and antimony. Arsenic vapour has definitely been characterized as As_4, and antimony possibly exists as Sb_4. Condensation of these vapours leads to metastable yellow 'non-metallic' solid forms, presumably containing As_4 and Sb_4 molecules. These readily transform to the stable grey 'metallic-like' forms, which are layer lattices.

(4) Bismuth. Bismuth is unknown in a non-metallic modification. The normal metallic form of bismuth, like the metallic forms of arsenic and antimony, resembles a layer lattice. Notice the trend from mainly covalent bonding in the arsenic layer to mainly metallic bonding in the bismuth lattice.

We can summarize by pointing out that, just as in group VI discrete S_8 molecules give place to infinite chains and layers in selenium and tellurium, so in group V discrete P_4 molecules may be compared with layer lattices of semi-metallic structure in arsenic and antimony. These elements, and to a lesser extent bismuth, represent a class of solid intermediate between true molecular and true metallic crystals.

27.2 FEATURES OF THE COMPOUNDS

The compounds formed by the joining of group V atoms with atoms of similar electronegativity exhibit predominantly covalent bonding. Covalent bonding is found in many discrete molecules, e.g. NH_3, $AsCl_3$, PCl_5, within ions such as NH_4^+, PCl_6^- and in giant lattice structures such as boron nitride, BN. This latter compound exists in both a 'diamond-like' and a 'graphite-like' form.

In nitrogen compounds, the maximum number of electron pairs around the nitrogen atom is four. With the other elements, some of the compounds have only four electron pairs around the group V atom, e.g. in $AsCl_3$, and similarly in the ion, PCl_4^+. However, more than four electron pairs can be accommodated in these later atoms. Thus five pairs are known in the penta-halides, such as PCl_5, and six pairs in ions such as PCl_6^-.

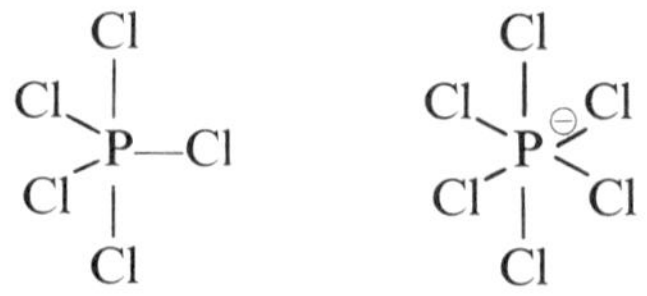

Fig 27.2 Valence structures of PCl_5 and PCl_6^-.

Compounds formed between atoms with greater electronegativity difference are predominantly ionic. Ionic structures are less common in group V than in the compounds of groups I, II, VI and VII as would be expected from the more intermediate electronegativity values. The wide range of behaviour in this group is illustrated by anion formation in nitrogen and cation formation in bismuth. Thus the nitride ion, N^{3-}, occurs in lithium nitride, Li_3N, and the bismuth cation, Bi^{3+}, is found in both BiF_3 and $Bi(NO_3)_3$. A few cations are known in aqueous solution: the ammonium ion, NH_4^+, and the arsenic, As^{3+}(aq), antimonyl, SbO^+(aq), and bismuthyl, BiO^+(aq), cations. A better representation for, say, the antimonyl cation would be $Sb(OH)_2^+$(aq); this dehydrates and gives rise to the SbO^+ ion in solids such as antimonyl chloride, SbOCl. The loss of five outer-shell electrons is not observed since no Z^{5+} cations are found. Presumably this would involve a prohibitively great energy expenditure.

27.3 THE HYDRIDES

The hydrides of formula ZH_3 are molecular in the gas, liquid and solid. Some physical properties are listed in Table 27.2.

Table 27.2

Physical properties of group V hydrides

Hydride	*Melting point (°C)*	*Boiling point (°C)*	*Standard heat of formation (ΔH_f° in kJ)*
Ammonia	−77.7	−33.3	−47.1
Phosphine	−132.5	−87.4	+9.25
Arsine	−111.2	−58.5	+41.0
Stibine	−88.5	−17.0	—
Bismuthine	—	+22.0	—

The steady increase in boiling point and melting point from PH_3 through the other members would be expected from the increasing value of the dispersion forces. The relatively high melting point and boiling point of ammonia are explicable in terms of hydrogen bonding between these molecules.

Ammonia is the most stable of these gases. The extent of decomposition to element and hydrogen increases progressively through the series. Phosphine and arsine are fairly stable, but stibine breaks up into antimony and hydrogen quite readily. For many years it was doubtful whether attempts to prepare

BiH_3 had been successful, for it decomposes into its constituent elements very readily. It can only be prepared in trace amounts, and it dissociates at quite low temperatures.

The molecules would be expected to be pyramidal in shape. Four charge clouds are arranged tetrahedrally about the group V atom, three of these charge clouds involving bonding electron pairs and one involving a non-bonding pair. The ammonia molecule has been discussed in some detail in chapter 9.

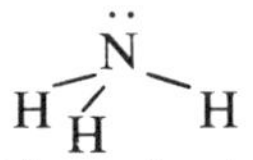

$$\ddot{N}(H)_3$$

Fig 27.3 Valence structure of NH_3.

Only nitrogen and phosphorus form tetrahedral ZH_4^+ cations. NH_4^+ is well known in ammonium salts, and the phosphonium ion, PH_4^+, exists in solid phosphonium iodide, PH_4I. No analogous cations are known for the other elements.

In aqueous solution NH_3 functions as a Lowry-Bronsted base.

$$NH_3 + H_2O \rightleftharpoons NH_4^+ + OH^-$$

To all practical intent the other hydrides exhibit no acid-base character in aqueous solutions.

The hydrides can all act as reductants, the reducing strength increasing down the group. Compare the $E°$s for the systems

$$P + 3H^+ + 3e \rightarrow PH_3 \qquad E° = -0.04 \text{ volt}$$

$$As + 3H^+ + 3e \rightarrow AsH_3 \qquad E° = -0.54 \text{ volt}$$

If phosphine is passed into a copper(II) sulphate solution, copper precipitates.

$$Cu^{2+} + 2e \rightarrow Cu$$

$$PH_3 \rightarrow P + 3H^+ + 3e$$

$$3Cu^{2+} + 2PH_3 \rightarrow 3Cu + 2P + 6H^+$$

27.4 THE HALIDES

Two series of simple halides are formed. Compounds of formula type ZX_3 have the group V element, Z, in an oxidation state of +3, and compounds of formula type ZX_5 have Z in an oxidation state of +5. These are called trihalides and pentahalides respectively. The related ionic species ZX_4^+ and ZX_6^- are known for some of the elements. Nitrogen appears in marked contrast to the other members, forming no pentahalides and only two trihalides.

(1) Trihalides. Nitrogen forms a very stable, unreactive, gaseous fluoride, NF_3, and an unstable, explosive, liquid chloride, NCl_3. Pure tribromide and tri-iodide compounds are unknown, but complex compounds of the type NI_3NH_3 are known. These latter compounds are very unstable and will detonate violently on impact.

Each of the other four elements forms each of the four possible trihalides. Generally they may be prepared by direct reaction of the elements in stoichiometric proportion, for example

$$P_4(s) + 6Cl_2(g) \rightarrow 4PCl_3(l) \qquad \Delta H = -321.8 \text{ kJ}$$

The reactions are all exothermic. In addition to the simple halides such as liquid PCl_3, liquid AsF_3, solid $SbBr_3$, etc., a number of mixed halides such as gaseous PF_2Cl, liquid $PFCl_2$, and solid $SbBrI_2$, are known. Only a small proportion of the possible mixed trihalides have been reported.

In the vapour the shapes of the trihalides are based upon the tetrahedral disposition of four charge clouds about a central atom, one of these charge clouds containing a non-bonding electron pair. They are thus pyramidal molecules.

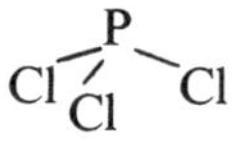

P
Cl Cl Cl

Fig 27.4 Valence structure of PCl_3.

In the solid state various structures occur depending upon both the group V element and the particular halogen. Structures range through a number of molecular crystals such as found in solid AsF_3 and PCl_3, to layer lattices in solid SbI_3 and BiI_3, to essentially ionic networks in BiF_3, comprising Bi^{3+} cations and F^- anions.

Of these compounds PCl_3 is perhaps best known. It hydrolyses in water to form phosphorous acid, H_3PO_3

$$PCl_3 + 3H_2O \rightarrow H_3PO_3 + 3HCl$$

It also reacts readily with oxygen to form the oxychloride, $POCl_3$. This latter molecule is tetrahedral and it has the valence structure shown in Figure 27.5.

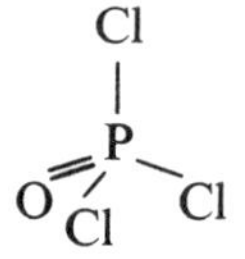

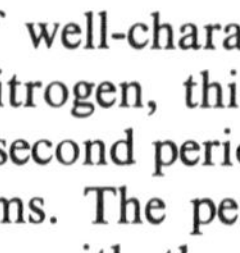

Fig 27.5 Valence structure of $POCl_3$.

The oxidation state of phosphorus in this molecule is +5.

(2) Pentahalides. The number of well-characterized pentahalides is limited. There are no pentahalides of nitrogen, this being consistent with the fact that none of the elements of the second period accommodate more than four charge clouds around their atoms. The pentahalides are formed either by direct reaction of excess halogen with the element, or by further reaction of the trihalide with the appropriate halogen. The formation of PCl_5 from PCl_3 and Cl_2 has been discussed in chapter 13.

PCl_5 in vapour and liquid states consists of discrete molecules of trigonal bipyramidal geometry, as described in chapter 9. This is to be anticipated from the accommodation of five charge clouds around a central phosphorus atom.

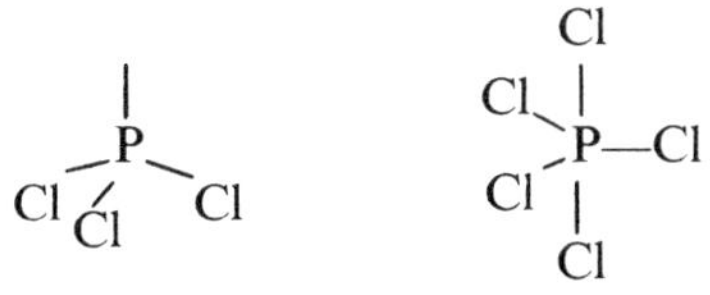

Fig 27.6 Valence structures of PCl_3 and PCl_5.

In the solid state, however, PCl_5 exists as a lattice structure of PCl_4^+ cations and PCl_6^- anions. As predicted by the charge cloud repulsion theory the cation is tetrahedral, and the anion octahedral. Note again that the non-conductivity in the molten state does not necessarily imply a molecular solid.

Fig 27.7 Valence structures of PCl_4^+ and PCl_6^-.

Hydrolysis of PCl_5 gives rise to the oxychloride, $POCl_3$:

$$PCl_5 + H_2O \rightarrow POCl_3 + 2HCl$$

but in excess water the following reaction producing phosphoric acid, H_3PO_4, is dominant

$$PCl_5 + 4H_2O \rightarrow H_3PO_4 + 5HCl$$

The corresponding pentabromide, PBr_5, has not been characterized in the vapour and is known only as an ionic solid containing PBr_4^+ cations and Br^- anions. Although six chlorine atoms surround phosphorus in PCl_6^-, it may be that the phosphorus atom is too small to accommodate six of the larger bromine atoms around itself. Notice the variation of structure type in such formally similar compounds. A mixed halide, PF_3Cl_2, is known both as a molecular solid of individual PF_3Cl_2 molecules, and as a solid ionic lattice of PCl_4^+ and PF_6^-.

A final example of the complexity of the chemistry of these compounds is that, despite the existence of PCl_5 and $SbCl_5$, no attempts to make $AsCl_5$ or $BiCl_5$ have been successful.

27.5 THE OXIDES

Two groups of oxides, of formula type Z_2O_3 and Z_2O_5, are known. Here the group V elements are showing oxidation numbers of +3 and +5, respectively. Because of the somewhat different chemistry of the nitrogen oxides these are treated separately from those of the other elements.

(1) Oxides of the +3 oxidation state. An empirical formula, Z_2O_3, may be assigned for oxides of each element. The structure of these oxides, however, varies. N_2O_3 is known in the gas phase from studies of the equilibrium system

$$NO + NO_2 \rightleftharpoons N_2O_3$$

At very low temperatures it exists as a solid of unknown structure. The trioxides of phosphorus, arsenic and antimony are similar, as in both the vapour and in one solid modification they exist as discrete Z_4O_6 molecules. Their structure, which is basically tetrahedral, is shown in Figure 27.8.

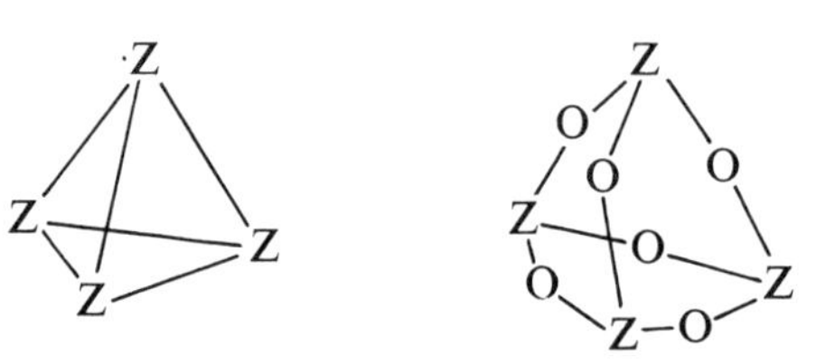

Fig 27.8 Valence structures of Z_4 and Z_4O_6.

In addition to these molecular crystals a layer lattice form of As_2O_3 and a chain lattice form of Sb_2O_3 occur, whilst Bi_2O_3 forms a network lattice.

These oxides show acidic and basic properties. Thus N_2O_3, P_4O_6, and As_4O_6 each react with water to form acids:

$$N_2O_3 + H_2O \rightarrow 2HNO_2 \quad \text{(nitrous acid)}$$

$$P_4O_6 + 6H_2O \rightarrow 4H_3PO_3 \quad \text{(phosphorous acid)}$$

$$As_4O_6 + 6H_2O \rightarrow 4H_3AsO_3 \quad \text{(arsenious acid)}$$

Sb_2O_3 is amphoteric, since it reacts with OH^- ions and with H^+ ions:

$$Sb_2O_3 + 2OH^- \rightarrow H_2O + SbO_2^- \quad \text{(antimonite ion)}$$

$$Sb_2O_3 + 2H^+ \rightarrow H_2O + 2SbO^+ \quad \text{(antimonyl ion)}$$

Bi_2O_3, however, is definitely a basic oxide. It reacts with acids and gives rise to cations (of complex structure) in solution. Thus the Z_2O_3 oxides show a clear trend from acidic to basic character down the group.

(2) Oxides of the +5 oxidation state. Just as the pentahalides were fewer in number and less well characterized than the trihalides, so our knowledge of the Z_2O_5 oxides is less clear than that of the trioxides.

Dinitrogen pentoxide, N_2O_5, exists as discrete molecules with this molecular formula in the vapour, and in the solid consists of an ionic lattice of nitronium cations, NO_2^+, and nitrate anions, NO_3^-.

Diphosphorus pentoxide exists as discrete molecules with the formula P_4O_{10}, both in the vapour and in one solid modification. Comparison of its valence structure with that of P_4O_6 and of P_4 shows that these three are related.

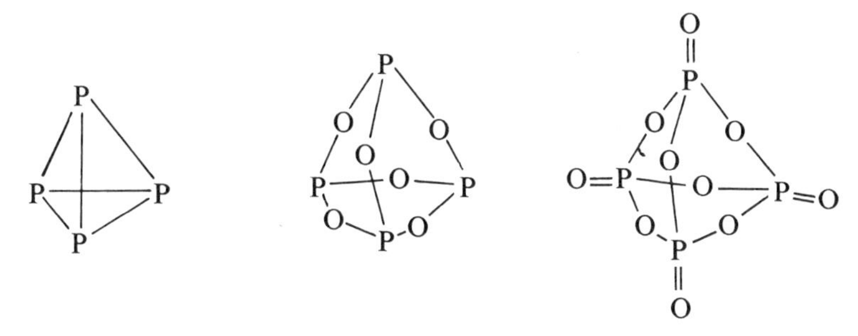

Fig 27.9 Valence structures of P_4, P_4O_6 and P_4O_{10}.

It is an acidic oxide, reacting with water to form phosphoric acid:

$$P_4O_{10} + 6H_2O \rightarrow 4H_3PO_4$$

It is an extremely effective drying agent because of its great avidity for water.

Whereas P_4O_{10} can be prepared simply by burning phosphorus in excess oxygen

$$P_4 + 5O_2 \rightarrow P_4O_{10}$$

the less stable As_2O_5 and Sb_2O_5 cannot be formed in this way as they decompose at high temperatures with loss of oxygen. They are very difficult to obtain in pure form and their structures have not been elucidated.

Bismuth pentoxide, Bi_2O_5, is indeed ill defined. It may well be that no pure Bi_2O_5 has ever been prepared.

(3) Nitrogen oxides. A large number of oxides of nitrogen are known and formal oxidation states may be assigned to each of them. The well characterized oxides are listed in Table 27.3.

Table 27.3

The oxides of nitrogen

Oxide	*Formula*	*Oxidation state of nitrogen*
'Nitrous oxide'	N_2O	+1
'Nitric oxide'	NO	+2
Dinitrogen trioxide	N_2O_3	+3
Dinitrogen tetroxide	N_2O_4	+4
Nitrogen dioxide	NO_2	+4
Dinitrogen pentoxide	N_2O_5	+5

Nitrogen dioxide and dinitrogen tetroxide are related by the equilibrium

$$N_2O_4 \rightleftharpoons 2NO_2$$

27.6 THE HYDROXY COMPOUNDS

The hydroxy compounds of nitrogen, phosphorus and arsenic are acids which form when the appropriate oxides are hydrolysed with water. Thus the oxides of these elements are acidic oxides. In Table 27.4 the oxides and the corresponding acids are listed, together with the acidity constants (K_a) of the acids, and the formula of the oxyion found in salts formed from the acids. Notice the formulation of the acids as hydroxy compounds to clearly indicate the basicity of the acid. Thus whereas H_3PO_4 is a tribasic (or triprotic) acid, H_3PO_3 is dibasic (or diprotic). This is reflected in their respective structures, viz. $PO(OH)_3$ and $HPO(OH)_2$.

Fig 27.10 Valence structures of H_3PO_4 and H_3PO_3.

Table 27.4

Some group V oxyacids

Oxide	*Oxyacid*	K_a	*Oxyion*
N_2O_3	$NO(OH)$	4×10^{-4}	NO_2^-
N_2O_5	$NO_2(OH)$	$1 \times 10^{+1}$	NO_3^-
P_4O_6	$HPO(OH)_2$	1×10^{-2}	$H_2PO_3^-$
P_4O_{10}	$PO(OH)_3$	8×10^{-3}	$H_2PO_4^-$
As_4O_6	$As(OH)_3$	6×10^{-10}	$H_2AsO_3^-$
As_4O_{10}	$AsO(OH)_3$	5×10^{-3}	$H_2AsO_4^-$

Acids such as the tribasic phosphoric, H_3PO_4, and arsenic, H_3AsO_4, acids form series of corresponding oxyions. Thus three orthophosphates and three orthoarsenates are known, exemplified in, say, the sodium salts, NaH_2PO_4, Na_2HPO_4 and Na_3PO_4. The nitrite and nitrate ions are well known in metallic salts, e.g. $NaNO_2$ and $NaNO_3$. The corresponding oxides of antimony and bismuth are less reactive with water, and their acids are not well characterized. The dominant species in antimony and bismuth solutions derive from X^{3+}(aq), but, because of hydrolysis, are better represented as $X(OH)_2^+$(aq). These ions form insoluble chlorides, e.g.

$$Bi(OH)_2^+ + Cl^- \rightarrow BiOCl + H_2O$$

Because of the formulae of the solids formed, the ions in the solution are usually written as SbO^+ (antimonyl) and BiO^+ (bismuthyl) cations.

27.7 SUMMARY

In this group, as in group VI, there is a marked trend towards metallic character in the elements as one moves down the group. Nitrogen is chemically and physically a non-metal. In phosphorus, arsenic and antimony we can recognize a trend towards increasing metallic character, whilst bismuth is definitely a metal. Nitrogen shows many unique differences from other members of the group. It can have no more than four electron pairs in its outer shell, whereas the other elements can exceed this number. As we move down the group we see that the +3 oxidation state becomes more stable relative to the +5 oxidation state, so that the chemistry of nitrogen is largely that of its covalent compounds, whilst that of bismuth is largely that of the bismuth ion (Bi^{3+}) or its hydrated form ($Bi(OH)_2^+$ or BiO^+). Again the trend from non-metallic to metallic character is stressed by this change in the character of the ions.

QUESTIONS

1. Draw a charge cloud diagram for the nitride anion N^{3-}, and compare it with charge cloud diagrams of the amide ion NH_2^-, the ammonia molecule NH_3, and the ammonium ion NH_4^+. State the formal oxidation state of nitrogen in these four species.
2. The formula of dinitrogen pentoxide, N_2O_5, is both an empirical and a molecular formula, but the formula P_2O_5 is only an empirical formula. Does this statement imply that there is no molecular phosphorus oxide in which the phosphorus shows an oxidation number of +5? Explain.
3. The decomposition of HNO_2 may be represented by

$$3HNO_2 \rightarrow H^+ + NO_3^- + 2NO + H_2O$$

   Explain clearly why this reaction may be regarded as one of oxidation and reduction. What is the oxidant? What is the reductant? State the oxidation number of nitrogen in each species containing it.
4. Write valence structures for NH_3 and H_2O, and hence predict the valence structures of hydroxylamine NH_2OH, hydrazine N_2H_4, and hydrogen peroxide H_2O_2.
5. Reduction of the nitrate ion NO_3^- may yield as products NO_2, NO, N_2 or NH_4^+. Write partial ionic equations for these possible reactions.
6. In what senses are the following gaseous reactions

$$4NH_3(g) + 3O_2(g) \rightarrow 2N_2(g) + 6H_2O(g)$$

$$4NH_3(g) + 6O_2(g) \rightarrow 4NO(g) + 6H_2O(g)$$

   instances of redox reactions?
7. By means of equations, illustrate the behaviour of NH_3 as
   (a) a base;
   (b) a reductant;
   (c) a complexing agent.
8. An aqueous solution of NH_4Cl contains the cation NH_4^+. This cation is a weak Bronsted acid ($K_a = 6.3 \times 10^{-10}$). Does it follow that the conjugate base, NH_3, is a strong base? Explain carefully.
9. $Ca_3(PO_4)_2$ is a sparingly soluble compound whose $K_s = 1 \times 10^{-25}$. Show algebraically how the molar solubility of the compound is related to its solubility product K_s.
10. What ions or molecules are present in 0.1 M solutions of each of (a) NH_3; (b) NH_4Cl; (c) NaH_2PO_4; (d) H_3PO_3?
    Indicate roughly the relative concentrations of the different species present (in terms of, say, moderate, low, and very low concentrations).
11. Comment on the fact that phosphorus has but one stable nuclide (isotope) of mass number 31, whereas the atomic weight of phosphorus is 30.97 and not 31.00.

12. For the tribasic acid H_3AsO_4, $pK_{a1} = 2.22$, $pK_{a2} = 6.98$ and $pK_{a3} = 11.53$. Decide whether solutions of NaH_2AsO_4 and Na_2HAsO_4 would be acidic (i.e. pH < 7) or alkaline (i.e. pH > 7).
13. Comment critically on each of the following:
    (a) PCl_3 is a covalent molecular substance in the pure state. Thus one might expect that, when dissolved in water, the solution would be a poor electrical conductor.
    (b) PBr_5 is a poor electrical conductor in the solid state. This suggests that the solid consists of a molecular lattice of discrete PBr_5 molecules.
    (c) Attempts to prepare NCl_5 and $AsCl_5$ have been unsuccessful to date. This is because the group V atoms can accommodate at most four pairs of electrons in their outer shells.
14. Consult at least three other textbooks and compare the values of electronegativities quoted for the group V elements. What generalizations could you make as to the variations of electronegativity in this group.
15. 'The group V elements exhibit, at room temperature and pressure, a wider range of structures than do the elements in group VII.' Discuss the validity of this generalization, and comment on the comparative similarities and differences between the structures of the elements of groups V, VI and VII.
16. 60 cm³ NH_3 and 60 cm³ O_2, measured at S.T.P., were reacted according to the equation

$$4NH_3(g) + 3O_2(g) \rightarrow 2N_2(g) + 6H_2O(g)$$

    After the reaction was completed the gas volumes were again measured at S.T.P.
    (a) What volume of oxygen was left unreacted?
    (b) What volume of nitrogen was produced?
    (c) What was the total volume change after the reaction?
17. A flask contains 50 cm³ of 0.1 M H_3PO_4. To this is added a solution of 0.2 M NaOH in successive stages. For each stage, state the principal ions which would be present in solution after the addition of
    (a) 25 cm³; (b) 37.5 cm³; (c) 50 cm³; (d) 67.5 cm³; (e) 75 cm³.
    Write down the formulae of the solids which would be formed in each case if the solution were evaporated to dryness.
18. Elemental phosphorus is formed from rock phosphate by reaction in an electric furnace with sand and carbon:

$$Ca_3(PO_4)_2 + 3SiO_2 + 5C \rightarrow 3CaSiO_3 + 5CO + 2P$$

    (a) What mass of rock phosphate would be needed to form one kilogram of phosphorus?
    (b) What volume of CO would be produced in (a), if the gas is measured at 400°C and 1 atm pressure?
19. A pure solid compound contains the elements phosphorus, sulphur and chlorine.
    (a) Its empirical formula was determined from the following analytical results: 0.6395 g of the compound was dissolved in distilled water when all the chlorine in the compound was converted to chloride ion, and the solution was made up to 250 cm³ with distilled water.
    25 cm³ aliquots from this sample solution required an average titre of 22.55 cm³ of 0.0500 M silver nitrate solution, 0.3081 g of the compound was dissolved in distilled water, the phosphorus in the compound was completely converted to orthophosphate ion which was precipitated as magnesium

ammonium phosphate ($MgNH_4PO_4$), and then finally ignited at high temperatures to give 0.2036 g of magnesium pyrophosphate ($Mg_2P_2O_7$).

$$2MgNH_4PO_4 \rightarrow Mg_2P_2O_7 + 2NH_3 + H_2O$$

From the analytical results calculate the empirical formula of the compound.

(b) The molecular weight of the compound was obtained from gas density measurements on the compound vaporized at elevated temperatures:
0.4705 g of the solid compound was placed in a 300 cm^3 vessel which was evacuated and then heated to 250°C, at which temperature the solid completely vaporized. The pressure in the vessel at 250°C was found to be 310.1 mmHg. From these measurements calculate:
(i) the molecular weight of the compound, and then
(ii) the molecular formula of the compound, using the empirical formula deduced above.
Discuss very briefly possible objections which could be raised to this method of measuring the molecular weight of the solid compound.

28

Group IV Elements

The elements in this group possess a common outer-shell configuration of four electrons (s^2p^2). Within this group there is a very marked change from the undoubted non-metals, carbon and silicon, through germanium showing metallic and non-metallic characteristics, to tin and lead which are undoubted metals. However, there is a striking difference in properties between carbon and silicon. Carbon shows many very unique properties and little of the chemistry of silicon can be inferred from that of carbon.

I	II											III	IV	V	VI	VII	VIII
1 H																	2 He
3 Li	4 Be											5 B	6 C	7 N	8 O	9 F	10 Ne
11 Na	12 Mg											13 Al	14 Si	15 P	16 S	17 Cl	18 Ar
19 K	20 Ca	21 Sc	22 Ti	23 V	24 Cr	25 Mn	26 Fe	27 Co	28 Ni	29 Cu	30 Zn	31 Ga	32 Ge	33 As	34 Se	35 Br	36 Kr
37 Rb	38 Sr	39 Y	40 Zr	41 Nb	42 Mo	43 Tc	44 Ru	45 Rh	46 Pd	47 Ag	48 Cd	49 In	50 Sn	51 Sb	52 Te	53 I	54 Xe
55 Cs	56 Ba	57 La	72 Hf	73 Ta	74 W	75 Re	76 Os	77 Ir	78 Pt	79 Au	80 Hg	81 Tl	82 Pb	83 Bi	84 Po	85 At	86 Rn
87 Fr	88 Ra	89 Ac															

Table 28.1
Physical properties of group IV elements

Element	*Atomic weight*	*Electron configuration*	*Outer-shell structure*	*Electro-negativity*	*Percentage abundance in earth's crust*
Carbon	12.0111	2.4	$2s^22p^2$	2.5	0.032
Silicon	28.086	2.8.4	$3s^23p^2$	1.8	27.72
Germanium	72.59	2.8.18.4	$4s^24p^2$	1.8	7×10^{-4}
Tin	118.69	2.8.18.18.4	$5s^25p^2$	1.8	4×10^{-3}
Lead	207.19	2.8.18.32.18.4	$6s^26p^2$	1.8	1.6×10^{-3}

The trend in properties for the elements from silicon to lead is more gradual. Carbon is not an abundant element in the earth's crust, but enormous quantities are found in the plant and animal kingdoms. Silicon is the second most abundant element in the earth's crust and is found in a wide range of silicate minerals. Germanium, tin and lead are very rare elements which are found in the form of sulphide or oxide minerals, e.g. galena, PbS, and cassiterite, SnO_2.

28.1 STRUCTURE OF THE ELEMENTS

Like the elements of groups V and VI, the elements of group IV show a wide variation in structure, and the gradation from the non-metallic element oxygen to the metallic element lead is clearly shown.

(a) *Carbon.* Carbon exists as two modifications. The first of these is diamond, which is a network lattice of carbon atoms covalently bonded together. Each carbon atom is at the centre of a tetrahedron bounded by four other carbons. The second modification is graphite, which is a layer lattice. Carbon atoms are covalently bonded to one another in sheets with weak dispersion forces acting between the sheets. Both forms have been discussed in detail in chapter 11.

(b) *Silicon.* Silicon exists as a network lattice, with silicon atoms bonded to one another covalently, as in diamond.

(c) *Germanium.* Germanium exists as a network lattice similar in structure to diamond and silicon.

(d) *Tin.* There are two modifications of tin. The stable form is a metallic lattice which conducts electricity and shows the properties of a true metal. The second form is a network lattice similar in structure to diamond, this form being metastable with respect to the metallic form.

(e) *Lead.* Lead exists as one form only, which is a true metallic lattice.

28.2 FEATURES OF THE COMPOUNDS

In most of their compounds the elements show oxidation states of +4 and of +2. An oxidation state of +2 is very rare in carbon and silicon but appears definitely in germanium and tin, and is the dominant oxidation state in lead. In fact, there is a gradual and steady increase in the stability of the lower oxidation state, with respect to the higher, as one moves down the group.

In compounds where an oxidation state of +4 is shown (and −4 in the hydrides), the bonding is almost entirely covalent, but it may assume intermediate character where the element is bonded to a very electronegative element, as in silicon tetrafluoride, SiF_4. Compounds of the higher atomic weight elements, where an oxidation state of +2 is shown, may exhibit ionic character, e.g. lead difluoride, PbF_2. Tin and lead in the +2 oxidation state are referred to as tin(II) and lead(II), whilst in the +4 oxidation state they are referred to as tin(IV) and lead(IV).

Carbon can have a maximum of four electron pairs in the outer shell, e.g. in carbon tetrachloride, CCl_4. The other elements may have more than four electron pairs in the outer shell, but in fact this is not common. In some complex ions, six electron pairs may be present in the outer shell, as in the fluorosilicate ion, SiF_6^{2-}.

F
F F
Si^{2-}
F F
F

Fig 28.1 Valence structure of SiF_6^{2-} ion.

Carbon atoms have the property of being able to bond one to another to form chains and rings. This property is called *catenation*. It is this property which results in carbon forming an enormous range of chemical compounds, the study of which forms the discipline of organic chemistry. Catenation is much less strongly developed in silicon, even less developed in germanium, and does not exist in tin and lead.

28.3 THE HYDRIDES

Structurally the hydrides are all discrete molecules. The simple hydrides, of formula MH_4, involve four electron pairs in the outer shell of the group IV atom and from the charge cloud repulsion theory we would predict the molecules to be tetrahedral. This is confirmed by experiment.

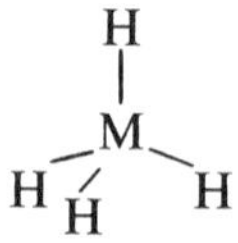

Fig 28.2 Valence structure of MH_4.

Carbon has a unique tendency to form a series of polymeric hydrides of general formula C_nH_{2n+2}, called the *alkanes*. There is an enormous number of these hydrides and their general properties have been discussed in chapter 19. Silicon forms a similar series of polymeric hydrides, involving up to eleven silicon atoms linked together. These are called the *silanes*. Germanium forms only three hydrides—the *germanes*. Tin and lead form one hydride each. The decrease in hydride stability down the group is reflected in the change in the standard heats of formation of the MH_4 compounds. For example, the values for methane, CH_4, and silane, SiH_4, are

$$C(s) + 2H_2(g) \rightarrow CH_4(g) \qquad \Delta H_f^\circ = -74.9\ kJ$$

$$Si(s) + 2H_2(g) \rightarrow SiH_4(g) \qquad \Delta H_f^\circ = -61.9\ kJ$$

Lead hydride, PbH_4, is so unstable that it is virtually impossible to isolate it.

The carbon and germanium hydrides do not react with water, but the others are hydrolysed in alkaline solution, Silane SiH_4, for example, is hydrolysed to silicic acid and hydrogen:

$$SiH_4 + 4H_2O \rightarrow Si(OH)_4 + 4H_2$$

28.4 THE CHLORIDES

There are two series of chlorides formed by the elements of group IV, viz. the dichlorides, MCl_2, in which the group IV element shows an oxidation state of +2, and the tetrachlorides, MCl_4, in which the group IV element shows an oxidation state of +4.

(1) Tetrachlorides. The group IV elements all form tetrachlorides which are liquids at room temperature, being composed of discrete molecules. From the charge cloud repulsion theory, we would predict tetrahedral molecules and this is confirmed experimentally.

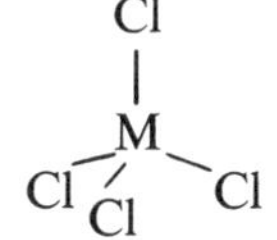

Fig 28.3 Valence structure of MCl_4.

It is found that CCl_4 does not hydrolyse, whereas the other tetrachlorides will hydrolyse. We will try to account for this difference. Let us consider the hydrolysis in basic solution of a tetrachloride such as $SnCl_4$. This is believed to proceed in stepwise fashion.

$$\begin{array}{llll} \text{I} & SnCl_4 & + & OH^- \rightarrow SnCl_3(OH) + Cl^- \\ \text{II} & SnCl_3(OH) & + & OH^- \rightarrow SnCl_2(OH)_2 + Cl^- \\ \text{III} & SnCl_2(OH)_2 & + & OH^- \rightarrow SnCl(OH)_3 + Cl^- \\ \text{IV} & SnCl(OH)_3 & + & OH^- \rightarrow Sn(OH)_4 + Cl^- \\ \hline \text{Nett reaction} & SnCl_4 & + & 4OH^- \rightarrow Sn(OH)_4 + 4Cl^- \end{array}$$

Tin(IV) hydroxide, $Sn(OH)_4$, is the almost insoluble, amphoteric hydroxy compound of tin in the +4 oxidation state. We can see how these hydrolysis steps might proceed. Tin atoms may accommodate more than four electron pairs in their outer shells, and hence step I in our hydrolysis will itself proceed in two stages, involving an unstable bipyramidal ion.

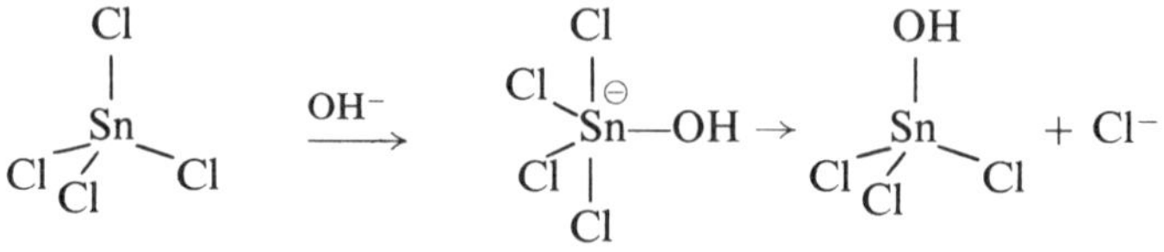

Cl Sn Cl Cl Cl $\xrightarrow{OH^-}$ Cl Cl ⊖ Sn—OH Cl Cl $\rightarrow$ OH Sn Cl Cl Cl + Cl^-

Fig 28.4 Mechanism of the hydrolysis of $SnCl_4$.

Such an easy pathway for this step is not available to carbon, however, since it is restricted to four electron pairs in its outer shell, i.e. no ion of composition $CCl_4(OH)^-$ is possible. A similar comparison may be drawn between the relative ease of hydrolysis of the mixed hydride-chlorides, methyl chloride, CH_3Cl, and silyl chloride, SiH_3Cl. Methyl chloride reacts relatively slowly with hydroxide ions to give methanol:

$$OH^- + CH_3Cl \rightarrow CH_3OH + Cl^-$$

whilst silyl chloride reacts virtually instantaneously with water, presumably yielding initially silanol, SiH_3OH. In the latter case the steps are shown in Figure 28.5:

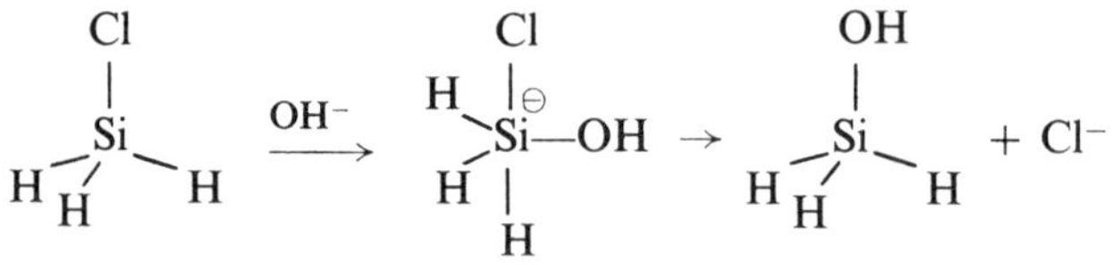

Cl Si H H H $\xrightarrow{OH^-}$ Cl H ⊖ Si—OH H H $\rightarrow$ OH Si H H H + Cl^-

Fig 28.5 Mechanism of hydrolysis of SiH_3Cl.

Here the bipyramidal ion, $SiH_3Cl(OH)^-$, acts as an easily formed intermediate. Carbon, however, is restricted to four electron pairs in its outer shell, and hence the chloride ion must be ejected whilst the hydroxide ion is moving in and no stable intermediate is possible.

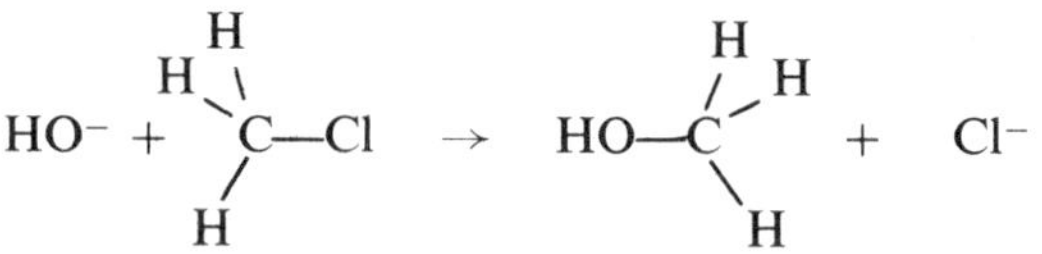

HO^- + H H C—Cl H $\rightarrow$ HO—C H H H + Cl^-

Fig 28.6 Mechanism of hydrolysis of CH_3Cl.

In this case the reaction proceeds quite slowly. A similar explanation probably holds for the difference in the ease of hydrolysis of SiH_4 and CH_4.

(2) Dichlorides. Dichlorides are formed by germanium, tin and lead. Germanium(II) chloride, $GeCl_2$, is rapidly oxidized by atmospheric oxygen:

$$2GeCl_2 + O_2 \rightarrow GeO_2 + GeCl_4$$

Ge^{2+} ions will reduce water to form GeO_2 and hydrogen gas. Tin(II) chloride $SnCl_2$ is readily soluble in water. If a concentrated solution is diluted the solution becomes turbid due to the precipitation of a basic salt, Sn(OH)Cl. Lead(II) chloride is sparingly soluble in cold water, but much more soluble in hot water. It shows little tendency to hydrolyse in solution.

There is a general tendency for the +2 oxidation state to become more stable relative to the +4 state as we proceed down the group. For germanium, tin, and lead this is well exemplified by the standard redox potentials for the systems

$$Ge^{4+} + 2e \rightarrow Ge^{2+} \qquad E^\circ = -1.6 \text{ volt}$$
$$Sn^{4+} + 2e \rightarrow Sn^{2+} \qquad E^\circ = +0.15 \text{ volt}$$
$$Pb^{4+} + 2e \rightarrow Pb^{2+} \qquad E^\circ = +1.80 \text{ volt}$$

showing the increasing stability of the reduced form relative to the oxidized form.

Tin(II) chloride is a common reductant through the action of the tin(II) ion. It will, for example, reduce potassium dichromate in acid solution:

$$Cr_2O_7^{2-} + 14H^+ + 6e \rightarrow 2Cr^{3+} + 7H_2O$$
$$Sn^{2+} \rightarrow Sn^{4+} + 2e$$
$$\overline{Cr_2O_7^{2-} + 14H^+ + 3Sn^{2+} \rightarrow 2Cr^{3+} + 3Sn^{4+} + 7H_2O}$$

28.5 THE OXIDES

There are two series of oxides formed by the elements of group IV, viz. the monoxides (MO) in which the group IV element shows an oxidation state of +2, and the dioxides (MO_2) in which the group IV element shows an oxidation state of +4.

(1) Dioxides. With the exception of lead dioxide, these oxides can all be formed by direct combination of element and oxygen. Lead dioxide can be prepared by electrolytic oxidation of the lead(II) ion, Pb^{2+}, in acidic solution.

Carbon dioxide in the solid, liquid and gaseous states is composed of discrete molecules. Solid carbon dioxide (dry ice) sublimes at −78.5°C under ordinary pressures. The other dioxides are network lattices involving bonds of varying degrees of polarity, ranging from almost pure covalent bonding in silicon dioxide (silica) to a partial ionic character in lead dioxide. Silicon dioxide (in the form of quartz) has a melting point of 1710°C and the other dioxides similarly have high melting points.

Lead dioxide is unstable when heated:

$$PbO_2 \rightarrow PbO + \tfrac{1}{2}O_2$$

The changing nature of the dioxides is illustrated by the decreasing acidic character of the oxides as we go down the group. Thus CO_2, SiO_2, and GeO_2 are acidic oxides which react with aqueous solutions of hydroxide ions giving the MO_3^{2-} ion, e.g.

$$GeO_2 + 2OH^- \rightarrow GeO_3^{2-} + H_2O$$

GeO_3^{2-} is called the germanate ion. In the case of CO_2 a bicarbonate ion, HCO_3^-, forms as well as a carbonate ion, CO_3^{2-}. Tin and lead dioxides are amphoteric. Neither of the pure anhydrous oxides is very soluble in aqueous solutions of acids or bases. They will, however, both react with fused hydroxides, e.g.

$$SnO_2 + 2OH^- \rightarrow SnO_3^{2-} + H_2O$$

SnO_3^{2-} is the anhydrous stannate ion. Lead dioxide forms an analogous, anhydrous plumbate ion, PbO_3^{2-}.

Both tin and lead dioxides may be prepared, in alkaline solution, in ill-defined hydrated forms, by the hydrolysis of tin and lead tetrachlorides:

$$SnCl_4 + 4OH^- \rightarrow Sn(OH)_4 + 4Cl^-$$

Tin(IV) hydroxide, $Sn(OH)_4$, is a poorly defined hydroxy compound that is frequently regarded as 'hydrated tin dioxide', $SnO_2.xH_2O$. It is only slightly soluble in water and readily loses water to yield SnO_2. Similarly the product of the hydrolysis of $PbCl_4$ is a hydrated lead dioxide, $PbO_2.xH_2O$, of indefinite composition. These hydrated oxides are fairly readily soluble in aqueous solutions containing hydroxide ions. They yield hydrated stannate and plumbate ions, $M(OH)_6^{2-}$

$$SnO_2\text{(hydrated)} + 2OH^- + 2H_2O \rightarrow Sn(OH)_6^{2-}$$

These hydrated oxides will also dissolve in acids yielding hydrated tin(IV) or lead(IV) ions which are themselves quite strong acids:

$$SnO_2\text{(hydrated)} + 4H^+(aq) \rightarrow Sn^{4+}(aq) + 2H_2O$$

(2) Monoxides. Carbon monoxide, CO, is a molecular compound with the probable valence structure $:\overset{-}{C}\equiv\overset{+}{O}:$ or $:C{=}\ddot{\underset{..}{O}}$

It may be formed by reduction of carbon dioxide, and is a gas at room temperature:

$$CO_2 + C \rightarrow 2CO$$

Silicon monoxide, SiO, apparently exists only in the gas phase at high temperatures, whilst germanium monoxide, GeO, is a black solid formed by reduction of GeO_2. Tin monoxide, SnO, and lead monoxide, PbO, are solids which form rather complex layer lattices. Both can be formed by heating the appropriate hydroxide, e.g.

$$Sn(OH)_2 \rightarrow SnO + H_2O$$

Carbon monoxide is a neutral oxide reacting with neither acid nor hydroxide ion solutions. Germanium monoxide shows weak acidic properties, reacting with hydroxide ions to form germanite ions, GeO_2^{2-}

$$GeO + 2OH^- \rightarrow GeO_2^{2-} + H_2O$$

It does not apparently show any basic properties, although, as we have seen, a Ge^{2+} ion does exist.

Both tin and lead monoxides are amphoteric, reacting with both acid and alkaline solutions, in the latter case yielding hydrated stannite or plumbite ions, $M(OH)_3^-$

$$SnO + 2OH^- + H_2O \rightarrow Sn(OH)_3^-$$

$$SnO + 2H^+ \rightarrow Sn^{2+} + H_2O$$

Both tin(II) and lead(II) ions react with hydroxide ion to form the appropriate hydroxide, e.g.

$$Sn^{2+} + 2OH^- \rightarrow Sn(OH)_2$$

These hydroxides are amphoteric, e.g.

$$Sn(OH)_2 + OH^- \rightarrow Sn(OH)_3^-$$

$$Sn(OH)_2 + 2H^+ \rightarrow Sn^{2+} + 2H_2O$$

28.6 SUMMARY

A number of discernible changes occur through this group:

(a) the marked tendency for the elements to become more metallic in structure as we go down the group;

(b) the general tendency for the +2 oxidation state to become more stable relative to the +4 state as we go down the group;

(c) the decrease in the acidic character of the oxides;

(d) the decrease in the acidity of the hydroxy compounds of both the $M(OH)_4$ and $M(OH)_2$ series. Decreasing electronegativity on passing down the group is undoubtedly an important factor in determining this trend. This decreasing electronegativity is also reflected in increasing ionic character, particularly of the +2 oxidation state, as we pass down the group.

QUESTIONS

1. Propose valence structures for carbon monoxide and for carbon dioxide.
2. Explain why potassium carbonate forms aqueous solutions which are alkaline. How may potassium bicarbonate be converted to potassium carbonate?
3. Compare the physical properties of carbon dioxide and silicon dioxide. Account for any differences in terms of differences in their crystal structures.
4. How may the lubricating and electrical conducting properties of graphite be explained in terms of its structure?
5. Compare the reaction with water of the chlorides of carbon and silicon. Account for any differences.
6. Oxalic acid ($H_2C_2O_4$) reacts with concentrated sulphuric acid as follows:

$$H_2C_2O_4 \xrightarrow{\text{conc. } H_2SO_4} CO + CO_2$$

   How could pure carbon monoxide be prepared using this reaction?
7. Explain what is meant by saying that 'carbon monoxide is isoelectronic with nitrogen.'
8. Write down valence structures for the following molecules and indicate their geometry:
   (a) CF_4; (b) CH_3F; (c) SiH_4; (d) SiH_3Cl; (e) CH_2Cl_2

Which molecule would be expected to be most polar? Which molecules would be expected to be non-polar?

9. For the systems

$$Sn^{4+} + 2e \rightarrow Sn^{2+} \qquad E^{\circ} = +0.15 \text{ volt}$$

and $$PbO_2 + 4H^+ + 2e \rightarrow Pb^{2+} + 2H_2O \qquad E^{\circ} = +1.46 \text{ volt}$$

which of the following reactions is most likely to occur:
(a) $PbO_2 + 4H^+ + Sn^{2+} \rightarrow Pb^{2+} + Sn^{4+} + 2H_2O$
(b) $Sn^{4+} + Pb^{2+} + 2H_2O \rightarrow Sn^{2+} + PbO_2 + 4H^+$

10. Explain why gaseous tin(II) chloride, $SnCl_2$, is an angular molecule rather than a linear one.
11. Suggest the steps that might take place when germanium tetrachloride, $GeCl_4$, reacts with water.
12. Oil paintings may contain lead pigments. Such paintings may darken due to the reaction of hydrogen sulphide, in the atmosphere, with the lead to form black lead sulphide, PbS. If such paintings are treated with dilute hydrogen peroxide solution, the painting is restored. Explain with the aid of equations the steps involved in these changes.
13. Calculate the solubility, in g dm^{-3}, of lead iodide, PbI_2, at room temperature, if the solubility product is 8.7×10^{-9}.
14. For the reaction

$$H_2CO_3 + H_2O \rightleftharpoons H_3O^+ + HCO_3^- \qquad K_{a1} = 4.5 \times 10^{-7}$$

whilst for the reaction

$$H_2CO_3 + 2H_2O \rightleftharpoons 2H_3O^+ + CO_3^- \qquad K_a = 2.0 \times 10^{-17}$$

Calculate K_a for the reaction

$$HCO_3^- + H_2O \rightleftharpoons H_3O^+ + CO_3^{2-}$$

15. Which solution has the higher pH, 0.2 M Na_2CO_3 or 0.02 M NaOH? Use the data given in question 14 and assume that only the hydrolysis of the CO_3^{2-} need be taken into account to calculate the pH of the Na_2CO_3 solution.
16. The standard heat of formation of silicon tetrachloride is: $\Delta H_f^{\circ}(SiCl_4(l)) = -640.1$ kJ. Calculate the heat evolved when the following react:
(a) 10.0 g Si and 110 g Cl_2;
(b) 10.0 g Si and 40.0 g Cl_2;
(c) 100 g Si and 17.0 g Cl_2;
(d) 100 g Si and 1.00 kg Cl_2.
17. The heat of combustion of cyclohexane (C_6H_{12}) at 25°C is

$$C_6H_{12}(l) + 9O_2(g) \rightarrow 6CO_2(g) + 6H_2O(l) \qquad \Delta H = -3929 \text{ kJ}$$

Given that the standard heats of formation of carbon dioxide and water are $\Delta H_f^{\circ}(CO_2(g)) = -393$ kJ and $\Delta H_f^{\circ}(H_2O(l)) = -286$ kJ, calculate the heat of formation of cyclohexane.
18. The boiling point of silicon tetrachloride is 57°C and at this temperature ΔH for vaporization is +151 joule g^{-1}. How much heat is evolved when 10 mole of silicon tetrachloride vapour condenses to liquid at the boiling point?

29

Group III Elements

The elements of group III have a common outer-shell configuration of three electrons (s^2p^1). Boron has a fairly high electronegativity and shows largely non-metallic characteristics. The other elements in the group are metals so that there is a distinct difference in properties between boron and the other elements. The chemistry of aluminium, gallium, indium and thallium is similar, but thallium shows some marked differences from other members of the group. Boron is not a common element but it is widely distributed in nature, usually combined with oxygen, e.g. as borax

I	II											III	IV	V	VI	VII	VIII
1 H																	2 He
3 Li	4 Be											5 B	6 C	7 N	8 O	9 F	10 Ne
11 Na	12 Mg											13 Al	14 Si	15 P	16 S	17 Cl	18 Ar
19 K	20 Ca	21 Sc	22 Ti	23 V	24 Cr	25 Mn	26 Fe	27 Co	28 Ni	29 Cu	30 Zn	31 Ga	32 Ge	33 As	34 Se	35 Br	36 Kr
37 Rb	38 Sr	39 Y	40 Zr	41 Nb	42 Mo	43 Tc	44 Ru	45 Rh	46 Pd	47 Ag	48 Cd	49 In	50 Sn	51 Sb	52 Te	53 I	54 Xe
55 Cs	56 Ba	57 La	72 Hf	73 Ta	74 W	75 Re	76 Os	77 Ir	78 Pt	79 Au	80 Hg	81 Tl	82 Pb	83 Bi	84 Po	85 At	86 Rn
87 Fr	88 Ra	89 Ac															

Table 29.1
Physical properties of group III elements

Element	*Atomic weight*	*Electron configuration*	*Outer-shell structure*	*Electro-negativity*	*Percentage abundance in earth's crust*
Boron	10.811	2.3	$2s^22p^1$	2.0	3×10^{-4}
Aluminium	26.9815	2.8.3	$3s^23p^1$	1.5	8.13
Gallium	69.72	2.8.18.3	$4s^24p^1$	1.6	1.5×10^{-3}
Indium	114.82	2.8.18.18.3	$5s^25p^1$	1.7	1×10^{-5}
Thallium	204.37	2.8.18.32.18.3	$6s^26p^1$	1.8	3.3×10^{-5}

$Na_2B_4O_7.10H_2O$. Aluminium is the third most abundant element in the earth's crust and is widely distributed, e.g. as bauxite, $Al_2O_3.xH_2O$, and in the form of silicate minerals (felspars and micas). The other elements are quite rare and are found as impurities in a number of ores, e.g. zinc ores.

29.1 FEATURES OF THE COMPOUNDS

Boron in many respects resembles beryllium, carbon, and silicon more closely than it does other members of group III. Thus it can form a series of hydrides (e.g. B_2H_6, B_4H_{10}, B_5H_9), whilst silicon also forms a series of hydrides (e.g. SiH_4, Si_2H_6, Si_3H_8). Aluminium forms only one hydride, a complex compound of empirical formula, AlH_3. It should be noted that the bonding in the boron hydrides shows some special features, whereas the bonding in the silicon hydrides involves the usual electron pair bonds.

Aluminium may be compared with the group I and II elements in the same period, viz. sodium and magnesium. From the standard redox potential for the system

$$Al^{3+} + 3e \rightarrow Al \qquad E^\circ = -1.67 \text{ volt}$$

we see that aluminium is a strong reductant. The rate at which Al is oxidized to Al^{3+} is often slow, however, because of the presence of an oxide film. It is for this reason that aluminium is resistant to atmospheric corrosion. The aluminium ion (Al^{3+}) has a smaller radius (0.5 Å) than that of sodium or magnesium, and is more highly charged. It readily forms hydrated ions which are more strongly acidic than those of sodium or magnesium.

$$Al(H_2O)_6^{3+} + H_2O \rightleftharpoons Al(H_2O)_5OH^{2+} + H_3O^+ \qquad K_{a1} = 1.4 \times 10^{-5}$$

$$Mg(H_2O)_6^{2+} + H_2O \rightleftharpoons Mg(H_2O)_5OH^+ + H_3O^+ \qquad K_{a1} = 4 \times 10^{-12}$$

If sodium hydroxide is added to a solution containing aluminium cations, a sequence of acid-base reactions can be written

$$Al(H_2O)_6^{3+} + OH^- \rightleftharpoons Al(H_2O)_5OH^{2+} + H_2O$$

$$Al(H_2O)_5OH^{2+} + OH^- \rightleftharpoons Al(H_2O)_4(OH)_2^+ + H_2O$$

$$Al(H_2O)_4(OH)_2^+ + OH^- \rightleftharpoons Al(H_2O)_3(OH)_3 + H_2O$$

$$Al(H_2O)_3(OH)_3 + OH^- \rightleftharpoons Al(H_2O)_2(OH)_4^- + H_2O$$

The neutral species $Al(H_2O)_3(OH)_3$ condense together, with the elimination of water molecules, to form a gelatinous precipitate of aluminium hydroxide. If excess of sodium hydroxide is added, the aluminate ion ($Al(OH)_4^-$) forms and the gelatinous precipitate dissolves. If acid is now added to this ultimate solution the above reactions are reversed:

$$Al(H_2O)_2(OH)_4^- + H_3O^+ \rightleftharpoons Al(H_2O)_3(OH)_3 + H_2O$$

Aluminium hydroxide first precipitates and then dissolves, the final product being the hydrated aluminium ion, $Al(H_2O)_6^{3+}$.

Complex ions, such as the aluminium fluoride ion (AlF_6^{3-}), are more readily formed by the aluminium ion than by the sodium or magnesium ion. Sodium and magnesium compounds usually involve sodium or magnesium ions as structural units, whereas in many aluminium compounds the bonding is intermediate in character between ionic and covalent.

Aluminium shows some resemblances, chemically, to beryllium. For example, the chlorides of both exist as dimeric molecules in the vapour.

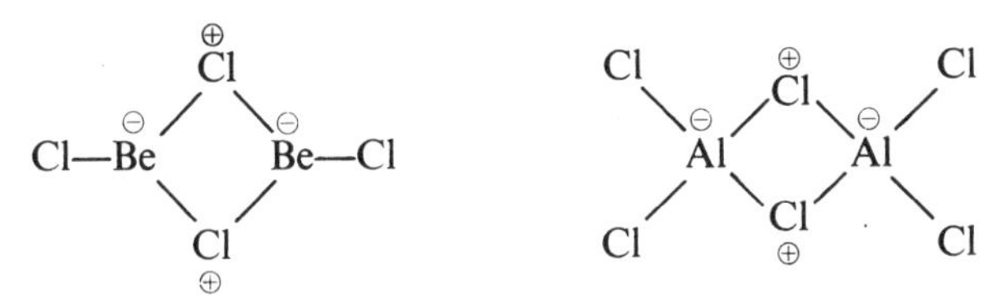

Fig 29.1 Valence structures of Be_2Cl_4 and Al_2Cl_6.

Gallium and indium are rare elements and their chemistry is far from completely known. Both form compounds involving Ga^+ and In^+, but these ions do not exist in aqueous solution, since they undergo oxidation to form the trivalent ion.

Thallium is well known in solution as the thallium(I) ion Tl^+, and many stable thallium(I) compounds are known, e.g. thallium(I) chloride, TlCl, thallium(I) hydroxide, TlOH. All members of group III form compounds in which the +3 oxidation state is shown, but this becomes less stable with respect to the +1 oxidation state as one moves down the group. Thus for the system

$$Tl^{3+} + 2e \rightarrow Tl^+$$

the standard redox potential is +1.25 volt, indicating a strong tendency for the thallium(III) ion to accept electrons and form the thallium(I) ion. In the solid state it is found that thallium(III) chloride loses chlorine at 40°C and forms thallium(I) chloride:

$$TlCl_3 \rightleftharpoons TlCl + Cl_2$$

29.2 THE CHLORIDES

Boron trichloride (BCl_3) is a planar, covalently bonded molecule involving three electron pairs in the outer shell of the boron atom. From the charge cloud repulsion theory we would predict that the chlorine atoms would occupy the corners of an equilateral triangle and this is found to be so.

```
      Cl
      |
      B
    /   \
  Cl     Cl
```

Fig 29.2 Valence structure of BCl_3.

Anhydrous aluminium chloride is a layer lattice involving intermediate bonding between ions in the layers and weak bonding between layers.

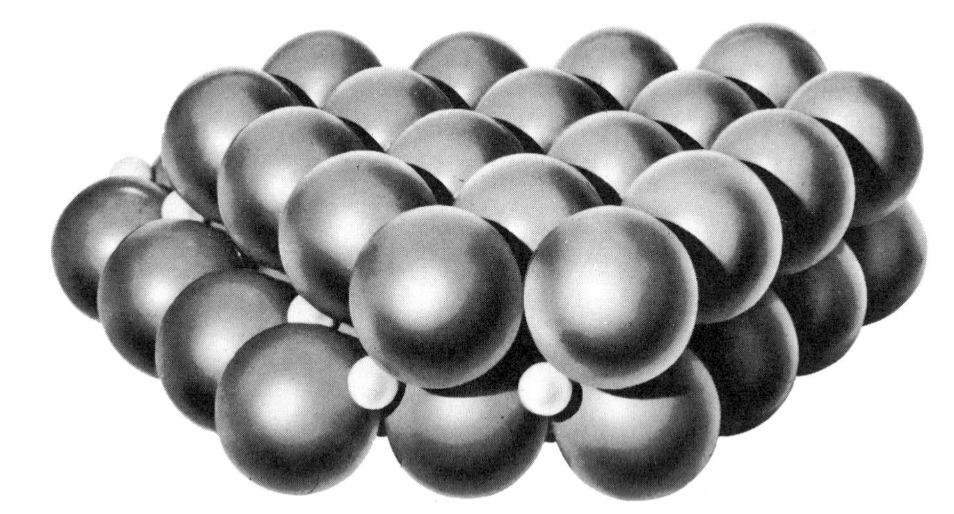

Fig 29.3 Aluminium chloride lattice.

If heated it is found that as the melting point is approached the electrical conductivity increases rapidly, and then at the melting point falls suddenly to nearly zero. At the melting point there is an unusually large decrease in density as the layer lattice changes to a melt containing Al_2Cl_6 molecules. The vapour also contains Al_2Cl_6 molecules.

```
Cl        ⊕        Cl
  \       Cl       /
   \ ⊖  /    \ ⊖  /
    Al          Al
   /    \    /    \
  /       Cl       \
Cl        ⊕        Cl
```

Fig 29.4 Valence structure of Al_2Cl_6 molecule.

The solid hydrate is a lattice of $Al(H_2O)_6^{3+}$ and Cl^- ions.

The melting point of aluminium chloride is 193 °C, whilst that of aluminium fluoride is 1290°C. The aluminium ion shows a co-ordination number of 6 towards the small fluoride ion and forms a network solid of aluminium and fluoride ions with a high melting point. With the larger chlorine atom or ion, a co-ordination number of 4 is preferred and a layer lattice is formed in the solid state and dimeric molecules form in the melt. In this case the melting point is lower.

Gallium, indium and thallium trichlorides resemble aluminium chloride in structure. Thallium(I) chloride (TlCl) has an ionic structure like that of caesium chloride, and chemically resembles silver chloride.

29.3 SUMMARY

In this group we see a number of marked characteristics:

(a) the striking difference in chemical properties between the largely non-metallic boron and the other members of the group;

(b) the similarity of aluminium to the group I and II elements, in that aluminium is a metal and a strong reductant;

(c) the decreasing stability of the +3 oxidation state down the group, and the increasing stability of the +1 oxidation state. This trend to greater stability of the lower oxidation state down a group is shown also in group IV, where the +2 state becomes more stable than the +4 state, and in group V where the +3 state becomes more stable than the +5 state.

QUESTIONS

1. Suggest valence structures for the following (show their stereochemistry):
   (a) AlF_6^{3-}
   (b) Ga_2Cl_6
   (c) BF_3
   (d) AlH_4^-
2. Boron nitride (BN) exists in the form of a solid very similar in properties to graphite. From your knowledge of the structure of graphite suggest a structure for boron nitride.
3. Aluminium is a very reactive element, yet it is used to make saucepans in which it comes into contact with many reactive materials. Explain why it may be so used.
4. When sodium carbonate is added to an aluminium salt solution, aluminium hydroxide precipitates. Explain.
5. Aluminium will not react with concentrated nitric acid, but will react readily with dilute nitric acid. Explain the difference in behaviour. Write an equation for the reaction of aluminium with dilute nitric acid assuming nitric oxide to be the chief gaseous product.
6. Thallium will dissolve in dilute sulphuric acid but not in dilute hydrochloric acid. Suggest a reason why.
7. The standard heat of formation of solid aluminium oxide is −1669 kJ. Calculate the quantity of heat evolved when one kilogram of aluminium oxide is formed from the elements.
8. 0.201 g of aluminium chloride occupy 18.1 cm³ at 15°C, 754 mmHg. Calculate

the observed mass of 1 mole of vapour. What is the structure of the molecule in the vapour?

9. 1.00 g of indium oxide (In_2O_3) was dissolved in 50.0 cm^3 of 0.480 M HCl. The excess acid was titrated with 0.102 M KOH and 23.5 cm^3 were required. Calculate the atomic weight of indium from these data, assuming the atomic weight of oxygen.
10. Calculate the pH of a 0.1 M solution of aluminium chloride. Assume that the acidity is entirely due to the first ionization of the aquo-cation.
(K_{a1} $Al(H_2O)_6^{3+}$ = 1.4×10^{-5}.)

30

Group VIII Elements

The group VIII elements are known as the *rare or noble gases.* They are called rare gases because they only exist in trace quantities in the atmosphere, and noble gases because they are chemically rather unreactive. Helium has an outer shell configuration of two electrons (s^2), whilst all the others have an outer-shell configuration of eight electrons (s^2p^6). Helium is formed as a product of radioactive decay in the form of α-particles, which are helium nuclei. Radon is a radioactive element which is a decay product of radium,

I	II											III	IV	V	VI	VII	VIII
1 H																	2 He
3 Li	4 Be											5 B	6 C	7 N	8 O	9 F	10 Ne
11 Na	12 Mg											13 Al	14 Si	15 P	16 S	17 Cl	18 Ar
19 K	20 Ca	21 Sc	22 Ti	23 V	24 Cr	25 Mn	26 Fe	27 Co	28 Ni	29 Cu	30 Zn	31 Ga	32 Ge	33 As	34 Se	35 Br	36 Kr
37 Rb	38 Sr	39 Y	40 Zr	41 Nb	42 Mo	43 Tc	44 Ru	45 Rh	46 Pd	47 Ag	48 Cd	49 In	50 Sn	51 Sb	52 Te	53 I	54 Xe
55 Cs	56 Ba	57 La	72 Hf	73 Ta	74 W	75 Re	76 Os	77 Ir	78 Pt	79 Au	80 Hg	81 Tl	82 Pb	83 Bi	84 Po	85 At	86 Rn
87 Fr	88 Ra	89 Ac															

Table 30.1
Physical properties of group VIII elements

Element	*Atomic weight*	*Electron configuration*	*Outer-shell structure*	*Electro-negativity*	*Percentage abundance in atmosphere*
Helium	4.0026	2	$1s^2$	—	5.24×10^{-4}
Neon	20.183	2.8	$2s^22p^6$	—	1.82×10^{-3}
Argon	39.948	2.8.8	$3s^23p^6$	—	0.934
Krypton	83.80	2.8.18.8	$4s^24p^6$	—	1.14×10^{-4}
Xenon	131.30	2.8.18.18.8	$5s^25p^6$	—	8.7×10^{-6}
Radon	—	2.8.18.32.18.8	$6s^26p^6$	—	—

but itself decays to polonium. The other elements exist in trace amounts in the atmosphere.

30.1 STRUCTURE OF THE ELEMENTS AND FEATURES OF THE COMPOUNDS

The noble gases are composed of single atoms. These persist as the structural units in the solid, liquid and gas phases. The melting points, boiling points, and heats of vaporization increase down the group. This is due to an increase in the strength of the dispersion forces as the size of the atoms increases (see chapter 10).

Table 30.2
Further physical properties of group VIII elements

Element	*Melting point (°C)*	*Boiling point (°C)*	*Heat of vaporization (ΔH in kJ mol⁻¹)*
Helium	—	−268.9	+0.092
Neon	−248.6	−246.0	+1.84
Argon	−189.4	−185.9	+6.28
Krypton	−157.2	−153.2	+9.66
Xenon	−111.9	−108.1	+13.68
Radon	−71	−62	+17.99

The noble gases form very few compounds. The electron affinities are probably close to zero and they have higher values for the ionization energy than other elements in the same periods. Thus we do not expect them to readily gain or lose electrons. Some of them can form compounds with elements of high electronegativity (such as fluorine and oxygen). In these compounds the bonding is probably covalent but it may possess some ionic character.

Neither helium nor neon would be expected to form compounds, as in both atoms the outer shell holds the maximum number of electrons permissible, viz. two in helium, and eight in neon. However, the others may have more than four electron pairs in the outer shell and hence may form compounds.

Compounds of argon have not been formed but compounds of xenon and krypton are known. In the case of radon the intense radioactivity tends to break up any chemical bonds as rapidly as they are formed.

30.2 THE FLUORIDES

When xenon and fluorine are heated in a nickel vessel at 400°C, they react to form a compound xenon tetrafluoride, XeF_4. At room temperature this is a stable, colourless, crystalline solid. It is isoelectronic with the ion IF_4^-, and its valence structure is shown in Figure 30.1. The charge cloud repulsion theory predicts that the molecule will be planar and this has been experimentally confirmed.

F F
Xe
F F

Fig 30.1 Valence structure of XeF_4 molecule.

Xenon tetrafluoride reacts violently with hydrogen:

$$XeF_4 + 2H_2 \rightarrow Xe + 4HF$$

Two other crystalline fluorides of xenon have been prepared, viz. xenon difluoride, XeF_2, and xenon hexafluoride, XeF_6. Xenon difluoride is isoelectronic with IF_2^- whilst XeF_6 might be expected to resemble IF_7 with a lone pair replacing one of the fluorine atoms in IF_7. The structure of XeF_6 is still uncertain.

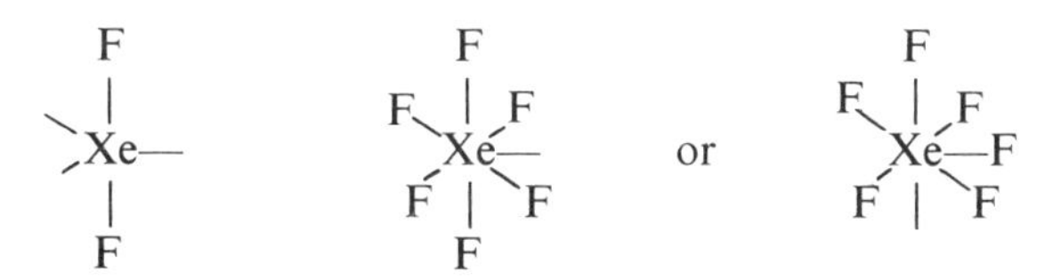

Fig 30.2 Valence structures of XeF_2 and $XeFe_6$ molecules.

Krypton forms a difluoride, KrF_2, and a tetrafluoride, KrF_4. These resemble the corresponding xenon fluorides but are less stable thermally.

All of these fluorides are good fluorinating agents, readily adding fluorine atoms to other elements and compounds. In this respect, as well as structurally, they resemble the interhalogen compounds.

30.3 THE OXIDES AND OXYACIDS

Both XeF_4 and XeF_6 may be reacted with water to produce xenon trioxide XeO_3. This is a colourless, crystalline solid which is violently explosive. It is a very strong oxidizing agent. Structurally the molecule is pyramidal as may be predicted from the charge cloud repulsion theory.

Xe
O O
O

Fig 30.3 Valence structure of XeO_3 molecule.

Xenon trioxide may be regarded as the derivative of xenic acid, H_6XeO_6 or $Xe(OH)_6$, which has been postulated but not definitely identified. By reacting XeF_6 with water and neutralizing the solution with barium hydroxide solution, a slightly soluble barium xenate, Ba_3XeO_6, may be obtained. In this compound xenon shows an oxidation state of +6.

If XeO_3 is reacted with sodium hydroxide solution the salt, sodium perxenate, Na_4XeO_6, may be obtained. This is a very powerful oxidant and shows xenon in the oxidation state +8.

$$4XeO_3 + 12OH^- \rightarrow Xe + 3XeO_6^{4-} + 6H_2O$$

Notice in this reaction that XeO_3 is both oxidized to XeO_6^{4-} (increase in oxidation number of Xe from +6 to +8) and reduced to Xe (decrease in oxidation number of Xe from +6 to 0). XeO_3 is functioning both as an oxidant (itself being reduced to Xe) and as a reductant (itself being oxidized to XeO_6^{4-}).

QUESTIONS

1. Suggest valence structures and stereochemistry for:
   (a) KrF_2
   (b) KrF_4
   (c) $XeOF_2$
   (d) $XeOF_4$
   (e) XeO_6^{4-}
2. Identify these two xenon fluorides from the following data:
   (a) 36.9 mg of the first fluoride yielded 13.5 mg of fluorine on decomposition, and
   (b) 0.409 g of xenon reacted with fluorine to form 0.758 g of the second fluoride.
3. 0.182 mole of xenon tetrafluoride reacted with 75.0 g of iodide ion in solution according to the equation

   $$XeF_4 + 4I^- \rightarrow 2I_2 + Xe + 4F^-$$

   What volume of xenon was evolved at 15°C, 768 mmHg pressure?
4. A 3000 cm^3 vessel contained xenon at partial pressure 395 mmHg, and fluorine at partial pressure 823 mmHg, at −30°C. The gases were reacted and 0.580 g of xenon difluoride formed. What was the total pressure in the vessel, at −30°C, after the reaction? (Ignore the volume taken up by the solid xenon difluoride.)

31

Transition Elements

In chapter 22 the structure of the periodic table was discussed in terms of the electron configurations of the elements. It was seen that in the fourth period (K to Kr), the first two elements, potassium and calcium, show the 4s subshell filling. With the next element, scandium, the next orbital of lowest energy lies in a 3d subshell, so that scandium has the electron configuration $1s^22s^22p^63s^23p^63d^14s^2$. A d-subshell contains five orbitals and hence ten electrons, so that following calcium we find a block of ten elements in which the 3d subshell is being filled. This feature results in these elements

I	II											III	IV	V	VI	VII	VIII
1 H																	2 He
3 Li	4 Be											5 B	6 C	7 N	8 O	9 F	10 Ne
11 Na	12 Mg											13 Al	14 Si	15 P	16 S	17 Cl	18 Ar
19 K	20 Ca	21 Sc	22 Ti	23 V	24 Cr	25 Mn	26 Fe	27 Co	28 Ni	29 Cu	30 Zn	31 Ga	32 Ge	33 As	34 Se	35 Br	36 Kr
37 Rb	38 Sr	39 Y	40 Zr	41 Nb	42 Mo	43 Tc	44 Ru	45 Rh	46 Pd	47 Ag	48 Cd	49 In	50 Sn	51 Sb	52 Te	53 I	54 Xe
55 Cs	56 Ba	57 La	72 Hf	73 Ta	74 W	75 Re	76 Os	77 Ir	78 Pt	79 Au	80 Hg	81 Tl	82 Pb	83 Bi	84 Po	85 At	86 Rn
87 Fr	88 Ra	89 Ac															

Table 31.1
Some physical properties of the transition elements

Element	*Electron configuration*	*Configuration of outer shells*	*Electronegativity*	*Percentage abundance in earth's crust*
Scandium	2.8.9.2	$3d^14s^2$	1.3	5×10^{-4}
Titanium	2.8.10.2	$3d^24s^2$	1.5	0.44
Vanadium	2.8.11.2	$3d^34s^2$	1.6	0.015
Chromium	2.8.13.1	$3d^54s^1$	1.6	0.02
Manganese	2.8.13.2	$3d^54s^2$	1.5	0.1
Iron	2.8.14.2	$3d^64s^2$	1.8	5.0
Cobalt	2.8.15.2	$3d^74s^2$	1.8	2.3×10^{-3}
Nickel	2.8.16.2	$3d^84s^2$	1.8	8.0×10^{-3}
Copper	2.8.18.1	$3d^{10}4s^1$	1.9	7.0×10^{-3}
Zinc	2.8.18.2	$3d^{10}4s^2$	1.6	0.0132

showing many properties in common and they are referred to as the *transition elements*. In the fifth and sixth periods we find the 4d and 5d subshells, respectively, filling in a similar way, giving rise to two more series of transition elements. In the case of the third transition series, the 4f subshell is of lower energy than the 5d subshell and a block of fourteen rare earth elements (or lanthanides) precedes the filling of the 5d subshell. In the rest of this chapter we will confine our attention to the first transition series (Sc to Zn). Iron is the most abundant of these elements, and some are quite rare. They occur largely in the form of the oxide or sulphide ores, e.g. haematite, Fe_2O_3, zinc blende, ZnS.

31.1 STRUCTURE OF THE ELEMENTS AND FEATURES OF THE COMPOUNDS

All of these elements have fairly low values for their electronegativities. They are, in fact, structurally all metallic crystals with generally high melting and boiling points, e.g. iron, m.p. 1535°C, b.p. 2735°C. Their densities are also generally high, e.g. iron, 7.87 g cm^{-3}. There are exceptions, e.g. zinc has an m.p. of 419°C, and titanium a density of 4.49 g cm^{-3}. The transition metals include some elements of great technological importance. The strength of iron, particularly in the form of its alloy steel, makes it a major engineering material. The high thermal and electrical conductivity of copper find a multitude of applications for this metal. The lustre of chromium and its apparent lack of reactivity find application in chrome plating. In this last instance it should be noted that chromium is, in fact, unstable with respect to atmospheric corrosion, but the presence of a thin surface film of oxide renders it apparently stable by reducing the rate of reaction. In the case of iron, the atmospheric corrosion, or rusting, occurs all too readily and represents a major economic problem.

The transition elements react with the halogens to form a wide range of halides, and with oxygen and sulphur to form oxides and sulphides. In these compounds we find a range of structures, but most commonly network

Table 31.2

Oxidation states of the transition elements

	Representative compounds							*Common oxidation numbers*	*Total 3d and 4s electrons*
Sc			$ScCl_3$					$+\mathbf{3}$	3
Ti		$TiCl_2$	$TiCl_3$	$TiCl_4$				$+2+3+\mathbf{4}$	4
V		VCl_2	VCl_3	VCl_4	V_2O_5			$+2+3+4+\mathbf{5}$	5
Cr		$CrCl_2$	$CrCl_3$			CrO_3		$+2+\mathbf{3}$ $+6$	6
Mn		$MnCl_2$	$MnCl_3$	MnO_2		K_2MnO_4	$KMnO_4$	$+2+3+\mathbf{4}$ $+6+7$	7
Fe		$FeCl_2$	$FeCl_3$			$BaFeO_4$		$+2+\mathbf{3}$ $+6$	8
Co		$CoCl_2$	$CoCl_3$					$+\mathbf{2}+3$	9
Ni		$NiCl_2$	Ni_2O_3					$+\mathbf{2}+3$	10
Cu	CuCl	$CuCl_2$						$+1+\mathbf{2}$	11
Zn		$ZnCl_2$						$+\mathbf{2}$	12

lattices or layer lattices. The bonding in these compounds can for most purposes be regarded as ionic, although in many cases it is intermediate in character between ionic and covalent.

There is a wide range of oxidation states found, as not only 4s but also 3d electrons may be involved in bonding. All the elements, except scandium, show an oxidation state of $+2$ and many show an oxidation state of $+3$ as well. The observed oxidation states are listed in Table 31.2, with the most common states shown in bold type. Notice that for the elements from Sc to Mn, the maximum oxidation number correlates with the total number of 3d and 4s electrons.

Many transition metal compounds and many transition metal complex ions are coloured. The colour is due to the absorption of photons of light energy by the electrons in the transition metal atoms or ions. The removal of certain components of visible light means that the compound appears to be coloured. The ability of electrons to absorb this light energy seems to be bound up with the presence of unfilled d-orbitals in these elements, In a compound or complex ion these orbitals do not in fact all occupy the same energy level, and it becomes possible for an electron to move from a lower to a higher energy level. Thus when the pale yellow iron(III) hydrated ion becomes associated with a thiocyanate ion (SCN^-), the new complex ion is an intense red colour:

$$Fe^{3+}(aq) + SCN^-(aq) \rightarrow Fe(SCN)^{2+}(aq)$$

In later sections we will see many other examples of coloured complex ions and coloured compounds.

31.2 COMPOUNDS OF VANADIUM, CHROMIUM, MANGANESE AND IRON

To illustrate the range of oxidation states shown by the transition elements it is convenient to consider in a little more detail four successive transition

Table 30.3

Oxides and derived ions of vanadium, chromium, manganese and iron

Oxide	*Formula*	*Derived ions*	*Colour of ion (hydrated)*
Vanadium(II) oxide	VO	V^{2+}	violet
Vanadium(III) oxide	V_2O_3	V^{3+}	green
Vanadium(IV) oxide	VO_2	VO^{2+} (vanadyl)	blue
Vanadium(V) oxide	V_2O_5	VO^{3+} (pervanadyl)	yellow
		VO_3^- (vanadate)	colourless
Chromium(II) oxide	CrO	Cr^{2+}	blue
Chromium(III) oxide	Cr_2O_3	Cr^{3+}	violet
		$Cr(OH)_4^-$ (chromite)	green
Chromium(VI) oxide	CrO_3	CrO_4^{2-} (chromate)	yellow
		$Cr_2O_7^{2-}$ (dichromate)	orange
Manganese(II) oxide	MnO	Mn^{2+}	pink
Manganese(III) oxide	Mn_2O_3	Mn^{3+}	violet
Manganese(IV) oxide	MnO_2	Mn^{4+}	—
		MnO_3^{2-} (manganite)	—
—	—	MnO_4^{2-} (manganate)	green
Manganese(VII) oxide	Mn_2O_7	MnO_4^- (permanganate)	purple
Iron(II) oxide	FeO	Fe^{2+}	green
Iron(III) oxide	Fe_2O_3	Fe^{3+}	yellow

elements, viz. vanadium, chromium, manganese and iron. The oxides formed by the four elements and the ions derived from them are listed in Table 30.3.

The table shows that for each element as the oxidation number rises the oxide becomes more acidic, basic oxides yielding positive ions and acidic oxides negative ions. For example, let us examine the case of chromium in detail. Chromium(II) oxide is a black powder. It is a basic oxide and if reacted with dilute sulphuric acid, in the absence of air it yields a blue solution containing chromium(II) ions:

$$CrO(s) + 2H^+(aq) \rightarrow Cr^{2+}(aq) + H_2O$$

The chromium(II) ion is a very powerful reductant (E° for the system Cr^{3+}/Cr^{2+} is -0.41 volt) and is readily oxidized to chromium(III) ion by atmospheric oxygen:

$$4Cr^{2+}(aq) + 4H^+(aq) + O_2(g) \rightarrow 4Cr^{3+}(aq) + 2H_2O$$

The hydrated chromium(III) ion is violet in colour. If sodium hydroxide solution is added, chromium(III) hydroxide precipitates as a green gelatinous solid:

$$Cr^{3+}(aq) + 3OH^-(aq) \rightarrow Cr(OH)_3(s)$$

The hydroxide is amphoteric, reacting with excess hydroxyide ion to form green chromite ions, and with acid to form chromium(III) ions:

$$Cr(OH)_3(s) + OH^-(aq) \rightarrow Cr(OH)_4^-(aq)$$

$$Cr(OH)_3(s) + 3H^+(aq) \rightarrow Cr^{3+}(aq) + 3H_2O$$

Chromium(III) oxide is a green powder and is also amphoteric, reacting with hydroxide ions to yield chromite ions, and with acid to yield chromium(III) ions.

In alkaline solution the chromite ion may be oxidized by an oxidant such as hydrogen peroxide to yield the yellow chromate ion

$$2Cr(OH)_4^-(aq) + 3H_2O_2(aq) + 2OH^-(aq) \rightarrow CrO_4^{2-}(aq) + 8H_2O$$

A valence structure for the chromate ion is shown in Figure 31.1.

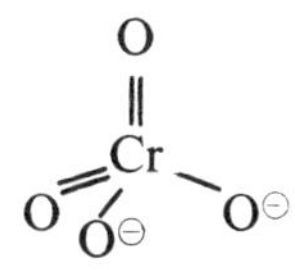

Fig 31.1 Valence structure of chromate ion, CrO_4^{2-}.

It is clear that this ion structurally resembles the sulphate ion, and indeed chromates and sulphates show many similarities.

If the solution containing chromate ions is acidified, the colour changes to orange as the following equilibrium is displaced, yielding a high equilibrium proportion of dichromate ion, $Cr_2O_7^{2-}$

$$2CrO_4^{2-}(aq) + 2H^+(aq) \rightleftharpoons Cr_2O_7^{2-}(aq) + H_2O \qquad K = 5 \times 10^{15}$$

In acidic solution the dichromate ion is most abundant and in alkaline solution the chromate ion is most abundant. The valence structure of the dichromate ion is shown in Figure 31.2.

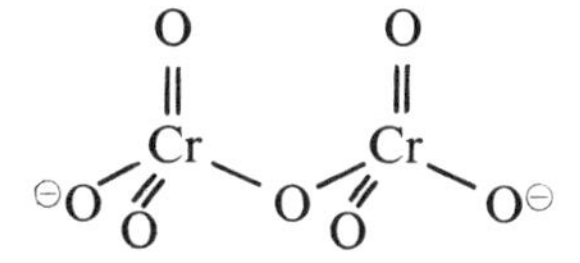

Fig 31.2 Valence structure of dichromate ion, $Cr_2O_7^{2-}$.

Chromium(VI) oxide is a deep red substance. It is an acidic oxide and when added to water it yields an equilibrium mixture of chromate ions, dichromate ions, and other still more complex species.

Both chromate and dichromate ions are oxidants. The dichromate ion is a powerful oxidant in acid solution, thus for the system

$$Cr_2O_7^{2-}(aq) + 14H^+(aq) + 6e \rightarrow 2Cr^{3+}(aq) + 7H_2O \qquad E^\circ = +1.33 \text{ volt}$$

whilst in alkaline solution the chromate ion can act as a much weaker oxidant

$$CrO_4^{2-}(aq) + 4H_2O(aq) + 3e \rightarrow Cr(OH)_3(s) + 5OH^- \qquad E^\circ = -0.13 \text{ volt}$$

In the case of vanadium and manganese a similar range of compounds exists and similar interconversions may be brought about. We will not consider these in detail, but confine our attention to one important ion of manganese, viz. the permanganate ion, MnO_4^-.

Manganese(IV) oxide is oxidized by atmospheric oxygen when fused with potassium hydroxide, forming the green manganate ion, MnO_4^{2-}

$$MnO_2(s) + \tfrac{1}{2}O_2(g) + 2OH^-(s) \rightarrow MnO_4^{2-} + H_2O$$

In acid solution the manganate ion reacts to yield the permanganate ion, MnO_4^-, which is an intense purple colour

$$3MnO_4^{2-}(aq) + 4H^+(aq) \rightarrow 2MnO_4^-(aq) + MnO_2(s) + 2H_2O$$

The permanganate ion has the valence structure shown in Figure 31.3 and clearly resembles the perchlorate ion, ClO_4^-. In these ions both the manganese and chlorine show an oxidation number of +7.

O⊖
|
Mn
O O O

Fig 31.3 Valence structure of permanganate ion, MnO_4^-.

The permanganate ion is a very powerful oxidant in acid solution, thus for the system

$$MnO_4^-(aq) + 8H^+(aq) + 5e \rightarrow Mn^{2+}(aq) + 4H_2O \qquad E^\circ = +1.51 \text{ volt}$$

The reaction of permanganate ion and a reducing agent in acidic solution appears to be catalysed by the manganese(II) ion, Mn^{2+}. Certainly this is true for the reaction with oxalate ion, $C_2O_4^{2-}$

$$2MnO_4^- + 5C_2O_4^{2-} + 16H^+ \rightarrow 2Mn^{2+} + 10CO_2 + 8H_2O$$

It may be noted that manganese(IV) oxide is itself quite a strong oxidant and in strongly acid solution can oxidize chloride ion to chlorine:

$$MnO_2(s) + 2Cl^- + 4H^+ \rightarrow Mn^{2+} + Cl_2 + 2H_2O$$

The common species of iron in aqueous solution are the iron(II) ion, Fe^{2+}, and the iron(III) ion, Fe^{3+}. The standard redox potential for the system

$$Fe^{3+} + e \rightarrow Fe^{2+} \qquad E^\circ = +0.77 \text{ volt}$$

shows Fe^{2+} to be a moderately strong reductant, and Fe^{3+} to be a moderately strong oxidant. Remember that the stronger an oxidant the weaker the conjugate reductant. Thus an ion such as MnO_4^- is a much stronger oxidant than Fe^{3+}, and in consequence the ion Mn^{2+} is a much weaker reductant than Fe^{2+}. We find that Fe^{2+} reduces MnO_4^- to Mn^{2+}, and reduces $Cr_2O_7^{2-}$ to Cr^{3+}, in acid solution.

$$MnO_4^- + 8H^+ + 5Fe^{2+} \rightarrow Mn^{2+} + 5Fe^{3+} + 4H_2O$$

$$Cr_2O_7^{2-} + 14H^+ + 6Fe^{2+} \rightarrow 2Cr^{3+} + 6Fe^{3+} + 7H_2O$$

On the other hand, Fe^{3+} oxidizes species such as I^- to I_2, and H_2S to S

$$2Fe^{3+} + 2I^- \rightarrow 2Fe^{2+} + I_2$$

$$2Fe^{3+} + H_2S \rightarrow 2Fe^{2+} + S + 2H^+$$

No redox reaction could occur if H_2S was passed into a Fe^{2+} solution, since both species are reductants. This is not to say, however, that *no reaction* could occur. In fact, if H_2S is passed into a slightly alkaline solution of Fe^{2+}, a black precipitate of FeS is formed.

$$Fe^{2+} + H_2S \rightarrow FeS(s) + 2H^+$$

Both the hydrated Fe^{2+} and the hydrated Fe^{3+} function as Bronsted acids in aqueous solution. The pK_a of the former is 9.5, indicating an acid strength comparable with the NH_4^+ ion, while the pK_a of the hydrated Fe^{3+} is 3.1, indicating this ion is a stronger acid than acetic acid. It is the strongest acid of the common aquated cations. In alkaline solution both species form insoluble hydroxides, the green iron(II) hydroxide and red-brown iron(III) hydroxide, generally represented for simplicity as $Fe(OH)_2$ and $Fe(OH)_3$ respectively. $Fe(OH)_2$ is readily and rapidly oxidized to $Fe(OH)_3$ in aqueous solution by atmospheric oxygen.

EXERCISE

Iron(II) hydroxide has a higher solubility product than iron(III) hydroxide. Does this *necessarily* imply that $Fe(OH)_2$ is more soluble than $Fe(OH)_3$?

In contrast to the behaviour of some other transition metal cations, neither Fe^{2+} nor Fe^{3+} form an ammine complex in solution.

EXERCISE

Two well-known compounds of iron are the ore minerals magnetite, Fe_3O_4, and iron pyrites, FeS_2. Comment briefly on the oxidation state of Fe in each of these compounds.

31.3 CO-ORDINATION COMPOUNDS

In chapter 15 the class of ions known as complex ions was discussed. Compounds containing these ions are known as *complex compounds* or *co-ordination compounds*. It was pointed out in chapter 15 that dissolved ions exist as hydrated ions and that it is possible to replace the water ligands with other molecules or ions. A familiar example of such a process is the class of complex ammine cations formed by treating a solution containing a cation with a solution of ammonia. Consider the reactions of the hydrated cations of silver, zinc, copper(II), and nickel(II) with concentrated ammonia solution. We may represent the equilibria as follows

$$Ag(H_2O)_2^+ + 2NH_3 \rightleftharpoons Ag(NH_3)_2^+ + 2H_2O \qquad K_{st} = 1.6 \times 10^7$$

$$Zn(H_2O)_4^{2+} + 4NH_3 \rightleftharpoons Zn(NH_3)_4^{2+} + 4H_2O \qquad K_{st} = 10^{10}$$

$$Cu(H_2O)_4^{2+} + 4NH_3 \rightleftharpoons Cu(NH_3)_4^{2+} + 4H_2O \qquad K_{st} = 4.7 \times 10^{12}$$

$$Ni(H_2O)_6^{2+} + 6NH_3 \rightleftharpoons Ni(NH_3)_6^{2+} + 6H_2O \qquad K_{st} = 10^9$$

In the first two cases all the species are colourless, but in the case of copper the aquo copper(II) ion is pale blue, whilst the tetrammine copper(II) ion is an intense blue: similarly the hexaquo nickel(II) ion is green, whilst the hexammine nickel(II) ion is blue.

What is the nature of the bonding in complex ions? In some complexes, such as FeF_6^{3-} the bonding may be regarded as attraction between positive Fe^{3+} and negative F^- ions. In the ammines and hydrated ions the bonding may be considered as due to the attraction between the polar ammonia or water molecules and the central positive ion. In other complexes, such as the ferricyanide ion, $Fe(CN)_6^{3-}$, the bonding between the central iron atom and the cyanide groups must be regarded as being appreciably covalent.

In aqueous solutions, water molecules will always fill any co-ordination sites around an ion which are not filled by other ligands. Thus Cu^{2+} can accommodate six ligands, so $Cu(NH_3)_4^{2+}$ should strictly be written $Cu(NH_3)_4(H_2O)_2^{2+}$ (with the four ammonia molecules in a plane). On the other hand, Zn^{2+} can only accommodate four ligands and hence $Zn(NH_3)_4^{2+}$ is tetrahedral.

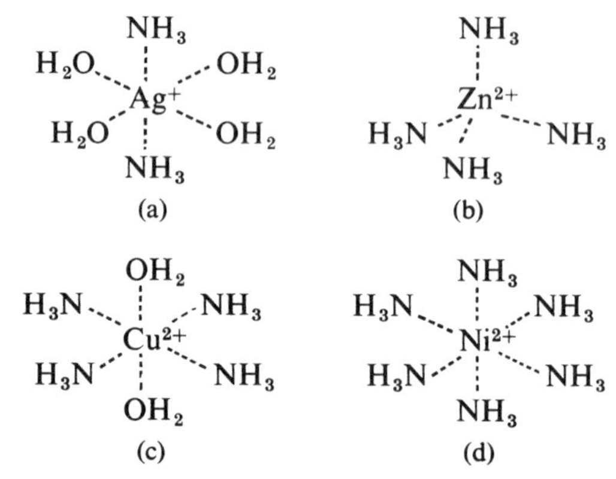

Fig 31.4 Structures of the complex ions: (a) $Ag(NH_3)_2^+$, (b) $Zn(NH_3)_4^{2+}$, (c) $Cu(NH_3)_4^{2+}$, (d) $Ni(NH_3)_6^{2+}$.

It should be noted that the complex ammine cations are associated with negative ions in solution and many of these complex compounds may be crystallized. Thus one can readily form solid tetrammine copper(II) sulphate which is an ionic lattice containing tetrammine copper(II) ions ($Cu(NH_3)_4^{2+}$) and sulphate ions (SO_4^{2-}).

Thus far we have discussed the ammonia ligand, but of course many other ligands may replace water in hydrated ions. Examples of suitable anions are OH^-, F^-, Cl^-, CN^- (cyanide ion), and SCN^- (thiocyanate ion). Some examples of the formation of complex ions involving these ligands are given

$$Al(H_2O)_6^{3+} + OH^- \rightleftharpoons Al(H_2O)_5(OH)^{2+} + H_2O \qquad K_{st} = 1.4 \times 10^9$$

$$Cu(H_2O)_6^{2+} + Cl^- \rightleftharpoons Cu(H_2O)_5Cl^+ + H_2O \qquad K_{st} = 1$$

$$Ag(H_2O)_2^+ + 2CN^- \rightleftharpoons Ag(CN)_2^- + 2H_2O \qquad K_{st} = 10^2$$
$$Fe(H_2O)_6^{3+} + SCN^- \rightleftharpoons Fe(H_2O)_5(SCN)^{2+} + H_2O \qquad K_{st} = 10^3$$

Replacement of water ligands may proceed sequentially, e.g. consider the case of the hexaquo chromium(III) ion:

$$Cr(H_2O)_6^{3+} + Cl^- \rightleftharpoons Cr(H_2O)_5Cl^{2+} + H_2O$$
$$Cr(H_2O)_5Cl^{2+} + Cl^- \rightleftharpoons Cr(H_2O)_4Cl_2^+ + H_2O$$

It is, in fact, possible to isolate and crystallize three chromium(III) chlorides, viz. a violet $[Cr(H_2O)_6]Cl_3$, a pale green $[Cr(H_2O)_5Cl]Cl_2.H_2O$, and a dark green $[Cr(H_2O)_4Cl_2]Cl.2H_2O$. A mole of each of these compounds would yield, respectively, 4, 3 and 2 moles of ionic species, e.g. $[Cr(H_2O)_5Cl]Cl_2.H_2O$ would yield a $[Cr(H_2O)_5Cl]^+$ cation and two Cl^- ions. The replacement process may continue, yielding a neutral species, $Cr(H_2O)_3Cl_3$, and eventually the species $CrCl_6^{3-}$. Notice that in this sequence we have complex ions with positive, negative and zero charges.

As discussed in chapter 15 some ligands may occupy more than one co-ordination position around a central ion. Such ligands are referred to as *chelating agents*, and the compounds containing such complex ions are referred to as *chelates*. Examples of chelating agents are the oxalate ion ($C_2O_4^{2-}$) and the ethylene diamine molecule ($H_2N.CH_2.CH_2.NH_2$) which is often symbolized as (en). Examples of complex ions involving chelating agents are the tris-oxalato chromium(III) ion and the tris-ethylenediamine cobalt(II) ion

$$Cr(H_2O)_6^{3+} + 3C_2O_4^{2-} \rightleftharpoons Cr(C_2O_4)_3^{3-} + 6H_2O$$
$$Co(H_2O)_6^{2+} + 3(en) \rightleftharpoons Co(en)_3^{2+} + 6H_2O$$

Some large molecules are examples of chelates. The molecule haem, a component of blood haemoglobin, is a chelate of iron(III), and the photosynthetic pigment chlorophyll is a chelate of magnesium.

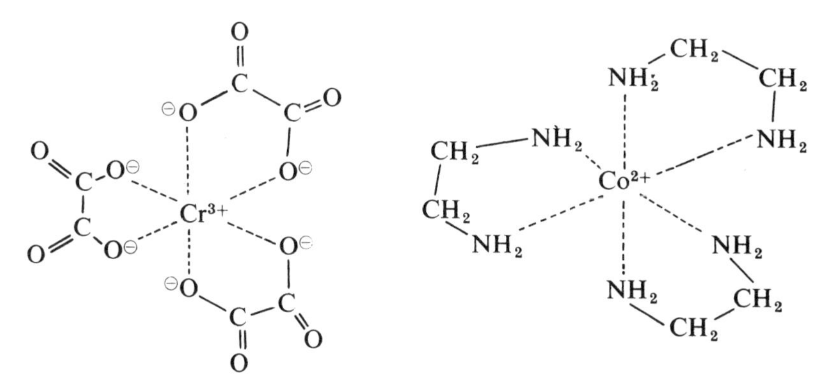

Fig 31.5 Structures of the complex ions (a) $Cr(C_2O_4)_3^{3-}$, (b) $Co(en)_3^{2+}$.

QUESTIONS

1. Why are the elements with atomic numbers 21 to 30 classified together, as a series, in the periodic table?
2. Write down the electron configurations of the elements: (a) scandium; (b) manganese; (c) zinc.
3. Complex ions are readily formed when H_2O molecules interact with cations, but not when hydrogen ions (H_3O^+) and cations interact. Suggest an explanation for this difference.
4. What is the oxidation number of chromium in: (a) CrO_4^{2-}; (b) Cr^{3+}; (c) CrO_3; (d) Cr_2O_3; (e) CrF_2; (f) $[Cr(H_2O)_5NH_3]Cl_3$; (g) $K_2SO_4.Cr_2(SO_4)_3 24H_2O$.
5. By assigning oxidation numbers show whether or not the reaction
$$2CrO_4^{2-} + 2H^+ \rightarrow Cr_2O_7^{2-} + H_2O$$
is an oxidation-reduction reaction.
6. Using values for standard redox potentials predict what might happen if a piece of zinc is added to a 1.0 M Mn^{2+} solution, and if a piece of manganese is added to a 1.0 M Zn^{2+} solution.
$$Zn^{2+} + 2e \rightarrow Zn; \quad E^\circ = -0.76 \text{ volt, and}$$
$$Mn^{2+} + 2e \rightarrow Mn; \quad E^\circ = -1.05 \text{ volt.}$$
7. Would one expect AgCl to be more or less soluble in 1.0 M NH_3 than in a mixture of 1.0 M NH_3 and 1.0 M NH_4NO_3?
8. A 0.1 M $ZnSO_4$ is acidic to litmus whilst a 0.1 M $MgSO_4$ solution is virtually neutral. Discuss this difference in behaviour.
9. For each of the following state in which of the two solutions the first substance is more soluble and why:
(a) AgCl(s) in 0.1 M Ag_2SO_4 or 0.1 M $AgNO_3$
(b) CrF_2(s) in 0.1 M NaF or 0.1 M HF
(c) $Cu(OH)_2$(s) in 1 M NaOH or 1 M NH_3
(d) MnF_2(s) in 0.08 M NaF or 0.1 M $MnSO_4$
10. Complete and balance the following equations:
(a) $MnO_4^- + H^+ + Fe^{2+} \rightarrow$
(b) $Cr_2O_7^{2-} + H^+ + NO_2^- \rightarrow$
(c) $MnO_2 + MnO_4^- + OH^- \rightarrow$
(d) $MnO_2 + H^+ + Br^- \rightarrow$
11. The standard redox potentials for the systems
$$V^{3+} + e \rightarrow V^{2+} \quad E^\circ = +0.20 \text{ volt}$$
and
$$Fe^{3+} + e \rightarrow Fe^{2+} \quad E^\circ = +0.77 \text{ volt}$$
are given, Predict what might happen if
(a) solutions of 1.0 M Fe^{3+} and 1.0 M V^{3+} are mixed;
(b) solutions of 1.0 M Fe^{2+} and 1.0 M V^{3+} are mixed;
(c) solutions of 1.0 M Fe^{3+} and 1.0 M V^{2+} are mixed.
12. An aqueous violet solution is treated with sodium hydroxide and a gelatinous blue-green precipitate forms. Excess hydroxide converts this to a clear green solution. When sodium peroxide (Na_2O_2) is added a yellow solution forms which becomes orange when treated with excess acid. Explain with the aid of equations this sequence of changes.
13. The colours of chromium(III) chloride solutions vary from violet to green depending on concentration and temperature. How is this variation explained?

14. How could one distinguish between $[Cr(H_2O)_5Cl]Cl_2.H_2O$ and $[Cr(H_2O)_4Cl_2]Cl.2H_2O$ other than by differences in colour?
15. It is possible to isolate two chlorides of formula $[Cr(H_2O)_4Cl_2].2H_2O$. By careful consideration of the structure of the complex ion explain why this is possible.
16. Starting with manganese(IV) oxide how could one prepare (a) $MnCl_2$; (b) K_2MnO_4; (c) $KMnO_4$?
17. A zinc ore is zinc spinel, $ZnAl_2O_4$. What mass of the ore is required to yield 1 kilogram of zinc?
18. Silver chromate, Ag_2CrO_4, has a solubility product of 1.7×10^{-12}. What is the solubility of silver chromate in 0.1 M K_2CrO_4 solution?
19. What is the pH of a 0.05 M $Cr(NO_3)_3$ solution, given that K_{a1} for the hexaquo chromium(III) ion ($Cr(H_2O)_6^{3+}$) is 2×10^{-4}, and assuming that the first dissociation is responsible for the acidity of the solution?
20. For the system

$$Fe^{3+}(aq) + SCN^-(aq) \rightleftharpoons Fe(SCN)^{2+}(aq) \qquad K = 10^3$$

In a solution it is found that, at equilibrium, $[SCN^-] = 10^{-2}$ M, and $[Fe^{3+}] = 3 \times 10^{-4}$ M. Find the equilibrium concentration of the complex ion.

21. A sample A is a green compound containing the elements manganese, carbon nitrogen and oxygen. Its empirical formula was determined from the following analytical results:

1.304 g of sample A was dissolved in distilled water and made up to 250 cm^3 with distilled water. 25-cm^3 aliquots from this sample solution were used in the analysis of manganese and nitrogen.

The manganese content was determined by converting the manganese to permanganate and titrating with iron(II) sulphate solution. 25 cm^3 aliquots of the sample solution required an average titre of 36.42 cm^3 of 0.1036 M iron(II) sulphate solution.

The nitrogen content was determined by converting nitrogen to ammonia, dissolving this in water and titrating the ammonia solution with hydrochloric acid. 25-cm^3 aliquots of the sample solution, when converted to ammonia, required an average titre of 21.93 cm^3 of 0.0987 M hydrochloric acid.

The carbon content of sample A was obtained by heating the sample in air and measuring the volume of carbon dioxide produced. It was found that 0.993 g of sample A gave 137.7 cm^3 of carbon dioxide, measured at 24°C and 755 mmHg pressure.

From these analytical results, calculate:

(a) the percentages of manganese, nitrogen and carbon in sample A. The percentage of oxygen can then be obtained by subtracting from 100% the combined percentages of manganese, nitrogen and carbon;
(b) the empirical formula of sample A;
(c) the molecular formula of sample A, given that the molecular weight is 173. (C = 12.01, N = 14.01, O = 16.00, Mn = 54.94.)

TABLE OF RELATIVE ATOMIC WEIGHTS 1969

Based on the Atomic Mass of ^{12}C = 12

Name	*Symbol*	*Atomic Number*	*Atomic Weight*
Actinium	Ac	89	——
Aluminium	Al	13	26.9815
Americium	Am	95	——
Antimony	Sb	51	121.75
Argon	Ar	18	39.948
Arsenic	As	33	74.9216
Astatine	At	85	——
Barium	Ba	56	137.34
Berkelium	Bk	97	——
Beryllium	Be	4	9.01218
Bismuth	Bi	83	208.9806
Boron	B	5	10.81
Bromine	Br	35	79.904
Cadmium	Cd	48	112.40
Caesium	Cs	55	132.9055
Calcium	Ca	20	40.08
Californium	Cf	98	——
Carbon	C	6	12.011
Cerium	Ce	58	140.12
Chlorine	Cl	17	35.453
Chromium	Cr	24	51.996
Cobalt	Co	27	58.9332
Copper	Cu	29	63.54
Curium	Cm	96	——
Dysprosium	Dy	66	162.50
Einsteinium	Es	99	——
Erbium	Er	68	167.26
Europium	Eu	63	151.96
Fermium	Fm	100	——
Fluorine	F	9	18.9984
Francium	Fr	87	——
Gadolinium	Gd	64	157.25
Gallium	Ga	31	69.72
Germanium	Ge	32	72.59
Gold	Au	79	196.9665
Hafnium	Hf	72	178.49
Helium	He	2	4.00260
Holmium	Ho	67	164.9303
Hydrogen	H	1	1.0080
Indium	In	49	114.82
Iodine	I	53	126.9045
Iridium	Ir	77	192.22
Iron	Fe	26	55.847
Krypton	Kr	36	83.80
Lanthanum	La	57	138.9055
Lead	Pb	82	207.2
Lithium	Li	3	6.94
Lutetium	Lu	71	174.97
Magnesium	Mg	12	24.305
Manganese	Mn	25	54.9380
Mendelevium	Md	101	——

The values for atomic weights given in the table apply to elements as they exist in nature, without artificial alteration of their isotopic composition, and, further, to natural mixtures that do not include isotopes of radiogenic origin.

Name	*Symbol*	*Atomic Number*	*Atomic Weight*
Mercury	Hg	80	200.59
Molybdenum	Mo	42	95.94
Neodymium	Nd	60	144.24
Neon	Ne	10	20.179
Neptunium	Np	93	——
Nickel	Ni	28	58.71
Niobium	Nb	41	92.9064
Nitrogen	N	7	14.0067
Nobelium	No	102	——
Osmium	Os	76	190.2
Oxygen	O	8	15.9994
Palladium	Pd	46	106.4
Phosphorus	P	15	30.9738
Platinum	Pt	78	195.90
Plutonium	Pu	94	——
Polonium	Po	84	——
Potassium	K	19	39.102
Praseodymium	Pr	59	140.9077
Promethium	Pm	61	——
Protactinium	Pa	91	——
Radium	Ra	88	——
Radon	Rn	86	——
Rhenium	Re	75	186.2
Rhodium	Rh	45	102.9055
Rubidium	Rb	37	85.467
Ruthenium	Ru	44	101.07
Samarium	Sm	62	150.4
Scandium	Sc	21	44.9559
Selenium	Se	34	78.96
Silicon	Si	14	28.086
Silver	Ag	47	107.868
Sodium	Na	11	22.9898
Strontium	Sr	38	87.62
Sulphur	S	16	32.06
Tantalum	Ta	73	180.947
Technetium	Tc	43	——
Tellurium	Te	52	127.60
Terbium	Tb	65	158.9254
Thallium	Tl	81	204.37
Thorium	Th	90	232.0381
Thulium	Tm	69	168.9342
Tin	Sn	50	118.69
Titanium	Ti	22	47.90
Tungsten	W	74	183.85
Uranium	U	92	238.029
Vanadium	V	23	50.941
Xenon	Xe	54	131.30
Ytterbium	Yb	70	173.04
Yttrium	Y	39	88.9059
Zinc	Zn	30	65.37
Zirconium	Zr	40	91.22

Appendix

SYSTEMATIC NOMENCLATURE

Chemists are not yet in complete agreement as to the best method of naming chemical compounds. However, the International Union of Pure and Applied Chemistry (IUPAC) has made detailed recommendations based on common usage and (hopefully) common sense, for both inorganic and organic substances. We have briefly summarized below some of these recommendations.

1 INORGANIC COMPOUNDS

(a) Naming of simple compounds. In simple compounds between non-metals, the first-named element retains its normal name, and the second has its name modified by the ending —ide. If desired, the numbers of atoms involved are indicated by the prefixes di-, tri-, tetra-, penta-, hexa-, hepta-, etc. Thus

Li_3N	lithium nitr*ide*
H_2S	hydrogen sulph*ide*
HCl	hydrogen chlor*ide*

These are three examples where there is no need to specify the *numbers* of atoms involved as there is only one known compound formed between the elements in each case. However, we would have

SF_2	sulphur *di*fluor*ide*
SF_4	sulphur *tetra*fluor*ide*
SF_6	sulphur *hexa*fluor*ide*

and

N_2O	*di*nitrogen ox*ide*
N_2O_4	*di*nitrogen tetrox*ide*
NO_2	nitrogen diox*ide*
Fe_2O_3	*di*-iron *trioxide*
Fe_3O_4	*tri*-iron *tetroxide*

(b) Naming of cations. For simple ions, that occur in one oxidation state only, the name of the element suffices, e.g.

Al^{3+}	the aluminium ion
Cd^{2+}	the cadmium ion
Li^{+}	the lithium ion

When a cation can exist in more than one oxidation state, the oxidation number (and hence the charge for a simple ion), is indicated with Roman numerals, e.g.

Cu^{+}	the copper(I) ion
Cu^{2+}	the copper(II) ion
Fe^{2+}	the iron(II) ion
Fe^{3+}	the iron(III) ion

(In the older trivial nomenclature, these were cuprous, cupric, ferrous and ferric respectively.)

Trivial names are retained for many common cations, e.g.

NH_4^+	ammonium
OH_3^+	oxonium or hydronium

And for polyatomic cations, where molecules or ions (ligands) are bonded to some central ion, the ion takes the name of the central ion, e.g.

$Fe(H_2O)_6^{3+}$	hexaquo-iron(III) ion
$Fe(CN)_6^{3-}$	hexacyano-iron(III) ion
$Co(NH_3)_6^{3+}$	hexammine cobalt(III) ion

(c) Naming of anions. For simple ions, the —ide ending is used, e.g.

H^-	hydr*ide*
O^{2-}	ox*ide*
P^{3-}	phosph*ide*
C^{4-}	carb*ide*

For many di- and polyatomic anions, well-entrenched trivial names are still used, e.g.

OH^-	hydroxide	SO_3^{2-}	sulphite
CN^-	cyanide	ClO_2^-	chlorite
O_2^-	superoxide	ClO^-	hypochlorite
N_3^-	azide	NO_2^-	nitrite
O_2^{2}	peroxide	NO_3^-	nitrate
SO_4^{2-}	sulphate		

2 ORGANIC COMPOUNDS

Virtually all naming of organic molecules is based on the carbon skeleton of the molecule. In naming any straight or branched chain alkane, the basic skeleton is drawn and the longest unbranched carbon chain is chosen to name the molecule.

All side-branches are then treated as substitutes on this longest chain, e.g.

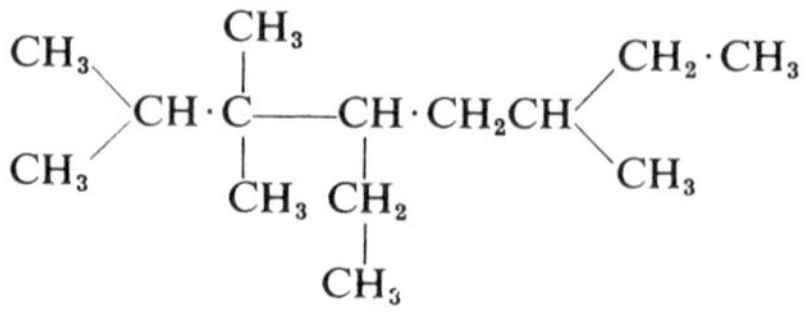

```
                 CH3
CH3              |                          CH2·CH3
   \CH·C————CH·CH2CH/
CH3/     |           |                      \CH3
         CH3   CH2
                     |
                     CH3
```

The skeleton may be represented

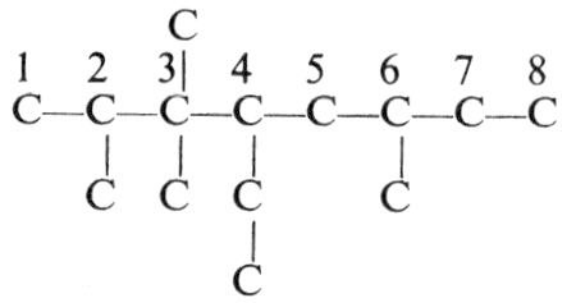

```
              C
1   2   3|  4   5   6   7   8
C—C—C—C—C—C—C—C
     |    |    |         |
     C   C   C        C
                |
                C
```

where the carbon atoms in the longest chain are numbered sequentially starting at the end that allows the substituents to be assigned the lowest possible numbers.

The compound is thus an octane (see Table 19.1) and would be named

2, 3, 3′, 6-tetramethyl-4-ethyl octane.

Unsaturated hydrocarbons with a carbon-carbon double bond are indicated by the suffix —ene, thus

$CH_2{=}CH_2$	ethene
4 3 2 1 $CH_3{\cdot}CH_2{\cdot}CH{=}CH_2$	1-butene
2 1 $CH_3CH{=}CH{\cdot}CH_3$	2-butene

A carbon-carbon triple bond is indicated by the suffix —yne, thus

$CH{\equiv}CH$	ethyne
4 3 2 1 $CH_3CH_2C{\equiv}CH$	1-butyne
4 3 2 1 $CH_3C{\equiv}CCH_3$	2-butyne

Cyclic hydrocarbons are indicated by the prefix cyclo-

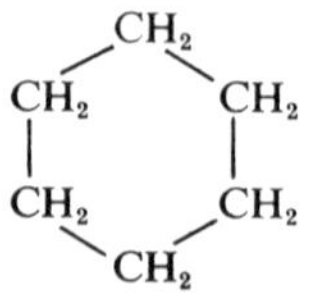

```
        CH2                      CH2
      /     \                  /     \
  CH2         CH2          CH2         CH2
   |           |             \         /
  CH2         CH2            CH2—CH2
      \     /
        CH2
  cyclohexane                cyclopentane
```

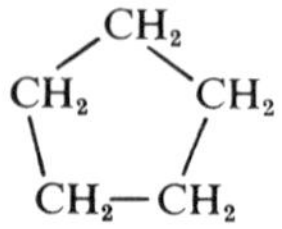

The various functional groups attached to alkyl chains are represented by appropriate prefixes or suffixes, e.g.:

Group	*Suffix*	*Prefix or group name*
-OH	—ol	-hydroxy
C=O	—one	
-CO·OH	—oic acid	-carboxy
-CHO	—al	
-Cl, -Br, -I		chloro-, bromo-, iodo-
$-NH_2$		amino
$-OCH_3$		methoxy
$-O\cdot CH_2CH_3$		ethoxy
$-CH_3$		methyl
$-CH_2CH_3$		ethyl
$-CH_2CH_2CH_3$		n-propyl
$-SO_2OH$		sulphonic acid

Compounds containing these groups are identified using the basic numbering system described earlier. Examples:

$\overset{1}{CH_3}\cdot\underset{\substack{\vert\\ OH}}{\overset{2}{CH}}\cdot\overset{3}{CH_2}\cdot\overset{4}{CH_3}$	2-butanol or butan-2-ol
$\overset{4}{CH_3}\cdot\overset{3}{CH_2}\cdot\overset{2}{CH_2}\overset{1}{CHOH}$	1-butanol
$CH_3\cdot CH_2\cdot CH_2\cdot CHO$	butanal
$CH_3\cdot CH_2\cdot CH_2\cdot CO\cdot OH$	butanoic acid
$CH_3\cdot\underset{\substack{\Vert\\ O}}{C}\cdot CH_2\cdot CH_3$	butanone
$CH_3\cdot\underset{\substack{\Vert\\ O}}{C}\cdot CH_2\cdot CH_2CH_3$	2-pentanone
$CH_3CH_2\underset{\substack{\Vert\\ O}}{C}\cdot CH_2CH_3$	3-pentanone
$CH_3\cdot\underset{\substack{\vert\\ Cl}}{CH}\cdot CH_3$	2-chloropropane
$CH_3CH_2CH_2CH_2NH_2$	1-aminobutane

Answers

CHAPTER 2

2. (a) Li^+, H^-; (b) F^-, Na^+; (c) N^{3-}, F^-, Mg^{2+}, Na^+, Al^{3+}
3. (a) 9; (b) 14; (c) 28; (d) 10; (e) 12; (f) 22; (g) 44; (h) $^{78}_{34}Se$; (i) 66; (j) $^{66}_{30}Zn$
4. (a) $^{16}_{7}N$; (b) $^{31}_{14}Si$; (c) $^{37}_{16}S$; (d) $^{85}_{36}Kr$
5. (a) 6×10^{13}g cm^{-3}; (b) 6×10^{7} ton cm^{-3}
6. 73.5% ^{85}Rb, 26.5% ^{87}Rb
7. (a) 6.95; (b) 24.3; (c) 63.6; (d) 28.1; (e) 69.7
8. $C_6H_5.NH_2^+$, $C_6H_5^+$, NH_2^{2+}, $C_6H_5^{2+}$, NH_2^+
9. (c) 12×10^{23} atoms of silver.
10. (a) 24; (b) 2; (c) 19.44 g
11. (a) 7.1 cm^3; (b) 1.2×10^{-23} cm^3; (c) 2.82×10^{-8} cm
12. 0.15; 2.4 g
13. 11.2 g
14. (a) 20.4 g; (b) 0.15, 0.45; (c) 6.2 g, 0.6 g; (d) 3×10^{23}; (e) 1.5×10^{23}, 4.5×10^{23}
15. (a) 0.435; (b) 0.3125; 0.039; 2.35×10^{22}
16. (a) 0.213; (b) 0.139
17. (a) 0.974; (b) 6.82; (c) 1.95
18. 0.00255
19. 196.9
20. 3×10^{-3} g
21. 6.0×10^{23}
22. (a) 6.66×10^{-23} g; (b) 3.44×10^{-22} g
23. 58.8
24. 58.9
25. 200.4
26. 184.0
27. 59.3
28. 6.84
29. 208
30. 65.82

CHAPTER 3

2. 8.8×10^{-5}
3. (a) increase; (b) decrease; (c) increase; (d) unchanged
4. 100 cm^3; 16 cm^3
5. 10.27 mmHg
6. 750.5 mmHg; 1876.3 mmHg
7. 1487.2 mmHg
8. 1.5×10^5 N m^{-2}
9. 0.628×10^5 N m^{-2}
10. 2.038×10^6 N m^{-2}
11. 4.6 mmHg
12. 2.84×10^{-3}
13. (a) 0.353; (b) 0.268; (c) 0.190
14. (a) 0.1, 0.2; (b) 6×10^{22}, 1.2×10^{23}; (c) 6.72 dm^3; (d) 8.03 dm^3
15. (a) 1.25; (b) 7.5×10^{23}; (c) 5; (d) 31.42 dm^3
16. (a) Nitrogen; (b) Hydrogen
17. 0.952 dm^3

18. Add 3.85×10^{-2} mole of nitrogen.
19. 6.38 cm^3
20. (a) same; (b) same; (c) mass of SO_2 is greater.
    (a) oxygen doubles; (b) same;
    (c) greater number in the oxygen.
21. 550 cm^3 sample.
22. 7.6×10^{10}
23. (a) 4.13 g dm^{-3}; (b) 3.80 g dm^{-3}
24. 65.58 g; 64.06 g
25. 42.3
26. 54.5
27. 34
28. 27.99; 28.01
29. 32.1 g
30. 104.5
31. 32.8
32. 4×10^{-8} cm; 6.4×10^{-23} cm^3
33. 5×10^{24}; 6×10^{23}
34. (a) 19.67 cm^3; (b) 18 cm^3;
    (c) 30610 cm^3
35. (a) 2.67×10^5 N m^{-2};
    (b) 5.13×10^5 N m^{-2}
36. 36.4°C

CHAPTER 4

3. (a) 52.92% Al, 47.08% O
   (b) 65.1% Cu, 32.82% O, 2.05% H;
   (c) 11.96% Mg, 34.87% Cl, 47.21% O, 5.96% H
   (d) 27.93% Fe, 24.02% S, 48.06% O
4. $C_2H_5.NO_2$
5. C_3H_8
6. P_2O_5, H_3PO_4
7. $CuCl_2.6NH_3$
8. C_6H_6
9. C_2H_2
10. C_2H_6O
11. SO_2Cl_2
12. $C_2H_5.NO_2$
13. 31.0
14. 631.9 g $BaSO_4$, 368.1 g $CaSO_4$
15. (a) 0.423 HNO_3; (b) 0.345 Zn;
    (c) 0.300 Zn
16. 2.43 g; yes; 4.01 g
17. (a) 5.77 g; (b) 5.45 cm^3
18. 74.1 g
19. 8.0 g
20. 15.5%
21. 2.84 g
22. (a) 15.0 g; (b) 60.6 g; (c) 74.7 cm^3
23. 14.35 cm^3
24. 5362 cm^3
25. (a) 2.154 g; (b) 0.0994
26. 16.2 dm^3
27. 3.61 dm^3
28. 32.9 g
29. 50%
30. 2.85 dm^3
31. O_2, 68.8 dm^3
32. (a) 4.88 dm^3; (b) 2.4 dm^3;
    (c) 6.15 dm^3
33. C_2H_6
34. 666.7 mmHg
35. C_2H_6
36. C_2H_6
37. 5 cm^3 CO_2, 5 cm^3 CH_4, 15 cm^3 H_2
38. (a) 24 g; (b) 49 g; (c) 325 g;
    (d) 17.2 g; (e) 572 g
39. (a) 110 g; (b) 78 g; (c) 278 g;
    (d) 0.371 g; (e) 8.63 g
40. (a) 5.84 g; (b) 546 g; (c) 0.78 g;
    (d) 4.69 g; (e) 84.8 g
41. 8.0 cm^3
42. 1.4 M KNO_3, 0.48 M K_2SO_4
43. 0.0075
44. (a) 4.57%, (b) 1.34 M
45. 5.6 M
46. (a) 16.6 cm^3; (b) 0.196 M; (c) 490 cm^3
47. 0.265 g
48. (a) 0.032, Mg; (b) 0.0016, Na_2CO_3;
    (c) 0.0034, H_2SO_4; (d) 0.0043, HCl
49. 134 cm^3
50. 70 cm^3
51. 223 cm^3
52. 0.396 M HNO_3, 0.423 M KOH
53. 122
54. 1.43 g
55. 1
56. 158 cm^3
57. 0.697 M
58. 0.0983 M
59. 33.7
60. 0.458
61. 26.3 g $BaCl_2$, 4.20 g $BaSO_4$
62. 43.13%
63. $C_3H_6Cl_2$, 59.0 cm^3, $C_3H_6Cl_2$
64. 0.67 dm^3
65. 1080 g
66. 26.4 cm^3
67. 0.332 M
68. 179 x
69. 59.4 cm^3
70. 0.16 M, 25.3 g dm^{-3}
71. 68.9%
72. Hg_2^{2+}

CHAPTER 5

1. (a) endothermic, lower; (b) endothermic, lower; (c) exothermic, higher; (d) endothermic, lower; (e) exothermic, higher.

2. (a) +3.5 kJ; (b) +1.75 kJ; (c) +1.75 kJ; (d) +1.75 kJ; (e) −1.11 kJ; (f) −1.11 kJ; (g) −1.11 kJ; (h) −1.11 kJ; (i) −6.19 kJ; (j) −5.53 kJ
3. (a) −0.485 kJ; (b) 6.25 g, 0.546 × 10^5 N m^{-2}
4. −9.8 kJ
5. +35.4 kJ
6. −4.04 kJ; 1.65 cm^3; yes
7. +13 kJ
8. +0.017 kJ, +0.34 kJ
9. (a) 112.5 g; (b) −3581 kJ
10. −827.2 kJ
11. (a) 110 kN m^{-2}; (b) −2043 kJ
12. 429.5 kJ
13. (a) 800 kJ; (b) 118 kJ; (c) 223 kJ
14. −1346.4 kJ
15. −985 kJ
16. −1126 kJ
17. (b) 243.4 kJ; (c) 23.9 kJ
18. −1254 kJ
19. −136.8 kJ, −28.1kJ
20. Super-hep, $\Delta H = -32.7$ kJ; Rocket-O, $\Delta H = -32.1$ kJ; 13.9 kJ
21. −201 kJ
22. +4.04 kJ
23. Benzene, $\Delta H = -4.06 \times 10^6$ kJ; diborane, $\Delta H = -3.65 \times 10^6$ kJ

CHAPTER 6

1. (a) $+2.306 \times 10^{-8}$ N; (b) -2.306×10^{-8} N; (c) -4.612×10^{-8} N; (d) $+9.224 \times 10^{-8}$ N; (e) $+9.224 \times 10^{-6}$ N.
2. (a) (i) $+2.306 \times 10^{-18}$ J, +1389 kJ mol^{-1}; (ii) -2.306×10^{-18} J, −1389 kJ mol^{-1}; (iii) -4.612×10^{-18} J, −2778 kJ mol^{-1}; (iv) $+9.224 \times 10^{-18}$ J, +5556 kJ mol^{-1}, (v) $+9.224 \times 10^{-17}$ J, +55,560 kJ mol^{-1}
3. (a) 1; (b) 2; (c) 2; (d) 3
8. +17,483 kJ mol^{-1}
10. Elements in the same group: (a) Z = 19, 11, 3; (b) Z = 8, 16, 34; (c) Z = 5, 31; (d) Z = 10, 2, 18
11. (a) Sodium; (b) Beryllium; (c) Potassium; (d) Aluminium; (e) Carbon.

CHAPTER 8

4. Na—Na; Rb—Rb; :Br—Br: ; :I—I: ; O=O ; :N≡N:
6. :Cl—Cl: ; K—K; H—Cl: ; H—I: ; :Br—Cl: ; ⊖:O—Cl:
7. ⊖:O—H; :S=O: ; :I—Cl: ; :S=S: ; H—S:⊖
9. :C≡O: (⊖ on C, ⊕ on O), :C=O
10. N=O
11. (b) Group VI; (c) Yes; Se_2, Se_8
13. NaCl, HCl, ClF, Cl_2
14. Smaller.

CHAPTER 9

1. (a) H, H, H, H around B⊖; H, H, H around O⊕
   (b) F, F, F, F, F, F around Al (3−); F, F, F, F, F, F around Si (2−); F, F, F, F, F, F around P⊖; F, F, F, F, F, F around S
4. (b) Cl—Be—Cl

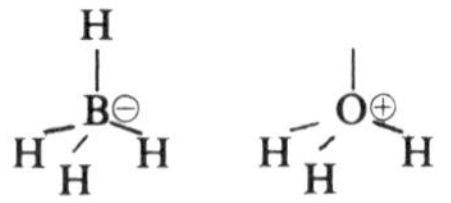

   N with Cl, Cl, Cl; B with Cl, Cl, Cl

   F—Mg—F

   S with H, H; Br—Cl

   Al with H, H, H; H—Br; C with H, Cl, Cl, Cl

   C with H, H, Cl, Cl; Si with F, F, F, F

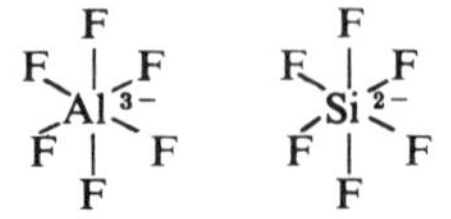

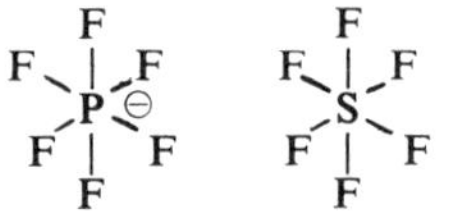

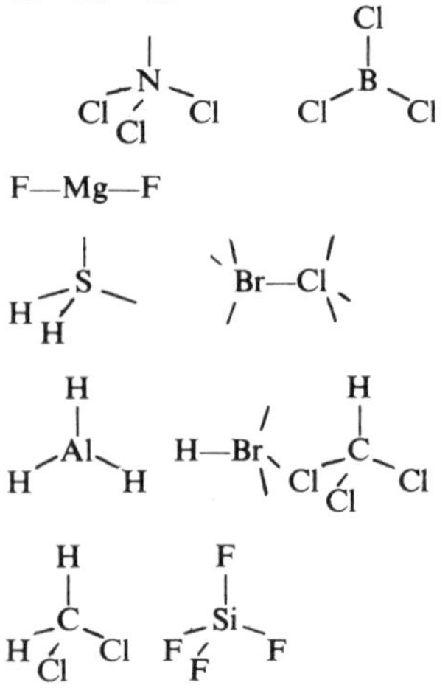

5. (b)
H H
Se As O
H H H H Cl

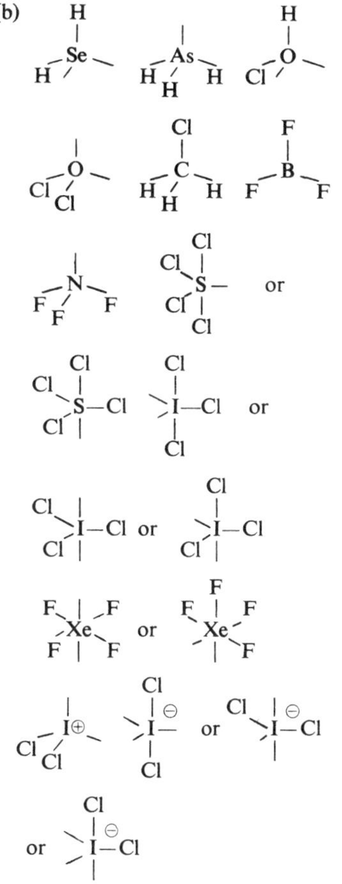

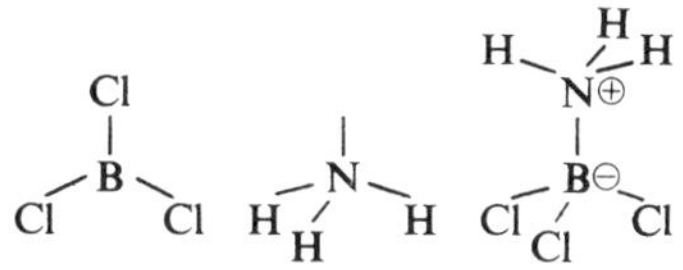

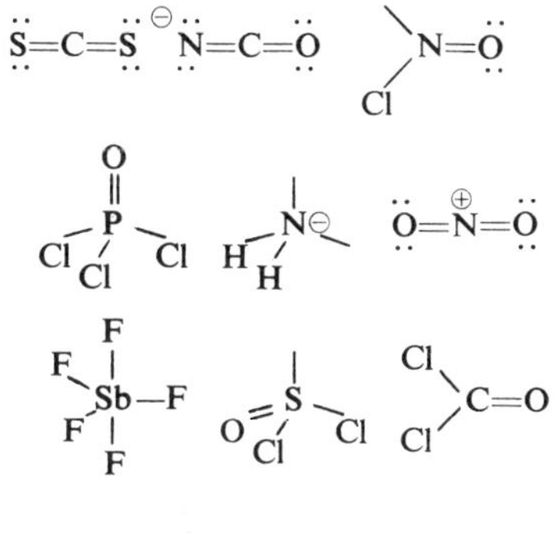

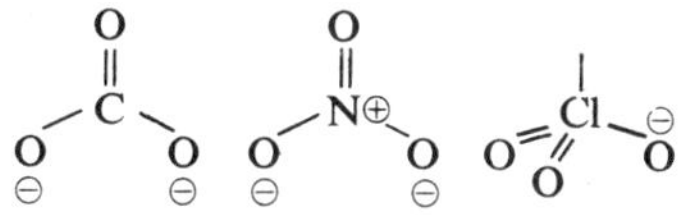

Cl F
O C B
Cl Cl H H H F F

Cl
N Cl S or
F F F Cl Cl

Cl Cl
Cl S—Cl I—Cl or
Cl Cl

Cl
Cl I—Cl or I—Cl
Cl Cl

F
F F F F
Xe or Xe
F F F

Cl
I⊕ I— ⊖ or Cl I—Cl ⊖
Cl Cl Cl

Cl
or I—Cl ⊖

6.
H H H
Cl N⊕
B N B⊖
Cl Cl H H H Cl Cl Cl

7. $\ddot{S}{=}C{=}\ddot{S}$ ⊖ $\ddot{N}{=}C{=}\ddot{O}$ N=O Cl

O
P N⊖ $\ddot{O}{=}N^{\oplus}{=}\ddot{O}$
Cl Cl Cl H H

F
F Sb—F S Cl
F F O Cl Cl C=O Cl

8.
O O
C N⊕ Cl ⊖
O O O O O O O
⊖ ⊖ ⊖ ⊖

9. :Ö=C=Ö: H—C≡N:

:N≡C—C≡N: ·N N·
:O: :O:
Ö

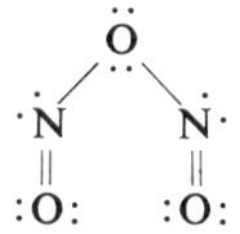

10.
O O⊖
P P
⊖O O ⊖O O
O ⊖ O
⊖ ⊖

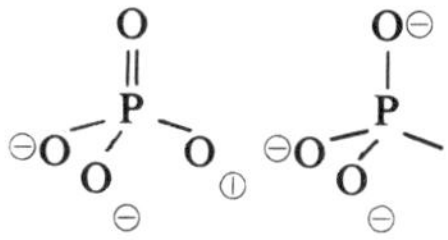

11.
O
Se
HO OH O

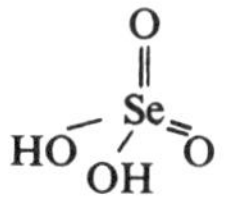

12.
N
⊖O O

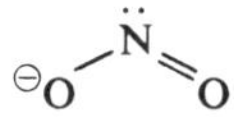

CHAPTER 10

4. NH_3; $CH_3.OH$; CH_3NH_2
7. (a) $Fe(H_2O)_6^{3+}$; (b) $Be(H_2O)_4^{2+}$
9. (a) N⊖
H H

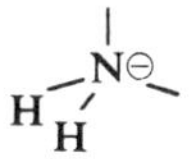

CHAPTER 11

1. $CaCl_2$, network lattice; PCl_3, discrete molecules; RbF, network lattice CCl_4, discrete molecules; MgO, network lattice; BrCl, discrete molecules; H_2O, discrete molecules; HBr, discrete molecules; SiC, network lattice.

2.
Cl O
C :Ö=C=Ö: C
H H H O O
⊖ ⊖

O⊖ H
Cl H—F: N⊕
O O O H H H

O⊖ O
N⊕ S
O O⊖ HO O O⊖

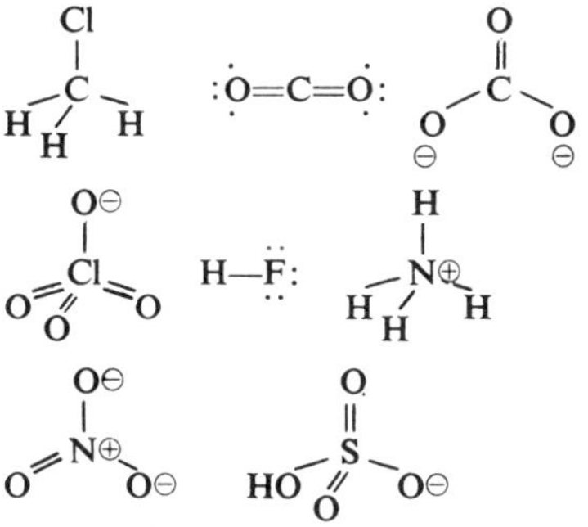

12. (a) HF; (b) Na; (c) KF; (d) He; (e) Si; (f) H_2S

CHAPTER 12

1. (a) 2R; (b) 4R; (c) $\frac{R}{4}$; (d) 16R
2. $R_{NO} = 2R_{O_2} = R_{NO_2}$
3. $R = k[A][B]$
4. (a) 1.75×10^{-6} M s^{-1}; (b) 8.7×10^{-7} M s^{-1}; (d) 0.175; (f) 1.75×10^{-5} M s^{-1}
5. 4.8×10^{7} M s^{-1}, 4.8×10^{7} M s^{-1}, 3.47×10^{-10} s
7. (a) 2.7×10^{-5} M s^{-1}; (b) 5.4×10^{-6} mol s^{-1}, 1.94×10^{-2} mol s^{-1}; (c) 2.5 dm^3; (d) 6.5×10^{-6} M
8. (a) R/9; (b) $\sqrt{2} \times$ [HI]; (c) M^{-1} s^{-1}
9. (a) $R = k[H_2O_2][I^-]$; (b) 1.75×10^{-2}; (c) M^{-1} s^{-1}
10. (a) 0.99%; (b) 4.94×10^{7} M s^{-1}; (c) 9.9×10^{-7} s^{-1}
11. 5.8×10^{-5} s^{-1}
12. 76%.

CHAPTER 13

1. (a) $\frac{p^2_{NO}}{p_{N_2}p_{O_2}}$ (b) $\frac{p^2_{ClF_3}}{p_{Cl_2}p^3_{F_2}}$ (c) $\frac{p_{CO_2}}{p_{CO}p^{\frac{1}{2}}_{O_2}}$ (d) $\frac{p^2_{O_3}}{p^3_{O_2}}$
2. 15
3. 0.24
4. 10 M
5. 2 M
6. 167.9, 100.7 atm
7. (a) 9.09×10^{-3}, 7.10×10^{-3}, 5.72×10^{-3}, 5.26×10^{-3}
   (b) 1.06×10^{-5}, 1.14×10^{-5}, 1.06×10^{-5}, 1.12×10^{-5}
8. (a) (i) 10; (ii) 0.316
   (b) (i) $[H_2] = 3.16$, $[Br_2] = 3.16$, $[HBr] = 1.0$ M
9. (a) No. One would need either the concentration of at least one species at equilibrium *or* the equilibrium constant at this temperature.
   (b) $K_{550^\circ} < K_{450^\circ}$
   (c) Proportions of HCl and O_2 increase, proportions of H_2O and Cl_2 decrease.
   (d) K remains unchanged.
   (e) $K = \frac{K^2_{(11)}}{K_{(1)}}$
10. (a) Remains unchanged; (b) decreases; (c) decreases; (d) increases; (e) remains unchanged.
11. Experiment 1: $K = 1.60$
    Experiment 2: $[H_2]_e = 0.44$, $[CO_2]_e = 0.44$, $[H_2O]_e = 0.56$, $[CO]_e = 0.56$ M
    Experiment 3: $[CO_2] = 2.0$
    Experiment 4: $[CO_2]_e = 0.88$, $[H_2O]_e = 1.12$, $[CO]_e = 1.12$ M
    Experiment 5: $[CO] = 2.0\,[H_2O]$
    Experiment 6: $[H_2O] = 1.0 = [CO_2]$
    Experiment 7: $[H_2]_e = 0.344$, $[CO_2]_e = 0.544$, $[CO]_e = 0.656$ M
12. (a) Endothermic
    (b) $p_{PCl_3} = 7.75$, $p_{Cl_2} = 7.75$, $p_{PCl_5} = 35.15$ atm.
    (c) $p_{PCl_3} = 18.2$, $p_{Cl_2} = 18.2$, $p_{PCl_5} = 196.2$ atm.
    (d) (i) $p_{PCl_3} = 7.75$, $p_{Cl_2} = 7.75$, $p_{PCl_5} = 35.15$ atm.
    (ii) $p_{PCl_3} = 1.55$, $p_{Cl_2} = 44.4$, $p_{PCl_5} = 41.35$ atm.

CHAPTER 14

1. (a) Na^+, OH^-, H_2O, H_3O^+; (b) 0.1 M; (c) 10^{-13} M; (d) 13
2. (a) 0.07 M; (b) 0.07 M; (c) 1.15; (d) 1.57 g
3. (a) 4; (b) 1.52; (c) 4.28
4. (a) $[H_3O^+] = 10^{-8}$, $[OH^-] = 10^{-6}$ M
   (b) 0.63, 1.6×10^{-14} M
   (c) 5×10^{-13}, 2×10^{-2} M
5. (a) 7; (b) 3.7; (c) 2.12
6. (a) 1.9×10^{-5}; (b) 4.8×10^{-10}; (c) 6.0×10^{-3}
7. (a) 1; (b) 2; (c) 3; (d) *ca.* 7 since $[H_3O^+]$ from water dissociation is at least 10^{-7} M
8. (a) Yes; (b) yes; (c) no.
9. (a) 2×10^{-4}; (b) 1.8×10^{-17}
10. (a) 3.6×10^{5}; (b) yes;
    (c) H.CO.OH, NH_3 low; NH_4^+ H.CO.O$^-$ moderate; H_3O^+, OH^- very low.
11. (a) $[H_3O^+] = 1.4 \times 10^{-3}$, $[HN_3] = 0.1$, $[N_3^-] = 1.4 \times 10^{-3}$, $[OH^-] = 7.2 \times 10^{-12}$ M;

(b) 1.4%; (c) 2.86

12. (a) $[H_3O^+] = 5.0 \times 10^{-14}$ M, pH = 13.3;
(b) 5×10^{-1}, 0.3;
(c) 0.069, 1.16;
(d) 8.1×10^{-14}, 13.09

13. (a) $[H_3O^+] = 1.78 \times 10^{-5}$ M, pH = 4.75;
(b) 1.78×10^{-5}, 4.75;
(c) 1.78×10^{-5}, 4.75

14. 15.68 g

15. (b) One would need K_a NH_4^+ and K_b CO_3^{2-} or K_a HCO_3^-;
(c) one would need K_a HS^- and K_b HS^- or K_a H_2S

16. (a) (i) 4.65; (ii) 0.04%
(b) (i) 10^{-6}; (i) 1%
(c) (i) 3.85; (ii) 1×10^{-6}

17. (a) 0.02; (b) 0.032

18. (a) 4.6×10^{-10}, 2.2×10^{-5}
(b) 4×10^{-10}

19. (a) 1.7×10^{-14}; (b) H_2Se moderately high; HSe^-, H_3O^+ low; Se^{2-}, OH^- very low; (c) $[H_3O^+] = 3 \times 10^{-3}$, $[Se^{2-}] = 10^{-10}$, $[H_2Se] = 4.7 \times 10^{-2}$, $[OH^-] = 3 \times 10^{-12}$ M $[HSe^-] = 3 \times 10^{-3}$

20. (b) 0.2 M H_3AsO_4: H_3AsO_4 moderate, $H_2AsO_4^-$ low, $HAsO_4^{2-}$ very low, AsO_4^{3-} extremely low, H_3O^+ low, OH^- very low.
0.2 M Na_2HAsO_4; Na^+ moderate, $HAsO_4^{2-}$ moderate, $H_2AsO_4^-$ low, H_3AsO_4 very low, AsO_4^{3-} low, H_3O^+ very low, OH^- very low.

21. (a) 5; (b) 10.7; (c) 8.48

22. (a) 3; (b) 8.1; (c) 122

CHAPTER 15

1. 1.25×10^{-8} M
2. 0.1 M
3. 0.1 M
4. (a) 0.1 M; (b) 0.1 M; (c) 0.0995 M $Cu(NH_3)_4^{2+}$, 4.55×10^{-4} M Cu^{2+}, 1.82×10^{-3} M NH_3; (d) 0.46%.

CHAPTER 16

1. (a) 2.24×10^{-6} M;
(b) 1.2×10^{-3} M;
(c) 1.08×10^{-4} M
2. (a) 3.03×10^{-13};
(b) 2.96×10^{-13};
(c) 1.35×10^{-18}
3. (a) 2.5×10^{-5}; (b) 6.4×10^{-6}
4. (a) $PbCrO_4$, $PbCO_3$, PbC_2O_4, $PbSO_4$
(b) $PbCrO_4$, $PbCO_3$, PbC_2O_4, $PbSO_4$
5. Ag_3AsO_4, Ag_3PO_4, AgN_3, Ag_2CrO_4, $AgBrO_3$
6. HgS, CuS, Ag_2S, Bi_2S_3
7. (a) 1×10^{-4} M;
(b) 1×10^{-7} M;
(c) 1×10^{-7} M
8. (a) 2.58×10^{-1} g dm^{-3};
(b) 2.67×10^{-2}g dm^{-3};
(c) 1.2×10^{-4} g dm^{-3}
9. (a) Precipitate will form;
(b) no precipitate;
(c) precipitate will form.
10. (a) 7×10^{-26} M;
(b) 1.7×10^{-1} M;
(c) 7×10^{-5} M
11. (a) 4×10^4; (b) 1×10^6;
(c) 3.48×10^7
12. $Fe(OH)_3$ precipitates first.
13. A range of $[OH^-]$ between 10^{-12} and 6×10^{-7} M
14. (a) 8×10^{-7} M;
(b) 7.9; (c) 2.5×10^{-2} M;
(d) 8×10^{-20} M;
15. (a) 1×10^{-22}; (b) 10^{-19} M;
(c) ZnS would precipitate;
(d) pH must be less than 0
16. (a) $PbCO_3$ precipitates if $[CO_3^{2-}]$ is greater than 10^{-11} M, whilst $MgCO_3$ precipitates if $[CO_3^{2-}]$ is greater than 10^{-3} M;
(b) 10^{-10} M
17. 1×10^{-7} M
18. (c), (d).
19. (a), (d).

CHAPTER 17

1. (a) +3, (b) +4, (c) +6, (d) +3, (e) +5, (f) +7, (g) +5, (h) +6, (i) +6, (j) +6, (k) +6, (l) +3, (m) +2
2. (a) +4, (b) +6 (c) +6 (d) +7 (e) +6 (f) +3 (g) +2 (h) +4
6. (c) 0.478 g
7. (b) (c) (d) (f) (g)
8. (c) (e) (g)
10. (a) Ag^+ strongest Mg^{2+} weakest;
(b) Mg strongest, Ag weakest;
(c) $AgNO_3$, $CuSO_4$;
(d) Zn, Mg;
(e) copper metal would form in the second vessel.
11. Zn metal.
16.

	Fe^{3+}	Fe^{2+}	Fe	Zn^{2+}	Zn	Cl^-
(a)	0.1	0.1	0.0	0.05	0.0	0.6
(b)	0.0	0.2	0.0	0.1	0.0	0.6
(c)	0.0	0.15	0.05	0.15	0.0	0.6

(d) 0.0 0.0 0.2 0.3 0.0 0.6
(e) 0.0 0.0 0.2 0.3 0.05 0.6

17. (a) 0.05 $SnCl_2$, 0.05 $SnCl_4$, 0.1 $FeCl_2$
(b) 0.02 $FeCl_3$, 0.08 $FeCl_2$, 0.04 $ZnCl_2$
(c) 0.15 $ZnCl_2$, 0.1 Fe, 0.25 Zn
18. (i) 6.4×10^{-11} M s^{-1};
(ii) 2.0×10^{-10} M s^{-1}

CHAPTER 18

8. (a) 4.8×10^{-4} M;
(b) 2.4×10^{-4}, 4.2×10^{-11} M

CHAPTER 22

6. (a) Ca^{2+}; (b) K^+; (c) F^-; (d) Cu^+
7. (a) Be; (b) Be; (c) Be; (d) Sr
11. (a) P; (b) Be; (c) Ca; (d) Cr; (e) I; (f) Ra

CHAPTER 23

2. (b) −64.02 kJ
5. (a) +4, +6
(b) I_2, Fe^{3+}
9. SiH_4 tetrahedral, PH_3 pyramidal, SH_2 angular, ClH linear.
10. (a) O=N⊕(O⊖)–O–N⊕(O⊖)=O
(b) O_3Cl–O–ClO_3
(c) O=S(=O)=O (d) F–O–F
11. 1.1×10^{-2}. Stronger (P more electronegative than B).
12. 5,431 kJ
13. 10.36
14. SF_4—distorted tetrahedron.
15. 15.8 RT

CHAPTER 24

1. (a) RaH_2; (b) Cs_2O, Cs_2O_2, CsO_2; (c) SrO, SrO_2
2. (a) Cs(Fr); (b) Cs(Fr); (c) Be; (d) Ra; (e) Cs(Fr)
3. (a) Tetrahedral; (b) octahedral;
(c) linear; (d) Cl—Be(μ-Cl⊕)₂Be—Cl
(e) linear; (f) tetrahedral
10. (a) Acidic; (b) alkaline; (c) neutral—slightly acidic; (d) neutral; (e) neutral.
12. 87.6
13. 62.65%
15. 12.9
16. (a) 0.043, (b) 3.2×10^{-4}; (c) 8.9×10^{-3}
17. (a) 0.067 M Ca^{2+}, 0.133 M Sr^{2+}, 0.4 M Cl^-
(b) 0.001 M Cl^-, 0.015 M OH^-, 0.008 M Ba^{2+}, 6.7×10^{-13} M H^+
(c) 0.067 M H^+, 0.067 M Cl^-, 8.7×10^{-4} M Sr^{2+}, 8.7×10^{-4} M SO_4^{2-}
18. (a) $Mg(OH)_2$ will precipitate, but not $Sr(OH)_2$

CHAPTER 25

3 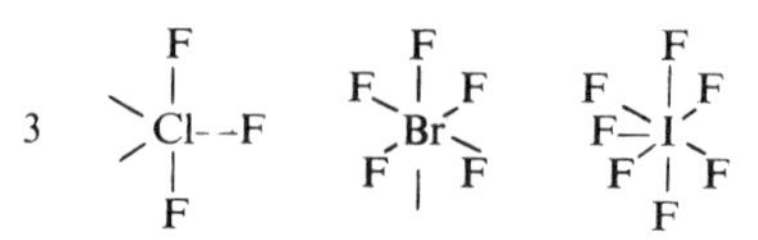

10. (a) Neutral; (b) alkaline; (c) acidic; (d) alkaline.
11. (a) T-shaped; (b) angular; (c) tetrahedral; (d) square-based pyramid; (e) linear; (f) square-based pyramid.
12. (a) 0.0289 M; (b) 0.559 g; (c) 0.024 M
13. No. Ionic product is 0.0025
14. 0.44 M, 8.76 dm^3
15. ClF_3
16. AgBr will precipitate, but not Ag_2SO_4
17. 6.6×10^{-3}
18. (a) 56.7%Ag, 17.95%NH_3, 25.4%SO_4
(b) $Ag_2(NH_3)_4SO_4$

CHAPTER 26

4. (a) SCl_2 (Cl–S–Cl) (b) Cl–S–S–Cl
(c) SO_2Cl_2 (d) $SeCl_4$

(e)  [structure: S with F, F, F, F, F, F] (f) [structure: S with O, Cl, Cl]

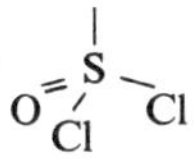

(g) 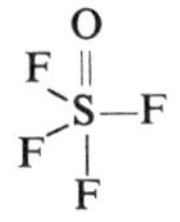 [structure: S with O, F, F, F, F] (ii) as (b) and [structure: S with S, F, F]

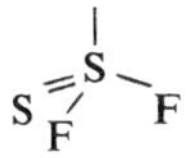

5. 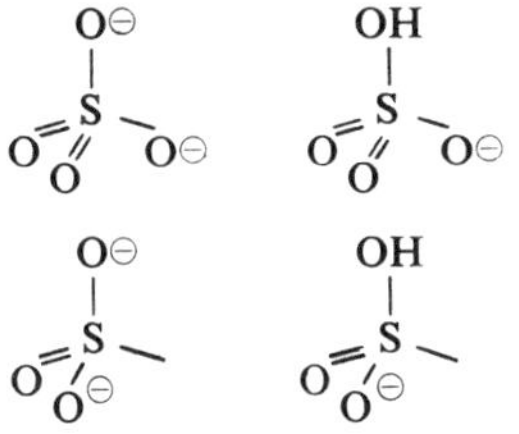 [structures: S with O⊖, O, O, O⊖; S with OH, O, O, O⊖; S with O⊖, O, O⊖; S with OH, O, O⊖]

14. Dilute 0.408 cm^3 of 98% acid to 1 dm^3 of solution.
15. 1752 g
16. (a) 4.17; (b) 0.5; (c) 4×10^{-7} M
17. 0.86×10^5 N m^{-2}
18. 70.0% $NaHSO_4$, 5.85% NaCl, 19.88% $NaSO_4$.
19. 23.7% S, 52.59% Cl, 23.7% O; SO_2Cl_2; 134.969

CHAPTER 27

1. Formal oxidation state −3

2. 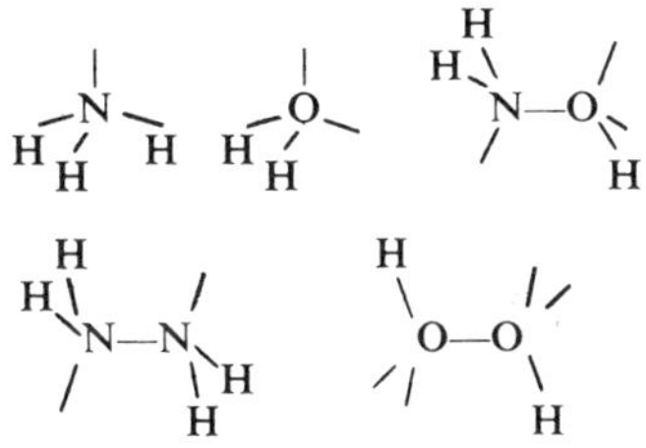 [structures: N with H, H, H; O with H, H; N with H, H and O with H; N—N with H, H, H, H; O—O with H, H]

9. 108 s^5
10. (a) NH_3 moderate; NH_4^+, OH^- low; H_3O^+ very low
    (b) NH_4^+, Cl^- moderate; H_3O^+ low; OH^- very low
    (c) Na^+, $H_2PO_4^-$ moderate; H_3O^+, HPO_4^{2-} low; H_3PO_4, PO_4^{3-}, OH^- very low
    (d) H_3PO_3 moderate; H_3O^+, $H_2PO_3^-$ low; HPO_3^{2-}, OH^- very low.
11. NaH_2AsO_4 acidic, Na_2HAsO_4 alkaline.
16. 15 cm^3 O_2; (b) 30 cm^3 N_2; (c) 75 cm^3
17. (a) Na^+, $H_2PO_4^-$ (NaH_2PO_4)
    (b) Na^+, $H_2PO_4^-$, HPO_4^{2-} (NaH_2PO_4, Na_2HPO_4)
    (c) Na^+, HPO_4^{2-} (Na_2HPO_4)
    (d) Na^+, HPO_4^{2-}, PO_4^{3-} (Na_2HPO_4, Na_3PO_4)
    (e) Na^+, PO_4^{3-} (Na_3PO_4)
18. (a) 5kg; (b) 4452 dm^3
19. (a) $PSCl_3$; (b) (i) 165, (ii) $PSCl_3$

CHAPTER 28

2. $\overset{\ominus}{C}\equiv\overset{\oplus}{O}$, O=C=O

8. (a) 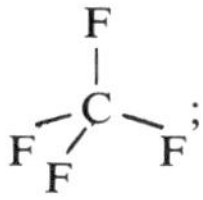 [C with F, F, F, F]; (b) [C with H, H, H, F]; (c) [Si with H, H, H, H]; (d) 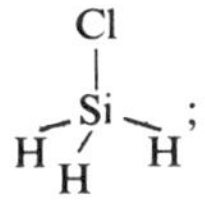 [Si with Cl, H, H, H]; (e) 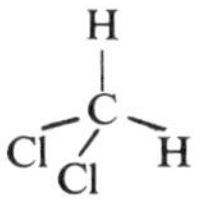 [C with H, Cl, Cl, H]

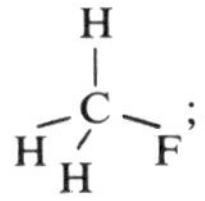

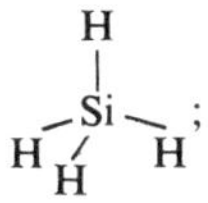

Most polar CH_3F, non-polar CF_4, SiH_4

13. 0.60 g dm^{-3}
14. 4.4×10^{-11}
15. pH Na_2CO_3 = 11.83, pH NaOH = 12.3
16. (a) −228 kJ; (b) −228 kJ; (c) −76.7 kJ; (d) −2278 kJ
17. −145 kJ
18. $\Delta H = -256.5$ kJ

CHAPTER 29

1. (a) 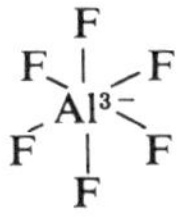 [Al^{3-} with F, F, F, F, F, F]
   (b) 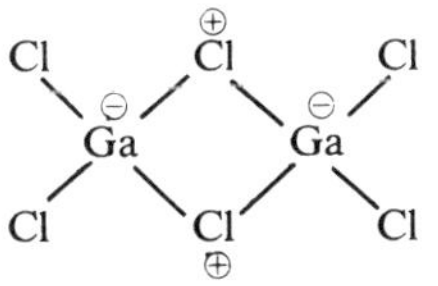 [$Cl_2Ga^{\ominus}(\mu\text{-}Cl^{\oplus})_2Ga^{\ominus}Cl_2$]
   (c) 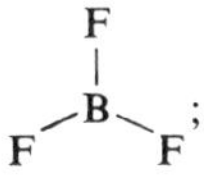 [B with F, F, F]; (d) 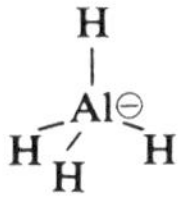 [$Al^{\ominus}$ with H, H, H, H]

7. $\Delta H = -16{,}360$ kJ
8. 264 g, Al_2Cl_6
9. 115
10. 2.9

CHAPTER 30

1. (a) F, Kr, F (b) F, F, Kr, F, F

(c) O, F, F, Xe, F, F (d) O, ⊖O, O⊖, Xe, ⊖O, O⊖, O

2. (a) XeF_4; (b) XeF_6

3. 3.45 dm^3
4. 1184 mmHg

CHAPTER 31

4. (a) +6; (b) +3; (c) +6; (d) +3; (e) +2; (f) +3; (g) +3
9. (a) 0.1 M $AgNO_3$; (b) 0.1 M HF; (c) 1 M NH_3; (d) 0.08 M NaF
11. (a), (b) no reaction; (c) Fe^{3+} reduced to Fe^{2+}, V^{2+} oxidized to V^{3+}
17. 2.81 kg
18. 6.8×10^{-4} g dm^{-3}
19. 2.5
20. 3×10^{-3} M
21. (a) 31.8% Mn, 23.2% N, 6.8% C, 38.2% O
    (b) MnN_3CO_4
    (c) MnN_3CO_4

Index

Acetaldehyde, 354, 360, 362
Acetamide, 367
Acetate ion, 363
Acetic acid, 253–4, 256–7, 354, 361–4, 367
— anhydride, 367
Acetone, 355, 361
Acetyl salicylic acid, 357
Acetylene, 163, 343–4, 347
Acetylide ion, 192
Acid amides, 367
Acidity constants, 253
— —, tables of, 256, 264, 363, 401, 408, 427, 431, 440, 444, 446, 460
Acids, 245–67
Actinide series, 114, 386
Activation energy, 378
Activity, 223–6
Addition reactions, 344–5
Adipic acid, 371
Alanine, 371
Alcohols, 351–5
—, primary, 351–2, 355, 361–2, 364
—, secondary, 352–3, 355, 361–2
—, tertiary, 352
Aldehydes, 355, 360–2
Alkali metals, 386, 412–19
— —, diatomic molecules of, 135–6
Alkaline earth metals, 386, 412–19
Alkanes, 336–43
—, combustion of, 342
Alkenes, 336, 343–6
Alkoxide ions, 355
Alkyl chlorides, 343
— groups, 338–9, 353
Alkynes, 343–6
Alloys, 183
Aluminate ion 405
Aluminium, and compounds, 399–403, 405–8 413–17
— chloride, 124, 191, 193, 475–7
— fluoride, 402–3
— hexaquo cation, *see* hexaquo aluminium(III) ion
— hydride, 400–2
— hydroxide, 284, 407–8
— oxide, 405–7
Amide ion, 153, 256
Amides, *see* acid amides
Amines, 334, 357–8
Amino acids, 371
Amino group, *see* amines
Aminoethane, *see* ethylamine
Ammonia, 87, 149–50, 153, 240–1, 245–6, 256, 400–1, 454–5
Ammonium acetate, 367
— chloride, 190–1
— hydrogen sulphide, 278–9
— ion, 151–3, 191, 255–6, 263–4
— nitrate, 190
— perchlorate, 247
— sulphide, 263–4
Ampholytes, 250, 256
Analytical concentration, 57
Angstrom unit, 21
Aniline, 358–9
Antimonite ion, 458
Antimony, and compounds, 451–60
— pentachloride, 457
— triiodide, 456
— trioxide, 458
Antimonyl ion, 454, 460
Argon, 119–20, 170, 399–400, 480
Aromatic hydrocarbon, 336, 346–8
Arsenic, and compounds, 451–60
Arsenic acid, 460
Arsenic pentachloride, 457
— pentoxide, 459–60
— trifluoride, 456
— trioxide, 458, 460
Arsenious acid, 458–60

Arsine, 454
Aspirin, *see* acetyl salicylic acid
Atomic number, 7
— orbital, *see* orbitals
— radius, 390–2
— structure, 6–8, 95–117, 383–4
— theory, 4–6
— weight, 9–12, 18
Atmosphere (pressure unit), 29
Avogadro's constant, 16–18
— law, 32

Bakelite, 356–7
Barium, and compounds, 413–19
— carbonate, 284
— chloride, 281
— chromate, 284
— sulphate, 283–4, 295
Bases, 245–67
Basicity constant, 254
Benzene, 340, 346–8, 358
Benzene sulphonic acid, 348, 356
Benzoic acid, 364–5
Beryllate ion, 404
Beryllium, and compounds, 399–408, 412–19
— chloride, 417
— fluoride, 402–3
— hydride, 152–3, 400, 402
— oxide, 403–4, 406–7
Bicarbonate ion, *see* hydrogen carbonate ion
Bimolecular step, 317, 381–2
Bismuth, and compounds, 451–4, 456–60
— nitrate, 454
— pentachloride, 457
— pentoxide, 459
— trifluoride, 454, 456
— triiodide, 456
— trioxide, 458
Bismuthine, 459
Bismuthyl ion, 459–60
Bisulphate ion, *see* hydrogen sulphate ion
Bisulphite ion, *see* hydrogen sulphite ion
Bomb calorimeter, *see* calorimeter
Bond, covalent, 134–6, 139, 146, 155–8, 181, 184–5
—, double, 141
— dissociation energy, 88–9, 130, 146 tables of, 89, 146
— energies, 88–9
—, single, 139–40
—, triple, 141–2
Bonding, between molecules, 168–79
—, ionic, 181, 186–90
—, metallic, 181–3
Boric acid, 407–8
Borohydride ion, 151–2
Boron, and compounds, 399–408, 473–6
— hydride, 400–1
— oxide, 403–4, 406–7
— trifluoride, 152, 402–3
Bromic acid, 431
Bromide ion, 186
Bromine, and compounds, 422–33
Bromobenzene, 346
Bromoethene, 344
Bronsted-Lowry concept of acids and bases, 249–50
Butanal, 360
n-Butane, 337–8
1-Butanol, 360
2-Butanol, 361
2-Butanone, 361
1-Butene, 344
2-Butene, 344
n-Butyl acetate, 365
n-Butyl group, 339
1-Butyne, 344
2-Butyne, 344
Butyraldehyde, *see* butanal
n-Butyric acid, 363

Caesium, and compounds, 413–19
— bromide, 432
— chloride, 187–9, 432
— fluoride, 144, 188, 432
Cadmium bromide, 191
— iodide, 125, 191, 194
— sulphide, 190
Calcium, and compounds, 413–19
— carbide, 343–4, *see also* acetylide ion
— carbonate, 279, 284, 287
— fluoride, 284, 288–9
— oxide, 188, 406
— sulphate, 389–90
— sulphite, 445
Calorie, 72–3
Calorimeter, bomb type, 75–6
—, electrical standardization of, 74, 76
— for solutions, 73–4
Canonical structure, 165–6
Carbon, and compounds, 86–7, 327–72, 399–404, 406–8, *see also* diamond, graphite
— dioxide, 80, 82–7, 122, 302, 332–3, 404, 406–7
— monoxide, 77, 79–80, 82–3, 87, 299, 302, 470
— tetrachloride, 328
— tetrafluoride, 402–3
Carbonate ion, 165–7, 256, 263
Carbonic acid, 408
Carbonyl group, 334, 360–2
Carboxyl group, *see* carboxylic acids
Carboxylic acids, 334, 362–5
Catalysis, 205–7, 239–40
Catalytic cracking, 340
— reforming, 340, 347
Catenation, 466
Cell reaction, 306–9
Celsius degree, 21
Chain lattices, *see* lattices
Charge cloud repulsion hypothesis, 138, 149
Charge clouds, 115–16, 137–9

Chelate compounds, 272, 491
Chemical bonding, *see* bond, bonding
— equations, 50–2
— formulae, *see* formulae
Chlorate ion, 192
Chloric acid, 431
Chloride ion, 166, 256
Chlorine, and compounds, 89, 399–409, 422–33
— heptoxide, 405–7
— trifluoride, 158–9, 402–3
Chloro group, 334
Chloroacetic acid, 363
Chlorobenzene, 348, 356
Chloroethane, 342, 344, 349, 352, 359
Chloroethene, *see* vinyl chloride
Chloroform, 328, 332
Chloromethane, 328–30
1-Chloropropane, 342, 352
2-Chloropropane, 342, 353
Chlorous acid, 257, 431
Chromate ion, 486–7
Chromium, and compounds, 484–91
—(III) oxide, 373
—(III) trisoxalato ion, *see* trisoxalto chromium(II) ion
Cobalt, and compounds, 484–5, 491
—(III) trisethylenediamine ion, *see* trisethylenediamine cobalt(III) ion
Competing equilibria, *see* equilibria
Complex ions, 271–6, 489–91
— ion equilibria, 271–6
Co-ordination compounds, 489–91
— number, 187
Copper, and compounds, 484–5, 489–90
—(I) fluoride, 190
—(II) chloride, 122
—(II) ion, 300, 304, 309
—(II) oxide, 190, 407
—(II) sulphate, 71
—(II) sulphide, 188, 284
—(II) tetrammine ion, *see* tetrammine copper(II) ion
—, use in alcohol oxidation, 354
Core charge, 145–6
Coulomb's law, 132
Cupric, cuprous, *see* copper(I), copper(II)
Cyanic acid, 256
Cyclobutane, 340
Cyclohexane, 340, 347
Cyclopentane, 340
Cyclopropane, 340

Dacron, 367
Daniell cell, 308
Decane, 338
Dehydrogenation 349
Diamond, 86–7, 123, 184–5
Diatomic molecule, 130–47
Diborane, *see* boron hydride
1,1-Dibromoethane, 344
1,2-Dibromoethane, 344
Dichloroacetic acid, 363
Dichloromethane, 328, 331
Dichromate ion, 305–6 486–7
— as organic oxidizing agent , 306 354–5
Diethyl ether, 359–60
Diffraction, electron, 122–7
—, X-ray, 128
Dinitrogen pentoxide, 198, 403–7, 409, 458–9
— tetroxide, 217, 459
— trioxidc, 458–9
Dipolar molecules, 143, 170, 172–5
Dipoles, electric, 168–70
Dispersion forces, 170–1

Electrolyte, 245–6
Electron affinity, 388
— diffraction, *see* diffraction
— volt, 104
Electronegativity, 144–6, 388–90
Electronic configurations, tables of, 110, 145
— —, structure of atoms, 107–10
Electrons, bonding, 134–6
—, charge on, 7
—, inner shell, 135
—, non-bonding, 137
—, outer shell, 136
Element, 7
Emission spectrum, 98
Endothermic reactions, 71, 73–74, 241
Endpoint (in titration), 59
Energy, conservation of, 78
—, ionization, 103–5, 387–8
— level, 97–100
—, units of, 71–3
Enthalpy, *see* heat content
Entropy, 119
Enzymes, 206
Equilibria, competing, 237–9
—, homogeneous, 212–42
—, heterogeneous, 277–95
Equilibrium constant, 216–23, 240–2, *see also* K_a, K_b, K_c, K_p, K_s, K_w
— —, dependence on temperature, 240–2, 295
Equivalence point (in titration), 58
Esters, 364–7
Ethane, 160–2, 337–8, 341–2, 344
Ethanoic acid, *see* acetic acid
Ethanol, 343, 352, 359–60, 362
Ethene, *see* ethylene
Ethers, 354–60
Ethoxide ion, 355
Ethyl acetate, 365
— benzene, 349
— chloride, *see* chloroethane
— formate, 365
— group, 339
— hydrogen sulphate, 344
— trichloroacetate, 365
Ethylamine, 358
Ethylene, 163, 341, 343–5, 357
— chlorohydrin, 357

Ethyne, *see* acetylene
Exothermic reactions, 70, 75–6, 80, 241
Extent of reaction, 212–14

Faraday, 16
Fats, 366–7
Ferric, Ferrous, *see* iron(II), iron(III)
Field emission microscope, 5
Fluoride ion, 186
Fluorine, and compounds, 87, 89, 122, 137–9, 422–33, 399–409
— monoxide, 403, 406
Fluoroborate ion, 192
Formal charge, 152
Formaldehyde, 162–3, 330–2, 356
Formalin, 331
Formamide, 368
Formate ion, 332–3, 363
Formic acid, 332–3, 363, 368
Formula weight, 12–14
Formulae, 46–50, 372–4
—, empirical, 46
—, molecular, 47
—, structural, 47
Francium, and compounds, 413
Functional groups, 327, 334, 351–73
— —, summary of reactions, 369–70

Gallium, and compounds, 473–7
Galvanic cell, 306–7
Gas constant, 32
— density balance, 34
Gases, 26–45
—, properties of, 26–7
Gay-Lussac's law, 56
General gas equation, 28–33
— — —, derivation from kinetic theory, 40–2
— — —, deviations from, 38–40
Germanate ion, 469
Germanes, 467
Germanite ion, 470
Germanium, and compounds, 464–70
—(II) chloride, 468
—(II) ion, 469
—(II) oxide, 470
—(IV) ion, 469
—(IV) oxide, 469–70
Glucose, 280
Glycerol, 257
Glyceryl tripalmitate, 366
Glycine, 371
Glycol, 357, 367
Graphite, 84–6, 190–1, 193
Group I, 412–19
— II, 412–19
— III, 473–7
— IV, 464–71
— V, 451–61
— VI, 436–48
— VII, 422–33
— VIII, 479–81

Half-cell, 306–9
Halogens and compounds, 136–40, 171, 422–33
Heat, and chemical reaction, 70–94
— changes, measurement of, 73–7
— changes, stoichiometry of, 77–8
— content, 78–81
— energy, 71–3
— of crystallization, 80
— of formation, 83–7, table of, 87
— of fusion, 81
— of hydration, 177
— of neutralization, 248
— of reaction, 74–80
— of solution, 73–4, 80
— of sublimation, 81
— of vaporization, 81
Helium, 170, 479–80
Heptane, 338
Heterogeneous catalysis, 206
— reactions, 197
Hexamethylene diamine, 371
Hexammine cobalt(III) ion, 191
Hexaquo aluminium(III) ion, 191, 193
Hexaquo magnesium(II) ion, 191
Homogeneous catalysis, 205
Homogeneous reactions, 197
Homologous series, 337, 344
Hydrated ions, 175–8, 264, 271
Hydration number, 178, 271
Hydrazine, 60–2
Hydride ion, 250, 303, 417–18
Hydrobromic acid, *see* hydrogen bromide
Hydrocarbons, 336–50
Hydrochloric acid, *see* hydrogen chloride
Hydrofluoric acid, *see* hydrogen fluoride
Hydrogen bond, 172–5
— bromide, 172–3, 221, 426–7
— carbonate, 192, 256, 263
— chloride, 172–3, 221, 245–6, 248–9, 251–2, 401–2, 426–7
— electrode, 310–12
— fluoride, 142, 151, 172–4, 221, 401–2, 426–7
— iodide, 172–3, 221
— ion, *see* hydronium ion
— molecule, 130–3
— molecule ion, 133–4
— peroxide, 161–2, 315, 441
— selenide, 172–3, 439–40
— sulphate ion, 192, 252–3, 256
— sulphide, 154, 172–3, 402, 439–40
— sulphite ion, 444–5
— telluride, 172–3, 438–40
Hydrolysis, 245, 262–4, 366
Hydronium ion, 151–3, 191, 246–7
— perchlorate, 191, 246–7
Hydroxide ion, 143, 153, 245–55
Hydroxy group, 334, 351–7
Hypobromous acid, 431
Hypochlorous acid, 257, 430–1
Hypoiodous acid, 431
Hypophosphorous acid, 257

Ice, 175–6
Indicators, 59, 266
Indium, and compounds, 473–7
Inert gases, *see* noble gases
Interhalogen compounds, 428–30
Internal energy, 78–9
Iodic acid, 430–1
Iodide ion, 186
Iodine, and compounds, 118–19, 422–33
Ionic bonding, *see* bonding
— equations, 51
— product, 289–91
— radii, 390–2
Ionization energy, *see* energy
Iron, and compounds, 484–90
—(II) hydroxide, 284
—(II) ions, 300, 309, 311–13
—(II) persulphide, 192
—(III) chloride, 191–4, 264
—(III) hydroxide, 284, 292
—(III) ions, 300, 309, 311–13
Isobutane, 337
Isobutyl group, 339
Isobutyric acid, 363
Isoelectric series, 151
Isomerism, 338
Isopropanol, *see* 2-propanol
Isopropyl acetate, 365
Isopropyl chloride, *see* 2-chloropropane
Isopropyl group, 339
Isotopes, 8–9

Joule, 72

K_a, *see* acidity constants, pK_a
K_b, *see* basicity constant, pK_b
K_c, 216, 218–19, 222–3
K_p, 222–3
K_s, *see* solubility product, pK_s
K_w, 251, 254–5, 257–62, 291
Kerosene, 339
Ketones, 355, 360–2
Kinetic theory of gases, 27–8, 40–2
K-shell, 102–3, 105–7, 134
Krypton, and compounds, 170, 479–81

Lattices, chain, 122, 126, 180
—, layer, 122, 124–5, 180, 190–4
—, network, 122–3, 180–90
Lattice energy, 187–8
Lauric acid, 366
Law of combining volumes, 56
Law of conservation of energy, 78
Law of conservation of mass, 54
Layer lattices, *see* lattices
Lead, and compounds, 464–71
—(II) chloride, 281–4, 468
—(II) chromate, 284
—(II) fluoride, 286
—(II) iodide, 284
—(II) ion, 469
—(II) oxide, 299
—(II) sulphate, 284
—(II) sulphide, 284
—(IV) oxide, 469
—(IV) hydride, 467
Ligands, 271
Liquids, 120
L-shell, 102–3, 105–7
Litre, 21
Lithium, and compounds, 135–6, 399–408, 413–19
— aluminium hydride, 362, 364
— diatomic molecule, 135–6
— fluoride, 402
— hydride, 402, 417
— hydroxide, 407–8
— oxide, 403, 406–7
— superoxide, 419
Lubricating oil, 339

Magnesium, and compounds, 399–408, 413–19
— carbonate, 284
— chloride, 191
— fluoride, 402–3
— hexaquo ion, *see* hexaquo magnesium (II) ion
— hydride, 400, 402
— hydroxide, 407–8
— oxide, 188, 405–6
Manganate ion, 488
Manganese, and compounds, 484–8
Mass number, 7
— spectrometer, 9–11
Mechanism of chemical reactions, 374–82
Metallic bonding, *see* bonding
Metals, 181–3
Metastable compounds, 395–6
Methane, 83, 149, 151, 328–9, 337–8, 341, 401–2, 453–4, 467
— diol, 331
—, stepwise oxidation of, 333
Methanol, 329–30, 344, 359, 364–5
Methoxide ion, 329, 359
Methyl acetate, 364
— acetamide, 368
— ammonium ion, 331
— benzoate, 365
— chloroacetate, 365
— chloride, *see* chloromethane
— cyclohexane, 340
— ethyl ketone, 361
— formate, 365
— group, 339
— red (indicator), 266
Methylamine, 330–1
Methylene dichloride, *see* dichloromethane
Molar volume, 39–40
Molarity, 57
Mole, 14–15
Molecular equations, 51
— formulae, 12, 47
— orbital, *see* orbitals
— weight, 12–14, 33–5

Molecules, 12
—, diatomic, 130–47
—, polyatomic, 148–67
M-shell, 102–3, 105–7
Myristic acid, 360

Natural gas, 339
Neon, 170, 480
Nernst equation, 315–17
Network lattices, *see* lattices
Neutralization, 264–6
Neutrons, 7
Nickel, and compounds, 484–5, 489
Nitrate ion, 164, 166, 458, 460
Nitric acid, 166, 460
— oxide, 87, 200–1, 206, 459
Nitride ion, 153, 453–4
Nitrite ion, 460
Nitrobenzene, 348, 358
Nitrogen, and compounds, 141–2, 399–409, 451–61
— dioxide, 217, 459
— pentoxide, *see* dinitrogen pentoxide
— tetroxide, *see* dinitrogen tetroxide
— trifluoride, 402–3, 455
— trioxide, *see* dinitrogen trioxide
Nitronium ion, 191, 348, 404, 458
Nitrous acid, 257, 458, 460
— oxide, 87, 459
Noble gases, 170, 386, 479–82
Nonane, 338
Nuclear binding energy, 11
Nucleons, 7
Nucleus, 7–8
Nuclides, 7
Nylon, 371

Octane, 338
Octet rule, 156
Orbitals, atomic, 101, 114–16
—, bonding, 137
—, d, 108
—, molecular, 133
—, non-binding, 137
—, p, 108
—, s, 108
Order of reaction, 202
Organic chemistry, 327–72
Orthophosphoric acid, 256–7, 460
Orthophosphorous acid, 257, 458, 460
Oxalate ion, 205, 272, 491
Oxidant, 300, *see also* redox potentials
Oxidation, 299–303
— number, 301–3
Oxidation-reduction reactions, 62–3, 299–322
Oxide ion, 153, 186, 416
Oxygen, and compounds, 141, 146, 399–406, 436–42
— difluoride, 402–3
Ozone, 166, 437

Palmitic acid, 366
Paraffin wax, 339
Partial pressure, *see* pressure
Pauli principle, 102, 109–10, 133–4, 136
Pentane, 338
Perchlorate ion, 192, 256
Perchloric acid, 256, 430–1
Perdisulphate ion, 196
Periodic acid, 431
Periodic classification, 110–14, 383–96
Permanganate ion, 205, 486, 488
Peroxide ion, 192, 418
Persulphide ion, 192
Perxenate ion, 482
Petroleum (Petrol), 339
Petroleum ether, 339
pH, 252
Phenol, 356–7
Phenolphthalein, 266
Phenoxide ion, 356
Phosphine, 154, 401–3, 454–5
Phosphoric acid, *see* orthophosphoric acid
Phosphorous acid, *see* orthophosphorous acid
Phosphorus, and compounds, 399–409, 451–60
— oxychloride, 456
— pentabromide, 477
— pentachloride, 227–9, 237–8, 454, 456–7
— pentafluoride, 150
— pentoxide, 406–7, 457–8
— trichloride, 227–9, 237–8, 456–7
— trifluoride, 153, 402–3
— trioxide, 458
Photons, 98
pK_a, 252–6
—, table of, 256
pK_b, 254–6
pK_s, *see* solubility product
Planck's constant, 89, 98
Plumbate ion, 470
Plumbic, Plumbous, *see* lead(IV), lead(II)
Plumbite ion, 471
Polonium, and compounds, 436–8
Polymers, 345–6
Polypeptide chain, 371
Polyprotic acids, 235–7
Polystyrene, 349
Polythene (Polyethylene), 345
Polyvinyl acetate, 345
— chloride, 345
Potassium, and compounds, 136, 413–19
— bromide, 281
— chloride, 144, 281
— chloride ion-pair, 144
— fluoride, 188
— hydroxide, 281
— oxide, 406
— superoxide, 419
Precipitates, dissolution of, 291–4
Precipitation reactions, 61, 289–91
Pressure, 28–9, 35–7, 42
—, partial, 35–7, 222

Propanal, 354, 360, 362
Propane, 337–8, 341–2
Propanoic acid, *see* propionic acid
1-Propanol, 352, 354, 360
2-Propanol, 353, 355, 361–2
2-Propanone, 355, 361
Propene, 344
Propionaldehyde, *see* propanal
Propionic acid, 354, 362–3
Propyl chloride, *see* 1-chloropropane
Propyne, 344
Proteins, 368
Protons, 7, 11, 246–7

Quantum mechanics, 100–2

Radium, 413
Radon, 170, 479–81
Rare earths, 114
Rare gases, *see* noble gases
Rate law, 198–205, 374
— constant, 198–205
— of reaction, 196–211, temperature dependence, 377–8
Redox potentials, 309–17
— —, concentration dependence of, 315–17
— —, stoichiometry of, 303–6
— —, tables of, 313, 424, 441, 469
Redox reactions, harnessing of, 306–9
— —, stoichiometry of, 303–6
Reductant, 300, *see also* redox potentials
Reduction, 299–303
Resonance, 164–7
— hybrid, 165, 347
Rubidium, and compounds, 413–19

Salicylic acid, 357
Salt, 247
— bridge, 308
Saturated hydrocarbons, *see* alkanes
Scandium, and compounds, 483–5
Secondary butyl group, 339
Selenous acid, 444
Selenium, and compounds, 436–46
Selenic acid, 446
SI units for physical quantities, 19–22
Sidgwick-Powell rule, 138, 149
Silanes, 401
Silanol, 468
Silicate ion, 401
Silicic acid, 407–8
Silicon, and compounds, 184, 399–408
— carbide, 184–5
— dioxide, 184, 405–7
— hydride, 401–2
— tetrafluoride, 402–3
Silver bromide, 190, 284, 433
— carbonate, 284
— chloride, 190, 284, 293, 295, 483
— chromate, 284, 290–1
— fluoride, 190, 281, 433
— iodide, 190, 284, 433
— oxide, 407
— sulphate, 284
Silyl chloride, 468
Soaps, 366–7
Sodium, and compounds, 399–408, 412–19
— bromide, 71, 281
— chloride, 123, 128, 144, 187–8, 246, 262–3, 281
— chloride ion-pair, 144
— fluoride, 402–3
— hydride, 400, 402
— hydroxide, 407–8
— oxide, 405–7
— palmitate, 367
— peroxide, 419
— thiosulphate, 71, 196–7, 303
Solid-gas equilibria, 277–80
Solids, 120
Solubility equilibria, 280–295
Solubility product, 282–7, 295
— —, table of, 284
— —, temperature dependence of, 295
Solvation of ions, 177–8
Spectrograph, 98
Stability constant, 273–5
Stability of chemical compounds, 394–6
Standard heat of formation, 84–7
— hydrogen electrode, *see* hydrogen electrode
— solution, 57
— state, 224
Stannate ion, 470
Stannic, Stannous, *see* tin(IV), tin(II)
Stannite ion, 471
Stearic acid, 366
Stibine, 454
Stoichiometry, 46–69
Strontium, and compounds, 412–19
Structural formulae, 47, 338
— isomerism, 338
Styrene, 349
Sublimation, 118, 120
Subshells, atomic, 107–9
Sulphate ion, 164–6, 256
Sulphide ion, 183, 263–4
Sulphite ion, 165, 445
Sulphur, and compounds, 399–409, 436–47
— difluoride, 219–20, 301
— dioxide, 206, 442–3
— hexafluoride, 155, 159, 402–3
— tetrafluoride, 402–3
— trioxide, 219–20, 405–6, 443
Sulphuric acid, 256, 394, 446–7
Sulphurous acid, 444–5
Superoxide ion, 419

Teflon, 346
Telluric acid, 394, 445
Tellurium, and compounds, 437–46
— tetrachloride, 159
Tellurous acid, 444–5
Temperature, 40–1
Terephthalic acid, 366

Termolecular step, 382
Tertiary butyl group, 339
Terylene, 367
Tetrafluoroethylene, 346
Tetrammine copper(II) ion, 272–3, 490
— zinc(II) ion, 272, 490
Thallium, and compounds, 473–7
Thiosulphate ion, 196–7, 303
Tin, and compounds, 464–71
—(II) chloride, 468
—(II) hydroxide, 470
—(II) ion, 469
—(II) oxide, 470
—(IV) chloride, 467–8
—(IV) ion, 469
—(IV) oxide, 470
—(IV) sulphide, 191
Titanium, and compounds, 484–5
Titration, 59
— curve (acid-base), 265
Toluene, 340, 364
Transition elements, 114, 483–91
Tribromide ion, 430
Trichloracetic acid, 363
Triiodide ion, 430
Trimethylamine, 330
Trisethylenediamine cobalt(III) ion, 491
Trisoxalatochromium(III) ion, 491
Tungsten, 120

Unimolecular step, 381–2
Units for physical quantities (SI), 19–22
Unsaturated hydrocarbons, *see* alkenes, alkynes

Valence structures, 47, 140, 153, 191–2
Van der Waals equation, 40
Vanadium, and compounds, 484–6
Vaseline, 339
Vinyl acetate, 345
— chloride, 345
Virial equation and coefficients, 40

Water, 81–2, 87, 148, 150–1, 172–4, 250–1, 254, 400–2
Work (thermodynamic), 78

Xenon, and compounds, 170, 479–82
X-ray diffraction, *see* diffraction

Zinc, and compounds, 484–90
— hydroxide, 289
— sulphide, 190
— tetrammine ion, *see* tetrammine zinc(II) ion
Zincate ion, 289